World Sustainability Series

Series Editor

Walter Leal Filho, Wissenschaften Hamburg, Hochschule für Angewandte, Hamburg, Germany

Due to its scope and nature, sustainable development is a matter which is very interdisciplinary, and draws from knowledge and inputs from the social sciences and environmental sciences on the one hand, but also from physical sciences and arts on the other. As such, there is a perceived need to foster integrative approaches, whereby the combination of inputs from various fields may contribute to a better understanding of what sustainability is, and means to people. But despite the need for and the relevance of integrative approaches towards sustainable development, there is a paucity of literature which address matters related to sustainability in an integrated way.

Notes on the quality assurance and peer review of this publication

Prior to publication, the works published in this book are initially assessed and reviewed by an in-house editor. If suitable for publication, manuscripts are sent for further review, which includes a combined effort by the editorial board and appointed subject experts, who provide independent peer-review. The feedback obtained in this way was communicated to authors, and with manuscripts checked upon return before finally accepted. The peer-reviewed nature of the books in the "World Sustainability Series" means that contributions to them have, over many years, been officially accepted for tenure and promotion purposes.

Archana Singh · Samuel Fosso Wamba ·
Megha Ojha · Vinay Kandpal · Seda Yıldırım
Editors

Human-Centered Sustainability: Bridging Well-Being, Gender Equity, and Technology

People-Centered Sustainability

Springer

Editors
Archana Singh
School of Management
Graphic Era Hill University
Dehradun, Uttarakhand, India

Megha Ojha
Department of Management Studies
Graphic Era (Deemed to University)
Dehradun, Uttarakhand, India

Seda Yıldırım
Tekirdag Namık Kemal University
Tekirdağ, Türkiye

Samuel Fosso Wamba
TBS Education
Toulouse, France

Vinay Kandpal
Department of Management Studies
Graphic Era (Deemed to University)
Dehradun, Uttarakhand, India

ISSN 2199-7373 ISSN 2199-7381 (electronic)
World Sustainability Series
ISBN 978-3-032-22019-6 ISBN 978-3-032-22020-2 (eBook)
https://doi.org/10.1007/978-3-032-22020-2

This Springer imprint is published by the registered company Springer Nature Switzerland AG
The registered company address is: Gewerbestrasse 11, 6330 Cham, Switzerland

Preface

With an increasing level of awareness and recognition of issues related to the environment, especially related to climate change, as well as the impact of technology on society, issues associated with social inequality (specifically gender inequality), and shifting patterns in the way political authority which is exercised globally, the complexity of the challenge to achieve sustainable development is also increasing. In this regard, the UN's Sustainable Development Goals (SDGs) provide broad guidelines for what is needed to address these multiple problems, but also provide an opportunity for international conversation on a wide range of issues concerning sustainability. It is with these goals in mind that the purpose of this edited book is to serve as a part of the global effort to improve how we understand sustainability by utilizing multiple disciplines, placing emphasis on the human element of sustainability, being more equitable, utilizing technology in ways that promote sustainability, and applying future-thinking to our understanding of sustainability.

The contents of the book have been divided into four parts that define the content for the 27 chapters that comprise this book. Each of the parts provides different perspectives from individuals who are working in academic, practitioner, or research roles at a national and/or international level. Each of the themes emphasises an aspect of Sustainable Development and supports the development of a common knowledge base for Global Policy, Institutional Transformation, and Community Empowerment.

The first part, "Human-Centred Sustainability, Well-Being and Social Development", focuses on the foundations of Human-Centered Sustainable Systems. Each of the chapters provides insight into how Human Dignity, Psychological Well-Being, and Behaviour Change contribute to building Human-Centered Systems. The chapters make clear the importance of Emotional Intelligence, Inclusive Design of Communities, and Human Dignity in the context of Sustainability, and they will significantly contribute to SDGs 3, 11, and 13. The second part of "Gender Equality, Inclusion and Empowerment in a Digital World" addresses the pressing need to combat many forms of inequality that affect women and other disadvantaged groups. Through an examination of Digital Education, Inclusive Financial Systems, Menstrual Health, Gendered Impacts of AI, and Public Policy, the chapters highlight critical opportunities to achieve SDGs 4, 5 and 10. They also emphasise the need for Inclusive Empowerment, using an Intersectional Approach, within a Digitally Mediated World.

The third part "Sustainable Business, Governance and Responsible Leadership" examines how organisations are able to incorporate Ethical Conduct, Transparency and Sustainability into business practices. The chapters in this part address issues related to CSR, Municipal Reforms, Leadership Models, Corporate Accountability, Human Resource Management—Corporate Social Responsibility (HRM-CSR) Integration and Governance. Together, these chapters reinforce the importance of Institutional Responsibility and Values-Based Leadership for achieving long-term sustainability within an organisation and the wider society.

The fourth part of "Digital Transformation, AI, FinTech Innovation and Future of Work" explores how new and emerging technologies, including AI, Blockchain, Digital HR and Fintech, are changing the nature of Economic Systems, Global Development and Workforce Dynamics. The chapters present the following insights that relate to responsible innovation, digital ethics, sustainable development goals (SDGs), aligned banking and technology inclusion. Collectively, these themes are consistent with SDGs 8, 9 and 17. These four themes form a complete picture of sustainable development. They connect theoretical assumptions to practical applications, global frameworks to local realities, and the potential of technology to meet the needs of people. This book aims to be a valuable resource for policy makers, academics, development professionals, industry leaders, and students interested in creating sustainable and inclusive futures.

We hope that the insights presented here will inspire further interdisciplinary research, innovative policymaking, and collaborative action as the world works collectively toward fulfilling the 2030 Agenda for Sustainable Development.

Dehradun, India — Dr. Archana Singh
Toulouse, France — Dr. Samuel Fosso Wamba
Dehradun, India — Dr. Megha Ojha
Dehradun, India — Dr. Vinay Kandpal
Tekirdağ, Türkiye — Dr. Seda Yıldırım

Acknowledgements

The editors would like to express their heartfelt appreciation for every individual whose commitment, backing and partnership made it possible for us to realise the creation of this edited volume. Each contributor to this volume provided a unique intellectual viewpoint, an extraordinary level of academic achievement and a desire to contribute to the continuous development of knowledge within the field by way of their area of scholarly expertise and practice. The subject areas that the different chapter contributors are dealing with are wide-ranging, with specific references to matters such as human-centric sustainability, gender equality, governance, digital transformation and responsibility in business. All of these areas will serve as a strong foundation to support the various chapters that will be compiled in this edited volume. The reviewers played a critical role in the process of this volume by providing expert advice, constructive criticism and academic guidance; thus, they greatly contributed to refining and ensuring the definition, clarity, coherence and scholarly level of each contributor's chapter.

The collaboration of experts and institutions, including universities, companies, and other types of organisations, has greatly contributed to this volume with their support of research projects and access to data and resources that went into this collaborative effort. In addition, to Springer Nature for recognizing the value of the work that has gone into this volume and the role of the editorial and production staff at Springer in helping to create this book and produce it to the highest standard at all stages of the proposal evaluation, editorial editing, and production processes. The expertise and technical competence of these editorial and production staff have been instrumental to the successful completion of this volume.

We are very grateful to the many international scholars, consultants and subject matter experts whose support and contributions to the book, their input and discussions, have contributed to the internationalization of the volume. Finally, we wish to thank our families and colleagues for their continued encouragement, understanding and support during this long and challenging but ultimately very rewarding pursuit. Their unwavering support provided us with the motivation and the ability to complete this book. This book represents collaboration and a shared commitment to advancing respectful, sustainable, inclusive, and human-centred development worldwide. We hope this book will provide value for scholars, policymakers, industry leaders, and practitioners.

Dr. Archana Singh
Dr. Samuel Fosso Wamba
Dr. Megha Ojha
Dr. Vinay Kandpal
Dr. Seda Yıldırım

Contents

About the Editors

Dr. Archana Singh is a distinguished academic and researcher in the field of Management, specializing in Human Resources, technology integration in management, and defense sector innovations. She holds a Doctorate in Management and has published extensively in Scopus-indexed journals and books. Her research focuses on e-recruitment, ethical AI in HRM, and the intersection of technology and sustainability in human capital management. With multiple books and chapters under publication, Dr. Singh continues to contribute significantly to global academic discourse on sustainable HR practices and digital transformation.

Dr. Samuel Fosso Wamba is a full professor of Information Systems and Data Science and Associate Dean of Research at TBS Education, France, and a distinguished visiting professor at the University of Johannesburg. He holds a Ph.D. in Industrial Engineering from Polytechnique Montréal and two master's degrees from HEC Montréal and the University of Sherbrooke. His research explores the business value and societal impact of digital technologies, including AI, big data, blockchain, and RFID. He has published widely and received multiple international awards (e.g., being recognized among the top 1% most-cited scholars by Clarivate and among the top 2% scientists by Stanford).

Dr. Megha Ojha is a dedicated professional and researcher in management and psychology, focusing on organizational behaviour, emotional intelligence, and sustainable HR practices. In addition to her academic interests, she emphasizes the importance of evidence-based approaches to understanding workplace dynamics and advancing human resource development. Her work reflects a commitment to integrating psychological principles with organizational frameworks to enhance employee well-being and foster effective leadership. She maintains a strong interest in contemporary issues shaping the modern workforce, including mental health, ethical management, and the evolving role of human capital in sustainable development. Dr. Megha continues to contribute thoughtfully to scholarly dialogue in these domains.

Dr. Vinay Kandpal (Postdoc and Ph.D. in Finance) possesses over 21 years of experience in Academics and Industry. He has published more than 85 research papers in leading refereed Journals. Published 15 books in the area of Management. He has presented several works at International Conferences on Smart City at the University of Nice Sophia Antipolis, France, and Oxford Brookes University, UK. He is a full-time member of ANAHEI, Florida, USA. He works as a professor in the Department of Management Studies, Graphic Era Deemed to be University, Dehradun, Uttarakhand, India. His areas of research are Sustainability, Fintech, Smart City, Risk Management, and Financial Inclusion.

Prof. Dr. Seda Yıldırım currently works at Tekirdag Namık Kemal University in Turkey. She has a doctoral degree (Ph.D.) in marketing. She is a full-time professor since 2017. She mostly studies sustainable development, 2030 SDGs, sustainable consumption patterns, environmental quality and food security. She is married and has two children. She has over two thousand citations by Google Scholar. She has many papers in journals indexed by Web of Science (SSCI, SCI-Exp, ESCI and others: h:13) and SCOPUS (h:17). She has published several books and book chapters by a reputable publishing house. She won the Emerald Literati Award-Outstanding Paper Award 2022.

Human-Centred Sustainability, Well-Being and Social Development

Framing Human-Centered Sustainability: Concepts, Theories, and Global Realities

Archana Singh, Girish Lakhera, Arvind Nain, and Jyoti Kumari

Abstract This chapter forms the conceptual and theoretical basis of human-centred sustainability, which is a model that incorporates environmental stewardship, social equity, technological responsibility, and human well-being. It is a synthesis of multidisciplinary viewpoints based on sustainability science, human development, behavioural studies, gender equity, and digital transformation to clarify the way people, not only systems, should be central to sustainable progress. The chapter under review identifies the most important theories related to sustainability outcomes that increase when designed towards human dignity, inclusiveness, and empowerment, by comparing such approaches as the Capability Approach, Ecological Modernisation, Triple Bottom Line, and Human-Centred Design. Case studies of global locations with different socio-economic backgrounds also highlight applications in the real world, showing that issues like digital inequalities, vulnerability to climate change, resource inequity, and gender inequalities remain unresolved. The chapter ends with the identification of research gaps, practical implications, and directions of creating a more sustainable model of the world based on equity, technological awareness, and human flourishing.

Keywords Human-centered sustainability · Well-being · Gender equity · Human-centered design · Social inclusion · Sustainable development

1 Introduction

During this era of fast digital mutation, escalating climate harm, and widening social disparities amid rich and poor, our concept of sustainability needs to transform and develop beyond our accustomed sense of its comfort-zoned limits. Sustainability was

A. Singh (✉)
Graphic Era Hill University, Dehradun, India
e-mail: Elvira.15singh@gmail.com

G. Lakhera · A. Nain · J. Kumari
Graphic Era Deemed to be University, Dehradun, India

A. Singh et al. (eds.), *Human-Centered Sustainability: Bridging Well-Being, Gender Equity, and Technology*, World Sustainability Series,
https://doi.org/10.1007/978-3-032-22020-2_1

initially a concept that concerned the conservation of the environment and stable economy. However, this must now be broader which means that it involves basic human rights, social justice, and ethical additions to technology. This change is an indication of the understanding that we cannot actually be sustainable without solving the core of the lives of people especially when it is they who end up being forgotten. A human-oriented approach to sustainability must propose three components of it: well-being, gender equality, and just technology that are three components which define how societies can be achieved and how they can govern and prosper and succeed. It is a strategy, which puts agency above assistance, which means rather than doing, it means accepting, instead of a small group being the face of a group (Sachs 2015; Raworth 2017).

With world institutions struggling with intersecting problems, including the digital divide and people losing their home through environmental issues, we must develop systems, policies, and technologies that would enable people to make the most out of their own capabilities without compromising the resources of the planet. This chapter initiates a very important multi-disciplinary discussion on the true meaning of sustainability that predisposes the remainder of the book, both conceptually and practically.

1.1 Rethinking Sustainability Through a Human Lens

The classic model of sustainability, known as the **Triple Bottom Line** (Elkington 1997), includes economic, environmental, and social dimensions. Yet, in practice, the social dimension often receives less attention or is reduced to peripheral interventions. Human-centered sustainability aims to correct this imbalance by focusing explicitly on qualitative human experiences—such as empowerment, psychological safety, cultural identity, and meaningful work. This approach is reported by critics after development that claims that development should be relevant and inspired by people. It also matches the UN's *Leave No One Behind* principle and says that development processes should be inclusive and justified from the beginning. By prioritizing living experiences, subjective welfare and social cohesion, this model transforms discourse by measuring development to improve the quality of life. Tools such as Social Progress Index and the OECD Better Life Index support the move, and support the move to evaluate human-centric results on the general economic matrix.

1.1.1 The Capability Approach (Sen 1999)

A basic principle of human-focused stability is Amartya Sen's capability, which reproduces the development of the real freedom of individuals, the ability of individuals and the ability to do what they do. Unlike a utilitarian or resource-based framework, this approach is the center to enable alternatives, speech and dignity. Practically, it is for health, education and investment, which is the primary indicator of social welfare. Martha Nussbaum later improved this structure to define the list of ten key human abilities to provide a rights-based approach to physical integrity, emotions and attachment development. Within sustainability debates, the Capability Approach fosters metrics that move beyond GDP or carbon emissions and instead evaluate how policy and technological interventions enhance human flourishing. For instance, when evaluating a clean energy project, a human-centered perspective would not only examine its emissions reductions but also whether it improves access to power for women-led households or indigenous communities.

1.2 Gender Equity: The Missing Pillar

Despite political frameworks such as the SDG and the Beijing Platform for action, gender is the lowest used column with equity stability. Very often, the woman's contribution is prepared as recipients instead of a change agent. Feminist scholars criticized the mainstream stability discourse for treating the penis as a check-out route instead of a structural question, ignoring the resource access to patriarchal norms, access to resource, work burden and political voice (Leach et al. 2016; Kabeer 2005). In addition, exclusion of LGBTQ+ and non-binary approaches further the definition of equity. A gender bus stability model should be responsible for unpaid care functions, reproductive rights, inter-judicial justice and gender effects of climate policy. Gender-written data and intersecting influence assessment in the stability scheme ensure that the guidelines are more responsive and transformative.

1.2.1 Feminist Political Ecology (FPE)

The feminist political ecology (FPE) provides an important lens to understand how environmental erosion and lack of resources affect women and marginalized societies. It challenges technical and market-based approaches to environmental rule by highlighting power differences in access, decision-making and knowledge production (Rocheleau et al. 1996). Women, especially in the global south, are often the primary salaries of natural resources—manages forests, water and agriculture—but still countries are often excluded from land ownership and formal political processes. FPE said that environmental law should be inseparable from gender law. FPE involves the principles of climate adaptation, land period and conservation strategies enable more fair and effective results.

1.2.2 Case Study: SEWA—India

The example of Self-scheduled Women's Association (SEWA) in India shows that gender-intelligent, socially powered methods may have long-term economic and ecological impacts. SEWA is an organization that exists at the border of economic justice, environmental control and power. Solar contractors, roofing and training of climate flexibility equipment Informal female workers with equipment to build green livelihoods are some of the programs.

Their collaborative model not only increases revenues, but also increases voice and visibility in political places. Research suggests that women involved in SEWA initiatives improve health, decision-making power and social leadership (Chen 2005). Sava's model is now repeated in South Asia and Africa, proving its scalability and transformative capacity.

1.3 Technology as a Bridge or Barrier?

Technology is often seen as a universal agent for sustainable development, but its placement without moral foresight can confuse existing differences or create new forms of exclusion. Human-focused stability requires technology to involve involvement, participation and only the results that are just results or profitability. Digital split, algorithm bias and monitoring of capitalism increase threats to equity in human agency and technology-based systems. UNESCO's recommendations such as moral technology structures, inclusive design, data transparency and innovation on ethics in AI (2021) argue. A nice approach admits that technology can be both an enabler and a barrier, depending on who controls it, who serves it, and how it is controlled.

1.3.1 Technological Determinism Versus Human-Centered Design

Technical provision considers a linear, autonomous progress of innovation that shapes society. In contrast, advocates for human-focused design (HCD) co-design with users, especially the historically marginalized as disabled, women in a low-income environment and indigenous community. HCD emphasizes sympathy, relevant understanding and relapse reaction. This is not just about the purpose, but about moral adaptation to social values. For example, inclusive fintech platforms now use voice commands and local dialects to complete illiterate users—describing how design options can democratization of access.

1.3.2 Responsible Innovation (Stilgoe et al. 2013)

According to Research and Innovation (RRI) framework, innovation processes should be forward-looking, of inclusion, refractor and accountable. This motivates the researchers, entrepreneurs and decision makers to make sure that technologies are aligned with social objectives so as to involve various populace in the cycle of innovation. This view holds significance when it comes to technologies (AI), biotechnology and renewable energy—where ethical, environmental and cultural concerns must be considered as part and parcel. RRI finds momentum in the EU research fund mechanism and offers an example to the world to adopt in SDG-related technical approaches.

1.3.3 Case Study: M-Pesa—Kenya

M-Pesa is a successful example of how technology can support inclusive development. It was launched in 2007 and enables mobile-based financial transactions even without internet access. There is much adopted-specially increased economic inclusion between women, rural communities and micro contractors, has reduced the cost of transactions, facilitating access to credit. From 2016, it was estimated that more than 194,000 houses were thrown out of poverty (Suri and Jack 2016). It is important that M-Pesa was successful due to its local adaptability, trust-based distribution networks and gender-sensitive outreach strategies that state that the inclusive technical design has similar results.

1.3.4 Case Study: AI Bias in Recruitment

On the other hand, Amazon's AI recruitment tool (2014–2017) suggests how the Algorithic system, when partially trained on historical data, can eliminate discrimination. This device resumed that included the term "woman" (like a "women's chess club"), which effectively harms female applicants. The case expressed global concern for algorithm bias and data moral (Eubanks 2018; O'Neil 2016). This outlines the need for diversity in the data cages, transparency in AI-declaration and regulatory inspections to ensure that new technologies do not strengthen the exclusion pattern.

1.4 Intersectional and Holistic Sustainability

Interactions launched by Crescow suggest that the interaction to produce gender, race, square, disability and migration as overlapping identity—to produce levels. When it comes to stability, it means that it is not equal to identify weaknesses for climate change, health risk or technical exclusion. For example, women of colors, disabled adolescents and stateless migrants increase the risk and reduce access to flexibility-building resources. A human-centered, interaction stability model ensures that strategies address these mixed realities instead of using a form-pass-up solutions. Overall guidelines require interregional cooperation between health, education, environment and digital management systems.

1.4.1 Case Study: Urban Poor Women and Climate Risks

Informal settlements such as Dhaka, Bangladesh, poor women meet land titles, water infrastructure and systemic exclusion from the health care system. Climate-sprouted floods not only destroy home, but also increase due to uneven conditions (UN Women 2020) due to gender-based violence and dangers. Projects such as Slum Dayler's International (SDI) and Women's Housing Trust (India) are working to equip climate data, safe recruitment rights and community-led adaptation solutions-affiliated that at the ground level, the interface level can get scalable models for inclusive stability.

1.5 The Doughnut Economy: A Framework for Balance

The Doughnut Economics model by Kate Raworth is an explicit synthesis of planetary challenges and social bases (Sen, Nussbaum) to establish a tangible structure of the safe and fair space to humanity. The inner circle includes human rights-based completely necessities such as access to food, education, shelter and equity, whereas the outer one denotes ecological boundaries that should never be transgressed. This variant requires circumstances of both neo-liberal boom tendencies and ecological centerpiece only views through inclusion of the regenerative, distributive economic model. The model of doughnut has been applied to cities such as Amsterdam to manual urban coverage with the inclusion of sustainability in procurement, housing, and employment. It gives a powerful, quantifiable, and flexible tool to assess the presence of a virtual man-centric development work.

2 Literature Review

2.1 Conventional Sustainability Paradigms

Triple Bottom Line (TBL) The model suggested by Elkington (1997) did not simply provide profit (financial) as outcomes of organizations, but also provided the planets (environment) and people (social) to evaluate their actions. Revolutionary critics at that time have argued that the social aspect is usually shallow or superficial. In case of the Corporate Social Responsibility (CSR) initiative, such social initiatives, even though sound on paper, do not usually go much deeper, much larger or lasting structural transformations. Moreover, the TBL takes into account three columns as single silos, and is not accountable to the inner existence of human welfare, equity and ecological flexibility. It has requested additional related contours that value social justice and human dignity as stability as valuable elements.

2.2 Human Development and Well-Being

The works of Amartya Sen, the Ability Approach (1999) picks up on the progress in the field of income calculations and concentrates on actual freedom in order to enable people lead to life they value.

This approach affected global policy, especially through the Human Development Index (HDI), which includes life expectancy, education and living standards. Sens work emphasizes the most important SDG indicators and has prepared the basis for understanding stability as natural as human cakes. It also challenged the most important models and resource distribution models by highlighting the qualitative dimensions of life as agency, autonomy and ability to participate in society. Expanding abilities instead of bare revenues became central to measuring social progress.

2.2.1 Gender and Intersectionality

The outline of traditional stability has historically been gendered in theory, but in the application gender blind (Kabeer 2005). The feminist scholars have criticized this inspection, emphasizing that without incorporating gender equality, the initiative of stability can unintentionally strengthening patriarchal norms. The concept of interaction initiated by Crenshaw (1989), a lot examines this criticism of how it recognizes—such as gender, race, class and disability—climate risk depends on shaping the differences between resources, opportunities and flexibility in the face of climate risk. The feminist political ecology (FPE) further reveals how women, especially marginalized or in indigenous communities, work as front-line environments, but often excluded from decision-making (Rocheleau et al. 1996). This insight requires a change to more inclusive and gender-provoking stability strategies.

2.2.2 Ethics and Technology

The digital age brings both opportunities and moral dilemmas. Emerging technologies such as AI, Blockchain and Big Data Analytics have transformational ability, but they also take into account existing inequalities if not carefully controlled (Eubanks 2018; O'Neil 2016). Algorithm Discrimination-which a biased recruitment software or future police technical provision, provokes the risk of technology, where technology is considered to be worthwhile. In response, responsible research and innovation (RRI) frameworks (Stilgoe et al. 2013) say for an active approach: fear of social influences, confusing stakeholders and morals built in design. This is especially important in contexts where technologies are crossed with human rights, such as digital ID systems, e-control or remote learning. Literature supports the integration of human-focused morality in all phases of innovation.

Table 1 Theoretical background

Theory	Description	Relevance to human-centered sustainability
Capability approach (Sen 1999)	Emphasis on people's real freedom to bring life, which focuses on opportunities and agencies rather than resource accumulation	In the context of human evidence, it acts as a moral and practical foundation for redefining development and stability, not just physical progress
Doughnut economics (Raworth 2017)	A social basis (e.g. health, equity) and organic roofs (e.g. co, level, biological diversity loss) imagine "safe and just space for humanity"	A system curing provides a framework that balances the boundaries of planets with social needs, and controls both macropolitics and societal action
Feminist political ecology (Rocheleau et al. 1996)	Examines the intersections of gender, environment, and power, often highlighting the role of women in managing ecological resources	Environmental policy lifts gender experiences and advocates to include local, indigenous peoples and women's knowledge systems in the stability scheme
Human-centered design (Norman 2013)	The focus on creating systems and technologies that meet users' actual needs and references, especially groups of margins	The inclusive participation encourages design processes that lead to adjusted equal and accessible innovations with stability goals
Intersectionality (Crenshaw 1989)	Analysis makes how many social identities (gender, race, class, capacity, etc.) are interested in producing complex forms of discrimination or privilege	This ensures that stability solutions are not a size-pass all, but diverse and often sequentially to solve unique challenges with overlapping weaknesses

2.3 *Theoretical Foundations*

See Table 1.

2.4 *Real-World Case Studies*

2.4.1 Gendered Climate Resilience in Bangladesh

In front of accelerating the rising sea level and monsoon patterns, women-led cooperatives in Bangladesh have ground breaking the floating garden, a form of hydroponic agriculture as innovative, low-technical flexibility strategies. The water hyacinth and bamboo-cooked gardens enable the families to cultivate vegetables and spices even in the face of protracted floods. Such initiatives are backed by groups like Oxfam and UNDP, and it is through this that local ecological knowledge and gender work are negotiated to generate sustainable solutions. Women do not just guarantee domestic food security, they also earn income, confront the conventional gender roles and establish solidarity in communities. This case shows how the human centered approach is based on agency, reference and equity can be used to promote climate adaptation in settings (UNDP 2021).

2.4.2 AI for Inclusive Health in India

Arvind Eye Care System is one of the exemplifications of moral innovation and inclusive health technology. Arvind has enhanced eye care to both rural and signed communities across India, with AI-controlled imaging, cloud computing and telemedicine. The high length, low cost model-a-tier price system and the committed with cross-safety disorder that no patient is defeated because of insufficient finances, in particular. The system is used to represent an ethos of humans where efficiency in the technical sense is enhanced rather than empathy. The Arvind model is a combination of stinginess, access and equity in health technology—how AI can preserve human dignity in the implementation of morally and inclusive (Prasad and Vasan 2020).

2.4.3 Gender Tech Gap: Digital Disparities in Africa

Although mobile penetration among Africans is improving, women are hardly evaluated in online engagement. Women 37% in Africa south of Sahara have a lower chance to use mobile Internet difference which is motivated by Maladness, literacy, cultural norms and security problems (GSMA 2023). This digital divide has a trickle down effect: Women are challenged in online education, e-health, financial services and civilian engagement platforms. Besides, the smart farming projects and new digital economies also pose the threat of copying pattern patterns whose penis is actively incorporated in the design of infrastructures and digital literature. Thus, digital gender differences should not be reduced only in terms of connection—but inclusive development, digital justice and economic power.

2.5 Current Issues and Emerging Trends

Digital Colonialism: Canary and Mejias have cautioned against the existence of a new type of colonialism, with data harvesting and technical addiction being analogous to the antique royal mobility. The Global South Nations tend to have no control over digital infrastructure, algorithms and platforms which guide their communities. The instantly posed questions on domination of Western technical giants in digital markets and train sets within AI domain bring up sovereignty concerns, cultural extinction and epidemic justice. Human centric stability needs a shift to decentralized open sources and locally controlled technical ecological which takes into account various values and priorities.

- **Climate-Induced Migration**: As climate change displaced millions of people globally, women are unevenly affected by disaster-linked and conflict areas. In temporary camps or urban slums, displaced women often have exclusion from gender-based violence, insufficient breeding health services and support programs. Nevertheless, their adaptive strategies and caring roles are rarely recognized in climate adaptation policies. Women's vulnerability asks for the Intermigration framework for safety and inclusion instead of physical transmission only instead of physical transfer.
- **Workplace Automation**: The fourth industrial revolution intensifies job shift in women and low-income workers, such as production, retail and administrative help. According to ILO (2021), automation can displace women inconsistently due to commercial insulation and skills that are not matching. Proactive apocillation, inclusive racing programs and gender-sensitive labor policy, elaboration of automation risk financial inequality. Forward-loving strategies should include technical equity in the heart of workforce infections.
- **Mental Health and Tech**: The prevalence of digital monitoring, productivity tracking tools and "allway-on" work cultures have increased mental health problems, especially burnout, anxiety and digital fatigue (WHO 2022). These effects are expanded to take care of, remove workers and playing job participants—many of whom are women. Algorine control systems often lack sympathy, ignore human variability, and the efficiency of welfare is preferable. A human-centric stability model should address mental health as the main element of digital morality and design in the workplace.

3 Chapter Objectives

1. To introduce the concept of human-centered sustainability and its relevance in the twenty-first century.
2. To critically review key theories linking sustainability, gender, well-being, and technology.
3. To explore real-world case studies that illustrate inclusive, equitable approaches to sustainability.
4. To identify emerging issues and tensions in the intersection of gender, tech, and sustainability.
5. To outline future directions and practical implications for policy, academia, and organizations.

3.1 Introducing Human-Centered Sustainability in the Twenty-First Century

The first goal of this chapter is to define and refer to the new paradigm for human-centric stability, which revived only profits or planetary boundaries in the heart of fellow stable development discourse. Traditional models often see environmental protection and economic development as competitive interests, and neglect human and moral dimensions of development. Human-focused stability, especially among the marginalized population, cure this imbalance by emphasizing welfare, dignity, equity and agency.

Models such as Donut Economics of Rorth describe this change by identifying a safe place where both organic roofs and social foundations are respected. Similarly, Sachs (2015) presents SDG as human-focused obligations, clearly equality, integrates education and reduces inequalities as well as inequalities.

Cities such as Copenhagen and Amsterdam simulate this integrated model by entering the reach for childcare, cheap mobility, public green places and participation in their climate action plans. These cities show that stability can exceed carbon extraction—it can be about making urban systems cheerful, inclusive and empowering. Thus, this goal determines the basis for seeing stability as a multi-dimensional and participation effort that lies in human escape and justice.

3.2 Synthesizing Theoretical Frameworks that Connect Sustainability, Gender, and Technology

The second goal is to seriously synthesize serious interdisciplinary principles that combine the extent of stability, gender law and digital changes. In advance is Amrit's ability to Sen's Ability (1999), which focuses development focus from income measurements to freedom, alternatives and capacity when analyzing penis access to education, health and digital units.

Construction of this, Crenshaw's (1989) The principle of renovation suggests how to identify individuals, experienced oppression based on identity—such as gender, race, square and disability. This is especially relevant in the digital age where the AI-operated systems risk strengthening structural inequalities unless they are designed with awareness of different user experiences.

In addition, Stilgoe has et al. It is consistent with feminist epistemology that criticizes critical neutrality and makes a lawyer for participation design.

Together, these theoretical lenses create a strong intellectual scaffolding to understand the mutual nature of sex and technology. They encourage us to reconsider innovation as a socially built-in procedure, not as a neutral good, but should be directed by values of equity, openness and justice.

3.3 *Highlighting Global Case Studies on Inclusive Sustainability*

This goal is focused on showing concrete, reference-specific models of human-focused stability, especially from different geographical regions, especially in the global south. An example is the floating garden of Bangladesh, which is an indigenous climate adaptation exercise, where women grow crops on the bed of floating beds in flood-exposed areas, supported by UNDP (2021), melts.

Another exemplary India is Arvind Eye Hospital, which provides affordable, AI-compatible eye care through a telemedicine model aimed at the infinitely rural population. The design emphasizes scalability, strength and sympathy, and shows how human-focused technologies can find health inequalities without giving up dignity (Prasad and Vasan 2020).

In addition, the Kenya M-Pesa platform has revolutionized financial inclusion for women by providing mobile-based banking services, getting millions out of poverty and strengthening female entrepreneurs. These cases of the case together confirm that inclusive, gender-covered and technology-accused practice are not only possible-they already change life, ecosystems and economies around the world.

3.4 *Identifying Emerging Tensions at the Intersection of Gender, Tech, and Sustainability*

While digital change promises progress, it also increases significantly moral stress, especially when technology is designed without sensitivity to gender, culture or class. This purpose examines these complications and starts with a digital gender department. According to GSMA (2023), women in low and moderate-or-in-the-country have become quite excluded from mobile internet use, digital literacy and technology-related job markets. This exclusion is not just technical—it refers to the extensive patriarchy systems, poverty and infrastructure neglect.

In addition, Eubanks (2018) said how algorithm decisions in public welfare systems can eliminate discrimination against breeds and gender. Automatic qualification equipment often abuses or punishes weaker home, causing unfair help. This data is complex of colonialism, a concept developed by Canary and Mejias, who criticizes how technical giants retrieve personal information from marginalized users without consent, openness or local benefits—creates a new form of digital abuse and addiction.

These examples reveal the unexpected results of exclusion design. By identifying such stresses, this chapter takes on CO design of technologies with affected communities, and ensures that innovation becomes a strength for liberation rather than exclusion.

3.5 *Envisioning Future Directions and Practical Implications*

The final goal explains a forward-looking vision to enter the human centenary in politics, practice and education. This asks for mission-driven innovation, as proposed by Majukato, where public investments actively form inclusive and durable markets instead of correcting the errors.

In education, experienced teaching models, interdisciplinary courses and gender-sensitive research are required to develop the next generation stability leaders. In workplaces, tools such as Microsoft Viva Insights and Wellbeing indexes begin to capture mental health, burn-out risk and balance between work and life, and emphasize that psychological welfare is integrated for social welfare stability (WHO 2022).

At the institutional level, the authorities ought to enforce shareholding digital policy, including obligatory algorithm revision and facilitate open data sovereignty and funding of inclusive design research. Meanwhile, the civil society organizations can give a voice to the ground to generate the localized stability models that are aligned with cultural values and human needs.

The bottom line is that this purpose suggests an explanation of the paradigm shift, Top-down system to participation, moral and inclusive future. Collective mandate must be human-oriented stability that will be founded on justice, innovation and flexibility.

4 Practical and Future Implications

The multi-stakeholder strategy required to attain human-centered sustainability must go beyond mere rhetoric to systems change. The cross-section between gender, technology, and sustainability cannot be looked at on a theory level alone but also in transformative practices in policymaking, institutional design, research, and community leadership. The section describes particular routes and advice that can guide the key actors to incorporate inclusive, ethical, and future-making principles into their spheres.

For Policymakers

- Incorporate impact assessment and digital rights of gender into every policy of climate and technology. The gender effect assessment (GIA) should become part of environmental and digital policies to be institutionally assessed by governments on how legislation, laws and financing currents can be differently impacted by women and marginalized societies. It aligns with gender mainstream in climate and digital rule on the part of the UN Women.

  Parallel, protection of digital rights—such as privacy, data ownership and algorithm transparency—by becoming a regulatory criteria, especially affects AI and surveillance technologies public service distribution and civilian freedom.

- **Promote gender-sensitive budgeting to fund inclusive innovation and healthcare** Gender-Northwest Public Finance is important to stop systemic intervals in health, education and technology. Governments should meet funds for female associated technological incubators, rural health programs and digital gender digital infrastructure. Such a budget is not only the same—it is financially strategic, as the evidence indicates that investment in women's education and digital inclusion provides high social returns.

For Organizations

- Build biased AI instruments and audits. Companies and institutions within the public sector that disseminate AI must transcend compliance and go into moral-for-design. This involves working with alternative training data sets, independent bias audits and allowing transparency on algorithm decision, particularly on employment, credit scoring and health care. Organizations can develop fair, transparent and socially accountable systems by aligning the standards, like the OECD AI principles (2019) and the EU AI Act.
- Introduce well-being models that are based on autonomy, belonging and purpose. It is not only productivity matrix that should be concerned with the stability of the workplace, but also the mental health of the employees, their dignity and contentment. The overall welfare model that provides flexible work, promotes psychological safety and facilitates objective-driven work cultures should be utilized by organizations. Such initiatives like Microsoft Viva Insights and SAPS Human Experience Management (HXM) demonstrate that technology may be utilized to monitor burn-out risk, facilitate the balance of work and life and enhanced real-time commitment.

For Academia and Research

- Develop transdisciplinary research and practice in the intersection of tech ethics, gender studies, and sustainability science. To address the challenges that occur at the intersection of gender, environment and technology, universities have to look beyond disciplinary silos. Other related research models integrating computer science, significant principles, feminists and system thinking may give more general and socially relevant knowledge. The transverse agenda is already in the priorities of financing agencies like Kshitij Europe and the Gates Foundation.

Democratizing research and innovation through open-access knowledge ecosystems. To guarantee that the knowledge economy has not marginalized the societies and global south researchers, institutions are supposed to facilitate open science, encourage civilian research and light linguistic publication forums. Such tools as the Open Science Framework of UNESCO and Scielo (in Latin America) describe how decentralized available knowledge could work equity in global scholarship and innovation.

In Case of Communities and Civil Society
Participation design and co-construction in health care, education and urban planning.

Social-centric solutions are most effective when they are crafted to be local, particularly women, youth and indigenous groups. Urban planning (e.g. Porto Alegre participation budget) and community school modalities and health co-operations assure not only increased efficiency of services, but also greater buses. These processes facilitate ground-level ownership, trust and flexibility.

Support technical governance, climate action and women management in collaboration economies.

Systemic changes are important by strengthening women in decision-making, new intellectuals and entrepreneurs. Civil society is able to spend on capacity programs, digital leadership academies and cross sectional mentors networks to influence AI policy, green infections and social enterprises. Scalable drawings can feed the inclusive leadership ecosystem by supplying the nourishment of platforms when he leads platforms such as he and women in AI.

5 Discussions

The human-oriented stability model holds the view that humans are active participants and not passive receptors of instructions since they are providers of knowledge and co-creators of remedies. This paradigm poses key questions unlike the traditional models of stability, which tend to favor carbon counting or GDP growth: Who benefits? Who decides? Who is excluded? These questions focus on structural disparities inherent in existing systems and shift the prism of technical efficiency to moral inclusion.

Such an approach is particularly required at the age of AI and climate adaptation, when technologies—uncontrolled otherwise, may either augment historical biases or disregard the weakest individuals to environmental chef. Every level of stability Built-in justice, morals and sympathy, collection to urban planning, course design to health services infrastructure, separating-pattern to which-has been long-term marginalized women, indigenous peoples and low-income communities.

The main conflict is evident: How do we accept the innovation and not to turn to technical socialism and think that everything can be decided by technology? It is not rejection they should answer, but radical redesign—via partnership regime, feminist AI framework, localized knowledge system and digital human rights protocol. The path also involves the necessity to engage with honor, a variety of uniformity in information. Things in this approach are not static but are built around them, which is the case in reality, cultural undertones and future aspirations.

6 Conclusion

Human-based stability is not merely a structure but a moral and strategic need. This compels us to re-initiate progress in manners that transgress existence, reinstating the same as a collective process towards mutual good, equity and harmony of worlds. It highlights this paradigm as an end and a means when we move into the twenty-first century—climate disorders, algorithms, growing inequality—and stress the urgency of this paradigm.

Individuals, particularly marginalized historically, are at the center of the stability endeavor, we clear space of transformation solutions that are just audio lists and social. The voices of women, indigenous people, young people and those who are marginalized are not marginal, but they are central to envision and actualize the flexible future. We should design not only human but skilled control systems, education, labor and health based on their leadership, innovations and experiences in life.

This chapter has given a platform through introducing the concepts to the crossroads of gender, welfare, technology and stability, the analysis of the case of the real world, and major stresses. The chapters which will go deeper into certain spheres reveal how human-centric and gendered e-based can transform the educational system, open the fields of work, redefine the health care allocation and democratize digital governance. At the bottom line, we should have stability as we do.

References

Chen M (2005) Rethinking the informal economy: linkages with the formal economy and the formal regulatory environment. UN DESA working paper series

Crenshaw K (1989) Demarginalizing the intersection of race and sex. Univ Chic Leg Forum 1989(1):139–167

Elkington J (1997) Cannibals with forks: the triple bottom line of 21st-century business. Capstone

Eubanks V (2018) Automating inequality: how high-tech tools profile, police, and punish the poor. St. Martin's Press

GSMA (2023) The mobile gender gap report 2023. Retrieved from https://www.gsma.com/

ILO (2021) World employment and social outlook—trends 2021

Kabeer N (2005) Gender equality and women's empowerment: a critical analysis of the third millennium development goal. Gender Dev 13(1):13–24

Leach M, Mehta L, Prabhakaran P (2016) Gender equality and sustainable development: a pathways approach. In: Leach M (ed) Gender equality and sustainable development. Routledge, pp 3–32

Norman DA (2013) The design of everyday things, revised and expanded edn. Basic Books

O'Neil C (2016) Weapons of math destruction: how big data increases inequality and threatens democracy. Crown

Prasad RM, Vasan S (2020) AI in eye care: redefining healthcare in rural India. Indian J Ophthalmol 68(12):2850–2855

Raworth K (2017) Doughnut economics: seven ways to think like a 21st-century economist. Chelsea Green

Rocheleau D, Thomas-Slayter B, Wangari E (1996) Feminist political ecology: global issues and local experiences. Routledge

Sachs JD (2015) The age of sustainable development. Columbia University Press

Sen A (1999) Development as freedom. Oxford University Press
Stilgoe J, Owen R, Macnaghten P (2013) Developing a framework for responsible innovation. Res Policy 42(9):1568–1580
Suri T, Jack W (2016) The long-run poverty and gender impacts of mobile money. Science 354(6317):1288–1292
UN Women (2020) Gender equality: women's rights in review 25 years after Beijing. Retrieved from https://www.unwomen.org
UNDP (2021) Promoting gender responsive climate action in Bangladesh. UNDP climate reports
WHO (2022) Mental health at work. Retrieved from https://www.who.int

Emotional Intelligence and Adolescent Adjustment: A Human-Centred Approach to Sustainable Well-Being

Sukanta Das

Abstract Adolescence is an early era marked by acute emotional, psychological, and social changes that often result in various types of adjustment issues. Emotional intelligence (EI) has emerged as an important aspect in helping people to understand and manage their feelings, as well as helping the youth interact with these issues. This study examines the impact of emotional intelligence on the adjustment capabilities of adolescents within a structure that gives importance to human-centred stability and total welfare. Research is inspired by increasing concerns about mental health issues of adolescents, as well as increasing concerns about long-term, individual-centric copying mechanisms. A quantitative technique was used to collect data from 102 teenagers using valid emotional intelligence and adjustment tests, which were complemented by qualitative interviews to help conclude. The study discovered an important inverse association between emotionally intelligent youth and adjustment problems, despite a high level of self-regulation, social ability, and adaptive imitation skills. Major findings suggest that EI dramatically helps reduce emotional, social, and educational adjustment difficulties. Conclusions also highlight the differences of gender and socio-economic status in EI and adjustment patterns, emphasizing the importance of teenagers. The importance of this study lies in addition to the discourse on long-term adolescent development by recommending the inclusion of emotional intelligence training in educational and community initiatives. This technique encourages emotional flexibility and helps adolescents to gain long-term personal and social benefits.

Keywords Emotional intelligence · Adolescent adjustment · Coping strategies · Human-centred approach · Sustainable wellbeing · Youth development

S. Das (✉)
Department of Geography, Dr. Bhupendra Nath Dutta Smriti Mahavidyalaya, Affiliated to The University of Burdwan, Hatgobindapur, West Bengal, India
e-mail: sukanta@dbndsm.ac.in

A. Singh et al. (eds.), *Human-Centered Sustainability: Bridging Well-Being, Gender Equity, and Technology*, World Sustainability Series,
https://doi.org/10.1007/978-3-032-22020-2_2

1 Introduction

Man is superior in all aspects of behaviour to any other in the universe, but humans are not identical; even an individual's growth pattern is unique and special. Every child develops at a different pace. Despite this, human life follows a systematic pattern. The helpless baby of today, who is an adult of tomorrow, has to cross various stages in the process of his development involving the development of his physical, intellectual, social, and emotional dimensions of personality. There are certain periods in the process of development, where certain characteristic features of behaviour stand out more prominently than in other periods of life; they are termed as a 'stage'. A stage is a time when specific changes take place. The development of the preceding stages serves as the foundation for each subsequent level. The activities that the youngster engages in throughout that time determine the designation of each stage. A person goes through four stages of development in their lifetime. They are the following stages: infancy (0–5 years), childhood (5–12 years), adolescence (12–19 years), and adulthood (after 19 years of age). Adolescence is a pivotal phase of human development characterized by profound physical, cognitive, emotional, and social transformations. It is a stage when individuals begin to form their identity, develop independence, and navigate complex interpersonal and societal expectations (Steinberg 2014; Sawyer et al. 2018). During this transitional period, adolescents face a range of adjustment challenges, including managing academic stress, peer relationships, family dynamics, emotional regulation, and societal norms (Eccles and Roeser 2011). Successful adjustment, defined as the individual's capacity to maintain psychological well-being, social harmony, and adaptive functioning in the face of these developmental tasks, is crucial for long-term mental health and societal integration (Arnett 1999; Compas et al. 2017). However, modern-day pressures ranging from academic competition to digital overload have exacerbated adjustment difficulties among adolescents, leading to increased rates of depression, anxiety, and social withdrawal (Twenge et al. 2017).

Amidst these challenges, the construct of Emotional Intelligence (EI) has gained significant attention in educational psychology and human development. Defined as the ability to perceive, understand, manage, and utilize emotions constructively (Salovey and Mayer 1990; Mayer et al. 2016), EI has been empirically linked with higher psychological resilience, better peer relationships, academic success, and reduced behavioural problems among adolescents (Brackett et al. 2012; Mavroveli et al. 2007). Emotional intelligence equips adolescents with the socio-emotional tools needed to navigate the complex landscape of adolescence, promoting self-regulation, empathy, and adaptive coping mechanisms (Goleman 1995; Petrides and Furnham 2001).

The novelty of this study lies in adopting a human-centered approach that integrates emotional intelligence into adolescent adjustment frameworks, not as a supplementary skill but as a foundational pillar of sustainable well-being. Unlike existing research that often isolates emotional intelligence or views adjustment as reactive behavior, this study proposes an integrative model that positions EI as a proactive, empowering process, enabling adolescents to thrive amidst uncertainty and change (Kumar and Muniandy 2012; Zeidner et al. 2009). Furthermore, the study emphasizes the role of context-sensitive EI interventions that account for cultural, gender, and socio-economic variables in adolescent development. The research problem centers on the limited implementation of emotional intelligence-based adjustment models within real-world educational and community settings. Despite the growing body of literature highlighting the benefits of EI, there is a significant gap in scalable, interdisciplinary, and culturally relevant programs targeting adolescent well-being (Durlak et al. 2011). From a business and policy perspective, this gap translates into rising costs in adolescent mental health care, lower productivity, school dropout rates, and underdeveloped human capital, especially in low-resource contexts (WHO 2021; OECD 2015). Aligned with the United Nations Sustainable Development Goals (SDGs), particularly SDG 3 (Good Health and Well-being), SDG 4 (Quality Education), and SDG 10 (Reduced Inequalities), this study contributes to designing inclusive and evidence-based adolescent development models. By promoting emotional intelligence as a tool for sustainable adjustment, the research advocates for systems-level change in education, mental health, and youth engagement policies, fostering emotionally resilient individuals capable of building cohesive and sustainable societies.

2 Literature Review

Emotional intelligence (EI) has developed as a fundamental psychological construct to explain the adaptive functioning of young people in different cultural, educational and socio-emotional contexts. International literature shows that emotional intelligence (EI) contributes significantly to school achievements, emotional regulation, relationships with peers and the general psychological adaptation of young people (Brackett et al. 2012; Mavroveli et al. 2007; Zeidner et al. 2009). Salovey and Mayer (1990) designed EI for the first time as a series of skills for the processing of emotional information that Goleman (1995) later made popular by linking it to life success. Studies in the USA and the UK underline that high emotional intelligence correlates with lower anxiety levels, better conflict management and improved social competence among young people (Mayer et al. 2016; Petrides et al. 2004). Durlak et al. (2011) through a meta-analysis that school-based social and emotional learning programmes (SEL) improve the social behaviour and emotional skills of students while reducing behavioural problems and emotional stress. In European contexts, emotional intelligence was associated with positive development of identity among young people and mental well-being (Sánchez-Álvarez et al. 2016; Extremera and Fernández-Berrocal 2005). Australian researchers like Parker et al. (2010) emphasised the importance of EI in reducing early school leaving rates and promoting engagement. In Asian contexts, especially in China and Japan, researchers have focused on culture-specific emotional competences and their influence on the mental health of young people (Li et al. 2020; Saklofske et al. 2003). In India, emotional intelligence has been increasingly investigated as a mediator between stress and adaptation in adolescents (Kaur and Kaur 2019; Singh and Sharma 2012). Indian studies emphasize the role of the family environment, cultural norms and gender differences in the design of emotional expression and regulation (Rani and Kumari 2018; Malhotra and Dutta 2019). Emotional intelligence was associated with better academic results, self-esteem and management mechanisms among Indian youth, especially in urban school environments (Mangal and Mangal 2015; Sharma 2011). In addition, Indian researchers have developed native instruments such as the Emotional Intelligence Scale (Hyde et al. 2001) to evaluate EI in different populations. Recent work in India also combines EI with social justice and inclusive education and advocates integration into the school curricula for holistic development (Garg and Rastogi 2009; Sharma and Bhardwaj 2021). EI was associated with sustainable well-being worldwide by providing young people with emotional resilience needed in rapidly changing socio-economic environments (Bar-On 2006; Schutte et al. 2007). Researchers agree that EI promotes empathy, ethical awareness, and social responsibility, key factors in the human-centered approach to youth development (Lomas et al. 2012; Lopes et al. 2003). A growing number of works brings research on emotional intelligence (EI) in line with the Sustainable Development Goals (SDGs), in particular SDG 3 (Health and Wellbeing) and SDG 4 (High Quality Education), and supports emotionally intelligent young people as actors of sustainable social change (UNDP 2015; OECD 2015). However, there are gaps in the contextualisation of EI interventions in various socio-cultural contexts, particularly in low-income and rural communities. This study therefore seeks to close the theoretical and practical gap by proposing a human-centered, emotionally intelligent model for adapting young people tailored to the Indian socio-cultural landscape.

3 Rationale of the Study

Adolescence is a crucial stage in a person's life since it prepares them for maturity. This phase, which starts with puberty, is characterized by identity vs. role confusion. Teenagers are forced to ask themselves questions due to the abrupt changes in their bodies and minds as well as the shifting expectations of society. The success of the nation in many facets of life rests on the appropriate upbringing of teenagers. Emotional intelligence (EI) plays a pivotal role in shaping adolescent adjustment, particularly in the context of their psychological, social, and academic well-being. Adolescence is recognized as a time of "emotional upheavals," with a variety of pressures, uncertainties, failures, and disappointments that can be handled more successfully with the right guidance and direction by using emotional intelligence as a tool. This study aims to address this gap by examining how emotional intelligence influences adolescent adjustment through a comprehensive, human-centric framework. The findings are expected to provide valuable insights for educators, psychologists, and policymakers in developing interventions that foster emotional competence, thereby contributing to the sustainable development of human capital.

Additionally, the study aligns with the broader objective of promoting mental health as a vital component of sustainable wellbeing, reinforcing its significance in academic and community settings.

4 Research Methodology

4.1 Objectives of the Study

The main objectives of this study are:

- To study the adjustment problems of adolescents based on gender.
- To study the ways of coping among adolescents based on gender.
- To study the effect of emotional intelligence on the adjustment problems of adolescents.
- To study the effect of emotional intelligence on ways of coping among adolescents.
- To study the interaction between sex and emotional intelligence for adjustment problems of adolescents.
- To study the interaction between sex and emotional intelligence for ways of coping among adolescents.
- To study the interaction among adjustment problems, ways of coping, and emotional intelligence among adolescents.

4.2 Hypotheses of the Study

In this study, the following hypotheses were taken to be adapted.

- There is no significant difference in the adjustment problems of adolescents based on gender.
- There is no significant difference in the ways of coping of adolescents based on gender.
- There is no significant difference in adjustment problems of adolescents with emotional intelligence.
- There is no significant difference in the ways of coping with emotional intelligence among adolescents.
- There is no interaction between sex and emotional intelligence for adjustment problems of adolescents.
- There is no interaction between sex and emotional intelligence for ways of coping among adolescents.
- There is no interaction among adjustment problems, ways of coping and emotional intelligence among adolescents.

4.3 Limitations of the Study

The study was limited to the following areas:

1. The study was limited to 102 adolescents only.
2. The study was limited to adolescents in the age group of 12–18 years of both sexes.
3. The study was conducted only within the 17 educational institutions of the Hooghly district in West Bengal.

4.4 Sample Collection

Simple random sampling has been used to collect the information from 17 educational institutions. A sample of 102 adolescents between the ages of 12 and 18 years (studying in classes seventh to 12th), consisting of 51 boys and 51 girls, 6 students from a single institution, consisting of 3 boys and 3 girls.

4.5 Tools Used

In the present study, the tools used for collecting data were:

1. To study the emotional intelligence of adolescents, an emotional intelligence test by Khera and Sarbjit.
2. To study the adjustment problems of adolescents, an adjustment inventory by Mittal.
3. To study the ways of coping among adolescents, a questionnaire by Folkman and Lazarus.

4.6 Design of Study

To study the role of emotional intelligence in adjustment problems and coping strategies, a factorial design involving a two-way analysis of variance (2×3) was used. For this purpose, sex (males and females) and emotional intelligence (high, average, and low) were taken as independent variables, whereas the variable of adjustment problems was kept as a dependent variable. The same was performed once more, using sex and emotional intelligence as independent variables and coping strategies as the dependent variable.

4.7 *Statistical Techniques Used*

1. Descriptive analyses, such as mean and standard deviation, were calculated to compare the average scores of various groups.
2. 'F' ratios were worked out to study the interaction of adjustment problems of adolescents and their ways of coping with sex and emotional intelligence.
3. 't' ratios were calculated to find out the significance of differences between means among various groups.

5 Results and Discussion

Adolescence is the most anxious and critical stage of a person's development. Poets have defined it as the spring of human life and a significant era in the overall lifespan. It is also known as the transitory era of life. Adolescence typically lasts from the ages of 12 to the early 20s. Adolescents become wiser, more sophisticated, and better in respect of making their own decisions. They are permitted to work, to get married, and to establish their different rights. They are expected to be able to support themselves financially and thus to adjust to society. Several studies have been conducted so far to understand the concept, nature, and problems of adolescence. Charlotte Pope made an extensive study of the problems of adolescence and reported the following areas of problems faced by them, like in the areas of finance, health adjustment, adjustment, physical development, sex, and occupation. Adjustment is an essential characteristic for a happy life in a society. It assists in keeping basic impulses at manageable levels, believing in one's talents, and achieving desired reinforcements. Thus, adjustment promotes self-directed growth and development in intellectual, emotional, social, physical, and vocational domains. It is the process of developing a pleasant psychological relationship as well as ongoing interaction between the individual and his surroundings. Emotional intelligence is a new concept, and it means making decisions and solving problems within the context of situations and interpersonal relations.

Emotional intelligence is the ability to perceive, appraise, and produce emotions to support ideas, comprehend emotions and emotional knowledge, and carefully control emotions in order to promote both intellectual and emotional development. Emotional intelligence, as opposed to IQ, is something that can be developed over time. Throughout a lifetime, emotional intelligence tends to increase as people learn to better manage difficult emotions by being more aware of their moods. In other words, as the person matures, he acquires certain emotional competencies that lead to outstanding performances. A human-centred approach to EI emphasizes the holistic development of adolescents, focusing not only on academic success but also on social and emotional well-being. After careful investigation and statistical analysis, the following results have emerged.

5.1. The effect of the independent variable of sex on the criterion variable of adjustment problems of adolescents was examined by employing the 't-test'.

The result seen (Table 1) that the 't' ratio (0.92) for the difference between means of boys and girls in adjustment problems is non-significant at the 0.05 and 0.01 level of significance, i.e., 1.96 and 2.58, respectively. Thus, the hypothesis I stated, "There is no significant difference in adjustment problems of adolescents based on gender," stands accepted. It might be concluded that both boys and girls, because of their education and parental guidance, don't face any adjustment problems at their emotional level.

5.2. The effect of the independent variable of sex on the criterion variable of ways of coping among adolescents was examined by employing the 't-test'.

It is revealed from Table 2 that the calculated value of 't' i.e., 0.875, is less than the tabular value of 't' i.e., 1.96 and 2.58 at both 0.05 and 0.01 level of significance. Therefore, Hypothesis II, indicating there is no significant difference between the means of boys and girls in ways of coping among adolescents based on gender, is accepted.

5.3. The main effect of the independent variable of emotional intelligence on the criterion variable of adjustment problems of adolescents was examined by employing the technique of "t-test".

Table 3 shows that the t-ratio for the difference between means of the low and average group (2.60) was found to be significant at 0.05 level of significant whereas the t-ratio for the difference between means of other two groups namely average and high (1.0) and low and high (1.2) were found to be non-significant at both 0.05 and 0.01 levels of significance. Table 3 clearly shows that adolescents with an average level of emotional intelligence face fewer adjustment problems than the other two groups, namely those with a low level and a high level of emotional intelligence. Thus, the

Table 1 Results of the adjustment problems of boys and girls

S. No.	Sex	Mean	S.D.	SE_D	t-ratio	Level of significance
1.	Girls	192.50	19.39	0.90	0.92	N.S.*
2.	Boys	193.33	22.63			

N.S.* → non-significant

Table 2 Showing the ways of coping of boys and girls

S. No.	Sex	Mean	S.D.	SE_D	t-ratio	Level of significance
1.	Girls	1.55	0.38	0.08	0.875	N.S.*
2.	Boys	1.62	0.44			

N.S.* → non-significant

hypothesis that "There is no significant difference in the adjustment problems of adolescents concerning emotional intelligence" stands rejected.

5.4. The main effect of the independent variable of emotional intelligence on the criterion variable of ways of coping among adolescents was made through 't-test'.

It is clear from Table 4 that the t-ratio for the difference between the means of low and average group (0.36) and the t-ratio of average and high group (1.10) were found to be non-significant at 0.05 level of significance, whereas the t-ratio for the difference between the means of high and low group (3.99) is greater than the table value of t-ratio, i.e., 1.96 and 2.58 at both level 0.05 and 0.01 level of significance. Further, it is clear from Table 4 that the adolescents having a high level of emotional intelligence, as compared to other groups, were proven to use maximum coping behaviour. Thus, the hypothesis that "There is no significant difference in ways of coping concerning emotional intelligence among adolescents" stands rejected.

5.5. The interactional effect of independent variables of sex and emotional intelligence on the criterion variable of adjustment problems of adolescents, the data was analyzed by employing the technique of two-way (2 × 3) analysis of variance (F-ratio).

It is seen from Table 5 that the interaction between sex and emotional intelligence is not significant because the computed F value, i.e. 0.41, is lower than the required value for significance at 0.05 levels. It shows that both sex and emotional intelligence have no significant interaction on adjustment problems.

Table 3 Showing adjustment problems and emotional intelligence

S. No.	Level of intelligence	Mean	S.D.	SE_D	t-ratio	Level of significance
1.	Low emotional intelligence	213.6	23.95	2.68	2.60	S# (low and average)
2.	Average emotional intelligence	220.66	12.69	3.02	1.0	N.S.* (average and high)
3.	High emotional intelligence	209.99	17.91	2.95	1.2	N.S.* (low and high)

S# → significant; N.S.* → non-significant

Table 4 Showing ways of coping and emotional intelligence

S. No.	Level of intelligence	Mean	S.D.	t-ratio	Level of significance
1.	Low emotional intelligence	1.54	0.58	0.36	N.S.* (low and average)
2.	Average emotional intelligence	1.79	0.34	1.10	N.S.* (average and high)
3.	High emotional intelligence	1.82	0.22	3.99	S# (high and low)

S# → significant; N.S.* → non-significant

Table 5 Summary of two-way (2 × 3) Analysis of variance (sex × emotional intelligence) on adjustment problems

S. No.	Source of variation	df	SS	MS 3638.05	F	Level of significance
1.	Sex (A)	1	3638.05		3.05	N.S.*
2.	Emotional intelligence (B)	2	2875	1437.5	1.20	N.S.*
3.	Interaction (A × B)	2	979.94	489.97	0.41	N.S.*
4.	Within	96	114,508.8	1192.80		
	Total	101				

N.S.* → non-significant

* Value of F to be significant for: df (1, 96) → at 0.05 = 3.94; at 0.01 = 6.90 df (2, 96) → at 0.05 = 3.09; at 0.01 = 4.82

It is seen from Table 6 that the 't' ratios for low and average group, i.e., 1.90; for average and high group, i.e., 1.33 and for high and low group, i.e., 0.12, are less than the table value of 't' i.e., 1.96 and 2.58 at both 0.05 and 0.01 level of significance. It implies that there exists no significant difference between the means of the three groups (namely, low and average; average and high and High and Low) in the case of girls.

It is seen from Table 7 that the 't' ratio for low and average group i.e. 1.66, 't' ratio for average and high group i.e. 1.89 and 't' ratio for high and low group i.e. 0.76 are less than the table value of 't' i.e. 1.96 and 2.58 at both 0.05 and 0.01 level of significance. It implies that there exists no significant difference between the means of the three groups (namely low and average; average and high and high and low) in case of boys. Thus, the hypothesis that "There is no interaction between sex and emotional intelligence for adjustment problems of adolescents" stands accepted.

5.6. The interactional influence of the independent variables of sex and emotional intelligence on the criterion variable of ways of coping among adolescents was investigated using the methodology of two-way (2 × 3) analysis of variance (F-ratio).

It is seen from Table 8 that the interaction between sex and emotional intelligence on ways of coping is not significant because the computed F value i.e. 0.44 is lower

Table 6 Adjustment problems and emotional intelligence (in case of girls)

S. No.	Level of intelligence	Mean	S.D.	t-ratio	Level of significance
1.	Low emotional intelligence	302.47	33.95	1.90	N.S.* (low and average)
2.	Average emotional intelligence	313.43	19.96	1.33	N.S.* (average and high)
3.	High emotional intelligence	305.47	38.93	0.12	N.S.* (high and low)

N.S.* → non-significant

Table 7 Table showing adjustment problems and emotional intelligence (in case of boys)

S. No.	Level of intelligence	Mean	S.D.	t-ratio	Level of significance
1.	Low emotional intelligence	313.46	19.96	1.66	N.S.* (low and average)
2.	Average emotional intelligence	319.29	15.66	1.89	N.S.* (average and high)
3.	High emotional intelligence	310.36	21.05	0.76	N.S.* (high and low)

N.S.* → non-significant

than the required value for significance at both 0.05 and 0.01 levels of significance. It shows that both sex and emotional intelligence have no significant interaction on ways of coping.

It is seen from Table 9 that the 't' ratios for low and average group i.e. 0.36; for average and high group i.e. 1.10 and for high and low group i.e. 1.90 and less than the table value of 't' i.e. 1.96 and 2.58 at both 0.05 and 0.01 level of significance. It implies that there exists no significant difference between means of three groups (namely low and average; average and high and high and low) in case of girls.

It is clear from Table 10 that the t-ratios for low and average group (0.349); for average and high group (1.09) and high and low group (1.93) are less than the table value of 't' i.e. 1.96 and 2.58 at both level 0.05 and 0.01 level of significance. It implies that there exists no significant difference between the means of the three groups (namely, low and average, average and high, and High and Low) in the case

Table 8 Summary of two way (2 × 3) analysis of variance (sex × EI on ways of coping strategies)

S. No.	Source of variation	df	MSV	F	Level of significance
1.	Sex (A)	1	0.25	2.21	N.S.*
2.	Emotional intelligence (B)	2	1.47	13.0	S#
3.	Interaction (A × B)	2	0.005	0.44	N.S.*
4.	Within	96	0.113		
	Total	101			

S# → significant; N.S.* → non-significant

* Value of F to be significant for: df (1, 96) → at 0.05 = 3.94; at 0.01 = 6.90 df (2, 96) → at 0.05 = 3.09; at 0.01 = 4.82

Table 9 Ways of coping and emotional intelligence (in case of girls)

S. No.	Level of intelligence	Mean	S.D.	t-ratio	Level of significance
1.	Low emotional intelligence	1.49	0.53	0.36	N.S.* (low and average EI)
2.	Average emotional intelligence	1.749	0.34	1.10	N.S.* (average and high EI)
3.	High emotional intelligence	2.240	0.18	1.90	N.S.* (high and low EI)

S# → significant; N.S.* → non-significant

of boys. Thus, the hypothesis that "There is no interaction between sex and emotional intelligence for ways of coping among adolescents" stands accepted.

5.7. The interactional effect of adjustment difficulties, coping strategies, and emotional intelligence was investigated using the 3-way analysis of variance (F-ratio) technique.

Table 11 depicts that:

1. The F-ratio of interaction between emotional intelligence and adjustment problems (A × B) is 1.0, which is less than the table value of F (18.41) at 0.05 level of significance. Thus, the connection between emotional intelligence and adjustment difficulty is revealed to be non-significant.
2. The F-ratio of interaction between adjustment problems and ways of coping (B × C) is 9.5, which is less than the table value of F (18.41) at 0.05 level of significance. So, interaction between adjustment problems and ways of coping is found to be non-significant.
3. Similarly, F ratio of interaction between emotional intelligence and ways of coping (A × C) is 1.23, which is less than the table value of F (18.41) at 0.05 level. So, interaction between emotional intelligence and ways of coping is found to be non-significant.
4. Again, there is no significant interaction effect between these three variables, i.e. emotional intelligence, adjustment problems and ways of coping (A × B × C) as the value of F (0.05) is not significant at both 0.05 and 0.01 level of significance.

Table 10 Ways of coping and emotional intelligence (in case of boys)

S. No.	Level of intelligence	Mean	S.D.	t-ratio	Level of significance
1.	Low emotional intelligence	1.594	0.54	0.349	N.S.* (low and average EI)
2.	Average emotional intelligence	1.756	0.37	1.09	N.S.* (average and high EI)
3.	High emotional intelligence	1.826	0.31	1.93	N.S.* (high and low EI)

N.S.* → non-significant

Table 11 Summary of three-way analysis of variance (emotional intelligence × adjustment difficulties × coping strategies)

S. No.	Source of variation	df	MSV	F	Level of significance
1.	Emotional intelligence (A)	2	360.00	54.88	N.S.*
2.	Adjustment problems (B)	1	90.00	13.72	N.S.*
3.	Ways of coping (C)	1	122.5	18.37	N.S.*
4.	A × B	2	2.50	1.0	N.S.*
5.	B × C	1	62.50	9.53	N.S.*
6.	A × C	2	2.50	1.23	N.S.*
7.	A × B × C	2	6.56	0.05	N.S.*

N.S.* → non-significant

Thus, the hypothesis that "There is no interaction among adjustment problems, ways of coping, and emotional intelligence among adolescents" stands accepted.

6 Major Findings of the Study

The major findings of the present study are:

1. Adolescents do not differ significantly in having adjustment problems on account of their sex.
2. Adolescents do not differ significantly in the ways of coping on account of their sex.
3. Emotional intelligence accounts for significant differences in the adjustment problems of adolescents.
4. Adolescents with an average level of emotional intelligence face less adjustment problems as compared to the high and low emotional intelligence groups.
5. The interactional effect of two variables (sex × emotional intelligence) does not account for a significant difference in adjustment problems among adolescents.
6. Emotional intelligence accounts for significant differences in the ways of coping among adolescents.
7. Adolescents with a high level of emotional intelligence show more coping behaviour as compared to average and low emotional intelligence groups.
8. There is no substantial connection between sex and emotional intelligence in coping strategies.
9. There is no interaction between emotional intelligence, adjustment problems and ways of coping among adolescents.
10. Trends of means show that:

    (a) Boys experience fewer adjustment issues compared to girls.
    (b) Boys use more coping behaviour as compared to girls.
    (c) Adolescents with average emotional intelligence face minimum adjustment problems.
    (d) Girls with low emotional intelligence use minimum coping behaviour.

7 Recommendations

Since the teenage stage is considered to be the most important stage of one's life. It's full of challenges, and as a result, they have problems adjusting to society, family, and themselves. Most of them are suffering from depression, conflict, complications, concerns, and concerns. They face difficulty in pursuing physical, emotional, and social coordination. Therefore, in order to support the normal growth and development of adolescents, they need to meet their needs before properly resolving the problems. Young people need a balanced diet, knowledge of health, personal hygiene, cleanliness and various diseases to maintain their physical and mental health. Sex plays a dominant role during adolescence, which requires adequate education about sex, hygiene and physiology. Promoting extracurricular activities and social situations can help to eliminate misunderstandings and promote healthy adaptation. Young people long for recognition and self-esteem, and teachers and parents should seek their opinions without offending them. Emotional education is crucial for young people, because they are restless, easily irritable and easily deceivable by political parties and opportunist politicians. Parents and teachers should convey emotional education, discipline their emotions and direct their resources to positive purposes. Young people have broad interests and skills, and discovering their unique interests and skills requires careful consideration. The curriculum should allow students to choose from a variety of subjects and activities based on their skills and personalities. Moral and religious education can greatly promote the growth of young people, but it should not be limited to imparting moral knowledge.

Based on the findings of this study, several basic recommendations are proposed to improve adolescents' adaptation through emotional intelligence within a human-centric context. Firstly, integrate emotional intelligence education into normal school programmes at all levels, with structured courses focused on self-awareness, emotion regulation, empathy, and conflict resolution. Teachers and school counsellors should get formal training in how to help students develop emotional intelligence so that they can better support them and make classes more emotionally welcoming. Second, interdisciplinary collaboration between psychologists, educators, community leaders, and policy makers in diverse socioeconomic contexts, especially in disadvantaged and rural areas, is critical to designing culturally responsive EI interventions. Thirdly, parents' involvement should be emphasized through laboratories and resources that allow families to support emotional learning at home, strengthening a consistent developmental environment. Fourthly, emotional intelligence should be incorporated into national youth policies and public health agendas, recognising its preventive role in reducing adolescents' mental health problems, substance abuse, intimidation, and school leaving rates. In addition, the indicators of emotional intelligence should be included in student evaluations, not just as academic indicators but as indicators of holistic well-being. Fifthly, mobile applications in digital platforms can be used to make education in emotional intelligence accessible to adolescents in remote areas, encouraging self-regulated and gambled learning experiences. Finally, future research should adopt long-term and intercultural approaches to examine the lasting effect of emotional intelligence on adolescents' development and the well-being of society, ensuring that interventions are based on evidence and adaptable. These integrated, multi-level strategies can effectively integrate emotional intelligence into adolescents' development, aligned with the Sustainable Development Goals (SDGs 3, 4, and 10) and ensure emotionally resilient, empathetic, and socially responsible future citizens who contribute essentially to sustainable and inclusive societies.

8 Implications

Emotional intelligence (EI) is a vital element of adolescent development, emphasizing its role in strengthening emotional regulation, decision-making, interpersonal relationships, and resilience. This finding encourages teachers and curriculum designers to review the traditional academically central education system and to incorporate socio-emotional learning as a key element, not just as an addition. By placing emotional intelligence as a tool for sustainable well-being, research also highlights its preventive potential in tackling the growing mental health challenges among adolescents, such as depression, anxiety, peer pressure, and behavioural disorders. For school systems and youth organisations, the study provides a strong empirical basis for implementing structured, culturally sensitive interventions based on emotional intelligence (EI) to support holistic development. In the context of public health and youth policy, research supports a shift towards human-centric contexts that treat adolescents as active factors in their emotional development, aligned with the Sustainable Development Goals 3 (Good Health and Well-being), 4 (Quality Education), and 10 (Reduction of Inequalities). In addition, it proposes that investing in the development of emotional intelligence can reduce long-term social costs by reducing juvenile delinquency, school dropouts, and youth unemployment. Finally, research has theoretical implications for promoting interdisciplinary models that incorporate emotional intelligence, developmental psychology, and sustainability studies, providing a comprehensive perspective on understanding and cultivating adolescents' adaptation into a rapidly changing world.

9 Conclusion

This study reinforces the importance of emotional intelligence in encouraging healthy adolescents' adjustments through a human-centered approach that promotes long-term well-being. Emotional intelligence provides teenagers with the abilities necessary for emotional regulation, empathy, interpersonal competence, and adaptive coping, allowing them to handle adolescence's complex emotional, social, and intellectual obstacles. This study bridges the gap between theory and practice by presenting EI as a fundamental construct rather than an auxiliary ability, providing a comprehensive framework that addresses both individual development and broader socio-educational goals. The incorporation of EI into school curricula, mental health programs, and community initiatives has revolutionary potential, particularly in culturally diverse and resource-constrained environments. The study supports the United Nations Sustainable Development Goals through promoting inclusive, emotionally supportive environments that foster resilient, empathetic, and socially responsible youth. Fostering emotional intelligence during adolescence is vital not just for personal growth but also for developing emotionally intelligent future citizens capable of contributing to a more equitable and sustainable society.

References

Arnett JJ (1999) Adolescent storm and stress, reconsidered. Am Psychol 54(5):317–326. https://doi.org/10.1037/0003-066X.54.5.317

Bar-On R (2006) The Bar-On model of emotional-social intelligence. Psicothema 18:13–25. https://doi.org/10.1037/t13989-000

Brackett MA, Rivers SE, Salovey P (2012) Emotional intelligence: implications for personal, social, academic, and workplace success. Soc Personal Psychol Compass 5(1):88–103. https://doi.org/10.1111/j.1751-9004.2010.00334.x

Compas BE, Jaser SS, Bettis AH et al (2017) Coping, emotion regulation, and psychopathology in childhood and adolescence: a meta-analysis and narrative review. Psychol Bull 143(9):939–991. https://doi.org/10.1037/bul0000110

Durlak JA, Weissberg RP, Dymnicki AB et al (2011) The impact of enhancing students' social and emotional learning: a meta-analysis of school-based universal interventions. Child Dev 82(1):405–432. https://doi.org/10.1111/j.1467-8624.2010.01564.x

Eccles JS, Roeser RW (2011) Schools as developmental contexts during adolescence. J Res Adolesc 21(1):225–241. https://doi.org/10.1111/j.1532-7795.2010.00725.x

Extremera N, Fernández-Berrocal P (2005) Perceived emotional intelligence and life satisfaction. Personal Individ Differ 39(5):937–948. https://doi.org/10.1016/j.paid.2005.03.012

Garg P, Rastogi R (2009) Effect of emotional intelligence on work values. Int J Bus Manag 4(3):147–153. https://doi.org/10.5539/ijbm.v4n3p147

Goleman D (1995) Emotional intelligence: why it can matter more than IQ. Bantam Books, New York

Hyde A, Pethe S, Dhar U (2001) Emotional intelligence scale (EIS). Vedant Publications, Lucknow

Kaur H, Kaur R (2019) Emotional intelligence and adjustment of adolescents. Int J Indian Psychol 7(3):763–769. https://doi.org/10.25215/0703.091

Kumar S, Muniandy B (2012) Emotional intelligence and academic achievement among adolescents. Int J Acad Res 4(3):193–199. https://doi.org/10.7813/2075-4124.2012/4-3/B.29

Li X, Wang Y, Wang Y (2020) Emotional intelligence and academic stress among Chinese students. Front Psychol 11:350. https://doi.org/10.3389/fpsyg.2020.00350

Lomas T et al (2012) The impact of emotional intelligence training in a UK University. Br J Guid Couns 40(4):395–410. https://doi.org/10.1080/03069885.2012.702074

Lopes PN et al (2003) Emotional intelligence and social interaction. Personal Soc Psychol Bull 29(8):1018–1034. https://doi.org/10.1177/0146167203253479

Malhotra S, Dutta R (2019) Gender differences in emotional intelligence among adolescents. Indian J Posit Psychol 10(2):134–137. https://doi.org/10.15614/ijpp.v10i02.19010

Mangal SK, Mangal S (2015) Emotional intelligence: theoretical and practical perspectives. PHI Learning

Mavroveli S, Petrides KV, Rieffe C, Bakker F (2007) Trait emotional intelligence, psychological well-being and peer-rated social competence in adolescence. Br J Dev Psychol 25(2):263–275. https://doi.org/10.1348/026151006X118577

Mayer JD, Caruso DR, Salovey P (2016) The ability model of emotional intelligence. Emot Rev 8(4):290–300. https://doi.org/10.1177/1754073916639667

OECD (2015) Skills for social progress: the power of social and emotional skills. https://doi.org/10.1787/9789264226159-en

Parker JDA et al (2010) Academic achievement and emotional intelligence. Learn Individ Differ 20(1):44–48. https://doi.org/10.1016/j.lindif.2009.10.003

Petrides KV, Furnham A (2001) Trait emotional intelligence: psychometric investigation with reference to established trait taxonomies. Eur J Personal 15(6):425–448. https://doi.org/10.1002/per.416

Petrides KV et al (2004) Emotional intelligence and psychological well-being. Personal Individ Differ 36(2):277–293. https://doi.org/10.1016/S0191-8869(03)00087-6

Rani R, Kumari P (2018) Emotional intelligence among adolescents: a gender-based study. J Psychosoc Res 13(2):431–438. https://doi.org/10.32381/JPR.2018.13.02.9

Saklofske DH et al (2003) Assessing emotional intelligence in Japan. Personal Individ Differ 34(4):707–721. https://doi.org/10.1016/S0191-8869(02)00063-X

Salovey P, Mayer JD (1990) Emotional intelligence. Imagin Cogn Pers 9(3):185–211. https://doi.org/10.2190/DUGG-P24E-52WK-6CDG

Sánchez-Álvarez N et al (2016) Trait emotional intelligence and subjective well-being. Personal Individ Differ 92:145–149. https://doi.org/10.1016/j.paid.2015.12.017

Sawyer SM, Azzopardi PS, Wickremarathne D, Patton GC (2018) The age of adolescence. Lancet Child Adolesc Health 2(3):223–228. https://doi.org/10.1016/S2352-4642(18)30022-1

Schutte NS et al (2007) A meta-analytic investigation of the relationship between emotional intelligence and health. Personal Individ Differ 42(6):921–933. https://doi.org/10.1016/j.paid.2006.09.003

Sharma A (2011) Role of emotional intelligence in academic achievement. Indian J Psychol Sci 2(1):78–83

Sharma R, Bhardwaj R (2021) Emotional intelligence and inclusive education. Indian J Educ Inf Manag 10(2):34–40

Singh D, Sharma R (2012) Emotional intelligence and stress in adolescence. J Indian Acad Appl Psychol 38(1):66–73

Steinberg L (2014) Age of opportunity: lessons from the new science of adolescence. Houghton Mifflin Harcourt

Twenge JM, Joiner TE, Rogers ML, Martin GN (2017) Increases in depressive symptoms, suicide-related outcomes, and suicide rates among U.S. adolescents after 2010 and links to increased new media screen time. Clin Psychol Sci 6(1):3–17. https://doi.org/10.1177/2167702617723376

UNDP (2015) Transforming our world: the 2030 agenda for sustainable development. United Nations. https://doi.org/10.18356/6e168f44-en

WHO (2021) Mental health of adolescents. World Health Organization. https://doi.org/10.4060/cb4199en

Zeidner M, Matthews G, Roberts RD (2009) What we know about emotional intelligence: how it affects learning, work, relationships, and our mental health. MIT Press. https://doi.org/10.7551/mitpress/7557.001.0001

Theorising Towards People-Centered Sustainability

Jashwini Jothishna Narayan

"The philosophical point is that our happiness and well-being is not based on income rising. This is not just the wisdom of sages but of ordinary people. Prosperity is more social and psychological: it's about identification, affiliation, participation in society and a sense of purpose." Tim Jackson

Abstract This paper aims to work towards theorising people-centered sustainability. At a broader level, people-centeredness requires incorporating the needs and views of all in various developments. Largely, research done thus far on this topic do not propose a wholesome framework for wider applicability. Much of the current research focus separately on specific areas only such as health, gender equality, education, solidarity, climate or impact of technology. This study is undertaken against this backdrop. A recent research highlights social sustainability as the 'forgotten' pillar of sustainability. Based on relevant secondary data reviews, this paper brings together a number of factors that may influence people-centered sustainability, keeping in mind the effect these will have on the general populace of any given country, leading to a wholesome conceptual framework and survey questionnaire. To this end, the chapter identified factors such as Equality, respect and inclusion; Solidarity and peace; Income redistribution/social responsibility; Universal healthcare; Justice; Strong institutions and Environmental protection. This study helps to better understand factors that may strongly influence people-centered sustainability across diverse country contexts. The study will prove effective for future bigger scale empirical research which can test these factors as variables. Future researchers can use these same factors to further this study by using a structural equation model. These factors

This chapter is dedicated to all those who are disadvantaged, treated unfairly and yearn for a better quality of life. It is also dedicated to those who tirelessly and genuinely continue to persist in people-centered sustainability initiatives.

J. J. Narayan (✉)
School of Business and Management, The University of the South Pacific, Laucala Campus, Suva, Fiji
e-mail: jashwini.narayan@usp.ac.fj

A. Singh et al. (eds.), *Human-Centered Sustainability: Bridging Well-Being, Gender Equity, and Technology*, World Sustainability Series,
https://doi.org/10.1007/978-3-032-22020-2_3

have not been discussed together, in the same manner, in any other prior study, to the best of the researcher's knowledge. Scholars, governments, bureaucrats, NGOs and policy makers will benefit from this chapter's recommendations.

Keywords People-centered sustainability · Framework

1 Introduction

The concept of sustainable development is no new phenomenon and is one that continues to evolve (Tilbury and Wortman 2004). Much has been discussed about what sustainable development constitutes or means (Tilbury and Wortman 2004). The most notable development on this concept materialised in 1990s, post the 1992 UNCED conference (United Nations 1992a, b), following which, numerous definitions emerged, one of the most prominent being, the triple-bottom-line approach that includes the dimensions of environment, economic and social factors (Elkington 1999).

Literature on indicators that influence sustainability, well-being and quality-of-life has no doubt been abound over the past 20 or so years including the amount of work that has been done to quantify as well qualify such concepts (Magee et al. 2012). Few scholars, such as Tilbury and Wortman (2004) likened people-centered sustainability to better quality of life enhanced through ecological sustainability, equity as well as fair distribution of wealth with access to resources.

The relevance of this study stems from the limitations of past research which this paper tries to address. For instance, despite much research, there is little consensus on what a better quality of life means. In other words, the way we interpret a better quality of life is open to different interpretations, in that it can be interpreted in various ways by different people and/or cultural segments (Tilbury and Wortman 2004). Magee et al. (2012) highlight that literature clearly informs that there are numerous divergent views and as such, a lack of consensus on what constitutes sustainability (see Bohringer and Jochem 2007; Parris and Kates 2003). Scholars like Atkinson (2000) report that among the various definitions of sustainability, some are strong while others are weak. According to Fraser et al. (2006), there are expert-driven as well as community-driven indicator developments. At a broader level, people-centeredness requires incorporating needs and views of all in various developments. This study is undertaken against this backdrop since current scholarly work lacks a wholesome framework. This is the research problem this paper addresses. Thus, the following research aims: (1) To highlight all the key factors that affect the sustainability of an entire population of a given country; and (2) to propose a framework that is wholesome and can be applied widely in any given country for sustainability-related studies. The aim of United Nations' seventeen SDGs is to overall attain decent lives for everyone by 2030 where the seventeen SDGs are: (1) no poverty, (2) zero hunger, (3) good health and well-being, (4) quality education, (5) gender equality, (6) clean water and sanitation, (7) affordable and clean energy, (8) decent work and economic growth, (9) industry innovation and infrastructure, (10) reduced inequality, (11) sustainable cities and communities, (12) responsible consumption and production, (13) climate action, (14) life below water, (15) life on land, (16) peace justice and strong institutions and (17) partnerships for the goals.

Whereas there is much research exploring people-centered sustainability, there still exists a research void in a wholesome framework that looks into this for wider applicability. Much of the current research focus on specific areas only such as health (Currie et al. 2024b), education, climate (Bianchi et al. 2024; Currie et al. 2024b; Wilkinson and Pickett 2020), Gender equality (Leach et al. 2016), solidarity and peace (Lopera-Arbeláez and Richter 2024) or impact of technology (Bianchi et al. 2024; Currie et al. 2024b). A recent research highlights social sustainability as the 'forgotten' pillar of sustainability.

This chapter is undertaken against this backdrop and seeks to propose a wholesome framework and survey questionnaire for wider applicability. Based on relevant secondary data reviews, this paper brings together a number of factors that may influence sustainability, keeping in mind the effect these will have on the general populace of any given country. To this end, the chapter has identified factors such as Equality (SDGs 5, 10); Respect and inclusion (SDGs 5, 10); Solidarity and peace (SDGs 11, 16); Income redistribution/social responsibility (SDGs 1, 2, 7); Universal healthcare (SDGs 3); Justice (SDGs 4, 8); Strong institutions (SDGs 9, 11, 16, 17) and, Environmental protection (SDGs 6, 11, 12, 13, 14, 15). The novelty of this chapter lies in the proposed new research framework and survey questionnaire. In this way, the chapter conceptually enhances and extends knowledge and research in this important area of people-centered sustainability.

The remainder of this chapter is divided into four parts. The following section presents methodology, followed by relevant review of literature and hypotheses development, leading to a conceptual model and survey questionnaire. The chapter then closes of with conclusion, relaying key points.

2 Methodology

This study is qualitative in nature and assumed the social constructionist paradigm. For social constructionists, the world is co-constituted, socially constructed and comprises a number of different realities of an individual and groups (Quilan et al. 2015), commonly utilizing qualitative methods (Mertens 2019).

This chapter made use of secondary data for write-up. Generally, there are three forms of secondary data—these being raw data, compiled data as well as survey data where raw data is directly sourced from websites, organisational databases, newspapers etc. Compiled data includes government publications, journals, books, industry statistics and reports which have some form of selection or summarizing while survey data includes survey strategies like census data.

This chapter, largely utilised the latter two forms of secondary data as follows. Compiled data included academic journal articles from various online sources and scholarly articles that were relevant. To this end, various e-resources were used to search for academic journal articles and theses such as Google Scholar, ProQuest, Science Direct, EBSCOhost among other academic databases. Academic journals used for this write up include Contemporary Social Science, Ecological Economics, Economics of Innovation and New Technology, Journal of the European Economic Association, Regional Studies, European Journal of Nuclear Medicine and Molecular Imaging, Radiography, Person-Centred Care in Radiology, Journal of Medical Imaging and Radiation Sciences, Philosophy and Public Affairs, Journal of Environmental Management, Society, Health and Vulnerability, Transactions of the Institute of British Geographers, Health Research Policy and Systems, Applied Research in Quality of Life, Regional Science, Annual Review of Environment and Resources, Journal of Clinical Medicine, International Journal of Sociology and Social Policy and Biological Sciences. Most recent sources were cited. In the absence of recent publications in certain areas, the most relevant earlier sources were then considered.

In addition, the author of this chapter reflected on her own position of that of an 'insider' since she has similar personal biography (Mercer 2007) such as gender (female), immigrant, ethnicity (Fiji born Indian whose ancestors were from India), of Hindu religion, generation X, middle-class and someone who had experienced resource/financial constraints earlier in life and work experience in retail, financial and educational sectors. As an insider, the author has an understanding of her own country (Fiji, a Global South, Pacific Island Country) which is multicultural, multilingual, a secular state and has natives, immigrants, issues of discrimination/gender, history and social life that helps her draw examples from her own home country to better understand the subject matter. The researcher has socialisation experience with the natives (ITaukeis'), Indo-Fijians as well as other ethnic groups in Fiji. Fiji's distinct sociocultural dynamics include close-knit social structures and is a male-dominated economy. Traditional gender roles are deeply embedded in this society, with men frequently holding positions of authority and economic power while women are typically given household jobs and obligations (Saliya 2024). Fiji fosters a communal culture that reinforces traditional gender roles, social cohesion and support thereby influencing educational experiences related to gender dynamics (Yabaki 2006). Pacific women continue to suffer systemic problems, owing to societal norms and male-centric gender policies (Boccuzzi 2021). This male-dominated societal structure creates inequalities in educational opportunities and outcomes, as women are often disadvantaged by limited access to resources and educational support (Saliya 2024). In Fiji, a common name 'Fijian' for all citizens has resulted in mixed reactions (Naidu 2013). The two major ethnic groups are the ITaukei and Indo-Fijians. Since 1987, Fiji experienced four military coups as well as a military mutiny that largely resulted from tensions between these two major ethnic groups (Humanists International 2024). The remaining smaller minority groups of Chinese, Banabans, Rotumans, Melanesians and others from other Pacific Islands are mostly politically, socially and economically, somewhat excluded (Naidu 2013). Fiji's economic and political developments and focus have created inequalities and divisions between the diverse ethnic groups (Humanists International 2024), leading to stereotypes and prejudices with both institutional and social racism being practiced, resulting in increasing socio-economic inequalities that created a sense of 'the affluent' and 'the poor' ethnicities (Naidu 2013).

To help generate a rich source of data and for validity and reliability reasons, the collected information between sources were cross-referenced to gain accurate insights. The use of multiple sources helped with data comparisons and in noting recent developments. After relevant data collection, manual thematic analysis technique was employed such as identifying, analysing, organizing, describing and reporting key themes which are the sub-topics of this chapter. This required familiarisation with the collected data through repeated readings of data in a way that helped to note what the key elements were and how and when these emerged. After this, the collected information was categorised into themes which are the sub-topics (factors) and lastly, the author went over the created themes to check how these fit into the overall chapter. Few initial overlapping themes were then reorganized into similar ones. The final write-up was then done to make it clearer, logical, non-repetitive within and across themes.

3 Literature Review and Development of Hypotheses

Based on relevant secondary data reviews, this chapter brings together a number of factors such as Equality, respect and inclusion; Solidarity and peace; Income redistribution/social responsibility; Universal healthcare; Justice; Strong institutions and Environmental protection that may influence sustainability, affecting the general populace of any given country. This helped create a wholesome conceptual framework, leading to a survey questionnaire that can be used for later empirical studies by future researchers. The following section elaborates on each of these factors.

3.1 Equality, Respect and Inclusion

Equality brings about declining polarised prosperity. Current increased use of technology has further exacerbated inequality. Inequality is commonplace—even the growing economies such as Asia, Latin America and South Africa have soaring inequality issues (Leach et al. 2016). As per Wilkinson and Pickett (2020), a campaign for sustainability is a campaign for enhanced equality. According to structuralist researchers, the categorisation of 'we and them' results in exclusion, inequality and segregation (Härenstam 2017). What is also true is that oftentimes, prioritising intergenerational equality, may adversely affect equality, [level of respect] among various generational groups, age, gender, ethnicity, nation (Leach et al. 2016), immigrant or native and differing sexual orientations—LGBTQ (lesbian, gay, bisexual, and transgender, queer and/or questioning, intersex, asexual). This is also evident at workplaces, which may neither be safe nor progressive for females. Categorizing jobs as male or female jobs creates unintended inequality and if nationwide, goals as well as agendas differ between males and females, younger and older generations, indigenous and immigrants, such agendas will then fail to enhance inclusivity or equality (Härenstam 2017). In this chapter, respect has been added since those who get little to no respect, given their past (such as jail term), disability, certain health conditions (e.g., HIV), poor education or being poor financially, may be treated unfairly or even excluded.

All key stakeholders, as well as communities, should be given equal opportunity to voice their opinions to influence sustainable development plans affecting them (Welsh Government 2024), including at the workplace. In their article, Leach et al. (2016) put gender equality on the spotlight, arguing that this should encompass all human rights (e.g., rights to basic needs fulfillment, livelihoods and integrity as well as security), claiming that some even 'promote sustainability at the cost of gender equality'. Essentially, education can be the key agent of transformation for sustainable development, leading to enhanced capacities and better decision making that considers long-term impacts promoting equity, ecology and economy for all communities (Tilbury and Wortman 2004). Hence, it is hypothesized that,

H1: Equality, respect and inclusion for all positively affect people-centered sustainability.

3.2 Solidarity and Peace

Social and Solidarity Economy (SSE) is a transformative approach that helps achieve peace-centred sustainable development (Lopera-Arbeláez and Richter 2024). Coraggio (2020) informs that SSE involves organizing production, its distribution as well as circulation, and consumption of products and services with cooperation from the society and state—this is a society focused on solidarity. In this way, SSE is not only a source for income but also enhances social cohesion (Mac Ginty and Richmond 2013) being both people-centered and nature-responsible, allowing for voluntary cooperation, collective products, community participative actions, as well as democratic governance and collective ownership (Lopera-Arbeláez and Richter 2024).

Solidarity and peace are interlinked and may mean everyone is working together for a better understanding to embrace diversity at all levels. It also promotes economic inclusion for those who are vulnerable like farmers, the indigenous, [disabled, the minority, immigrants, street dwellers] and females, who are at times excluded from the formal economy (OECD 2023). SSE democratizes governance, empowering the vulnerable (UN Inter-Agency Task Force on Social and Solidarity Economy 2022), encouraging them to take control of their own economic activities, thus creating opportunities for them to actively shape their own destinies (Murtagh 2016). Despite its significance, it is often not given much attention by academics or practitioners (UN Inter-Agency Task Force on Social and Solidarity Economy 2022). With equality comes stronger community life, encouraging all to willingly and jointly work together for the common good, creating a community that focuses on equal societies, recycling waste with business leaders who are supportive of environmental policies also (Wilkinson and Pickett 2020). Therefore, it is hypothesized that,

H2: Solidarity and peace positively affect people-centered sustainability.

3.3 Income Redistribution/Social Responsibility

Earlier studies have reported the widening of opportunities as well as income disparities (Capello and Lenzi 2023; Corradini et al. 2021), raising early warnings on longer-term social inequalities which, while could lead to economic growth may polarise prosperity (Bianchi et al. 2024). The attributes of such polarisation can also be technologies (Bianchi et al. 2024; Chancel and Piketty 2021) and even regional economic divergence (Martin 2021). For a fairer distribution of wealth and to prevent the irreversible divergence of income, adequate mechanisms need to be in place for a just and equitable transition (Bianchi et al. 2024). In this chapter, income redistribution/social responsibility is defined as progressive taxation, not taxing individuals up to a particular threshold of income and allowing for voluntary contribution by richer individuals and corporations to assist governments in helping those on or below poverty line. Hence, it is hypothesized that,

H3: Income Redistribution/social responsibility positively affect people-centered sustainability.

3.4 Universal Healthcare

This chapter defines Universal healthcare as fair, safe and timely access to affordable healthcare services and medicine for all citizens, irrespective of background. Social sustainability in healthcare system improves well-being and quality of life by maintaining beneficial health outcomes and enhanced processes that ensure such benefits continue (Lennox et al. 2020), decreasing morbidity, healthcare spending, conserving resources across a large populace (Currie et al. 2024b). This requires emotionally and culturally safe environments to address the lack of diversity within the healthcare staff and unequal accessibility to healthcare services (Currie et al. 2024b). Sometimes, professionals, even those who are emotionally and culturally aware, may be biased and prejudiced given the stressful environment they work in (Currie et al. 2024b). Time, [doctor to patient ratio; long working hours] can be a barrier in genuinely offering person-centered care, suggesting the need for Artificial Intelligence (AI) for improvements in patient care (Currie et al. 2019, 2020, 2024a; Currie and Hawk 2021). However, AI may also deepen healthcare inequities (Currie et al. 2024a) for those who cannot afford expensive treatments. Political will, [budget, charges] and bureaucracy of complex healthcare processes are not the same across nations (Currie et al. 2024b) and requires commitment from key stakeholders such as the national regulator and state health departments (Currie et al. 2024b). Thus, it is hypothesized that,

H4: Universal healthcare positively affects people-centered sustainability.

3.5 Justice

Justice is defined as fair treatment to all citizens irrespective of age, gender, ethnicity, qualification, financial status in every aspect including job opportunity/education, the freedom to practise own religions and legal justice. Questioning sustainability also raises questions on gender and various inequality issues as well as justice (Leach et al. 2016). Social justice and sustainability are natural allies and what is more concerning is that while the richer nations produce more carbon footprint, its the poor who are adversely affected by the negative climate change effects (Wilkinson and Pickett 2020). According to the ILO (2015), an all inclusive environmentally sustainable economy and society, requires distributional and also procedural justice (Bianchi et al. 2024). There are generally three types of justice: (1) distributive justice that reflects at the distribution of entitlements and opportunities, related to any economic transition (Rawls 1971); (2) procedural justice that involves inclusion of all citizens and stakeholders in the decision-making process, allowing them to activate their capabilities, enhancing their access to goods and rights; and (3) substantive justice that is the result of the above two which are similar to the concepts of 'substantive equality' of Hay (1995) and 'equality of welfare' of Dworkin (1981). Social justice requires sensitivity towards economic growth and its societal and environmental impacts, assessing consumption at personal and societal levels (Tilbury and Wortman 2004). Therefore, it is hypothesized that,

H5: Justice positively affects people-centered sustainability.

3.6 Strong Institutions

Strong institutions are described as capable, effective, accountable and transparent institutions advocating and ensuring peace, stability, human rights, equal participation, environment conservation as well as effective governance, abiding by the law. Such institutions are not only the ones owned and run by the state but also the NGOs that can question and assist with policy formulation, seeking greater participation. Further, states/governments need to beef up their capacity for service delivery that are centered on sustainability, gender (Leach et al. 2016), the disabled, the disadvantaged, the retirees and minority to ensure equality. Additionally, the role of social institutions in change and development, with democratic participation in both research and the policy making (Ward 2024) systems such as in selection of governments and permitting expression of opinion in seeking consensus while addressing differences are crucial (Tilbury and Wortman 2004). Such expressions of feelings and experiences must not be reductionist but plural and inclusive (Madgin and Howcroft 2024). Global institutions, the state and/or all co-produced local institutions should effectively work together to promote: equality for gender, [the disabled, the disadvantaged, LGBTQ, the retirees, immigrants, indigenous and minority] and, sustainability, allowing these segments to also have a voice in overall planning and implementation of such policies (Leach et al. 2016). Thus, it is hypothesized that,

H6: Strong institutions positively affect people-centered sustainability.

3.7 *Environmental Protection*

Environmental protection reflects at the preservation, pollution controls and restoration of natural resources to ensure resource availability for all and in the future.

For general well-being, a third of world's populace rely on natural resources, threatening sufficient resource availability for future generations, giving rise to unsustainable patterns that exacerbate poverty and inequalities (Leach et al. 2016) even leading to political oppositions to policies that could curb carbon emissions, spreading denial of climate change impacts (Wilkinson and Pickett 2020). Sustainable development not only includes equality and human rights [at all levels] but also adequate social and economic policies for the environment (Leach et al. 2016) that raises awareness on the fragile environment and how this is affected by human activities and decisions (Tilbury and Wortman 2004). Equity is also a thorny issue in sustainable technologies to address climate change impact that end up adding to the workload of females in contexts that already have existing discrimination and inequality (Leach et al. 2016). Such technologies should be deliverable, efficient and at the same time environmentally sustainable, enhancing productivity (Bianchi et al. 2024; Bellandi and De Propris 2021). As per Currie et al. (2024b), other sectors within countries such as sustainable healthcare practices may also help lower carbon footprint and improve environmental impacts with greener practices. As per global estimates, the healthcare sector contributes to 4.4% of carbon dioxide emissions (which is 5% in UK, 7% in Australia and much higher—10% in US), where medical imaging contributes to 10% of the healthcare quota (between 0.4 and 1% of total global footprint) (McAlister et al. 2022; Picano et al. 2022). Overall, campaigning for the environment should translate into radical movements for social justice to all (Wilkinson and Pickett 2020).

Therefore, it is hypothesized that,

H7: Environmental protection positively affects people-centered sustainability.

In line with the aim of this chapter and, in keeping consistency with the literature reviewed, a conceptual model is drawn as shown in Fig. 1, depicting the association between the seven factors and people-centered sustainability.

Based on the aforementioned extant literature, scales were created from the key points discussed under each factor for the following five-point Likert scale questionnaire with responses ranging from "Strongly Disagree" (1) to "Strongly Agree" (5)

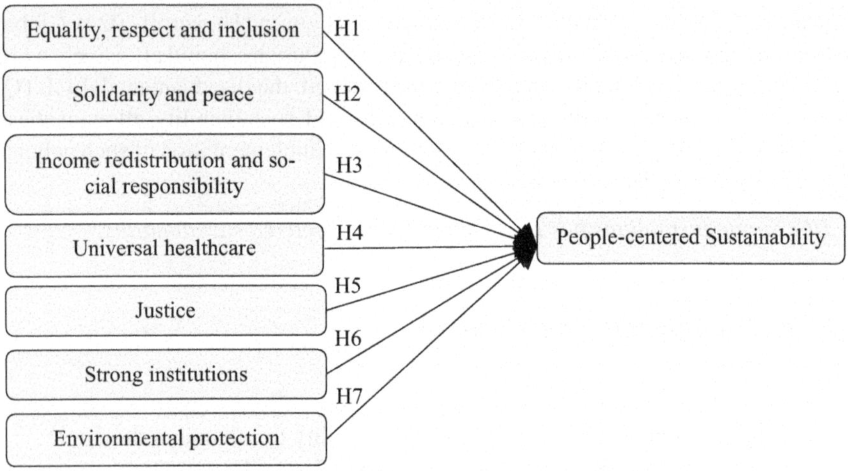

Fig. 1 Factors affecting people-centered sustainability. H1: Equality, respect and inclusion positively affect people-centered sustainability. H2: Solidarity and peace positively affect people-centered sustainability. H3: Income redistribution/social responsibility positively affect people-centered sustainability. H4: Universal healthcare positively affects people-centered sustainability. H5: Justice positively affects people-centered sustainability. H6: Strong institutions positively affect people-centered sustainability. H7: Environmental protection positively affects people-centered sustainability

for future empirical research. The scales being the factors of Equality, Respect and Inclusion (ERI); Solidarity and Peace (SP); Income Redistribution/Social Responsibility (IRSR); Universal Healthcare (UH); Justice (J); Strong Institutions (SI); and Environmental Protection (EP) and People-Centered Sustainability (PCS).

Questionnaire
TOPIC: Factors influencing people-centered sustainability

Objective:
This research aims to identify factors that may influence people-centered sustainability. Collected information from this questionnaire will remain highly confidential, and will only be used for academic purpose. It will take about 10 minutes to answer this questionnaire. Thank you for your participation. Please read each statement carefully for better understanding.
Participation:
We seek your voluntary participation, however, you may exit this survey or refuse to participate.
Contact Information:
For questions relating to this survey, please contact ________________________________
Do we have your informed consent for this survey? ☐ ***Yes*** ☐ ***No***
If yes, kindly proceed to answer questions below. If not, thank you.

Section A: Personal Information

For this section, please tick "✓" or put "X" to select an option for each question below.

1. Age					
☐ 18–20	☐ 21–30	☐ 31–40	☐ 41–50	☐ 51 and above	
2. Gender					
☐ Male		☐ Female		☐ Other	
3. Level of Study					
☐ Primary school	☐ High school	☐ Undergraduate level		☐ Postgraduate level	
4. Residential Area					
☐ Urban	☐ Semi-urban	☐ Semi-rural		☐ Rural	
5. Status in Country					
☐ Immigrant			☐ Native		
6. Sexual Orientation					
☐ Heterosexual	☐ Homosexual	☐ Bisexual		☐ Asexual	
7. Sector Employed					
☐ Tourism	☐ Agriculture	☐ Manufacturing	☐ Service	☐ Fisheries	
☐ Forestry	☐ Infrastructure	☐ Health	☐ Education	☐ Energy	☐ IT
8. Income Level					
0–30,000	30,001–50,000	50,001–70,000	70,001–90,000	Above 90,000	

Section B: People-Centered Sustainability

For this section, please tick "✓" or indicate "X" to select an option that best reflects at how you feel about each statement below. Options provided are: Strongly disagree (1); Disagree (2); Neutral (3); Agree (4); Strongly agree (5).

People-Centered Sustainability					
Statement:					
(A) Equality, Respect and Inclusion (ERI)					
1. Females and males are treated equally and fairly in my country	1	2	3	4	5
2. The disabled are treated equally and fairly in my country	1	2	3	4	5
3. Immigrants are treated equally and fairly in my country	1	2	3	4	5
4. The minorities are treated equally and fairly in my country	1	2	3	4	5
5. The poor/disadvantaged are treated equally as well as fairly in my country	1	2	3	4	5
6. People with different sexual orientation—LGBTQ are treated equally as well as fairly in my country	1	2	3	4	5
7. Despite different sexual orientation, appearance, status and health condition, people are respected in my country	1	2	3	4	5
8. People generally feel safe, have equal opportunities and have a voice at their workplaces in my country	1	2	3	4	5

(continued)

(continued)

(B) Solidarity and Peace (SP)					
1. Everyone in my community willingly work together	1	2	3	4	5
2. Everyone in my country willingly work together	1	2	3	4	5
3. Everyone in my community embrace diversity	1	2	3	4	5
4. Everyone in my country embrace diversity	1	2	3	4	5
5. Relationship among diverse people in my country is harmonious	1	2	3	4	5
6. Everyone makes an effort to be at peace with one another	1	2	3	4	5
(C) Income Redistribution/Social Responsibility (IRSR)					
1. Social welfare benefits are sufficient to address income disparity in my country	1	2	3	4	5
2. The corporate sector/rich contribute adequately towards their social responsibility in my country	1	2	3	4	5
3. Unemployment benefits are adequate in my country	1	2	3	4	5
4. During emergency situations, government of my country comes up with effective policies to assist those who are adversely affected	1	2	3	4	5
(D) Universal Healthcare (UH)					
1. Healthcare facilities in my county are well decentralized	1	2	3	4	5
2. Healthcare facilities in my county are adequate	1	2	3	4	5
3. Healthcare facilities in my county are safe	1	2	3	4	5
4. I have ready access to all basic healthcare facilities in my country	1	2	3	4	5
5. Healthcare facilities in my county are affordable	1	2	3	4	5
6. Healthcare staff in my country are empathetic	1	2	3	4	5
7. Healthcare staff in my country are well qualified	1	2	3	4	5
8. There are enough healthcare staff in my country	1	2	3	4	5
9. There are enough healthcare products in my country	1	2	3	4	5
(E) Justice (J)					
1. People of different age groups are given job opportunities in my country	1	2	3	4	5
2. People of different age groups are given educational opportunities in my country	1	2	3	4	5
3. People of different age groups have the right to seek legal counsel in my country	1	2	3	4	5
4. Both males and females are given equal job opportunities in my country	1	2	3	4	5
5. Both males and females are given educational opportunities in my country	1	2	3	4	5
6. Both males and females have the right to seek legal counsel in my country	1	2	3	4	5
7. The disabled are given job opportunities in my country	1	2	3	4	5

(continued)

(continued)

8. The disabled are given educational opportunities in my country	1	2	3	4	5
9. The disabled have the right to seek legal counsel in my country	1	2	3	4	5
10. All ethnic groups are given equal job opportunities in my country	1	2	3	4	5
11. All ethnic groups are given educational opportunities in my country	1	2	3	4	5
12. All ethnic groups have the right to seek legal counsel in my country	1	2	3	4	5
13. People with no/low qualification level are also given job opportunities in my country	1	2	3	4	5
14. People with no/low qualification level are given educational opportunities in my country	1	2	3	4	5
15. People with no/low qualification level have the right to seek legal counsel in my country	1	2	3	4	5
16. The poor are given equal job opportunities in my country	1	2	3	4	5
17. The poor are given educational opportunities in my country	1	2	3	4	5
18. The poor have the right to seek legal counsel in my country	1	2	3	4	5
19. Immigrants are given equal job opportunities in my country	1	2	3	4	5
20. Immigrants are given educational opportunities in my country	1	2	3	4	5
21. Immigrants have the right to seek legal counsel in my country	1	2	3	4	5
22. Native are given equal job opportunities in my country	1	2	3	4	5
23. Native are given educational opportunities in my country	1	2	3	4	5
24. Native have the right to seek legal counsel in my country	1	2	3	4	5
25. Everyone has the freedom to practise their own religion in my country	1	2	3	4	5
26. Those with different sexual orientation are given equal job opportunities in my country	1	2	3	4	5
27. Those with different sexual orientation are given educational opportunities in my country	1	2	3	4	5
28. Those with different sexual orientation have the right to seek legal counsel in my country	1	2	3	4	5
(F) Strong Institutions (SI)					
1. In my country, the government agencies advocate and ensure peace, stability, human rights, equal participation, environment conservation as well as effective governance, as per law	1	2	3	4	5
2. In my country, the government agencies are capable and effective	1	2	3	4	5
3. In my country, the government agencies are accountable	1	2	3	4	5
4. In my country, the government agencies are transparent	1	2	3	4	5
5. In my country, the government agencies are well resourced and funded	1	2	3	4	5
6. In my country, the government agencies allow for democratic participation	1	2	3	4	5

(continued)

(continued)

7. In my country, the government agencies practise equal treatment	1	2	3	4	5
8. NGOs in my country are very active and vocal	1	2	3	4	5
(G) Environmental Protection (EP)					
1. In my country, there are adequate policies to preserve and conserve the natural environment	1	2	3	4	5
2. In my country, there are adequate policies on pollution control	1	2	3	4	5
3. In my country, there are adequate focus and actions taken to restore the natural environment	1	2	3	4	5
4. In my country, there are adequate penalties for not abiding by the environment legislation	1	2	3	4	5
(H) People-Centered Sustainability (PCS)					
1. Equality, respect and inclusion in my country will improve my quality of life and well-being	1	2	3	4	5
2. Strong communal relationships will improve my quality of life and well-being	1	2	3	4	5
3. Effective income redistribution initiatives of the government and business sector's/rich's social responsibility in my country will improve my quality of life and well-being	1	2	3	4	5
4. Universal healthcare facilities in my country will improve my quality of life and well-being	1	2	3	4	5
5. Fair and readily accessible justice system in my country will improve my quality of life and well-being	1	2	3	4	5
6. Competent governmental institutions and NGOs in my country will improve my quality of life and well-being	1	2	3	4	5
7. Appropriate environmental protection initiatives in my country will improve my quality of life and well-being	1	2	3	4	5

Thank you for your time in answering this questionnaire. Your responses will help us better understand the area of research.

4 Implications

This study largely makes a few theoretical contributions to current literature on the subject of people centred sustainability. First, while most previous studies have singled out certain factors, this research takes a wholesome approach to overall sustainability that is aligned to all 17 SDGs. Second, it is among the first few studies that proposes a ready wholesome framework that future studies can use and test empirically.

5 Conclusion

The aim of this research was to investigate into the factors that may influence people-centered sustainability to help prepare a wholesome conceptual framework and survey questionnaire. It has identified factors such as Equality, respect and inclusion; Solidarity and peace, Income redistribution/social responsibility, Universal health-care, Justice, Strong institutions and Environmental protection. With this proposed research framework and a detailed Likert scale survey questionnaire, the chapter aims to conceptually enhance and extend knowledge and research in this important area of people-centered sustainability. The next step is empirical research to validate the importance and relevance of these factors to people-centered sustainability. Future researchers can use the same or modify the scale factors of ERI, SP, IRSR, UH, J, SI, EP, and PCS and the measured variables that are indicated as questions under each scale to further this research by using a structural equation model to better understand the association of these factors to people-centered sustainability. Scholars can also conduct multi-group analysis for comparative studies such as comparisons between the male/female group, native/immigrant group, in terms of employment sector, qualification levels, income levels and age groups. These factors have not been discussed together, in this same way in any of the previous studies, to the best of the author's knowledge.

References

Atkinson G (2000) Measuring corporate sustainability. J Environ Plann Manag 43(2):235–252. https://doi.org/10.1080/09640560010694

Bellandi M, De Propris L (2021) Local productive systems' transitions to Industry 4.0+. Sustainability 13(23):13052. https://doi.org/10.3390/su132313052

Bianchi P, De Propris L, Labory S (2024) People-centred policies for a just transition (digital, green and skills). Contemp Soc Sci 19(1–3):262–282. https://doi.org/10.1080/21582041.2024.2351479

Boccuzzi E (2021) The future of work for women in the Pacific Islands. The Asia Foundation. https://asiafoundation.org/wp-content/uploads/2021/02/The-Future-of-Work-for-Women-in-the-Pacific-Islands.updateMarch1.pdf

Bohringer C, Jochem P (2007) Measuring the immeasurable—a survey of sustainability indices. Ecol Econ 63(1):1–8. https://doi.org/10.1016/j.ecolecon.2007.03.008

Capello R, Lenzi C (2023) 4.0 technological transformations: heterogeneous effects on regional growth. Econ Innov New Technol 33(6):1–20. https://doi.org/10.1080/10438599.2023.2204523

Chancel L, Piketty T (2021) Global income inequality 1820–2020: the persistence and mutation of extreme inequality. J Eur Econ Assoc 19(6):3025–3062. https://doi.org/10.1093/jeea/jvab047

Coraggio JL (2020) Economía social y economía popular: Conceptos básicos. Ser Consejeres 1:1–16

Corradini C, Santini E, Vecciolini C (2021) The geography of industry 4.0 technologies across European regions. Reg Stud 55(10–11):1667–1680. https://doi.org/10.1080/00343404.2021.1884216

Currie G, Hawk KE (2021) Ethical and legal challenges of artificial intelligence in nuclear medicine. Sem Nucl Med 51(2):120–125. https://doi.org/10.1053/j.semnuclmed.2020.08.001

Currie G, Hawk KE, Rohren E, Vial A, Klein R (2019) Machine learning and deep learning in medical imaging: intelligent imaging. J Med Imaging Radiat Sci 50(4):477–487. https://doi.org/10.1016/j.jmir.2019.09.005

Currie G, Hawk KE, Rohren EM (2020) Ethical principles for the application of artificial intelligence (AI) in nuclear medicine. Eur J Nucl Med Mol Imaging 47:748–752. https://doi.org/10.1007/s00259-020-04678-1

Currie G, Hawk KE, Rohren EM (2024a) The role of artificial intelligence in enhancing person-centred care. In: Chau S, Hyde E, Knapp K, Hayre C (eds) Medical imaging in practice. CRC Press, pp 343–362. https://doi.org/10.1201/9781003310143-29

Currie GM, Hawk KE, Rohren EM (2024b) Challenges confronting sustainability in nuclear medicine practice. Radiography 30(1):1–8. https://doi.org/10.1016/j.radi.2024.04.026

Dworkin R (1981) What is equality? Part 1: equality of welfare, philosophy & public affairs. Philos Public Aff 10(3):185–246. https://www.jstor.org/stable/2264894

Elkington J (1999) Cannibals with forks: the triple bottom-line of 21st century business. Capstone, Oxford. ISBN: 9781841120843, 1841120847

Fraser EDG, Dougill AJ, Mabee WE, Reed M, McAlpine P (2006) Bottom up and top down: analysis of participatory processes for sustainability indicator identification as a pathway to community empowerment and sustainable environmental management. J Environ Manage 78(2):114–127. https://doi.org/10.1016/j.jenvman.2005.04.009

Härenstam A (2017) Inclusion, sustainability, and equality: how can research contribute? Soc Health Vulnerabil 8(1):1332856. https://doi.org/10.1080/20021518.2017.1332856

Hay AM (1995) Concepts of equity, fairness and justice in geographical studies. Trans Inst Br Geogr 20(4):500–508. https://doi.org/10.2307/622979

Humanists International (2024) The freedom of thought report—Fiji. https://fot.humanists.international/countries/oceania-melanesia/fiji/

International Labour Organization (2015) Guidelines for a just transition towards environmentally sustainable economies and societies for all. Geneva

Leach M, Mehta L, Prabhakaran P (2016) Gender equality and sustainable development: a pathways approach. The UN women discussion paper, 13

Lennox L, Linwood-Amor A, Maher L, Reed J (2020) Making change last? Exploring the value of sustainability approaches in healthcare: a scoping review. Health Res Policy Syst 18(120):1–24. https://doi.org/10.1186/s12961-020-00601-0

Lopera-Arbeláez I, Richter S (2024) Transformative approaches for peace-centred sustainable development: the role of social and solidarity economy. World Dev Perspect 34:100593. https://doi.org/10.1016/j.wdp.2024.100593

Mac Ginty R, Richmond OP (2013) The local turn in peace building: a critical agenda for peace. Third World Q 34(5):763–783. https://doi.org/10.1080/01436597.2013.800750

Madgin R, Howcroft M (2024) Advancing people-centred, place-based approaches. AHRC place-based research programme report series, No. 2. University of Glasgow. https://doi.org/10.36399/gla.pubs.342111

Magee L, Scerri A, James P (2012) Measuring social sustainability: a community-centred approach. Appl Res Qual Life 7:239–261. https://doi.org/10.1007/s11482-012-9166-x

Martin R (2021) Rebuilding the economy from the COVID crisis: time to rethink regional studies? Reg Stud Reg Sci 8(1):143–161. https://doi.org/10.1080/21681376.2021.1919191

McAlister S, McGain F, Breth-Petersen M, Story D, Charlesworth K, Ison G, Barratt A (2022) The carbon footprint of hospital diagnostic imaging in Australia. Lancet Reg Health West Pac 24. https://doi.org/10.1016/j.lanwpc.2022.100459

Mercer J (2007) The challenges of insider research in educational institutions: wielding a double-edged sword and resolving delicate dilemmas. Oxf Rev Educ 33(1):1–17. https://doi.org/10.1080/03054980601094651

Mertens DM (2019) Research and evaluation in education and psychology: integrating diversity with quantitative, qualitative, and mixed methods. Sage

Murtagh B (2016) Economics: neoliberal peace and the politics of social economics. In: Richmond OP, Pogodda S, Ramovic J (eds) The Palgrave handbook of disciplinary and regional approaches to peace. Palgrave Macmillan UK, pp 110–122. https://doi.org/10.1007/978-1-137-40761-0_9

Naidu V (2013) Fiji: the challenges and opportunities of diversity. Minority Rights Group International. Accessed 28 Apr 2025. https://books.google.com.fj/books?hl=en&lr=&id=jk0sEQAAQBAJ&oi=fnd&pg=PA2&ots=NIOBEveeUk&sig=8PqReeURb0xi7jmk1FL0z-a6iWI&redir_esc=y#v=onepage&q&f=false

OECD (2023) Policy guide on social impact measurement for the social and solidarity economy. https://doi.org/10.1787/270c7194-en

Parris TM, Kates RW (2003) Characterizing and measuring sustainable development. Annu Rev Environ Resour 28(1):559–586. https://doi.org/10.1146/annurev.energy.28.050302.105551

Picano E, Mangia C, D'Andrea A (2022) Climate change, carbon dioxide emissions, and medical imaging contribution. J Clin Med 12(1):215. https://doi.org/10.3390/jcm12010215

Quinlan C, Babin BJ, Carr J, Griffin M, Zikmund WG (2015) Business research methods, 1st edn. Cengage Learning, Andover, UK. ISBN-13: 9781473704855, ISBN-10: 1473704855

Rawls J (1971) A theory of justice. Belknap Press of Harvard University Press, Cambridge, Massachusetts. Accessed 28 Apr 2025. https://books.google.com.fj/books?hl=en&lr=&id=AjrXZIlbK1cC&oi=fnd&pg=PA1&dq=Rawls,+J.+(1971).%C2%A0A+theory+of+justice.+Cambridge,+Massachusetts:+Belknap+Press+of+Harvard+University+Press,&ots=sesmTCk-fo&sig=9mMJGO8CK6M-1rHnEYIhyB2i7VQ&redir_esc=y#v=onepage&q&f=false

Saliya CA (2024) Exploring gender inequality and sociocultural factors in Fiji: unveiling the "pink unicorns." Int J Sociol Soc Policy 44(5/6):516–536. https://doi.org/10.1108/IJSSP-08-2023-0184

Tilbury D, Wortman D (2004) Engaging people in sustainability. Accessed 26 Apr 2025. https://books.google.com.fj/books?hl=en&lr=&id=F-ypMpjqPv0C&oi=fnd&pg=PR5&dq=Tilbury,+D.,+%26+Wortman,+D.+(2004).+Engaging+people+in+sustainability.+IUCN.&ots=v3lKXUhlsp&sig=dHySvfY8wMLDmg0WMPSF_mvgwsc&redir_esc=y#v=onepage&q=Tilbury%2C%20D.%2C%20%26%20Wortman%2C%20D.%20(2004).%20Engaging%20people%20in%20sustainability.%20IUCN.&f=false

UN Inter-Agency Task Force on Social and Solidarity Economy (2022) Advancing the 2030 agenda through the social and solidarity economy. Position paper. ILO—International Labour Organization, p 126. ISBN: 978-92-2-037431-3 (web PDF)

United Nations (1992a) Agenda 21. United Nations Conference on Environment and Development (UNCED). UN General Assembly, Rio de Janeiro. Accessed 20 Apr 2025. https://sustainabledevelopment.un.org/outcomedocuments/agenda21

United Nations (1992b) Rio declaration on environment and development. United Nations Conference on Environment and Development (UNCED). UN General Assembly, Rio de Janeiro. Accessed 15 Apr 2025. https://www.un.org/en/development/desa/population/migration/generalassembly/docs/globalcompact/A_CONF.151_26_Vol.I_Declaration.pdf and https://www.un.org/en/conferences/environment/rio1992

Ward D (2024) Prospectus our work and offer, 2024–25. Policy Lab UK, HM Government. Accessed 10 Apr 2025. https://openpolicy.blog.gov.uk/wp-content/uploads/sites/35/2024/07/PUBLIC_Policy_Lab_Prospectus_2024-25-1.pdf

Welsh Government (2024) Consultation document infrastructure (Wales) bill requirements for pre-application consultation and methods of engaging stakeholders and local communities, 13 May 2024. Number: WG49599. Accessed 19 Apr 2025. https://www.gov.wales/sites/default/files/consultations/2024-05/infrastructure-wales-bill-requirements-for-pre-application-consultation.pdf

Wilkinson R, Pickett K (2020) The struggle for equality and sustainability. In: Costanza R, Erickson JD, Farley J, Kubiszewski I (eds) Sustainable wellbeing futures. Edward Elgar Publishing, pp 179–187. https://doi.org/10.4337/9781789900958.00020

Yabaki T (2006) Women's life in a Fijian village. Doctoral dissertation, University of Canberra. Accessed 19 Apr 2025. https://doi.org/10.26191/tcp3-tt09

Beyond Survival: Designing Sustainable Systems that Prioritize Human Dignity and Well-Being

Megha Ojha and Vinay Kandpal

Abstract This Chapter explores the crossroad of the United Nations Sustainable Development Goals (SDGs) and Mortal-Centred Design (HCD) as synergistic fabrics for creating sustainable, indifferent, and inclusive results to global challenges. The SDGs, espoused in 2015, offer a comprehensive design to address poverty, inequality, climate change, environmental degradation, peace, and justice by 2030. Still, rephrasing these high-position pretensions into effective, real-world interventions demands design approaches prioritizing moral quality, community engagement, and original applicability. Mortal-Centred Design, emphasizing empathy, iterative problem-working, and stakeholder involvement, complements the SDGs by ensuring that results are adapted to the specific requirements, societies, and boundaries of different communities. This paper examines how HCD can enhance the impact of SDG-related enterprises across sectors such as health, education, civic planning, and environmental stewardship. Through a critical analysis of SDG 1–17, the Chapter highlights how mortal-centric approaches can strengthen strategies to make sustainable metropolises, promote responsible consumption, advance climate action, conserve biodiversity, foster peaceful institutions, and develop global meetings. The discussion underscores that sustainability isn't solely a specialized or environmental issue but a mortal bone. Results that neglect the voices and lived experiences of the people they aim to serve are likely to fail. Again, integrating HCD into SDG strategies bridges the gap between global ambition and original action, fostering inventions that aren't only sustainable but also inclusive, just, and empowering. Eventually, the confluence of these paradigms offers a path forward toward achieving a thriving, flexible, and indifferent future for all.

Keywords Human-centred · SDG · Sustainability · Well-being

M. Ojha (✉) · V. Kandpal
Graphic Era deemed to be University, Dehradun, India
e-mail: Meghaojha96@gmail.com

A. Singh et al. (eds.), *Human-Centered Sustainability: Bridging Well-Being, Gender Equity, and Technology*, World Sustainability Series,
https://doi.org/10.1007/978-3-032-22020-2_4

1 Introduction

For too long, mortal systems—profitable, political, and environmental—have been structured around bare survival. In a world struggling with issues like climate change, inequality, and resource reduction, the traditional fabrics no longer serve. To truly thrive, we must revise sustainability not just as an ecological necessity but as a mortality-centred imperative. Sustainable systems should prioritize the natural quality of every existent, ensuring that communities are not just surviving but flourishing. This shift demands an integration of ethical design principles, community-led invention, and policy fabrics that fete well-being as a critical measure of success. Beyond access to food, water, and shelter, the future must guarantee openings for education, emotional health, artistic expression, and independent participation. Erecting similar systems requires collaboration across disciplines, centering the voices of the marginalized, and embracing adaptability as a dynamic, inclusive process. This vision goes beyond specialized results; it calls for a profound artistic metamorphosis that honors life in all its confines. It's only by moving once that survival that humanity can achieve a sustainable, stable future for all. In a period marked by enhanced climate change, widening social inequalities, and rapid-fire technological shifts, the traditional models of sustainability are proving inadequate. Historically, sustainability enterprise has concentrated primarily on resource conservation and environmental protection, frequently sidelining the abecedarian mortal need for quality, autonomy, and holistic well-being (Raworth 2017).

Survival, while critical, is the utmost introductory standard of societal success; a sustainable future must be erected on fabrics that promote flourishing lives, not simply enduring actuality. Arising education emphasises that systems designed without integrating social equity and moral quality eventually immortalize structural vulnerabilities and ecological decline (Ostrom 2009; IPCC 2022). Moving beyond survival demands reimagining sustainability as a dynamic, inclusive process that titleholders' adaptability, justice, and the full consummation of mortal eventuality.

Reconsidering sustainability involves admitting that mortal well-being is multidimensional, encompassing not only physical health but emotional, social, and spiritual fulfilment (WHO 2023). This broader understanding shifts the focus from minimizing detriment to laboriously creating systems that support thriving communities. Raworth's (2017) conception of" doughnut economics" exemplifies this shift, proposing a model where profitable exertion operates within ecological limits while ensuring that all people meet their basic requirements and live lives of quality. Also, the United Nations' Sustainable Development Goals (SDGs) explicitly link environmental sustainability with poverty reduction, gender equality, and access to education, affirming those moral rights and ecological rights. Designing sustainable systems that prioritize quality requires an interdisciplinary approach, combining perceptivity from ecology, economics, sociology, and indigenous knowledge systems. Ostrom's (2009) exploration of common-pool resource governance highlights the significance of community-driven, participatory models that empower original stakeholders rather than imposing top-down regulations. In civic surrounds, scholars endorse “just sustainability,” which ties environmental planning directly to issues of social justice, aiming to produce metropolises that are not only greener but also more equitable and inclusive (Agyeman 2013). Without such an intertwined approach, sweatshops threaten buttressing exacerbate power imbalances and fail the very populations most vulnerable to environmental harms. The necessity for this paradigm shift is further underlined by contemporary global challenges. The Intergovernmental Panel on Climate Change (IPCC) warns that climate adaptation strategies must prioritize vulnerable populations, noting that measures enhance rather than undermine moral quality and agency (IPCC 2022). Likewise, during global health crises like the COVID-19 epidemic, systems that demanded adaptability and equity aggravated suffering and strengthened inequalities (WHO 2023). These assignments make clear that adaptability and well-being cannot be afterthoughts; they must be foundational design principles.

Innovation must be ethical, participatory, and guided by the principle of regenerative design, where mortal exertion contributes appreciatively to ecosystems and societies rather than simply mollifying damage (Mang and Reed 2012). Technologies, programs, and architectures must be estimated not solely for their effectiveness but for their capacity to uphold moral quality and promote collaborative well-being. Designing for flourishing requires a value-grounded approach that places empathy, justice, and care at the core of decision-making, making structure sustainable systems that prioritize moral quality is not simply an aspirational thing, it is a survival strategy for the twenty-first century. A narrow focus on specialized fixes without addressing social injuries will be inadequate to navigate the complex challenges ahead. By centering the full diapason of mortal requirements, sustainable design can foster societies that are not only flexible to shocks but also vibrant, inclusive, and joyous. In doing so, we move beyond survival toward a truly sustainable and staid mortal future. The design field is regarded as an exertion or a process sphere rather than a scientific discipline. It’s the confluence of scientific, on-technical and social processes. In diurnal communication, the term design is substantially associated with casting and/or constructing vestiges; still, it’s a study process for creating palpable and impalpable products, services and processes. Depending on the field of interest, there are prominent differences along the axes (Holmes 1995). Design may encompass a wide variety of environments, knowledge and gestures from different disciplines. Design issues may range from vehicles, machines, consumer products and cultural particulars at one end of the diapason, to software, interactive systems and allowed processes at the other end. This broad perspective brings the need to consider both specialized and non-specialized aspects, which include form and functionality, as well as means of integration with advancements in technology, engineering, cultural trends, lore, society, policy, timber, education, sustainability, and business education.

2 Human Element and Sustainability in Product Design

In contemporary product design, two critical imperatives have surfaced as essential: the integration of the mortal element and the commitment to sustainability. Traditional design practices that prioritized aesthetics or functionality alone are no longer sufficient in the face of mounting social and environmental challenges (Brown 2009). The mortal element emphasizes a compassionate, stoner-centred approach, focusing on the cerebral, emotional, and artistic requirements of different individualities. Mortal-centered design (HCD) styles engage deeply with druggies' lived experiences, promote inclusivity and ensure that products resonate meaningfully across different capacities, genders, and artistic backgrounds (Norman 2013). This approach fosters invention by resting design opinions in empathy, leading to results that aren't only functional but also empowering and emotionally engaging.

Contemporaneously, sustainability in product design demands a holistic retrospection of every stage in a product's lifecycle—from material sourcing to manufacturing, use, and eventual disposal (McDonough and Braungart 2002). Sustainable design strategies concentrate on continuity, order, energy effectiveness, and the use of renewable or recycled accoutrements. Indirect frugality models further advocate for products that can be completely disassembled, reused, or biodegraded, minimising waste and environmental detriment (Ellen MacArthur Foundation 2013). Ethical considerations, including fair labour practices and minimizing carbon vestiges, are also pivotal to authentically sustainable product development. Importantly, the mortal element and sustainability must not be treated as separate disciplines but understood as deeply connected. Products designed for sustainability must still prioritize moral quality, availability, and emotional connection; else, they risk rejection by druggies or failure to produce lasting behavioral change (Manzini 2015). Companies like Patagonia and Fairphone illustrate this integration by producing durable, ethically sourced products that empower consumers to make environmentally responsible choices without sacrificing usability or values. Therefore, by bedding both empathy and ecological stewardship at the core of product design, we move toward a future where invention serves not just requests, but the broader pretensions of mortal flourishing and planetary health.

3 Human-Centered Design

Mortal-Centred Design (HCD) is a creative and problem-solving approach that prioritizes the requirements, gestures, and values of the people for whom a product, service, or system is intended. At its core, HCD seeks to understand druggies deeply by empathizing with their gestures, observing their actions, and involving them directly in the design process (Brown 2009). Unlike traditional design styles that may emphasize technological capabilities or request trends, HCD begins with a compassionate disquisition of real-world challenges faced by different communities. It follows an iterative process of alleviation, creativity, and perpetration, where nonstop feedback from druggies refines results at every stage (IDEO.org 2015). By centering design around mortal requirements, HCD fosters products and services that aren't only more usable but also more meaningful and emotionally reverberative. Crucially, HCD embraces diversity and inclusivity, recognizing that people vary extensively based on their physical capacities, artistic backgrounds, genders, periods, and socioeconomic conditions. This perspective challenges contrivers to suppose beyond the" average" stoner and produce adaptable, accessible gestures that cater to a wide range of requirements (Norman 2013). In fields similar to healthcare, education, technology, and public services, applying HCD has led to further successful issues because results are acclimatized to fit into the everyday lives of the people they aim to serve (Buchanan 2001). For example, in healthcare settings, bias designed through HCD principles frequently enhances patient autonomy and quality, making treatment processes less intimidating and more stoner-friendly. Also, HCD contributes to ethical and socially responsible invention. As Buchanan (2001) emphasized, HCD isn't only a system but also a gospel that affirms the quality and agency of individuality in shaping their surroundings. In a period where technology and globalization frequently alienate or marginalize vulnerable populations, HCD offers a frame to ensure that design processes empower rather than count. It encourages interdisciplinary collaboration, drawing on perceptivity from anthropology, psychology, engineering, and business to produce holistic results. Eventually, mortal-centred design aligns technological advancement with mortal well-being. As global challenges—from climate change to public health heads become more complex, the need for compassionate, inclusive, and adaptable results grows more critical. HCD's commitment to harkening, repeating, and co-creating with real people ensures that invention remains predicated on humanity, not abstraction. In this way, mortal-centred-centred-Centered Design not only improves the usability of products and services but also advances the larger thing of social good.

4 Defining Sustainability

Sustainability is the guiding principle for meeting present requirements without compromising the capability of unborn generations to meet their own (World Commission on Environment and Development (WCED) 1987). Firstly defined by the Brundtland Commission as "development that meets the requirements of the present without compromising the capability of unborn generations," the conception has since expanded to encompass environmental integrity, profitable viability, and social equity—the three pillars of the triadic nethermost line (Elkington 1997). Environmental integrity calls for the protection and restoration of ecosystems through practices such as biodiversity conservation, pollution reduction, and resource effectiveness. Profitable viability demands that businesses and husbandry operate in ways that induce substance without depleting natural capital. Social equity insists on fair distribution of benefits and burdens, ensuring that all communities have access to introductory coffers, openings, and a healthy terrain (United Nations 2015a, b).

Enforcing sustainability requires a systemic, life-cycle approach. Life Cycle Assessment(LCA) evaluates the environmental impacts of a product or service from raw-material birth through product, use, and disposal, relating hotspots where advancements yield the topmost benefits (ISO 2006). indirect frugality models make on LCA by closing material circles designing products for continuity, order, and recyclability—and shifting value chains from direct "take - make - dispose" systems to regenerative overflows of accoutrements and energy(Ellen MacArthur Foundation 2013). Policy instruments such as carbon pricing, extended producer responsibility(EPR), and green public procurement further incentivise sustainable product and consumption patterns (OECD 2020). Beyond specialised tools and programs, sustainability is deeply intertwined with social and artistic metamorphosis. Behavioural change—at individual, organisational, and societal situations—is critical for reducing overconsumption and fostering stewardship. Education for Sustainable Development(ESD) empowers people with the knowledge, chops, and values demanded to make informed opinions and take responsible conduct(UNESCO 2017). Inversely important is inclusive governance, which engages stakeholders, particularly marginalized communities, in decision - timber processes to ensure that sustainability strategies address different requirements and uphold social justice(Agyeman 2013). Despite significant progress, major challenges remain settled reactionary—energy dependences, widening socioeconomic inequalities, and the accelerating pace of biodiversity loss and climate change. Achieving the Sustainable Development Goals (SDGs) by 2030 demands unprecedented cooperation among governments, businesses, civil society, and individuals (United Nations 2015a, b). Eventually, sustainability is not a fixed target but a nonstop trip of invention, equity, and watch for both people and the earth.

5 Sustainable Development Goals

The Sustainable Development Goals (SDGs) are a universal call to action espoused by all 193 United Nations Member States in 2015, forming the core of the 2030 Agenda for Sustainable Development. This docket provides a participatory design for peace and substance, for people and the earth, now and into the future. At its heart are 17 SDGs, which address the most burning global challenges, from poverty and hunger to health, education, gender equality, clean energy, and climate action. These pretensions were designed as a critical call for action by both developed and developing nations, sustained by a commitment to "leave no one behind" and achieve inclusive development through global cooperation. The SDGs rest on three interlinked pillars—environmental integrity, profitable viability, and social equity—frequently appertained to as the "triadic nethermost line." Environmental integrity demands the protection and restoration of ecosystems through pollution reduction, biodiversity conservation, and resource effectiveness. Profitable viability calls for business models and programs that induce substance without depleting natural capital, creating lasting wealth and stability. Social equity insists on fair distribution of benefits and burdens, guaranteeing that all communities have access to essential services, opportunities, and a healthy terrain. This holistic frame moves beyond traditional development paradigms that prioritized growth at the expense of people and the earth. The 17 pretensions themselves are far-reaching, ambitious, and interdependent. Goal 1 seeks to "End poverty in all its forms far and wide," while Goal 2 aims to "End hunger, achieve food security and better nutrition and promote sustainable agriculture." Goal 3 commits to "insure healthy lives and promote well - being for all at all periods," and Goal 4 to "insure inclusive and indifferent quality education and promote lifelong literacy openings for all." pretensions 5 through 17 continue this holistic vision achieving gender equivalency; securing clean water and sanitation; icing affordable, dependable, sustainable energy; promoting decent work and profitable growth; erecting flexible structure and invention; reducing inequalities; fostering sustainable metropolises and communities; icing responsible consumption and product; taking critical action on climate change; conserving abysses and marine coffers; guarding terrestrial ecosystems; promoting peaceful and inclusive societies; and revitalizing global meetings for sustainable development. Despite the clarity of these objects, global progress has been uneven and, in numerous areas, inadequate. According to the Sustainable Development Report 2024, only 16% of the 231 SDG targets are on track to be met by 2030, while 84% show limited progress or outright reversal since 2020. Particularly intimidating are SDG 2 (Zero Hunger), SDG 11(Sustainable metropolises and Communities), SDG 14(Life Below Water), SDG 15(Life on Land), and SDG 16(Peace, Justice and Strong Institutions), all of which are off track. The report warns that without accelerated action, the 2030 Deadline will pass with numerous core pretensions unmet. Ending these gaps requires addressing a stunning funding gap; an estimated USD 4.2 trillion periodic funding gap stands between current investments and what's demanded to achieve the SDGs. In response, the United Nations has proposed an SDG encouragement Plan—promoting debt relief, providing backing, and reforming the global fiscal architecture to more support developing countries. Contemporaneously, the private sector is being prompted to introduce and invest in measurable impact, shifting from expenditure-based backing to outgrowth—grounded models that directly tie commercial benefits to concrete advancements in livelihoods, education, and health. Stakeholder prospects and new nonsupervisory fabrics are driving companies to align ESG targets with SDG issues, creating problems that can rally coffers at scale. Eventually, the SDGs represent both a moral imperative and a realistic roadmap for mortal and planetary well-being. Their success hinges on collaborative action—governments, businesses, civil society, and individuals must unite in unknown ways. As underlined in the UN's Global Sustainable Development Report 2023, transformative progress requires systems-position change across sectors, using technological advances, indigenous knowledge, and innovative backing mechanisms. Only through loyal political will, inclusive governance, and sustained meetings can we hope to fulfil the pledge of the 2030 Agenda, achieving a sustainable, inclusive, and prosperous future for all.

- **SDG 1: End Poverty**

Sustainable Development Goal 1 aims to annihilate extreme poverty in all its forms by 2030, targeting those living below the international poverty line of USD 1.90 per day(2011 PPP) and extending social protection to vulnerable groups(United Nations, n.d.- a). Central to this Goal are seven targets that include reducing at least by half the proportion of men, women, and children living in poverty; enforcing nationally applicable social protection systems; and ensuring equal rights to profitable coffers, land, and introductory services. Progress over the once two decades has been significant between 2002 and 2012, the global proportion of people living under USD 1.90 a day fell from 26 to 13% (United Nations, n.d.- a). Nonetheless, the COVID-19 epidemic, rising inequality, and climate-related shocks have reversed earnings in numerous regions, pushing an estimated 100 million people back into extreme poverty in 2020 (World Bank 2022). Sweats to accelerate SDG 1 bear integrated strategies. Expanding social protection bottoms can cushion homes against income loss due to illness, severance, or disasters, while inclusive fiscal services similar as microcredit and digital banking—enable the poor to make means and invest in education or small enterprises(United Nations, n.d.- b) also, icing affordable access to quality introductory services(education, healthcare, water, sanitation, and energy) is critical for breaking poverty traps. Strengthening adaptability through climate-smart husbandry and disaster threat reduction also protects pastoral populations from climate volatility, which disproportionately affects smallholder growers. Despite these measures, crucial challenges persist, backing faults hang the scale-up of social safety nets in low-income countries, and systemic inequalities, grounded in gender, race, and disability, frequently leave the most marginalized behind. To meet SDG 1 by 2030, global meetings must rally fresh coffers, foster pro-poor programs, and promote data disaggregation to target interventions effectively. Only by combining profitable addition, social protection, and adaptable structure can the world truly end poverty in all its forms.

- **SDG2: Zero Hunger**

SDG 2 aspires to end hunger and all forms of malnutrition by 2030, ensuring that everyone, especially children and the most vulnerable, has year-round access to sufficient, safe, and nutritional food(United Nations). Eight targets bolster this Goal, including doubling agrarian productivity and inflows of small-scale food producers, increasing sustainable food product systems, and maintaining the inheritable diversity of seeds and cultivated crops. Progress has been mixed globally. Undernourishment fell from 15% in 2000–2002 to 11% in 2014–2016, yet over 790 million people still faced hunger pre-epidemic (FAO 2020). Recent heads including COVID-19 dislocations, climate axes, and conflicts have since reversed earnings, pushing food instability to its loftiest situations in decades (FAO et al. 2023). Achieving SDG 2 demands a holistic approach. Sustainable agrarian practices similar as agroecology, climate smart husbandry, and effective water operation—boost productivity while guarding soils and biodiversity (United Nations, n.d.- c), Inversely important is strengthening pastoral structure (roads, storehouse, and cold chains), which reduces post crop losses and connects growers to requests (UN Women 2021). Social safety nets, similar to academic feeding programs and temporary cash transfers, ameliorate salutary diversity and adaptability among the poorest homes. Investments in exploration and invention—for failure-tolerant crops, sustainable diseases, and digital premonitory services—can further empower smallholders.

The major obstacles remain. Gender difference in land power and access to finance hamper women growers, who produce up to 80% of food in some regions, yet contribute a bit to the coffers (UN Women 2021). Conflict and forced relocation disrupt agrarian livelihoods, while unpredictable food prices complicate vulnerability. To meet SDG 2, programs must integrate nutrition education, equal resource access, and emergency response mechanisms. By aligning agrarian development with social protection and environmental stewardship, the global community can transform food systems to nourish all people sustainably.

- **SDG 3: Good Health and Well-being**

SDG 3 seeks to insure healthy lives and promote well being for all at all periods, addressing a diapason of challenges from motherly and child mortality to noninfectious conditions(United Nations, n.d.- d) Thirteen targets include reducing the global motherly mortality rate below 70 per 100,000 live births; ending preventable deaths of babe and children under five; combating HIV/AIDS, tuberculosis, malaria, and neglected tropical conditions; and achieving universal health content(World Health Organization (WHO) 2023) Despite progress—under five mortality declined by 47% between 2000 and 2016—an estimated 5.6 million children under five still failed in 2016, with deaths heavily concentrated in sub Saharan Africa and South Asia (WHO 2023). Achieving SDG 3 hinges on strengthening health systems. Universal health care requires accessible, affordable, and quality primary care services, including essential drugs and vaccines (WHO 2023). Robust backing mechanisms blending domestic backing, transnational aid, and innovative instruments like health bonds can close gaps in resource-poor settings. Integrated service delivery, bolstered by trained health workers, telemedicine, and community health programs, enhances outreach to remote and marginalised populations. Preventative measures(e.g., immunisation juggernauts, health education, and tobacco control) are inversely critical to check the growing burden of noninfectious conditions and support internal health (WHO 2023). Global health faces new pitfalls, afflictions, antimicrobial resistance, and climate-related health extremities that challenge system adaptability. The COVID-19 epidemic exposed fragilities—diverting coffers, dismembering routine services, and aggravating injuries (WHO 2023). Addressing these requires intertwined preparedness fabrics, early warning systems, and multi-sectoral collaboration across health, terrain, and social protection sectors. By investing in universal health content, strengthening primary care, and prioritising equity, nations can ensure that SDG 3's pledge of health and well-being for all becomes a reality.

- **SDG 4: Quality Education**

SDG 4 strives to “ensure inclusive and equitable quality education and promote lifelong literacy opportunities for all” by 2030, featuring education as a foundational motorist of sustainable development (United Nations). Ten targets encompass free primary and secondary education; equal access to quality pre-primary, specialized, vocational, and advanced education; knowledge and numeracy for youth and grown-ups; and training of good preceptors (UNESCO 2023). Although global primary academy registration rose from 83% in 2000 to 91% in 2015, 262 million children and youth remained out of school, with learning issues affecting over half of scholars in low-income countries fail to achieve minimal proficiency in reading and mathematics(UNESCO 2023). To accelerate progress, SDG 4 calls for systemic reforms. Investing in early childhood education yields the loftiest returns, perfecting cognitive development and long-term issues. Equipping preceptors through nonstop professional development and perfecting pupil–schoolteacher rates enhance learning quality. Structure upgrades safe classrooms, inclusive installations for learners with disabilities, and digital connectivity—address barriers to participation. Lifelong literacy enterprise, including adult knowledge programs and chops training, empowers individuals to acclimate to labour market shifts (UNESCO 2023). Still, inequality persists, gender gaps, geographic differences, and socioeconomic status determine educational access and attainment. The COVID-19 pandemic widened these divides, with academy closures impacting 1.6 billion learners and disproportionately harming those without access to remote literacy (UNESCO 2023). Bridging these gaps demands targeted outreach, education programs, and community engagement. By aligning public education strategies with inclusive programs, adequate backing, and data-driven monitoring, policymakers can ensure that every learner benefits from quality education and lifelong literacy opportunities.

- **SDG 5: Gender Equality**

SDG 5 trials to “achieve gender equivalency and empower all women and girls,” feting gender equivalency as a abecedarian mortal right and a foundation for sustainable development(United Nations, n.d.- f) Nine targets include ending all forms of demarcation and violence against women and girls; barring dangerous practices similar as child marriage and womanish genital mutilation; feting and valuing overdue care work; icing women’s full participation in leadership and decision timber; and universal access to sexual and reproductive health and rights(UN Women 2023) While legislative fabrics have bettered—over 140 countries now guarantee women’s legal equivalency in the plant—the pace of change remains slow. Encyclopedically, 1 in 3 women experience physical or sexual violence in their lifetime, and women hold only 26% of administrative seats as of 2023 (UN Women 2023). Accelerating SDG 5 requires multi-rounded strategies. Legal reforms must be executed with responsibility mechanisms, and gender responsive budgeting should allocate coffers to women’s commission programs. Social morals can be shifted through community discourses, media juggernauts, and engaging men and boys as abettors. Profitable commission enterprise, like access to credit, property rights, and entrepreneurship training, enables women’s fiscal independence (United Nations). Also, expanding childcare services and maternal leave programs reduces the disproportionate burden of overdue care work on women patients' walls, including discriminatory morals, conflict, climate shocks, and the digital peak—bear targeted interventions. Data disaggregation by coitus, age, and other factors is essential for covering progress and acclimatizing programs. Strengthening meetings with civil society and grass-roots movements ensures that the voices of marginalized women inform program design. By achieving gender equity across all sectors and situations of governance, the global community can unleash the full potential of half the world's population and realise SDG 5 by 2030.

- **SDG 6: Clean Water and Sanitation:**

Sustainable Development Goal 6 aims to "ensure the availability and sustainable operation of water and sanitation for all" by 2030. Universal access to safely managed drinking water, sanitation, and hygiene (marshland) services is abecedarian for health, education, and profitable productivity, yet current progress is oppressively off track, As of 2022, 2.2 billion people demanded safely managed drinking water services and 3.5 billion demanded safely managed sanitation services; nearly half of the world's population endured severe water failure for at least part of the time.

Achieving universal content by 2030 requires accelerating progress sixfold for drinking water and fivefold for sanitation compared to current rates (UN Stats Division 2024a, b, c, d). Core strategies include expanding investment in flexible marshland structure, promoting community-led total sanitation, and integrating hygiene education into academic classes. Technological inventions similar as decentralized water treatment, low- cost detectors, and nature-grounded results offer fresh pathways to extend services to remote and informal agreements., financing gaps remain dispiriting; an estimated USD 114 billion per year is demanded to reach SDG 6 targets, with low-income countries bearing the mass of the space (UN Sustainable Development Report 2024). Injuries in service provision persist along pastoral–civic lines and between socioeconomic groups, demanding pro-poor backing models, gender sensitive programming, and stronger governance to ensure equitable resource allocation. Without critical, cooperative action to rally finances, strengthen institutions, and empower communities, the Goal of safe water and sanitation for all will remain elusive.

- **SDG 7: Affordable and Clean Energy:**

SDG 7 seeks to "ensure access to affordable, dependable, sustainable, and ultra-modern energy for all" by 2030, citing energy as essential for poverty eradication, health, education, and profitable growth. Significant strides have been made in electrification, but progress on clean cuisine and renewable energy remains uneven. Worldwide, the electrification rate has risen steadily over the past decade, yet 770 million people still lack access to electricity, and nearly 2.4 billion people cook with contaminated energies. To achieve SDG 7, global stakeholders must accelerate deployment of off-grid and mini-grid systems, expand clean cuisine enterprises, and enhance energy effectiveness across sectors (IRENA 2024). Integrating smart grid technologies and demand-side operation can further optimise resource use and reduce losses. Backing remains a hedge, periodic investment in renewable energy and energy effectiveness needs to triple to USD 2.6 trillion by 2030, with a particular focus on emerging markets (ESMAP 2024). Policy fabrics such as subvention reform, feed-in tariffs, and carbon pricing are critical to creating enabling surroundings for private capital. Transnational cooperation, technology transfer, and capacity structure must be geared up to support developing countries in the transition. Without this combined conduct, the pledge of universal, clean, and affordable energy will remain out of reach.

- **SDG 8: Decent Work and Economic Growth:**

SDG 8 promotes "sustained, inclusive, and sustainable profitable growth, full and productive employment and decent work for all." Profitable expansion alone is inadequate; the quality and equity of growth determine its social impact.

Global GDP growth recovered to 4.6% in 2022 and is projected at 5.5% in 2024, yet these rates fall suddenly of the 7% demanded in least developed countries(LDCs) to annihilate poverty (UN Stats Division 2024a, b, c, d), Youth severance remains stubbornly high—at 12.9% encyclopedically in 2023—and informal employment accounts for nearly 60% of total jobs, aggravating vulnerable work conditions, programs to achieve SDG 8 include strengthening labor request institutions, expanding vocational training, and promoting green jobs through investments in renewable energy and sustainable structure. Social protection benefits and pay envelope subvention programs can buffer shocks and grease formalization. Also, supporting small and medium-sized enterprises (SMES) with access to finance and technology enhances productivity and job creation. Challenges persist despite policy advances, numerous countries warrant the financial space to invest in job-rich growth strategies, and rising robotization threatens to displace low-skilled workers. Global shocks—like the COVID-19 epidemic and force chain dislocations—emphasise the need for flexible, inclusive labour requests. Ending the decent work gap requires coordinated action across governments, businesses, and transnational associations.

- **SDG 9: Industry, Innovation and Infrastructure:**

SDG 9 aspires to "make flexible structure, promote inclusive and sustainable industrialization and foster invention." Robust structure and manufacturing are machines of profitable development, job creation, and technological progress. Post epidemic, global manufacturing value added per capita rose by 16% from 2015 to 2023, reaching USD 1922—but periodic growth has slowed at 2.7% since 2022, indicating the need for revitalized artificial programs. Digital structure expansion has accelerated, yet nearly 37% of the world's population remains offline, limiting invention prolixity. Crucial strategies include scaling up public–private meetings to finance transport and energy networks, incentivising R&D through duty credits, and fostering invention capitals that link academia, industry, and government. Emphasising sustainable artificial processes, like indirect manufacturing and green hydrogen, can uncouple growth from resource reduction. Indigenous difference is stark, manufacturing in sub–Saharan Africa remains at USD 163 per capita, compared to USD 4986 in Europe and Northern America, emphasizing the need for targeted capacity structure and technology transfer. To fulfil SDG 9, countries must integrate structural planning with climate adaptability and adaptation, ensuring that innovation serves all parts of society.

- **SDG 10: Reduced Inequalities:**

SDG 10 targets "reducing inequality within and among countries" by empowering the poorest 40% of income earners, promoting social, economic, and political inclusion irrespective of age, sex, disability, race, or other status. In 2022, further than half of 124 countries achieved briskly income growth for the poorest 40% of the population, yet indigenous differences persist in only 30% of countries in Central and Southern Asia saw similar earnings compared to 78% in North America and Europe. Overseas development backing remains concentrated in many sectors, and remittance costs average 7% of transferred quantities, limiting income mobility for low-income homes. Strategies to advance SDG 10 include progressive duty and transfer systems to redistribute wealth, scalable social protection programs, and targeted fiscal addition enterprise for marginalized groups. Strengthening anti-demarcation laws and fostering inclusive governance fabrics can amplify underrepresented voices in decision-making; nevertheless, global migration, climate shocks, and digital divides pose new challenges to equality. Rising geopolitical pressures and profitable nationalism hang cross-border cooperation on migration and backing. To bridge patient gaps, policymakers must align domestic programs with global commitments, ensuring that growth, social cohesion, and environmental sustainability advance hand in hand.

- **SDG 11: Sustainable Cities and Communities:**

Goal 11 seeks to "make metropolises and mortal agreements inclusive, safe, flexible and sustainable" by 2030. With over 4.4 billion people, further than half the world's population—formerly living in rural areas, the rapid pace of urbanization presents both opportunities and challenges. Core targets include furnishing access to acceptable, safe and affordable housing; enhancing public transport and green public spaces; and strengthening disaster threat reduction strategies. Despite progress, roughly one billion civic residents still live in slums or informal settlements, lacking basic services such as water, sanitation, and a flexible structure. Achieving SDG 11 demands sixfold acceleration of progress on casing and transport, robust civic planning that centres the requirements of vulnerable groups, and integrated waste operation systems. Only by aligning programs across casing, mobility, and environmental adaptability can metropolises achieve equity and sustainability.

- **SDG 12: Responsible Consumption and Production:**

Goal 12 aims to "ensure sustainable consumption and product patterns". It emphasises divorcing profitable growth from resource use, promoting indirect frugality models, and reducing waste at all stages of the product life cycle. Crucial targets include achieving sustainable operation of chemicals and waste, halving per capita global food waste, and encouraging companies to adopt sustainability reporting practices. Yet, the global material footprint continues to rise, reaching over 100 billion tonnes in 2020, while periodic external solid waste covers 2 billion tonnes. To meet SDG 12, governments must apply extended product responsibility schemes, incentivize eco-design, and foster consumer mindfulness through education and labeling. Businesses, in turn, need to shift from direct "take make dispose" models to systems that prioritize exercise, form, and recycling, icing resource security for current and unborn generations.

- **SDG 13: Climate Action:**

Goal 13 calls for critical action to combat climate change and its impacts, recognised as the topmost threat to development. Climate change has formerly raised global temperatures by roughly 1.1 °C above preindustrial levels, enhancing extreme rainfall events and disproportionately burdening the poorest communities (IPCC 2022). Targets under SDG 13 include integrating climate change measures into public programs, perfecting adaptability and adaptive capacity, and marshalling USD 100 billion annually by 2020 to support developing countries (UN Stats Division 2024a, b, c, d). Despite pledges, the 100 billion thing remains unmet, and hothouse gas emissions continue to climb. Accelerated mitigation and adaptation bear phasing out reactionary energy subsidies, scaling up renewable energy, and erecting early warning systems. Cross-border cooperation, backed by robust finance and technology transfer, is essential to guard lives, livelihoods, and ecosystems from rising climate pitfalls.

- **SDG 14: Life Below Water:**

Goal 14 aims to "conserve and sustainably use the abysses, swell and marine coffers for sustainable development". Abysses give livelihoods for over 3 billion people and absorb around 30% of anthropogenic CO_2, yet they face severe pitfalls from overfishing, pollution, and acidification (UN Stats Division 2024a, b, c, d). Targets include precluding and significantly reducing marine pollution by 2025, sustainably managing littoral ecosystems, and establishing marine protected areas covering at least 10% of littoral and marine zones. Presently, only 7% of the ocean is protected. Meeting SDG 14 demands spanning up ecosystem grounded operation, bridling plastic and nutrient runoff, and administering sustainable fisheries. Transnational agreements similar to the UN Decade of Ocean Science must be completely enforced to restore ocean health, ensuring that marine biodiversity and services remain available for unborn generations.

- **SDG 15: Life on Land:**

Goal 15's accreditation is to "cover, restore and promote sustainable use of terrestrial ecosystems, sustainably manage timbers, combat desertification, halt and rear land degradation and halt biodiversity loss" (United Nations, n.d.- Goal 15). Terrestrial ecosystems support the livelihoods of 1.6 billion people and house 80% of all known terrestrial species, yet one million species face extinction due to niche loss and unsustainable land use (IPBES 2019). Specific targets include halving the rate of land degradation, restoring 15% of demoralized ecosystems, and ensuring that all agro-biodiversity and wild species are sustainably managed. Achieving these objectives requires integrated geography operations, reforestation enterprise, and measures to combat illegal logging and wildlife trafficking. Strengthening community rights and indigenous stewardship is critical, as original land druggies frequently give effective, culturally applicable conservation. Only through coordinated restoration and protection can the world safeguard vital terrestrial ecosystems and the services they provide.

- **SDG 16: Peace, Justice and Strong Institutions:**

Goal 16 aspires to "promote peaceful and inclusive societies for sustainable development, give access to justice for all and make effective, responsible and inclusive institutions at all situations" (United Nations, n.d.- Goal 16) Targets include significantly reducing all forms of violence and related death rates, ending abuse and trafficking of children, icing equal access to justice, and developing transparent institutions. Yet, in 2023, homicide rates varied tenfold between regions, and an estimated 71% of people worldwide warrant legal identity attestation (UN DESA, 2024). Realising SDG 16 demands strengthening the rule of law, investing in police and judicial capacities, and fostering communal participation. Anti-corruption measures and freedom of information laws can enhance governmental responsibility, while inclusive policymaking—engaging women, youth, and non-age groups ensures that institutions serve all citizens fairly.

- **SDG 17: Partnerships for the Goals**

Goal 17 underscores the need to "strengthen the means of perpetuation and revitalize the Global Partnership for Sustainable Development". It emphasizes finance, technology, capacity structure, trade, and systemic consonance. Crucial targets include marshalling USD 100 billion in climate finance, enhancing North–South cooperation, and promoting policy consonance for sustainable development. Still, developing countries face a persistent financing gap of roughly USD 4.2 trillion for SDG implementation (UN CTAD 2024). To bridge this peak, innovative backing instruments—green bonds, amalgamated finance, and debt for nature barters must be scaled up, alongside reforms to global duty and trade architecture. Inversely vital are South–South and triangular meetings, knowledge sharing, and data collaboration enterprises that leave no country or community behind.

6 Conclusion and Discussion

In the end, the Sustainable Development Goals and Human-Centred Design represent two reciprocal fabrics that together chart a path toward a further just, flexible, and flourishing world. The 17 SDGs stimulate the global imperatives—from ending poverty and hunger, to securing ecosystems and fostering inclusive institutions—that must be met to secure the well-being of people and the earth by 2030 (United Nations 2015a, b). Yet achieving these intentions requires more than high-position targets and policy instruments; it demands design processes that deeply understand and admire the lived experiences, requirements, and burdens of those whom these pretensions are meant to serve (Brown, 2009). Mortal-Centred Design brings empathy, replication, and co - co-creation to the forefront of problem-solving, icing that results aren't only technically sound but socially meaningful and culturally applicable (Norman 2013). By engaging communities in every stage—from requirements assessment and prototyping, through testing and refinement—contributors can uncover environment-specific walls and influence original strengths. This approach fosters power, builds trust, and increases the likelihood that interventions—whether in health, education, energy, or governance—will be espoused and sustained. Moreover, HCD's emphasis on inclusivity aligns directly with the SDGs' "leave no one behind" principle. Inclusive design practices ensure that marginalized groups—women and girls, persons with disabilities, indigenous peoples, and pastoral communities are active actors in results that affect their lives, thereby advancing both social equity (SDG 10 and 5) and the quality of life (SDG 3 and 4). Contemporaneously, bedding sustainability criteria like lifecycle thinking and indirect-frugality models within HCD processes help attune mortal well-being with environmental boundaries (SDG 12 and 13). Eventually, the emulsion of SDG precedence and moral design creates an important community global pretension predicated on original realities, and design processes informed by universal imperatives. By marrying the macro-position vision of the 2030 docket with micro-position empathy and co - co-co-creation, stakeholders can co-construct results that are formerly ambitious and attainable. In doing so, we can move beyond bare survival toward a future in which every person can lead a staid, empowered, and sustainable life.

References

Agyeman J (2013) Introducing just sustainability's: policy, planning, and practice. Zed Books

Brown T (2009) Change by design: how design thinking creates new alternatives for business and society. Harper Business

Buchanan R (2001) Human-centred design: philosophy and practice. In: Laurel B (ed) Design research: methods and perspectives. MIT Press, pp 55–65

Elkington J (1997) Cannibals with forks: the triple bottom line of 21st century business. Capstone

Ellen MacArthur Foundation. (2013). Towards the circular economy: economic and business rationale for an accelerated transition. https://ellenmacarthurfoundation.org/towards-the-circular-economy

ESMAP (2024) Tracking SDG 7: progress towards sustainable energy. The World Bank. Retrieved from https://trackingsdg7.esmap.org/

Food and Agriculture Organization (2020). The State of Food and Agriculture 2020. FAO

FAO, IFAD, UNICEF, WFP & WHO (2023) The state of food security and nutrition in the world 2023. Food and Agriculture Organization of the United Nations

Holmes WP, Azam MA, Hills PC (1995) A preliminary study of the relationship between industrial design and engineering design. In: DIS '95: Proceedings of the 1st conference on Designing interactive systems: processes, practices, methods, & technique, pp 105–114

IDEO.org (2015) The field guide to human-centered design. IDEO

Intergovernmental Panel on Climate Change (IPCC) (2022) Climate change 2022: impacts, adaptation, and vulnerability. https://www.ipcc.ch/report/ar6/wg2/

International Organization for Standardization (ISO) (2006) ISO 14040: environmental management—life cycle assessment—principles and framework

IPBES (2019) Global assessment report on biodiversity and ecosystem services. Intergovernmental Science-Policy Platform on Biodiversity and Ecosystem Services

IPCC (2022) Climate change 2022: impacts, adaptation and vulnerability. Intergovernmental Panel on Climate Change

IRENA (2024) Tracking SDG 7: the energy progress report 2024. International Renewable Energy Agency, IRENA

Mang P, Reed B (2012) Designing from place: a regenerative framework and methodology. Build Res Inf 40(1):23–38. https://doi.org/10.1080/09613218.2012.621341

Manzini E (2015) Design, when everybody designs: an introduction to design for social innovation. MIT Press

McDonough W, Braungart M (2002) Cradle to cradle: remaking the way we make things. North Point Press

Norman DA (2013) The design of everyday things (revised and expanded edition). Basic Books

OECD (2020) Environmental policy tools and evaluation: feedback on EPR and carbon pricing. OECD Publishing

Ostrom E (2009) Beyond markets and states: polycentric governance of complex economic systems. Am Econ Rev 100(3):641–672. https://doi.org/10.1257/aer.100.3.641

Raworth K (2017) Doughnut economics: seven ways to think like a 21st-century economist. Chelsea Green Publishing

UN CTAD (2024) Financing for sustainable development report 2024: financing for development at a crossroads. United Nations Conference on Trade and Development

United Nations Department of Economic and Social Affairs (2024). *Global Sustainable Development Report 2024: Times of Crisis, Times of Change – Science for Accelerating Transformations to Sustainable Development*. Retrieved from https://desapublications.un.org/publications/global-sustainable-development-report-2024

United Nations Development Programme (2024). Human Development Report 2024. UNDP

UN Stats Division (2024a) Goal 10: progress and info. United Nations Statistics Division. Retrieved from https://unstats.un.org/sdgs/report/2024/Goal-10/

UN Stats Division (2024b) Goal 6: SDG indicators. United Nations Statistics Division, UNSD. Retrieved from https://unstats.un.org/sdgs/report/2024/Goal-06/

UN Stats Division (2024c) Goal 8: SDG indicators. United Nations Statistics Division, UNSD. Retrieved from https://unstats.un.org/sdgs/report/2024/Goal-08/

UN Stats Division (2024d) The sustainable development goals report 2024. United Nations, UNSD

UN Women (2021) Women and the SDGs: gender snapshot 2021. United Nations Entity for Gender Equality and the Empowerment of Women

UN Women (2023). Progress on the sustainable development goals: the gender snapshot 2023. UN Women

UNESCO (2017) Education for sustainable development goals: learning objectives

UNESCO (2023) Global education monitoring report 2023/4. United Nations Educational, Scientific and Cultural Organization

United Nations (2015a) Transforming our world: the 2030 agenda for sustainable development. https://sdgs.un.org/2030agenda

United Nations (2015b) Transforming our world: the 2030 agenda for sustainable development. United Nations Department of Economic and Social Affairs

United Nations (n.d.-a) Goal 1: end poverty. SDG knowledge platform. https://sdgs.un.org/goals/goal1
United Nations (n.d.-a) Goal 10: reduced inequalities. UN SDG knowledge platform. Retrieved from https://sdgs.un.org/goals/goal10
United Nations (n.d.-a) Goal 16: peace, justice and strong institutions. UN SDG Knowledge Platform
United Nations (n.d.-b) Goal 11: sustainable cities and communities. UN SDG Knowledge Platform
United Nations (n.d.-b) Goal 17: partnerships for the goals. UN SDG Knowledge Platform
United Nations (n.d.-c) Goal 2: zero hunger. SDG knowledge platform. https://sdgs.un.org/goals/goal2
United Nations (n.d.-c) Goal 12: responsible consumption and production. UN SDG Knowledge Platform
United Nations (n.d.-c) Goal 6: clean water and sanitation. UN SDG Knowledge Platform. Retrieved from https://sdgs.un.org/goals/goal6
United Nations (n.d.-d) Goal 3: good health and Well-being. SDG knowledge platform. https://sdgs.un.org/goals/goal3
United Nations (n.d.-d) Goal 13: climate action. UN SDG Knowledge Platform
United Nations (n.d.-d). Goal 7: affordable and clean energy. UN SDG Knowledge Platform. Retrieved from https://sdgs.un.org/goals/goal7
United Nations (n.d.-e) Goal 4: quality education. SDG knowledge platform. https://sdgs.un.org/goals/goal4
United Nations (n.d.-e) Goal 14: life below water. UN SDG Knowledge Platform
United Nations (n.d.-e) Goal 8: decent work and economic growth. UN SDG knowledge platform. Retrieved from https://sdgs.un.org/goals/goal8
United Nations (n.d.-f) Goal 5: gender equality. SDG knowledge platform. United Nations. https://sdgs.un.org/goals/goal5
United Nations (n.d.-f) Goal 15: life on land. UN SDG Knowledge Platform
United Nations (n.d.-f) Goal 9: industry, innovation and infrastructure. UN SDG Knowledge Platform. Retrieved from https://sdgs.un.org/goals/goal9
United Nations Statistics Division (2024) The sustainable development goals report 2024. United Nations. Retrieved from https://unstats.un.org/sdgs/report/2024/The-Sustainable-Development-Goals-Report-2024.pdf
WCED (World Commission on Environment and Development) (1987) Our common future. Oxford University Press
WHO (2023) World health statistics 2023: monitoring health for the SDGs. World Health Organization
World Bank (2022) Poverty and shared prosperity 2022: correcting course. World Bank Publications
World Health Organization (WHO) (2023) World health statistics 2023: monitoring health for the SDGs. https://www.who.int/data/gho/publications/world-health-statistics

Economic Empowerment and Menstrual Hygiene: A Comparative Study of Working and Non-working Women in Panchla and Mathabhanga

Lopamudra Ganguly, Subhadip Mondal, and Saptadipa Chakraborty

Abstract The study examines women's economic empowerment in relation to menstrual hygiene management (MHM) as part of a comparison of working and non-working women's MHM behavior. The work is based in two areas of West Bengal with distinguished climatic and socio-spatial features. They are Panchla (peri-urban) and Mathabhanga (rural). The study also considered the relationship between employment status, income generation, infrastructural availability and climate variability on MHM. A total of 95 respondents were included in the study using structured interviews and data analysed using statistical and spatial tools (SPSS and QGIS). The findings suggest that women workers particularly engaged in SHG-led tailoring in Mathabhanga demonstrated better use (and associated hygiene) of sanitary pads than non-working women especially in Panchla, who did not use pads because of economic barriers and insufficient infrastructure. The study demonstrated a clear role of education and access to adequate sanitation facilities also influenced menstrual behavior. Climatic context of monsoon flooding in Mathabhanga and water scarcity in Panchla also affected MHM and will affect the use of re-usable cloths. Mathabhanga has the advantage of active NGO and ASHA outreach, which is absent in Panchla. The research findings suggest economic empowerment is important but should also be advanced by infrastructure, health interventions, and climate responsive policies. Improving menstrual hygiene therefore requires an integrated approach to the public's health, gender equity, and sustainability in resource-poor settings.

Keywords Women empowerment · Menstrual hygiene · WASH · Sustainable development goals · Climatic zones · Economic independence · Comparative study

L. Ganguly (✉) · S. Mondal · S. Chakraborty
Department of Geography, Vedanta College, Kolkata, West Bengal, India
e-mail: ganguly.geography@gmail.com

A. Singh et al. (eds.), *Human-Centered Sustainability: Bridging Well-Being, Gender Equity, and Technology*, World Sustainability Series,
https://doi.org/10.1007/978-3-032-22020-2_5

1 Introduction

Menstrual hygiene management (MHM) is not simply a personal hygiene issue, but a fundamental element of gender justice, public health and social equity. MHM refers to the capacity to access safe and hygienic menstrual products, have access to adequate and appropriate sanitation and privacy to change menstrual materials (Sommer et al. 2016). MHM interacts with diverse development issues and is persistently overlooked within mainstream health and development dialogues. In India, but especially in rural and peri-urban areas, menstrual health is highly stigmatized, and culture taboos, lack of infrastructure and economic issues create obstacles to MHM for millions of women and girls. While adolescent girls and their menstrual needs are starting to gain increased attention in schools, there continues to be a lack of knowledge on how adult women, in particular economically active or disadvantaged women manage their menstrual hygiene as a lived experience in the real-world particularly in informal settings or working outside institutional contexts (UNICEF 2019). This study intends to address this gap. The current research is novel because it investigates the interactions of economic empowerment and environmental context for MHM practices in two regions of West Bengal that are unique in geo-climatic and socio-economic conditions: Panchla (peri-urban, industrial fringe) and Mathabhanga (rural, SHG-reliant economy). The comparative approach is particularly innovative, since it looks at women's experiences across employed status, climate risks (water shortage vs inundation), and their infrastructural support systems, exposing the multi-faceted impact on MHM practices. Moreover, while existing literature tends to examine health or income alone, this research combines the socio-economic environment, consideration of access and infrastructure, and climate resilience; it combines quantitative data sources, inferential statistics, and geographical Info Systems (GIS) based spatial analysis.

Two research problems are being constructed in this investigation. First, economic empowerment is often viewed as a means of improving women's control over health decisions (Kabeer 1999), but there is little empirical evidence indicating how this control plays a role in menstrual hygiene practices of women in resource-poor settings. Does having an income make women more inclined to have hygienic menstrual practices? Does economic activity, as with SHG engagement, encourage awareness via peer networks or community outreach? These are real issues facing policy makers, but remain underexplored. Second, economic studies very rarely include the environmental or infrastructure considerations of MHM. For example, well-earning women may have increased incomes, but limited capacity to follow hygienic menstrual practices due to inadequate infrastructure such as exposure to poor sanitation, seasonal water shortages, or inadequate disposal facilities, all of which may impact menstrual health (Caruso et al. 2017). Therefore, research not only illustrates the behavioural opportunities of working versus the behaviours of non-working women, it demonstrates that these behaviours are also made in the context of climatic and infrastructural limitations. In addition, this study addresses a business and productivity issue. Poor MHM has been cited to contribute to absenteeism, lower productivity at the workplace, and mental stress that leads to lesser economic growth. Women who cannot access to safe and dignified menstruation may steer clear of work altogether, especially in informal sectors or destination micro-industries where sanitation does not exist. As such, menstrual hygiene is not only a health issue, but also a workforce readiness and economic participation issue. If this connection is disregarded, the female labor force goes markedly underutilized, and the earning gap continues, and in particular, rural women in India.

This study has geater alignment with the United Nations Sustainable Development Goals (SDGs) and intends to advance three separate targets. First, SDG 3 (Good Healthy and Well-being) actively engages with target 3.7, which seeks to reduce health loss from menstrual hygiene-related risks such as infection and psychological stress. Second, SDG 5 (Gender Equality) directly engages targets outlining women's economic and social empowerment with respect to the ways in which it can guide a woman's autonomy to make decisions impacting her health, and by extension her ability to manage menstruation safely. Finally, SDG 6 (Clean Water and Sanitation) is directly integrated into this analysis through study of the water infrastructure in local and disposal systems that informs menstrual behavior, primarily in climate vulnerable areas such as Mathabhanga. Additionally, the comparative nature of this research lends itself well to suggesting interventions that are climate-adapted and economically inclusive as well as culturally relevant. Essentially, this research resituates menstrual hygiene not as a distinct health issue, but as a multi-sectoral development issue. It embeds MHM into the broader discussions of sustainable development, gender justice, climate adaptation, and economic empowerment. By investigating adult women in different workplace contexts, the research addresses a largely neglected population in the MHM literature. The outcomes were aimed at informing menstrual product subsidy policies, climate-resilient WASH infrastructure, and community pathways with the goal of arriving at menstrual dignity for all women, irrespective of their income, their occupation and/or their geography.

2 Literature Review

Over the past few decades, menstrual hygiene management (MHM) has emerged as a significant area of interdisciplinary study, reflecting its interactive linkages with public health, gender equity, sanitation, and economic development. Menstruation is a natural biological process, but it remains one of the most stigmatized and least talked about aspects of women's health across the globe, especially in developing countries like India. Lack of access to sanitary and hygienic menstrual practice not only contributes to compromising the health of women but also discourages women's education, livelihood, mobility, and dignity (Sommer et al. 2016). Consequently, understanding socio-economic, cultural, infrastructural, and environmental aspects that motivate MHM is important for the planning of effective policy making and intervention initiatives. The aim of this review of literature is to explore the existing literature relevant to women's economic empowerment and its influence on MHM, particular to the contextual location of rural and peri-urban India. The World Health Organization (WHO) states that MHM is a method of using clean menstrual material to absorb or store menstrual blood, which can be replaced in a private space as often as needed, with ready availability of soap and water for body washing and facilities for the safe disposal of soiled materials (WHO and UNICEF 2013). In India, a number of studies have reported the constraints that many women experience in using such minimum hygiene facilities. According to the National Family Health Survey (NFHS-5, 2019–21), there has been an increase in menstrual product use among Indian women aged 15–24 years; however, significant gaps exist, both urban-rural and between social-economic categories. Rural women, especially those from economically poor households, are still likely to use unhygienic alternatives, such as cloths, husk, or sand, since it is often too expensive, taboo, and sanitation infrastructure is poor (Chandra-Mouli and Patel 2017).

Economic empowerment is known to be a determinant of women's autonomy and decision-making ability, including in health and hygiene matters. Kabeer (1999) describes women's empowerment as the widening of their capacity to make strategic life choices in contexts where this capacity was once denied. Educated, employed, or self-help group participating women are more likely to have access to information, break taboos, and pay for hygienic menstrual products like sanitary pads. For instance, a research conducted by Rajagopal and Mathur (2017) revealed that working women in Tamil Nadu who were enrolled in SHG programs had greater knowledge and improved menstrual hygiene practices than non-working women. The same is supported by Gopalakrishnan and Dhanalakshmi (2014), who found that economic independence has a positive relationship with health-seeking behavior and maintenance of personal hygiene among women. Cultural taboos and beliefs remain strong deterrents to MHM throughout India. Most girls learn little or nothing about menstruation prior to menarche and receive fragmented or inaccurate information from peers or mothers (Dasgupta and Sarkar 2008). Such unawareness tends to be more exaggerated in non-working women and economically poor women, perpetuating silence and misinformation cycles. But economic involvement can slowly transform these norms. Women involved in income-generating groups or cooperatives often get exposed to awareness campaigns by NGOs, government health teams, or corporate social responsibility projects. A case study in Jharkhand found that women participating in weaving cooperatives were more likely to go for MHM awareness sessions and practice hygienic habits (WaterAid India 2019).

The function of infrastructure, particularly access to WASH facilities, is key to menstrual hygiene. The Swachh Bharat Mission (SBM), which was initiated in 2014, was a significant national movement towards better sanitation in India. Yet, the quality, privacy, and gender sensitivity of sanitation infrastructure are patchy, particularly in rural and peri-urban settings (Caruso et al. 2017). Poor toilet facilities, water unavailability, and non-existence of waste disposal channels render menstruation management problematic, especially for those using reusable cloths. Climatic factors generally add to the problems of infrastructure. During floods, drying cloths is a health risk, whereas during droughts, unavailability of water restricts body cleansing (Sahoo et al. 2015). Such environmental pressures disproportionately fall on economically marginalized women, who cannot purchase disposable sanitary items or substitute coping strategies. The integration of geospatial technologies such as GIS (Geographic Information Systems) into MHM research is a fairly new but emerging trend. GIS helps show and assess spatial variations in WASH services and interventions to Impacts of MHM. Pradhan and Subramaniam (2021) used GIS mapping to show the areas vulnerable to menstrual health in Odisha based on vulnerabilities of access to water and sanitation, as well as population density. For the purpose of this research QGIS 3.26 will allow me to access the spatial patterns in Mathabhanga and Panchla in order to identify trends indicating patterning of infrastructure disadvantage and how they are related to economic and environmental influences on menstrual health. Many researchers have highlighted the need for focused policy interventions to deal with the complex barriers to menstrual hygiene. UNICEF (2019) suggests a life-cycle approach to MHM involving policy support, education, infrastructure, and access to products. Even though, the Menstrual Hygiene Scheme (MHS) and Rashtriya Kishor Swasthya Karyakram (RKSK) are attempting to provide access to menstrual products and menstrual education, the barriers for implementation stem from geographic inequities and lack of coordination across departments (UNFPA 2021). In addition, the schemes tend to have substantial neglect towards adult women and are more focused on adolescent girls, so they do not adequately reach the needs of economically active women (of reproductive age).

Gender-sensitive development frameworks promote the mainstreaming of MHM as a part of overall strategies for women's empowerment. Scholarly research has shown that community-led initiatives, including SHG production of inexpensive sanitary pads, can achieve improvements in menstrual hygiene, create livelihoods, and destroy cultural silences on menstruation (Balamurugan and Bendigeri 2012). Such a comprehensive approach is entirely relevant to communities such as Mathabhanga which already have SHGs formed and can be theoretically applied to the regions of Panchla, given appropriate policies and training. The literature consistently identifies menstrual hygiene as not merely a health or individual issue, but also a very structural and systemic issue. It is linked to economic status, livelihoods, climate, infrastructure, and social values and norms. Because of their ability to earn, their increased visibility, and their ability to participate in social networks, working women are better positioned to manage menstruation in a hygienic manner. However, systemic barriers—affordability, access to water and sanitation, and ongoing taboos—still impair progress. The literature calls for consolidated, multi-sector programming that situates MHM in the context of women's dignity, health, and empowerment. This research contributes to those understandings by contributing a comparative study of two contexts—Panchla and Mathabhanga-and addressing the evidence gap in localized studies, and providing policy alternatives.

3 Methodology

This study utilizes a cross-sectional study design to examine the relationship between women's economic empowerment and menstrual hygiene management (MHM) in two geographical settings with different socio-economic contexts, as separate studies, Panchla, Howrah district and Mathabhanga, Cooch Behar district in West Bengal. A cross-sectional design is appropriate for this study because it allows for data collection at a single point in time reflecting current behaviours and practices and factors impacting menstrual hygiene among working and non-working women at the research sites.

The target sample population for this study consisted of 95 participants: 33 females in Panchla and 62 in Mathabhanga. A purposive sampling strategy was utilized in order to incorporate both working and non-working women from each area with different income levels, ages and educations. The larger sample size in Mathabhanga reflects the larger pool of the SHG participants as well as practical considerations of easy access in the rural environment. The questionnaires were developed in the local dialect, and were administered using face-to-face interviews so that quantitative and qualitative information could be obtained. All surveys were patterned on demographic questions, several questions on menstrual hygiene practices, product type and frequency of use, sanitation availability, and knowledge and sources of knowledge.

The data analysis was performed using SPSS version 20.00. Descriptive statistics were calculated to illustrate the basic characteristics of the respondents. The descriptive statistics provided were frequencies, means, and percentages. The inferential tests that were run were chi square tests and t-tests for independent samples to see if there was significance related to the relationships between variables, such as, employment status and menstrual hygiene product used, or difference between geographical regions. These analyses provided a strong indication of whether or not statistically significant differences exist between work and non-work women, with regard to their menstrual hygiene practice, in both Panchla and Mathabhanga. The research also had a spatial component to it, in addition to statistical analysis. The research used QGIS version 3.26 (https://www.qgis.org/en/site/) to geometrically map the study locations, assess spatial variation in terms of water availability, sanitation access, and locations experiencing flood or drought spatially. While conducting a mapping exercise via GIS, it allows researchers to visualize how the climatic and developmental scenarios intersect with MHM, giving an ancillary layer of meaning to the reported analysis. Maps were made to illustrate distribution patterns in access to WASH infrastructure and environmental challenges in terms of the seasonal shortage of water and flooding during monsoons. Being able to complete the geospatial analysis allows for locating the quantitative findings within the physical space, and highlights the spatial disparities that shape menstrual hygiene outcomes. By effectively using quantitative, inferential, and geospatial methods, this approach provides a comprehensive exploration of the complex and dynamic interrelationship between empowerment, environmental context, and menstrual hygiene practice. Using SPSS-based analysis of quantitative data, as well as QGIS-based spatial analysis, allows this research to go beyond anecdote to provide evidence that has implemented meaning to practice, policy and future research.

4 Analysis and Results

4.1 Study Area-1

This map (Fig. 1) presents Panchla the study site's geographical setting as it exists inside Howrah district West Bengal India. The top section of the map displays Howrah's districts through different Community Development (CD) blocks which feature Panchla block as pink colored against adjacent blocks including Uluberia, Jagatballavpur, Domjur and Sankrail and more. An inset below demonstrates detailed administration boundaries showing the settlements of the Panchla block. This detailed map reveals how densely settled Panchla functions as a peri-urban space near Kolkata as it develops rapidly in the urbanizing fringe of India. Your study on economic empowerment and menstrual hygiene benefits from the vicinity of urban facilities which provide better employment opportunities and health and educational services. The inset shows the location map of Satgharia which researchers believe exists either in or close to Panchla block. Satgharia's distinctive spatial formation deFigted in cyan color suggests that it follows small-scale community configuration. The area functions both as a specific research location for on-site observations and as an exemplary demographic section in Panchla's regional landscape. These present maps demonstrate the clear distinction between Panchla's urbanized character and the distant Mathabhanga territory which supports your study's main finding on geographic patterns that impact women's health and economic conditions.

A chi-square analysis (Table 2) determined how educational attainment affects the selection of materials for pad or cloth disposal (Table 1). The statistical data analysis showed an important match between education level and disposal practices ($\chi^2 = 20.512$, df = 8, p = 0.009). A significant Linear-by-Linear Association (p = 0.001) shows that more educated individuals tend to use better disposal methods for sanitary waste. The chi-square test's assumptions are violated by the fact that 93.3% of cells displayed expected counts lower than 5. Although the results have merit they need further verification because they stem from studies with meager sample numbers that require validation through larger research projects.

The research failed to establish a meaningful statistical connection between menstrual waste disposal methods at home or work sites and environmental concerns (Table 3) but the negative outcome (r = – 0.288, p = 0.104) indicates an area worth extra investigation. The understanding of appropriate sanitary waste disposal techniques seems to influence individuals' likelihood to perform actions which pollute the environment. These results suggest that environmental awareness fails to create a negative relationship with behavior even though statistical evidence shows limited

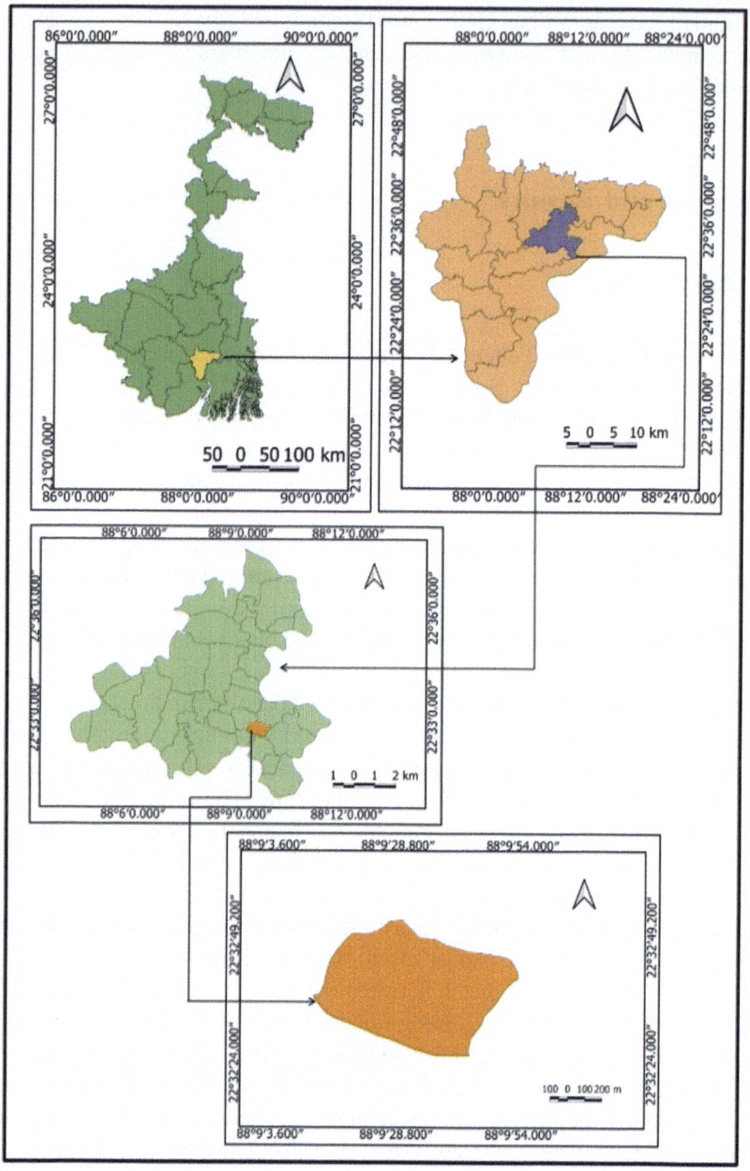

Fig. 1 Study area map of Panchla, Howrah. *Source* Author's work

Table 1 Descriptive statistics of the selected variables

			Wrapping material used during dispose of sanitary pad/cloths					Total
			WITh a cloth	Plastic bag	NO WRAP	No wrap in school (at home plastic bag)	At home plastic bag	
Education status	LITERATE	Count	7	3	2	0	0	12
		Expected count	4.7	4.0	2.2	0.7	0.4	12.0
	< CLASS 8	Count	6	7	1	0	0	14
		Expected count	5.5	4.7	2.5	0.8	0.4	14.0
	> CLASS 8	Count	0	1	3	2	1	7
		Expected count	2.8	2.3	1.3	0.4	0.2	7.0
Total		Count	13	11	6	2	1	33
		Expected count	13.0	11.0	6.0	2.0	1.0	33.0

Source Author's work

Table 2 Chi-square tests between education status and wrapping material used to dispose sanitary pad/cloths

	Value	df	Asymp. Sig. (2-sided)
Pearson chi-square	20.512[a]	8	0.009
Likelihood ratio	20.994	8	0.007
Linear-by-linear association	10.963	1	0.001
N of valid cases	33		

[a] 14 cells (93.3%) have expected count less than 5. The minimum expected count is 0.21
Source Author's work

effects. The findings reveal an insufficient connection due to three potential reasons: shortfalls in menstrual education, limited access to disposal guidance, and cultural customs that restrict open discussions about waste management. The research findings urge awareness initiatives and menstrual hygiene education despite observed weak statistical data. Focused educational efforts and public awareness programs must reach schools and workplaces and rural residents to reduce environmental damage stemming from improper disposal of menstrual waste.

Table 3 Pearson correlation between knowledge of disposing sanitary wastes and environment pollution caused due to it

		Disposal of menstrual waste in working place/at home	Do you think sanitary waste disposal create environment problem
Disposal of menstrual waste in working place/at home	Pearson correlation	1	− 0.288
	Sig. (2-tailed)		0.104
	N	33	33
Do you think sanitary waste disposal create environment problem	Pearson correlation	− 0.288	1
	Sig. (2-tailed)	0.104	
	N	33	33

Source Author's work

A one-way Analysis of Variance (ANOVA) revealed statistical differences in menstrual material choice for various participant groups which matched education level, employment and geographic place of residence (Table 4). Results indicated the groups showed a significant change in menstrual material preferences $F(2, 30) = 6.945$ at $p = 0.003$. The between-groups sum of squares was 2.839 and the within-groups sum of squares was 6.131. According to statistical significance with $p = 0.003$ this confirms that group mean differences cannot be explained through random chance. The results show that the specific classification variable (education or employment) significantly affects which menstrual materials people select. Education level and employment status determine a woman's ability to access sanitary products as women who have both tend to use commercial products whereas women with fewer resources often rely on cloth pads or traditional methods.

The ANOVA reveals there are important differences between groups yet fails to show how the individual groups relate to each other. Moving forward to identify which groups differ, researchers should conduct post-hoc analysis using Tukey's HSD test. Further analysis will reveal the exact group relationships which produce results such as educated versus uneducated women or Panchla versus Mathabhanga differences in menstrual products and working or non-working women distinctions.

The chart (Fig. 2) shows the mean of menstrual material use during menstruation by education status. It shows educational level has a positive significant relationship with the use of menstrual hygiene materials. Among the four categories of women, the illiterate women show the lowest mean usage, suggesting that they are not aware of the right products to use during their menstrual periods. The overall respondents whose formal education level is below class 8 have relatively improved usage level and this implies that even the basic level of education enhances the usage of improved MHCs. However, the largest increase is noted among women with education level beyond class 8, with the mean value of usage rising to 1. This implies that all the women in this category of users utilize the menstrual materials during the menstrual period. Aging reveals the importance of education in informing and promoting safer practices on menstruation, this form of information goes a long way in supporting

Table 4 One way ANOVA between education level and use of material during menstruation

Descriptives

Use of material during menstruation

	N	Mean	Std. deviation	Std. error	95% confidence interval for mean		Minimum	Maximum
					Lower bound	Upper bound		
LITERATE	12	0.0833	0.28868	0.08333	− 0.1001	0.2667	0.00	1.00
< CLASS 8	14	0.2143	0.42582	0.11380	− 0.0316	0.4601	0.00	1.00
> CLASS 8	7	0.8571	0.69007	0.26082	0.2189	1.4953	0.00	2.00
Total	33	0.3030	0.52944	0.09216	0.1153	0.4908	0.00	2.00

Test of homogeneity of variances

Use of material during menstruation

Levene statistic	df1	df2	Sig.
3.082	2	30	0.061

ANOVA

Use of material during menstruation

	Sum of squares	df	Mean square	F	Sig.
Between groups	2.839	2	1.419	6.945	0.003
Within groups	6.131	30	0.204		
Total	8.970	32			

Source Author's work

the education of girls since that is the only way they will be knowledgeable on how to maintain their hygiene during those times of the month.

Significance analysis (Table 5) using a Tukey's Honest Significant Difference (HSD) test after the one way ANOVA test will show that there is a significant difference in education level based on the menstrual hygiene practices among the respondents. When making the comparison based on the level of education between the literate women and those with less than class 8 education their mean difference is − 0.131 with $p = 0.744$ this implies that there is no significant difference between the two groups. In more detail, literate women used sanitary pads more than does (Mean Diff = −0.774, $p = 0.003$) women with education below class 8 whom they also used more cloth as alternatives to modern MHHs.

Thus, the research underlines the education's importance in promoting the proper behavior during menstruation. First, higher figures of modern menstrual products' use can be a sign that more educated women are embracing such products. Better MCH not only leads to the better health of women but also to low truancy in school and work, as well as increase in productivity. Also, the so needed sanitary products enhance the markets in case of demand and employ people on production and distribution of the products. Consequently, education and literacy especially beyond the primary level is empowering women with better option to make healthier choices for themselves and at the same time the society and market gains a better and healthier environment to expand.

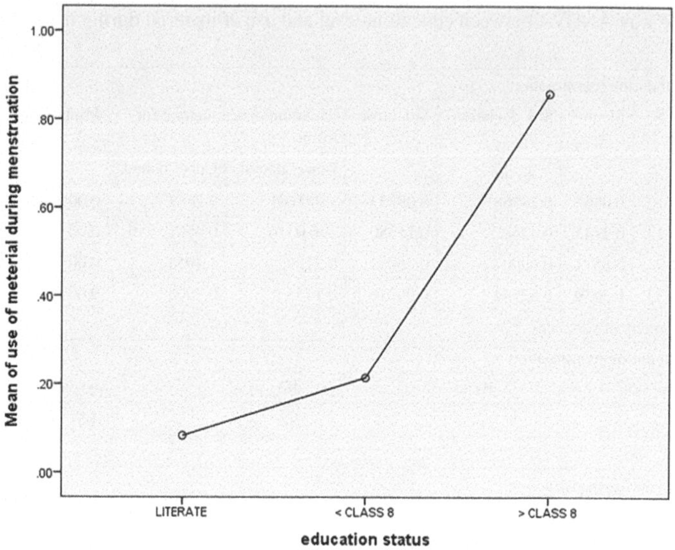

Fig. 2 Relationship between education status and use of menstrual materials. *Source* Author's work

Table 5 Post hoc tests (Tukey's HSD test) of selected variables

Dependent Variable: use of material during menstruation Tukey HSD

(I) Education status	(J) Education status	Mean difference (I–J)	Std. error	Sig.	95% confidence interval	
					Lower bound	Upper bound
LITERATE	< CLASS 8	– 0.13095	0.17784	0.744	– 0.5694	0.3075
	> CLASS 8	– 0.77381[a]	0.21500	0.003	– 1.3038	– 0.2438
< CLASS 8	LITERATE	0.13095	0.17784	0.744	– 0.3075	0.5694
	> CLASS 8	– 0.64286[a]	0.20927	0.012	– 1.1588	– 0.1270
> CLASS 8	LITERATE	0.77381[a]	0.21500	0.003	0.2438	1.3038
	< CLASS 8	0.64286[a]	0.20927	0.012	0.1270	1.1588

[a]The mean difference is significant at the 0.05 level
Source Author's work

4.2 Study Area-2

The map shown above (Fig. 3) portrays the geographical location of the study area, which is the Cooch Behar district of West Bengal, India. The above part of the map delineates the district administration, where different blocks like Dinhata-I, Dinhata-II, Tufanganj-I, Tufanganj-II and Mathabhanga-I, etc. The letter 'M' uniquely marked in Mathabhanga-I signifies it as the directly affected block which is going to be the area of the study. The middle pan of the map zooms into the Mathabhanga I block; it shows administrative divisions and inhabited localities within it. This quick reference map is useful in as far as giving details on the spatial distribution and access to the block. The third and the rightmost bar of the map further zooms into two particular gram panchayats which are shown green in colour namely Golokganj and Suterkuti. These are the designed data collection area or survey site that may have been identified depending on their demographic, socio-economic or health importance. It has also graphic features such as coordinate axes, scales, and north arrow for the enhancement of geographical referencing. In totality, it makes this visual layout really helpful in situating the research location from a broad to the specific level in order to provide clarity and full appreciation in field-based research and analysis.

In order to establish whether there is a significant relationship between the education status and the method of wrapping that was used during the disposal of sanitary pads (Table 6) or cloths a chi-square test of independence was used (Table 7). The angiographic data analysis showed the existence of a statistically significant relationship between the two studied variables, $\chi^2(25, N = 62) = 45.305, p = 0.000008$. This means that the way the respondents dispose of their articles is strongly linked to their level of education. Cross tabulation finding out that whereas there were women respondents who disposed of their articles by using hygiene such as, the usage of the paper or the plastic bags, more women with master's degree were likely to dispose of their articles by the said method. While respondents with less education (up to completion of secondary or graduation level) in contrast to those responding to the survey on use of improved toilets only to dispose human waste, used diverse and sometimes less hygienically acceptable methods such as no wrapping at all or use of cloths.

However, looking at the summary below it should be noted that all the cells of the two tables had expected frequencies with values less than 5 in 91.7% of the cases, which goes contrary to one of the assumptions of the chi-square test. This limitation points to the fact that although the result is statistically significant it should not be conceived as absolute. It research is a small sample size of a short time, and some categories have little data, so the test may not be very good. Further studies using a larger sample of population with uniform distribution are recommended to affirm these results. The study underlines importance of education regarding hygienical utilization of the menstrual waste and the need for disseminating awareness programmes for the women with low levels of education.

The Pearson correlation coefficient (Table 8) showing any relationship between the place of menstrual waste disposal (either in the household or at the workplace) and the view on problems created through improper disposal of sanitary waste was found to be very low and insignificant ($r = 0.141, p = 0.273$, N 62). This virtually negates that only the place where these wastes are disposed influences the perception of whether such wastes cause environmental issues. Consequently, if people are disposing of menstrual waste in suboptimal environments, they may not be aware of the negative influences of such a decision. On this note, there is an urgent need for more education and sensitization regarding better and proper ways of removing the wastes. This means that increased knowledge about the consequences of improper disposal might lead to less littering, even if the disposal is happening elsewhere. Such awareness helps in addressing issues of pollution and proper protection of the environment.

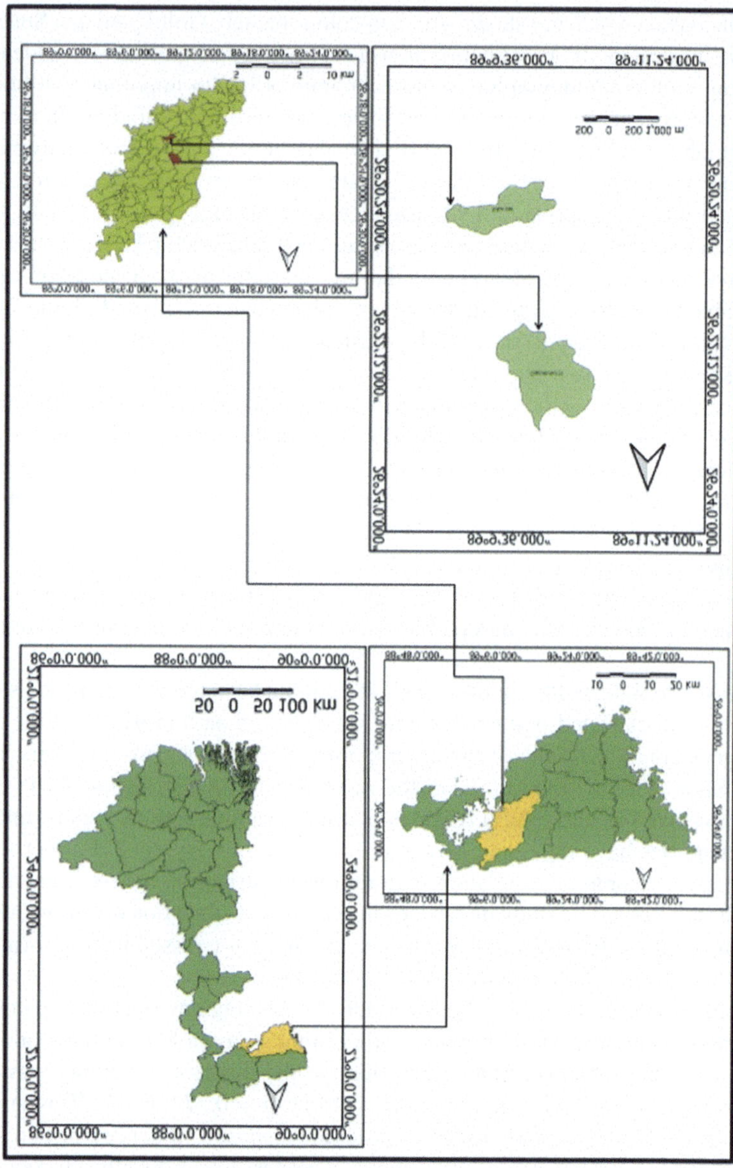

Fig. 3 Study area map of Mathabhanga, Coochbihar. *Source* Author's work

Table 6 Descriptive statistics of the selected variables

			Wrapping material during dispose of sanitary pad/cloths						Total
			Paper bag	N.A	Plastic WRAP OR bag	Without wrap	With a cloth	No wrap in school	
Education status	Upto secondary	Count	1	5	7	3	1	0	17
		Expected count	1.1	3.6	7.7	1.9	0.8	1.9	17.0
	Upto graduation	Count	0	2	9	3	2	0	16
		Expected count	1.0	3.4	7.2	1.8	0.8	1.8	16.0
	Upto higher secondary	Count	0	2	4	1	0	0	7
		Expected count	0.5	1.5	3.2	0.8	0.3	0.8	7.0
	Masters	Count	2	3	3	0	0	0	8
		Expected count	0.5	1.7	3.6	0.9	0.4	0.9	8.0
	4.00	Count	0	0	1	0	0	0	1
		Expected count	0.1	0.2	0.5	0.1	0.0	0.1	1.0
	5.00	Count	1	1	4	0	0	7	13
		Expected count	0.8	2.7	5.9	1.5	0.6	1.5	13.0
Total		Count	4	13	28	7	3	7	62
		Expected count	4.0	13.0	28.0	7.0	3.0	7.0	62.0

Source Author's work

Table 7 Chi-square tests between education status and wrapping material used to dispose sanitary pad/cloths

	Value	df	Asymp. Sig. (2-sided)
Pearson chi-square	45.305[a]	25	0.008
Likelihood ratio	43.818	25	0.011
Linear-by-linear association	5.648	1	0.017
N of valid cases	62		

[a]33 cells (91.7%) have expected count less than 5. The minimum expected count is 0.05
Source Author's work

Table 8 Pearson correlation between knowledge of disposing sanitary wastes and environment pollution caused due to it

		Disposal of menstrual waste in working place/at home	Do you think sanitary waste disposal create environment problem
Disposal of menstrual waste in working place/at home	Pearson correlation	1	0.141
	Sig. (2-tailed)		0.273
	N	62	62
Do you think sanitary waste disposal create environment problem	Pearson correlation	0.141	1
	Sig. (2-tailed)	0.273	
	N	62	62

Source Author's work

Exploratory research (Table 9) was undertaken to find out if there is any correlation between educational status and the type of material used at the time of menstruating. Thus, analysis of the goals by descriptive statistics showed a clear and apparent trend. When education level raised, the mean score related to material usage goes down which is indicative of the improved level of sanitation. For instance, respondents with a master's degree had the lowest numerical mean score of 0.125 (SD = 0.35), which indicates that most of them use sanitary napkins only; the scores were higher for respondents with lower education level, meaning that they frequently use other or less hygienic products.

To test the homogeneity of variance the Levene's test was carried out and the results as follows: F = 4.171 while the p value was 0.005, this means this there was a significant difference in the variance between the educational groups. However, in the one-way ANOVA test, the result did not show any statistical significance for the group F = 1.154 and p = 0.343). This has an implication that differences between material usage across the different education levels are not very much different, even when results were analyzed using the chi-square test, the significance level used was 5% but the difference in the usage of material in the different education levels did not quite reach this level. However, there is evidence in descriptive results showing the relationship between increased education level and improved menstrual hygiene practices. This agrees with the hypothesis that education has a significant influence on the utilization of safer and hygienic products during the period, although the results in this study did not provide a significant chi-square value to support the model significantly.

The line graph (Fig. 4) depicts mean usage of the menstrual materials based on the level of education. The usage in this case remains relatively constant and below a mean of 1 for those educated up to the secondary, higher secondary and graduation level. However, there is a decline to the means for the individuals who hold a master's degree, where the means have the lowest values. Unexpectedly after that the highest mean of 2.00 is noticed in the category labeled as "4.00". It then decreases once

Table 9 One way ANOVA between education level and use of material during menstruation

Descriptives

Use of material during menstruation

	N	Mean	Std. deviation	Std. error	95% confidence interval for mean		Minimum	Maximum
					Lower bound	Upper bound		
Upto secondary	17	0.7647	0.83137	0.20164	0.3373	1.1922	0.00	2.00
Upto graduation	16	0.6875	1.01448	0.25362	0.1469	1.2281	0.00	3.00
Upto higher secondary	7	0.7143	0.95119	0.35952	− 0.1654	1.5940	0.00	2.00
Masters	8	0.1250	0.35355	0.12500	− 0.1706	0.4206	0.00	1.00
4.00	1	2.0000	.	.	.	.	2.00	2.00
5.00	13	0.7692	0.92681	0.25705	0.2092	1.3293	0.00	2.00
Total	62	0.6774	0.88288	0.11213	0.4532	0.9016	0.00	3.00

Test of homogeneity of variances

Use of material during menstruation

Levene statistic	df1	df2	Sig.
4.171[a]	4	56	.005

ANOVA

Use of material during menstruation

	Sum of squares	df	Mean square	F	Sig.
Between groups	4.441	5	0.888	1.154	0.343
Within groups	43.108	56	0.770		
Total	47.548	61			

Source Author's work

[a]Groups with only one case are ignored in computing the test of homogeneity of variance for use of material during menstruation

more at "5.00". This irregular trend tends to indicate the presence of potential errors in the data or inaccuracies of education levels above the master's level.

To determine this, Tukey's HSD test's (Table 10) were used in order to get a feel of the pattern of material usage during menstruation based on the education levels of the females. The findings also revealed a tendency that is quite evident showing that women with higher education use better hygiene during menstruation Respondent-masters had the least mean score of 0.125 (SD = 0.35) signifying the use of hygienic products like the sanitary napkins. On the other hand, the literacy group categorized as "upto secondary level education" (n = 115, M = 0.76, SD = 0.83).) toward "upto graduation," M = 0.69, SD = 1.01) to indicate more frequent use of other or less clean absorbents, such as cloth or mixed method.

The results show the "upto graduation" group scored the highest possible value of 3, inform of wider spreadsheet software using. The corresponding result for the respondent in category 4.00 (only one individual) was a mean of 2.00, which also may be considered as a less hygienic practice, but it shouldn't be generalized because of the one-case sample. All in all, the overall mean score for all levels of education

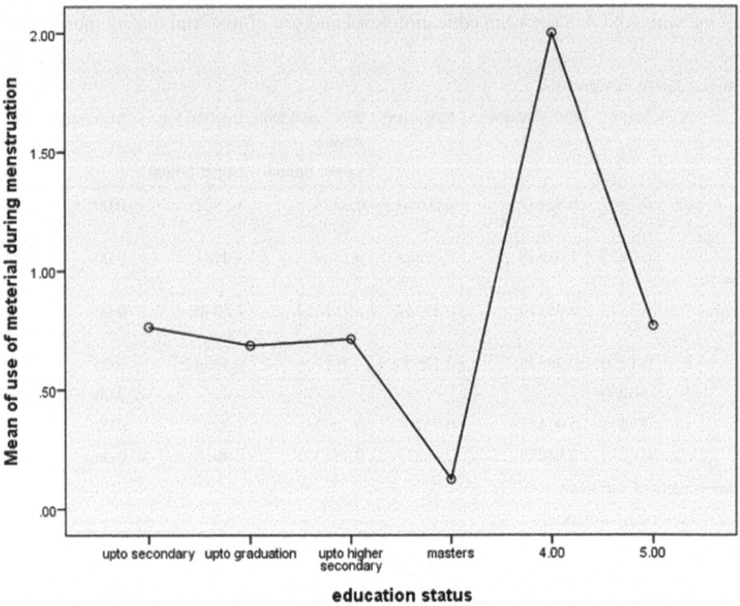

Fig. 4 Relationship between education status and use of menstrual materials. *Source* Author's work

Table 10 Post hoc tests (Tukey's HSD test) of selected variables

Use of material during menstruation

	N	Mean	Std. deviation	Std. error	95% confidence interval for mean		Minimum	Maximum
					Lower bound	Upper bound		
Upto secondary	17	0.7647	0.83137	0.20164	0.3373	1.1922	0.00	2.00
Upto graduation	16	0.6875	1.01448	0.25362	0.1469	1.2281	0.00	3.00
Upto higher secondary	7	0.7143	0.95119	0.35952	− 0.1654	1.5940	0.00	2.00
Masters	8	0.1250	0.35355	0.12500	− 0.1706	0.4206	0.00	1.00
4.00	1	2.0000	.	.	.	.	2.00	2.00
5.00	13	0.7692	0.92681	0.25705	0.2092	1.3293	0.00	2.00
Total	62	0.6774	0.88288	0.11213	0.4532	0.9016	0.00	3.00

Source Author's work

encompassed 0.68 (SD = 0.88), thus it was identified that, although most of the respondents followed rather a hygienic approach, some variations are still remained especially in cases of the respondents with lower levels of education. These findings confirm the education factor in the promotion of the use of safer MHM practices.

5 Comparative Discussion

The current study makes a comparative examination of two geographically isolated regions—Panchla, a peri-urban area of Howrah district, and Mathabhanga, a far-off rural subdivision in Cooch Behar district. The contrasts in geographical landscape have a direct bearing on the rates of economic empowerment and menstrual hygiene management (MHM) among working and non-working women in these locations.

1. Geographical Setting and Economic Opportunities:

   Panchla, has a better position in proximity to Kolkata and industrial centers and is better connected comparatively to employment, educational, medical and financial services. This urban fringe has created increased economic engagement for women, particularly in small industry and service units. In contrast, Mathabhanga has rural disbursements, it has less infrastructure, and there are fewer opportunities to earn; women's economic participation in Mathabhanga—especially by any women who don't work—refers exclusively to the household division of labor, or to very low wages in the agriculture-sector. It appears that proximity to an urban area (like Panchla) is significant and contributes to women's economic agency and ultimately women's economic empowerment more so than the remoteness of Mathabhanga or its rural context.
2. Menstrual Hygiene Management (MHM) Practices:

   Menstrual hygiene practices also exhibit geographical differences. In Panchla, there are medical stores, NGO interventions, and improvements in sanitation infrastructure, which facilitate awareness and use of sanitary products, particularly for working women. Mathabhanga's dispersed habituation and lack of a proper health care infrastructure continues to suffer from the absence of unsanitary items, inadequate water supply, and inadequate toilets. While income improves accessibility in both locations, geographic isolation in Mathabhanga restricts availability and knowledge flow, particularly for women who are not working.
3. Education and Health Awareness:

   What is true for Panchla also applies to Mathabhanga. In Panchla the accessibility to education is better because of its proximity to schools, colleges, and urban outreach programs. This results in a noteworthily better awareness of women's concerns over menstrual hygiene and reproductive health issues. However, the rural nature of Mathabhanga makes access more limited, meaning that Mathabhanga women have lesser levels of health awareness. Non-working women, especially, have it worse owing to lower literacy levels and corresponding cultural taboos.
4. Social Norms and Family Support:

   Urban influences in Panchla are responsible for a slow transition of gender roles so that working women can access support from family members for handling work-life and menstrual management. Traditional norms in rural Mathabhanga are more strict, and working as well as non-working women tend to be left with no support for household chores. Social and cultural environment influenced by the geographical setting strongly impacts women's autonomy and well-being.

5. Sanitation Infrastructure:

   Geographical inequality is glaring as far as sanitation is concerned. Panchla enjoys improved road connectivity, electricity, and sanitation facilities—while not ideal, they are relatively better than those available in Mathabhanga. Mathabhanga's rural landscape has no basic hygiene infrastructure, and women find it challenging to deal with menstruation hygienically and with dignity, whether at work or not.

6 Conclusion

The comparison of Panchla and Mathabhanga will highlight how the geographical location factor determines the status of women's economic development, as well as their menstrual hygiene status. Compared to other villages and due to its Urban-fringe characteristic and better connectivity to infrastructure; women of Panchla have comparatively better economic independencies, awareness, and facilities as constraints to access menstrual products those who are working employees. On the other hand, the locational constraint of Mathabhanga adds multiple barriers due to weak infrastructural development, scarce employment facilities, and continuation of disparaging cultural perceptions. While, employment brings increase in purchasing and decision making power in two respects for women, the extent of these effects is larger in Panchla because of availability of such factors as information and health facilities along with family support. The non-working women and especially the women form Mathabhanga, the most vulnerable group of women are dependent on men and they do not have the opportunity to properly manage their menstruation. It reveals the significance of anchored programs and strategies to cater to the regional differences like infrastructure development in the developing areas, sexuality and knowledge enhancement campaigns in the concerned geographic areas, and women no matter their employment status to provide them dignity and overall development.

7 Implications

The impacts of this research are widespread and complex, touching on the domains of public health, gender equity, economic development, and climate adaptation. Importantly, this study provides evidence to reinforce the position that menstrual hygiene management (MHM) cannot be understood in a vacuum or solely as a health- or behavior-related issue. MHM must be conceptualized as a systemic development challenge, meaning women's experiences with MHM are shaped by women's economic situations, infrastructure access, education, and environmental vulnerabilities. For example, the inequities between working and non-working women, particularly the geographical differences of locations in Panchla and Mathabhanga, exemplify how women given economic power are able to make better choices about their menstrual health in combination with other educational and health factors; however, their choices may not be realized due to infrastructure deficiencies or climate challenges like tap water access, drought, and flooding. The point here is that MHM must be interwoven with broader agenda items for policymakers related to lives, economic development, climate-resilient WASH programming, and public/community health. Targeted investment in sanitation infrastructure, which is particularly lagging in some of the most climate vulnerable and economically marginalized areas, is required to place women in a position to manage menstruation openly and with dignity. The results also indicate that community-based platforms similar to Self Help Groups/ SHGs could be valuable for menstruation product distribution and peer-led health education (especially in a rural context). The scaling of such models could provide a socially embedded and economically viable menstrual health ecosystem and simultaneously generate local jobs. For businesses and employers and particularly for the informal sector the study highlights productivity and retention losses linked to inadequate MHM practices; for businesses, there is an economic case for rationalizing menstrual hygiene facilities at workplaces. At the national and global policy levels, the research is directly connected to Sustainable Development Goals (SDGs) 3 (Good Health and Well-being), 5 (Gender Equality) and 6 (Clean Water and Sanitation) through ground [ed] and evidence-based extrapolations of how menstrual hygiene is relevant to women's empowerment, environmental justice and access to basic services. Thus, the work calls for transformational cross-sectoral approaches to menstrual health, which can address short-term hygiene needs and challenge structural inequities that create gendered vulnerabilities in low-resource settings.

8 Scope for Future Work

Although the study has examined the relationship between economic empowerment and menstrual hygiene across two socio-climatic contexts in detail, there is still plenty of room for the discussion to become deeper and wider. Future research should use large-scale, longitudinal studies, across districts with local variation in socio-economic and ecological contexts in order to create generalizable evidence, and to unpack the regional specificities. While we endeavored to analyze patterns of MHM across adult women, it is apparent more research is needed to understand the lived experience, the views, and the practices of menstruation, when it comes to other marginalized groups, especially adolescents, elderly women, and persons with disabilities, whose menstrual needs tend to be overlooked in the health discussion. In a world that is ever more digitized, and where there is widespread engagement in digital platforms, the use of mHealth tools and menstrual tracking apps to promote awareness-raising and behaviour change, challenge social taboos and stigma associated with MHM, would be useful, especially in rural and semi-literate contexts etc. Future research can also investigate the economics of menstrual product accessibility by evaluating the cost, availability, and supply chain logistics of SHGs-led distribution channels. In a similar realm, environmental assessments should be conducted on the increasing use and disposal of non-biodegradable menstrual products as part of overall approaches to include MHM into climate-resilient WASH policies and waste management approaches. Finally, the institutional aspects of playing a role in menstrual dignity schools, workplaces, or healthcare providers would all serve to develop collaborations across institutions and inform approaches for policy development for cross-institutional policies. A multidisciplinary, systems-based approach would have much to offer the advancement of inclusive, sustainable MHM solutions.

References

Balamurugan SS, Bendigeri ND (2012) Community-based study on menstrual hygiene among adolescent girls in a rural area of Tamil Nadu, India. J Indian Med Assoc 110(6):314–317

Caruso BA, Clasen TF, Yount KM, Cooper HL, Hadley C, Haardörfer R (2017) Assessing women's negative sanitation experiences and concerns: the development of a novel sanitation insecurity measure. Int J Environ Res Public Health 14(7):755. https://doi.org/10.3390/ijerph14070755

Chandra-Mouli V, Patel SV (2017) Mapping the knowledge and understanding of menarche, menstrual hygiene and menstrual health among adolescent girls in low- and middle-income countries. Reprod Health 14(1):30. https://doi.org/10.1186/s12978-017-0293-6

Dasgupta A, Sarkar M (2008) Menstrual hygiene: how hygienic is the adolescent girl? Indian J Community Med 33(2):77–80. https://doi.org/10.4103/0970-0218.40872

Gopalakrishnan S, Dhanalakshmi D (2014) A study on menstrual hygiene and related personal hygiene practices among adolescent girls in rural Puducherry. Int J Commun Med Public Health 1(2):145–150

Kabeer N (1999) Resources, agency, achievements: reflections on the measurement of women's empowerment. Dev Chang 30(3):435–464. https://doi.org/10.1111/1467-7660.00125

Pradhan S, Subramaniam V (2021) Mapping menstrual health vulnerabilities using GIS: a study in rural Odisha. J Public Health Res 10(2):211–220

Rajagopal S, Mathur K (2017) Breaking the silence around menstruation: experiences of adolescent girls in an urban setting in India. Gend Dev 25(3):393–407. https://doi.org/10.1080/13552074.2017.1379783

Sahoo KC, Hulland KR, Caruso BA, Swain R, Freeman MC, Panigrahi P, Dreibelbis R (2015) Sanitation-related psychosocial stress: a grounded theory study of women across the life-course in Odisha, India. Soc Sci Med 139:80–89. https://doi.org/10.1016/j.socscimed.2015.06.031

Sommer M, Hirsch JS, Nathanson C, Parker RG (2016) Comfortably, safely, and without shame: defining menstrual hygiene management as a public health issue. Am J Public Health 105(7):1302–1311. https://doi.org/10.2105/AJPH.2014.302525

UNFPA (2021) Menstrual health is a public health and human rights issue. United Nations Population Fund. https://www.unfpa.org/menstrual-health

UNICEF (2019) Guidance on menstrual health and hygiene. United Nations Children's Fund. https://www.unicef.org/documents/guidance-menstrual-health-and-hygiene

WaterAid India (2019) Menstrual hygiene in India: status and way forward. https://www.wateraid.org/in/publications

WHO, & UNICEF (2013) Progress on sanitation and drinking-water: 2013 update. World Health Organization. https://www.who.int/publications/i/item/9789241563956

Gender Equality, Inclusion and Empowerment in a Digital World

Bridging Gender Disparities Through Digital Education: A Systematic Literature Review

Geetanjali C. Achrekar, Chirra Baburao, D. Rajyalakshmi, and Priya V. Nagvekar

Abstract The digitalization of education has expanded global learning opportunities, but has not effectively addressed entrenched gender disparities, particularly in developing countries like India. The 78th round of the NSSO highlights a widening digital divide, with rural women and older populations facing significant digital inadequacies. These gaps hinder progress toward Sustainable Development Goals (SDG 4—Quality Education, and SDG 5—Gender Equality). This systematic literature review analyzes 31 peer-reviewed publications (articles, book chapters, and conference papers) published between January 2015 and February 2025 and indexed in Scopus to examine the barriers, patterns, and interventions in digital education for women. Thematic analysis reveals critical challenges such as limited access to devices, poor connectivity, insufficient digital skills, and sociocultural barriers. The findings underscore the need for community learning programs, women-focused digital literacy initiatives, and policies that improve rural digital infrastructure and affordability. A suggestive framework connects structural disparities and inclusive educational ecosystems. The research provides an integrated evidence based review of how digital learning can empower women in closing the gender digital divide. It provides practical guidelines to stakeholders, including educators, policymakers and community leaders to facilitate inclusive, equitable, and sustainable education

G. C. Achrekar (✉)
Department of Economics, Goa Vidyaprasarak Mandal's Gopal Govind Poy Raiturcar College of Commerce & Economics (affiliated to Goa University), Ponda, Goa, India
e-mail: geetanjali@gvmcommercecollege.ac.in

C. Baburao
Sir CR Reddy Educational Institutions, Eluru, AP, India
e-mail: bchirra@gnlu.ac.in

D. Rajyalakshmi
Psychology, Department of Social Sciences & Humanities, Vignan's Foundation for Science, Technology & Research (deemed to be university), Guntur, AP, India

P. V. Nagvekar
Department of Computer Science, Goa Vidyaprasarak Mandal's, Gopal Govind Poy Raiturcar College of Commerce and Economics (affiliated to Goa University), Ponda, Goa, India
e-mail: priyanagvekar@gvmcommercecollege.ac.in

A. Singh et al. (eds.), *Human-Centered Sustainability: Bridging Well-Being, Gender Equity, and Technology*, World Sustainability Series,
https://doi.org/10.1007/978-3-032-22020-2_6

systems. This review points out that women education and empowerment in the digital sphere can be viewed as not only an educational convergence but also the route towards creating overall gender equality and social integration. Digital education is a channel to reduce the gender inequity as specific barriers may be reduced by changing the socio-economic environment of women through digital education.

Keywords Digital education · Gender digital divide · Women empowerment · Digital literacy · Community-based learning · Policy interventions

1 Introduction

Across the globe, educational sectors have undergone significant change due to rapidly growing information and communication technologies (ICTs) (UNESCO 2020). The combination of online courses with mobile learning and blended pedagogies through digital education offers unprecedented access to learning materials that go beyond physical classrooms (Johnson et al. 2016). In low- and middle-income countries, women lag behind men in internet usage by 25% (International Telecommunication Union [ITU] 2021). The National Sample Survey Office (78th round) in India reported that 29% of females use digital devices compared to 37% of males, highlighting an increasing gap in digital activity involvement (MoSPI 2021). The combination of societal cultural norms that favor patriarchal beliefs and safety uncertainties and low digital literacy competencies and limited economic resources and poor infrastructure suppresses women from engaging in digital learning, according to Helsper and Van Deursen (2017) and Srivastava and Pathak (2019). The Government of India launched Digital India (2015) together with Skill India (2015) to provide digital infrastructure and vocational training as well as programming for women's digital up-skilling (Ministry of Electronics & Information Technology [MeitY] 2022). Results demonstrate weak adoption behavior among rural women which stems mainly from patriarchal tendencies in household choices (Agarwal and Malathi 2020).

Digital education is slowly proving to be a transformative agent in reducing various socio-economic barriers. Its promise to confront prolonged gender gaps—especially in education—is underserved. The present study supports the narrative of the positive role of digital education in closing the gender based inequalities, a step towards building a more inclusive learning environment. Online education, with its scalability and accessibility, offers a compelling solution to transcend the physical, cultural and economic limitations that have historically impeded women. If we take advantage of digital platforms, we can democratize access to quality education and thus empower people, especially women, creating sustainable development.

What this review adds that is new is a much-needed synthesis of existing research on the intersection of digital education and gender equality, going beyond anecdotes to offer a more comprehensive, evidence-based understanding of what works and what challenges remain. Existing work has addressed elements of digital learning or gender in education, but a unified, comprehensive account of how digital education actively bridges these gaps is missing. Our research seeks to fill this crucial gap, guiding best practices, technological interventions, and policy recommendations that can help hasten the path forward.

Additionally, this issue maps directly to many UN SDGs, most prominently SDG 4 (Quality Education) and SDG 5 (Gender Equality). Digital education, in fostering inclusive and equitable quality education and lifelong learning opportunities for all, makes a direct contribution towards achieving gender equality and empowering all women and girls. This review will emphasise how digital education projects can be a trigger for such global goals, a path to a more just and sustainable future in which everyone, regardless of gender, has the chance to flourish.

A systematic literature review structure has been implemented in this chapter to synthesise empirical research alongside theoretical studies about digital education as a tool for addressing gender inequalities according to Antonio and Tuffley (2014a, b). A review of academic research between 2015 and 2025 outlines obstacles and impactful solutions and proposed programs to create equal digital learning access for women (Qazi 2022; Sánchez-Prieto et al. 2020; Wickens and Miller 2020; Yañez et al. 2023).

Objective of the study

- To examine digital education's role in bridging gender gaps by analysing barriers, evaluating interventions, and proposing inclusive policy reforms

2 Literature Review

2.1 Theoretical Framework

2.1.1 Gender Digital Divide

The gender digital divide produces multiple dimensions that go beyond technology access to show differences in skills development, together with use patterns and digital outcome production between genders (Sánchez-Prieto et al. 2020). The digital divide exists at three levels according to van Dijk (2020): the first level targets hardware devices and connection networks, and the second level touches on digital learning ability as well as skills, while the third level observes distinctive advantages and effects from technology usage.

The first-level divide is an ongoing obstacle, especially in rural areas and developing countries (Habibi et al. 2023). Women encounter substantial obstacles to obtaining digital devices and internet connectivity, primarily due to financial limitations and social traditions and cultural demands that create barriers to their inclusion (Singh 2017). The lack of access maintains established gender inequalities because it restricts women from obtaining education and employment opportunities in addition to social participation across the digital realm (Esteban et al. 2021).

The second-level discrimination exists because people need basic digital literacy skills. Women who obtain access to technology might struggle to leverage this technology effectively since they lack appropriate skills as stated in Mariscal et al. 2019. Women are less likely to obtain essential digital skills mainly because of education-based gender bias and limited early exposure to technology and because societal beliefs discourage women from entering STEM professions. A new UNESCO (2025) report highlights significant gender gaps, with fewer women accessing the internet and only 35% of women doing graduation in STEM. To address this, specific efforts should be implemented in educational programs to teach women digital skills, thereby bridging the developing mismatch (O'Brien et al. 2014).

According to Hilbert (2011a, b) third-level digital divides analyse the different advantages and consequences from technology utilisation between men and women (Hilbert 2023). The analysis of this level explores how people of different genders use digital technology and the effect it creates. Several social and cultural elements prevent women from using technology to expand their economic potential and professional prospects (Sobieraj and Krämer 2019). Women bear a higher risk of harassment and cyberbullying in digital spaces, leading to inhibitions regarding their participation there.

A detailed strategy should be put in place to effectively reduce the digital gender divide. The main approach involves improving infrastructure while lowering costs to help women overcome initial barriers, combining targeted education and training sessions with efforts to increase digital literacy, and creating welcoming digital spaces that encourage female participation. The digital gender gap can be effectively narrowed through policy measures that promote gender-responsive ICT policies, along with initiatives aimed at increasing the number of women in the tech industry (Bertaux et al. 2007; Borgonovi et al. 2018; Hilbert 2011a, b; Leveraging Digital Technologies to Advance Women's Economic Empowerment 2024; World Bank Document 2024; Mariscal et al. 2019; Naga et al. 2020; Trigueros and Aldecoa 2021).

Long-term change requires dealing with the fundamental social beliefs and prejudices against women in technology (Mackey and Petrucka 2021). Educational institutions must challenge gender stereotypes while actively finding technology-oriented female role models to create an environment where digital empowerment of women thrives and have women user-friendly digital technology (Kuteesa et al. 2024; Sánchez Peláez et al. 2023). All levels of the gender digital divide need comprehensive solutions that require joint efforts between governments, private sector organisations and civil society organisations (Mariscal et al. 2019). Digital technology dominance in modern life requires immediate action to address the gender digital divide because achieving gender equality and inclusive development depends on it (Mariscal et al. 2019). Societies can create advanced, innovative digital futures through total gender equality by providing equal opportunities and skills across the digital space (Borgonovi et al. 2018).

2.1.2 Empowerment through Digital Education

The powerful tool of digital education serves as an empowering mechanism that strengthens human capital and social capital as well as economic agency in women (World Bank 2018). The adaptability of digital education enables better inclusion for women to obtain educational opportunities and develop necessary skills in the face of gender disparities. Making the most of digital education potential requires both the solution of multiple obstacles and the establishment of specific approaches to overcome these barriers (Kuteesa et al. 2024).

The gender digital divide continues to exist as a persistent barrier through three distinct levels, including technology access and digital literacy skills, together with different benefits from technology usage (van Dijk 2020). Indian women exhibit significantly less digital participation compared to men by an 8% margin and this disparity worsens when looking at advanced digital activities (Ministry of Statistics & Programme Implementation 2021). Few women participate in digital learning due to sociocultural elements such as patriarchal customs as well as safety issues and inadequate technical literacy and infrastructural barriers (Helsper and Van Deursen 2017; Srivastava and Pathak 2019).

A solution that covers all elements must be developed to solve these problems. The process of resolving first-level access barriers depends on better infrastructure alongside reduced costs. Das and Joshi (2018) establish that the lack of electricity and insufficient broadband networks, and limited personal devices, act as the primary obstacles preventing women from participating in digital educational programs. Rural women typically share smartphones between each other thus their study time suffers from reduced privacy along with maintaining male dominance within the home digital space (Srivastava and Pathak 2019).

The importance of developing digital proficiency along with digital education competencies is identical. The Skill India programs have exhibited moderate achievement through their training programs yet they face limitations from standard educational curricula and predominantly male classroom environments (Agarwal and Malathi 2020). The combination of specialised educational programs and separate training sessions for women prove superior in both completion rate results and participants' progress scores (Rao et al. 2022) and justify women-specific digital skills education models.

The effectiveness of digital education reaches higher levels when educational techniques develop innovative strategies specifically for women. Community learning centres managed by female facilitators provide safe educational environments for students in rural areas (Malhotra and Gupta 2023). The execution of localized mobile-based micro learning applications promotes knowledge retention in addition to better practical usage (Desai 2021). The teaching approaches comply with feminist pedagogies, which establish learner-focused collaborative rooms that dismantle conventional power structures (Lorde 1984; Hooks 1994).

Gender-responsive learning practices implemented in e-learning systems create better student engagement, together with improved educational results (UN Women 2019). The design of content should focus on women's desired pursuits, including entrepreneurial success as well as health concerns and civic engagement opportunities, to improve their interest in learning. The use of existing local institutions, consisting of schools and self-help groups helps both build social trust and reduce digital education program reluctance (Malhotra and Gupta 2023).

The successful reduction of the gender digital divide depends heavily on well-developed policies which receive essential institutional backing. The adoption of three key recommendations consists of providing discounted hardware devices to women as well as setting gender requirements for digital skills programs and including gender-informed evaluations for digital education policies (MeitY 2022 and World Bank 2018). The recommended actions require synergy with initiatives aimed at changing social beliefs that deter female digital space participation (Trigueros and de Aldecoa 2021).

Implementing effective solutions at all levels requires partnership among public institutions and private companies and non-profit organisations, Adeniyi et al. (2024). A multi-sectoral partnership enables solutions to rectify all gender digital divide elements starting from basic infrastructure development to education-related gender bias elimination and technology field model creation (Klingen et al. 2024).

The importance of digital education empowerment for resolving the digital gender gap assumes fundamental significance because modern life becomes increasingly digital while development and gender equality need full activation of all gender potential (Sánchez-Prieto et al. 2020). Equal digital opportunities with access to training and resources enable societies to achieve maximum potential from everyone regardless of gender thus creating an inclusive and innovatively better digital future.

Researchers need to study how digital education initiatives will affect women's social and economic circumstances throughout multiple years (Krieger-Boden and Sorgner 2018). The development of flexible models for digital education requires investigation to enable its wider adoption among women from varied socio-economic backgrounds.

3 Research Methodology

3.1 Search Strategy

A Boolean search of Scopus database was performed in February 2025 to investigate ("digital education" OR "e-learning" OR "online learning") AND ("gender disparity" OR "gender divide" OR "women empowerment") AND ("India" OR "developing countries") (Antonio and Tuffley 2014a, b; Mohamed et al. 2023; Talukdar 2019). A language restriction for English with final publication requirements served as the search filtering conditions. A total of 124 records were identified through the search but 31 documents satisfied the inclusion criteria after the title and abstract screening process.

3.2 Inclusion and Exclusion Criteria

Research focusing on gender and digital learning that was published between January 2015 and February 2025 as well as theoretical analysis and review articles qualified for inclusion when they addressed relevant topics for policy and practical implementation within India or comparable contexts.

The criteria excluded research about ICT outside education and results without peer-reviewing and no gender data separation from the analysis.

3.3 Data Extraction and Analysis

Research authors obtained their names from their published work as well as the year of publication and location. The study methods received attention together with participant information and explained interventions and results along with proposed recommendations. The analysis identified five main recurring concepts: Access Barriers, Digital Literacy, Sociocultural Norms, Pedagogical Innovations, and Policy Measures. Table 1 highlights significant studies related to these concepts.

Table 1 Summary of reviewed studies (2015–2025)

Author(s)	Year	Country/region	Focus area	Key findings
Antonio and Tuffley	2014	Developing countries	Gender digital divide, digital literacy	Highlighted deep-rooted gender gaps in access and skills; called for targeted interventions
Srivastava and Pathak	2019	India (rural)	Household decisions, device sharing	Found that shared device access reinforces gender-based control in households
Rao et al.	2022	India (Maharashtra)	Women-only digital skills training	Women-only cohorts had higher completion and confidence levels
Esteban et al.	2021	Global	Businesswomen and digital inclusion via social media	Emphasised role of informal networks in promoting digital entrepreneurship
UNESCO	2025	Africa (generalised)	Girls' access to STEM and internet usage	Reported only 35% of STEM graduates are women; urged stronger early digital exposure

4 Analysis and Results

4.1 Access Barriers

As shown in Table 2, women encounter significant barriers to participating in digital education, primarily due to access issues, particularly in developing countries and rural areas (Bertaux et al. 2007). The lack of electricity, slow internet, and shortage of personal devices create infrastructural obstacles that prevent women from accessing digital education, according to Das and Joshi (2018) and Kaur and Mahajan (2020). Gender inequality in access to household resources and technology often results in greater disadvantages for women.

Most rural women share smartphone access rather than owning individual devices. This situation limits students' learning time and puts their privacy at risk when using digital tools (Patel et al. 2021). According to Srivastava and Pathak (2019), shared device use serves two purposes: it restricts access time and reinforces male control over household online domains (p. 45). The current communication system maintains existing gender barriers that prevent women from gaining education or skill development opportunities.

Remote areas face the most significant challenge because of their unreliable power supply system. The irregular power outages in the region create problems for online learning by stopping virtual classes and blocking women from finishing their work or self-teaching (Sigama and Kalema 2022). The availability of dependable internet access remains essential because inconsistent or poor internet quality blocks women from accessing digital teaching materials and interactive chat sessions. Poor broadband infrastructure leads to sluggish internet speeds and video buffering

Table 2 Key barriers to digital education for women

Barrier category	Description	Supporting sources
Access barriers	Limited access to smartphones, computers, internet, and electricity, especially in rural areas. Shared device usage restricts privacy and time	Das and Joshi (2018), Srivastava and Pathak (2019), Frederick (2022)
Digital literacy gaps	Lack of basic ICT skills and limited exposure to technology, especially among women from traditional or underprivileged backgrounds	Antonio and Tuffley (2014a, b), Rao et al. (2022)
Sociocultural norms	Patriarchal beliefs, gendered responsibilities, and safety concerns (e.g., cyberbullying) discourage women's participation in online learning	Singh (2017), Citron and Penney (2019), Gupta (2024)
Infrastructure limitations	Poor broadband connectivity and frequent power outages hinder consistent participation in digital education, particularly in rural regions	Sigama and Kalema (2022), Kuteesa et al. (2024)
Economic constraints	High cost of digital devices and data plans; women receive lower priority for access to household digital resources	Patel et al. (2021), Chakraborty (2017)

and disconnected sessions while participating in real-time educational activities for women.

The high expense of purchasing devices and data solutions proves to be an obstacle that stops many people from being techno-inclined. A scarcity of financial resources as well as lower incomes makes smartphone, tablet and computer ownership unattainable for many women in digital learning according to Frederick (2022). Women typically receive a lower priority in device access than the male members of their household even when devices exist within their home environment.

Several obstacles prevent women from fully accessing digital education because of limited digital skills together with technological access problems. People with insufficient knowledge of technology and digital platforms avoid utilising digital educational resources even when they have physical access to them (Wickens and Miller 2020). Social gender prejudices and reduced technology exposure beginning at an early stage create a skills deficit for women which impedes their digital participation. Bipasha et al. (2025) identified lack of access to e-resources, insufficient digital training, fast technological upgradations and cultural and gender stereotyping as major impediments for digital literacy among female students from higher education.

A solution must deal with multiple points of access limitations. Rural electrification improvement with added broadband infrastructure development represents part of this approach (Kuteesa et al. 2024). The implementation of device-financing or subsidy initiatives specifically aimed at women would help break financial obstacles. Reliable internet access in learning centres that provide devices allows women to study in safe environments through digital platforms (Borgonovi et al. 2018).

Well-designed digital literacy programs for women will teach them essential skills needed to study on online platforms successfully. Programs designed with gender-responsive pedagogical approaches should possibly conduct classes for women only to foster comfortable learning environments according to Sixl-Daniell and Wong (2015).

Policy interventions function as essential tools to eliminate accessibility problems that women face during their educational journey. ICT policies require gender sensitivity along with digital inclusion focus on women according to Greener (2021) and Liang et al. (2022). The initiative provides two key elements to encourage gender equality in digital education: it gives educational content at reduced prices through data plan subsidies while establishing minimum quotas for female trainees in specific digital programs.

The path to digital education equity for women needs complete solutions that simultaneously eliminate both physical connectivity limitations and learning cost obstacles while reducing social hurdles. Digital education becomes a strong instrument for women's empowerment when obstacles are removed which creates better equality between genders regarding educational and economic possibilities.

"Shared device usage not only limits time but reinforces male control of digital space within households" (Srivastava and Pathak 2019, p. 45).

4.2 Digital Literacy and Skills

The shortage of basic ICT skills causes women to encounter major difficulties when they attempt to access and use digital education through available devices. The digital literacy gap creates resistance for women to achieve equal access to online education settings (Antonio and Tuffley 2014a, b). Under the Skill India program the training programs produced limited improved outcomes for digital literacy among female students. These programs achieve minimal results because their learning methods ignore the particular requirements of women and the male-dominated learning environments.

The challenges that individual women encounter in their digital skill development cannot be solved by conventional homogeneous training methods. The standardised training modules typically do not recognise how varying technology knowledge exists among students along with individual learning styles and the essential digital skills that match women's life needs and career goals (Babu et al. 2013). Women in majority male classrooms face discomfort because the environment feels threatening to women who are either from traditional families or have little technological background experience.

Studies prove that digital literacy training programs developed specifically for women produce better results (Esteban et al. 2021). Research shows female educational programs succeed better in achieving their goals when they include dedicated courses designed for women's needs which lead participants to show better completion results and self-assurance improvements. Gender-sensitive training methods establish secure educational settings that enable women to feel protected through asking questions and testing new technology while developing their competencies without intimidation or judgment (Sixl-Daniell and Wong 2015). The effective closure of the digital literacy gap requires delivering training programs that address the special contexts and requirements of women learners (Antonio and Tuffley 2014a, b). Educational programs must develop a curriculum that demonstrates actual scenarios from women's everyday life by utilizing digital platforms for business startups and medical information lookup and virtual community networking. Gaining better learning results can be achieved through female facilitators and instructors working to establish environments where students connect with the material better (Meherali et al. 2021).

Digital literacy training through community-based programs demonstrates effective methods to teach women learners especially in rural communities. Women can develop their digital skills through safe accessible facilities at local learning centres which have female facilitators. The centres provide adaptable scheduling alongside peer learning groups which enable women to balance their responsibilities (Krithika et al. 2019).

Studies show that mobile learning applications delivering short educational contents targeted for specific areas improve digital literacy skills among female learners. The apps send fast educational lessons in local languages which make learning more practical and relevant to how women spend their time each day

(Schneegass et al. 2018). Women can adapt mobile learning platforms to fit their individual speed and schedule requirements thus enabling them to overcome some common barriers that prevent traditional training participation.

Raising digital literacy skills among women needs multiple strategies and exceeds basic hardware distribution since it requires targeted educational programs designed for women's needs. Gender-sensitive training programs need to be designed exclusively for the target audience using innovative educational strategies and supportive learning facilities (Gilster 1997). Institutions must invest in women's digital literacy to make digital education serve as an empowerment tool for eradicating the gender divide across the digital domain.

"Tailored modules and women-only cohorts resulted in higher completion rates and confidence scores" (Rao et al. 2022, p. 112).

4.3 Sociocultural Norms

The way women should behave in society depends heavily on patriarchal societal standards that lead women to stay away from both digital technologies and online learning platforms (Antonio and Tuffley 2014a, b; Singh 2017). Societal norms have deep roots that frequently cut off women from using electronic devices by seeing screen hours as contradictory with household duties. Traditional societal norms introduce both psychological and physical obstacles that prevent women from joining digital education programs.

Women together with adolescents tend to avoid extensive internet-based learning due to safety-related concerns particularly the risk of receiving harassing online messages. The risks of encountering cyberbullying along with stalking behavior and encountering inappropriate material lead women and their families to refrain from extensive online activities (Citron and Penney 2019; Hazlyna et al. 2021). The fear is particularly strong in younger women and girls which reduces their opportunities to benefit from digital resources and educational content for skill development.

Multiple sociocultural hurdles together generate and escalate additional barriers which include insufficient device availability and inadequate connectivity standards and inadequate digital competence. The availability of technology access does not guarantee educational resource utilisation because females face mental barriers against these activities (Antonio and Tuffley 2014a, b). Digital education participation shows greater gender inequality due to online safety anxieties which might force users to develop self-censoring behaviours or minimise their digital platform activities.

A comprehensive solution needs to be deployed because of multiple obstacles. Educational institutions should launch programs which both destroy gender limitations through education and display digital benefits for female participation (Kuteesa et al. 2024). Implementing safe online spaces for women goes hand in hand with effective mechanisms for addressing internet harassment to protect virtual communities. Community-based programs which includes family members and village leaders can transform public opinion to generate better digital educational environments for women (Gupta 2024).

The implementation of gender-sensitive educational approaches into digital programs will reduce the obstacles that women encounter in such platforms. The education program should provide women-only online courses alongside female instructors while creating instructional materials that address their specific needs(Borgonovi et al. 2018). The addition of digital skills training to current women's empowerment programs uses established trust networks to decrease the resistance women face when participating digitally.

The key function of policy measures consists of resolving sociocultural barriers that prevent implementation. Public authorities must establish complete plans to enhance digital systems and accessibility while actively eliminating the social factors preventing women from participating digitally (Antonio and Tuffley 2014a, b). Media campaigns for protective online rights should be created to support women while educational institutions need to construct formal curricula which teach digital literacy and governments need to create policies ensuring safe digital environments.

Digital education will become a truly inclusive empowerment tool for women when we combine efforts to resolve their sociocultural barriers and safety issues with other current barriers to their learning. Digital learning platforms require a complete solution to bridge gender digital divides which enables women to access their full potential from these platforms.

4.4 Pedagogical Innovations

Technology-based education initiatives achieve successful outcomes by closing gender gaps and enhancing learning opportunities for women, especially among rural students. Community learning centres with female staff members serve as effective digital learning environments that create safe, protective spaces where women feel comfortable to learn (Krithika et al. 2019). These facilities provide specialised areas that enable women to learn technology and improve their digital skills, while ensuring they are protected from intimidation and judgment. When learners engage with female facilitators, they develop trust and a sense of identification, which helps them connect better with the instructional content.

Mobile micro learning applications through smartphones show their ability to enhance digital education retention and literacy skills for women learners across different settings. Mobile learning applications provide local language-based educational content which enables women to learn more effectively and apply their knowledge to their everyday situations (Panicker 2020). Mobile learning provides women with the agility to study educational content whenever they want and wherever they wish which exceeds barriers that traditional educational programs impose on their enrolment. Content targeted at women that mirrors their life perspectives builds their drive to learn and demonstrates compatibility with their interests (Beegle and Rubiano-Matulevich 2023).

Studies show that blended educational formats, which combine virtual learning techniques with direct learner assistance, are effective for overcoming the various barriers women face when engaging with digital education. This educational approach uses digital systems while providing essential human guidance, including peer assistance (Tzavara and Wilczek 2019). Learning through blended models offers direct support for technology use, along with virtual coursework to help close digital literacy gaps. These programs foster networking among women learners and create supportive learning environments where they can form meaningful community connections (Lazem 2019).

The educational strategies match feminist pedagogical principles which focus on active students and cooperative learning experiences that disrupt standard educational power structures (Hayes 1989). The purpose of these innovative approaches is to develop specific methods which align with women's needs and experiences, thus helping them break through social barriers while developing digital technology confidence (Peña et al. 2021).

Integration of digital education within schools and self-help groups succeeds in leveraging community trust networks while minimising opposition to digital participation by women (Nedungadi et al. 2018). This method allows digital learning to become part of familiar settings through which access and cultural acceptance increase for educational opportunities. This system enables the combination of digital skills training with other women development programs while providing a comprehensive approach to their empowerment (Borgonovi et al. 2018).

Digital education becomes more relevant by developing content that addresses women's business goals together with their health needs and their involvement in civic matters. The clear identification of digital skills' potential in helping women achieve personal and professional objectives produces increased motivation levels together with better engagement. The tailored approach enables women to see direct advantages of digital literacy which exist in their everyday lives (Borgonovi et al. 2018).

There are two essential factors for maintaining and expanding modern techniques through official support and governmental actions. The recommendations highlight the need to give women reduced-cost devices for digital learning and enforce gender-based goals during training and include gender perspectives in all digital education planning. World Bank Document 2024 shows that these measures assist in solving institutional barriers which establish effective conditions for digital empowerment among women.

These innovations require a unified method that deals with access challenges together with literacy and cultural barriers which must be solved in a system-wide manner. The strategies to enhance digital access and use for women require upgrades in basic facilities and network expansion into rural regions alongside specialised digital training for women and public programs that strive to unsettle traditional gender-based technology biases (Terry and Gómez 2010).

Future research needs to assess the extended economic and social effects that come from these modern educational methods which target women. The expansion of digital education programs that can scale to different socioeconomic contexts demands exploration to maximise their effectiveness when empowering women from all backgrounds (Sterling et al. 2020).

Digital education empowered through pedagogical innovations and supportive policies will develop into an effective solution to reduce gender disparities in education and increase women's inclusive learning possibilities. The digital age demands full female participation in education through technology since this is needed to attain true gender equality and inclusive development.

4.5 Policy Measures and Institutional Support

The digital participation of women in education requires supportive policies and institutional backing to establish connections between gender groups and inhibit digital inequalities. There are several recommendations in this study which emphasise:

- Women's digital education can benefit from programs that provide affordable smartphone or table devices or computers exclusively for female students. Through collaborative agreements between device manufacturers and telecom companies women learners would receive minimal price plans.
- The government should enforce gender quotas which require minimal female enrolment in all digital skills training activities it organises. Such initiatives help reduce disadvantaged groups and establish learning environments that include everyone.
- Digital education policy development should use gender analysis to complete extensive assessments which will show barriers along with unintended consequences affecting women learners.
- The expansion of reliable electricity and broadband internet access as a priority for underserved locations which would serve women learners.
- Community learning centres act as safe digital training facilities supported by female instructors who deliver digital education through local centres which serve women learners.
- Mobile learning platform development consists of creating gender-sensitive mobile learning applications that serve women learners in their specific locations.
- The organisation will execute media-based and community outreach initiatives as a part of their campaign to break down gender barriers in technology usage while fighting for digital rights for women.
- The system needs to develop solid online protection frameworks and operating processes that establish secure spaces for female educational students.
- Integration with existing programs: Incorporating digital skills training into established women's empowerment initiatives to leverage existing trust networks.
- Educational institutions need to receive guidance through resource provision for implementing teaching approaches which specifically cater to the learning requirements and circumstances of women participating in digital education activities.

- The funding of assessments through research should evaluate digital education's enduring effects on gender-related economic growth and social development for policy improvement.
- Multiple stakeholder entities should work together as partnerships between government institutions and private companies and educational centres also civil society organisations for the implementation of complete solutions.

Government institutions alongside organisations can establish an enabling digital environment for women through proper policy implementation while building institutional support to close gender discrepancies in educational digital accessibility and results.

5 Discussion

This study shows digital education acts as a transformative approach for closing gender differences especially within areas like India, which experiences minimal female digital participation in relation to male users. The national initiatives supporting digital infrastructure expansion under the Digital India and Skill India programs have failed to eliminate multiple barriers which prevent women from participating digitally. The core obstacles preventing women from digital access are derived from structural barriers affecting network speed and lack of personalised learning curricula tailored for women.

Studies show that digital gender differences exist at various levels. Initial barriers to access devices while maintaining connectivity exist because of economic status and structural problems. The assessment at the second level shows overall training methods and male-biased classroom environments stop women from effectively using digital tools. The third category demonstrates how technology leads to different advantages as well as effects yet women continue to be restricted from digital platforms because of social beliefs along with concerns about safety such as internet harassment.

Productive pedagogical procedures have emerged as beneficial educational solutions. Learning centres that serve communities with instructors who are women and teaching groups consisting entirely of women and smartphone applications containing local educational content help women learners build higher engagement and self-assurance. Gender-responsive educational strategies establish protective and encouraging digital environments by designing learning programs that face real-life situations of women and direct towards their personal ambitions.

State policies together with institutional backing serve as fundamental requirements for success. The combination of strategic policy recommendations including digital device financing schemes for women along with gender quota requirements in training programs and gender analysis system evaluation in digital learning initiatives brings the potential to weaken structural obstacles. To successfully implement

Table 3 Strategic interventions and policy recommendations

Barrier addressed	Strategic intervention	Policy recommendation	Supporting sources
Access barriers	Community learning centres with female facilitators; mobile learning hubs	Government-sponsored device distribution or financing schemes for women	Krithika et al. (2019), MeitY (2022), World Bank (2018)
Digital literacy gaps	Gender-sensitive training programs; women-only digital literacy cohorts	Integrate digital skills into women empowerment schemes; set gender quotas in training	Rao et al. (2022), Agarwal and Malathi (2020), UNESCO (2025)
Sociocultural norms	Local awareness programs; inclusion of family and community leaders in educational initiatives	Gender-responsive ICT and education policies; safe digital environments for women	Gupta (2024), Antonio and Tuffley (2014a, b), Srivastava and Pathak (2019)
Infrastructure limitations	Development of local broadband networks; solar-powered digital kiosks in remote areas	Rural electrification with digital education in mind; telecom infrastructure upgrades	Sigama and Kalema (2022), Kuteesa et al. (2024)
Economic constraints	Subsidised internet/data plans for women; shared community digital resources	Public-private partnerships to reduce digital access costs for disadvantaged women	Frederick (2022), Borgonovi et al. (2018)

a digital empowerment strategy, a comprehensive solution is essential. This solution should address infrastructure issues, enhance digital literacy education, and challenge traditional societal norms to achieve gender equality in educational development, as illustrated in Table 3, which outlines Strategic Interventions and Policy Recommendations.

6 Conclusion

Digital education demonstrates powerful capabilities in reducing long-standing gender inequalities between men and women in digital spaces that show lower female interest. The initiatives of Digital India and Skill India have not overcome the major barriers which include socio-cultural norms along with limited device access and insufficient digital literacy among the population. The reviewed evidence demonstrates that combination approaches of community-based interventions along with gender-sensitive teaching methods and supportive policies which include funding electronic devices and expanded rural internet access efficiently guide women progress in education. The transformation of education through technology demands an integrated strategy focusing on digital access together with skills education and policy refinement for developing inclusive learning systems. Social advancement transforms into new possibilities when these actions are implemented.

7 Implications and Scope for Future Work

This systematic review provides valuable insights into how digital education can act as a transformative force for bridging gender disparities, particularly in developing contexts such as India. The findings carry several important implications for policymakers, educators, and community stakeholders.

- **Policy Implications**: The review underscores the urgent need for gender-responsive digital education policies. Government and institutional initiatives should prioritise subsidised digital devices, ensure equitable internet access in rural areas, and incorporate digital literacy into national skill development frameworks. Setting enrolment quotas and monitoring gender-disaggregated data in digital programs will enhance inclusivity.
- **Educational and Institutional Implications**: Educational institutions must adopt inclusive, feminist pedagogical approaches that address the unique learning needs of women. Creating safe, community-based learning centres staffed by female instructors can provide psychologically secure environments. Integrating digital literacy into the curricula of women's empowerment programs can further enhance reach and effectiveness.
- **Social and Economic Implications**: Bridging the digital gender divide will boost women's participation in the digital economy, entrepreneurship, and skill-based employment sectors. Closing this gap can enhance household decision-making power for women and improve the socio-economic status of communities.
- **Scope for Future Research**: Future research should explore longitudinal impacts of digital education programs on women's economic independence, career progression, and social mobility. There is also a need to develop adaptive, scalable digital learning models that consider diverse socio-cultural contexts and user behaviours. Additionally, emerging technologies like AI and mobile micro learning platforms should be studied for their potential to advance gender-equitable education.

In summary, a collaborative, multi-stakeholder approach is essential to convert digital education into a sustainable vehicle for achieving gender equality and inclusive development.

References

Adeniyi IS, Hamad NMA, Adewusi OE, Unachukwu CC, Osawaru B, Onyebuchi CN, Omolawal SA, Aliu AO, David IO (2024) Gender equality in the workplace: a comparative review of USA

and African practices [review of gender equality in the workplace: a comparative review of USA and African practices]. World J Adv Res Rev 21(2):763. https://doi.org/10.30574/wjarr.2024.21.2.0491
Agarwal P, Malathi S (2020) Assessing the impact of ICT training programs on women's digital literacy in rural India. Int J Educ Dev 76:102–115. https://doi.org/10.1016/j.ijedudev.2019.102115
Antonio A, Tuffley D (2014a) Digital literacy in the developing world: a gender gap. http://eprints.usq.edu.au/27633/
Antonio A, Tuffley D (2014b) The gender digital divide in developing countries. Future Internet 6(4):673. https://doi.org/10.3390/fi6040673
Babu SC, Ferguson J, Parsai N, Almoguera R (2013) Open distance learning for development: lessons from strengthening research capacity on gender, crisis prevention, and recovery. Int Rev Res Open Distr Learn 14(5). https://doi.org/10.19173/irrodl.v14i5.1611
Beegle K, Rubiano-Matulevich E (2023) Five ways to make skills training work for women. https://blogs.worldbank.org/jobs/five-ways-make-skills-training-work-women
Bertaux N, Okunoye A, Rashed JA (2007) Information technology education for women in developing countries: benefits, barriers and policies. Glob Bus Econ Rev 9(4):353. https://doi.org/10.1504/gber.2007.015099
Bipasha NJ, Islam MS, Bhuyan MM (2025) Bridging the digital gender divide: is digital literacy truly an equalizer for female students across different demographic factors? Digit Libr Perspect 41:497–517
Borgonovi F, Centurelli R, Dernis H, Grundke R, Horvát P, Jamet S, Keese M, Liebender A-S, Marcolin L, Rosenfeld D, Squicciarini M (2018) Bridging the digital gender divide. http://repositorio.ciem.ucr.ac.cr/jspui/bitstream/123456789/307/1/RCIEM271.pdf
Chakraborty L (2017) Gendered perceptions of technology in Indian households. Gend Technol Dev 21(2):135–150. https://doi.org/10.1080/09718524.2017.1342950
Citron DK, Penney JW (2019) When law frees us to speak. Fordham Law Rev 87(6):2317. https://ir.lawnet.fordham.edu/flr/vol87/iss6/2/
CSIS (2024) Leveraging Digital Technologies to Advance Women's Economic Empowerment. https://www.csis.org/analysis/leveraging-digital-technologies-advance-womens-economic-empowerment
Das S, Joshi A (2018) Connectivity constraints and their effect on women's e-learning participation. J Rural Stud 59:112–121. https://doi.org/10.1016/j.jrurstud.2018.03.005
Desai R (2021) Microlearning in local languages: enhancing digital education for rural women. Educ Technol Res Dev 69(4):2345–2362. https://doi.org/10.1007/s11423-021-09990-1
Esteban PG, Miranda F d PR, Delgado SC, Rincón MLH (2021) Digital inclusion of businesswomen and women entrepreneurs through social networks in the informal environment. Multidisciplinary J Gend Stud 10(3). https://doi.org/10.17583/generos.7146
Frederick DE (2022) The new feminine face of the digital divide. Library Hi Tech News 39(9):6. https://doi.org/10.1108/lhtn-04-2022-0050
Gilster P (1997) Digital literacy. https://www.goodreads.com/work/editions/2346785-digital-literacy
Greener S (2021) Developing learning technology as inclusive practice. Interact Learn Environ 29(3):343. https://doi.org/10.1080/10494820.2021.1927494
Gupta K (2024) Transforming women safety with information technology: a Mobile real-time intelligence framework. Deleted J 20:1193. https://doi.org/10.52783/jes.1430
Habibi A, Sofyan S, Mukminin A (2023) Factors affecting digital technology access in vocational education. Sci Rep 13(1). https://doi.org/10.1038/s41598-023-32755-6
Hayes E (1989) Insights from women's experiences for teaching and learning. New Dir Adult Contin Educ 1989(43):55. https://doi.org/10.1002/ace.36719894307
Hazlyna HN et al (2021) Awareness about cyber bullying on social media among female students in a Malaysian public university. Türk Bilgisayar ve Matematik Eğitimi Dergisi 12(3):1592. https://doi.org/10.17762/turcomat.v12i3.972

Helsper E, Van Deursen A (2017) Digital skills in Europe: gender divides and policy responses. New Media Soc 19(8):1315–1336. https://doi.org/10.1177/1461444816638788

Hilbert M (2011a) Digital gender divide or technologically empowered women in developing countries? Electron J Inf Syst Dev Countries 47(7):1–16. https://doi.org/10.1002/j.1681-4835.2011.tb00436.x

Hilbert M (2011b) Digital gender divide or technologically empowered women in developing countries? A typical case of lies, damned lies, and statistics. Womens Stud Int Forum 34(6):479. https://doi.org/10.1016/j.wsif.2011.07.001

Hilbert M (2023) Digital gender divide or technologically empowered Women in developing countries?. https://www.semanticscholar.org/paper/Digital-Gender-Divide-or-Technologically-Empowered-Hilbert/702b8e496faea2d49cb72843a439c15f45bf2ab0

Hooks B (1994) Teaching to transgress: education as the practice of freedom. Routledge

International Telecommunication Union (2021) Measuring digital development: facts and figures 2021. ITU

Johnson L, Adams Becker S, Estrada V, Freeman A (2016) NMC horizon report: 2016 higher education edition. The New Media Consortium

Kaur G, Mahajan R (2020) Gender differences in digital device ownership in northern India. Gender Manag Int J 35(5):423–437. https://doi.org/10.1108/GM-01-2019-0006

Klingen N, Hammond A, Srinivasan S (2024) Putting women and girls at the center of digital development. https://blogs.worldbank.org/en/digital-development/putting-women-and-girls-center-digital-development

Krieger-Boden C, Sorgner A (2018) Labor market opportunities for women in the digital age. Economics 12(1). https://doi.org/10.5018/economics-ejournal.ja.2018-28

Krithika S et al (2019) Study on economic and technological empowerment of rural Women through'digital literacy and digital livelihood enhancement programme in Tamil Nadu. Int J Agric Sci Res 9(3):51. https://doi.org/10.24247/ijasrjun20197

Kumar N, Singh P (2021) Online safety concerns among young women learners in India. Comput Educ 165:104–118. https://doi.org/10.1016/j.compedu.2021.104118

Kuteesa KN, Akpuokwe CU, Udeh CA (2024) Gender equity in education: addressing challenges and promoting opportunities for social empowerment. Int J Appl Res Soc Sci 6(4):631. https://doi.org/10.51594/ijarss.v6i4.1034

Lazem S (2019) On designing blended learning environments for resource-challenged communities. Int J Emerg Technol Learn (iJET) 14(12):182. https://doi.org/10.3991/ijet.v14i12.10320

Liang M, Lim CP, Park J, Mendoza NB (2022) A review of ICT-enabled learning for schoolgirls in Asia and its impacts on education equity [review of a review of ICT-enabled learning for schoolgirls in Asia and its impacts on education equity]. Educ Technol Res Dev 71(2):267. https://doi.org/10.1007/s11423-022-10178-w

Lorde A (1984) Sister Outsider: Essays and Speeches. Crossing Press

Mackey A, Petrucka P (2021) Technology as the key to women's empowerment: a scoping review [review of technology as the key to women's empowerment: a scoping review]. BMC Womens Health 21(1):78. https://doi.org/10.1186/s12905-021-01225-4

Malhotra N, Gupta A (2023) Community learning centers as safe digital spaces for rural women. Commun Dev J 58(2):189–207. https://doi.org/10.1093/cdj/bsac011

Mariscal J, Mayne G, Aneja U, Sorgner A (2019) Bridging the gender digital gap. Economics 13(1). https://doi.org/10.5018/economics-ejournal.ja.2019-9

Meherali S, Rahim KA, Campbell S, Lassi ZS (2021) Does digital literacy empower adolescent girls in low- and middle-income countries: a systematic review [review of does digital literacy empower adolescent girls in low- and middle-income countries: a systematic review]. Front Public Health 9. https://doi.org/10.3389/fpubh.2021.761394

Ministry of Electronics & Information Technology (2022) Annual report 2021–22. Government of India

Ministry of Statistics & Programme Implementation (2021) Household social consumption: education, 78th round. Government of India

Mohamed H, Ismail A, Sutan R, Rahman RA, Juval K (2023) A scoping review of digital technologies in Antenatal Care: recent progress and applications of digital technologies [review of a scoping review of digital technologies in Antenatal Care: recent progress and applications of digital technologies]. Res Sq (Res Sq). https://doi.org/10.21203/rs.3.rs-3492596/v1

Naga JF, Catindig MAC, Caparida LT (2020) Development of E-learning module for ICT skills of marginalized Women and girls for ICT4D. Int J Emerg Technol Learn (iJET) 15(16):94. https://doi.org/10.3991/ijet.v15i16.14929

Nedungadi P, Menon R, Gutjahr G, Erickson L, Raman R (2018) Towards an inclusive digital literacy framework for digital India. Educ Train 60(6):516. https://doi.org/10.1108/et-03-2018-0061

O'Brien LT, Blodorn A, Adams G, Garcia DM, Hammer ED (2014) Ethnic variation in gender-STEM stereotypes and STEM participation: an intersectional approach. Cultur Divers Ethnic Minor Psychol 21(2):169. https://doi.org/10.1037/a0037944

Panicker P (2020) Critical pedagogy in the implementation of educational technologies. arXiv (Cornell University). https://doi.org/10.48550/arXiv.2006.00245

Patel Y, Bansal R, Verma S (2021) Shared smartphones and women's learning time in rural India. Asian J Distance Educ 16(1):58–74. https://doi.org/10.5281/zenodo.1234567

Peña J, Hanrahan BV, Rosson MB, Cole C (2021) After-hours learning. ACM Trans Comput Educ 21(2):1. https://doi.org/10.1145/3446964

Qazi A (2022) Gender differences in information and communication technology use & skills: a systematic review and meta-analysis. https://link.springer.com/content/pdf/10.1007/s10639-021-10775-x.pdf?error=cookies_not_supported&code=742e2905-bc44-4cb7-a18f-f2821e4fbcff

Rao S, Mehta K, Joshi L (2022) Evaluating women-only digital skills cohorts in Maharashtra. Comput Hum Behav Rep 6:110–118. https://doi.org/10.1016/j.chbr.2022.100115

Sánchez Peláez I, Reyes GCE, Morales GDL (2023) Gender digital divide in education 4.0: a systematic literature review of factors and strategies for inclusion. Future in Educ Res 1(2):129–146. https://doi.org/10.1002/fer3.16

Sánchez-Prieto J, Torres JMT, García MG, García GG (2020) Gender and digital teaching competence in dual vocational education and training. Educ Sci 10(3):84. https://doi.org/10.3390/educsci10030084

Schneegass C, Terzimehić N, Nettah M, Schneegaß S (2018) Informing the design of user-adaptive mobile language learning applications, p 233. https://doi.org/10.1145/3282894.3282926

Sigama K, Kalema BM (2022) A model for MOOC implementation in areas of low bandwidth in developing countries. Int J Distance Educ Technol 20(1):1. https://doi.org/10.4018/ijdet.312182

Singh S (2017) Bridging the gender digital divide in developing countries. J Child Media 11(2):245. https://doi.org/10.1080/17482798.2017.1305604

Sixl-Daniell K, Wong A (2015) Examining students' perceptions of E-learning in a Women-only environment. Int J Adv Corp Learn (iJAC) 8(4):4. https://doi.org/10.3991/ijac.v8i4.4835

Sobieraj S, Krämer NC (2019) Similarities and differences between genders in the usage of computer with different levels of technological complexity. Comput Hum Behav 104:106145. https://doi.org/10.1016/j.chb.2019.09.021

Srivastava S, Pathak A (2019) Household decision-making and women's access to digital learning. J Gend Stud 28(1):40–56. https://doi.org/10.1080/09589236.2018.1428176

Sterling R, Grubbs L, Koutsky T (2020) Breaking through the gender digital divide: technology, social norms, and the Women Connect challenge. SSRN Electron J. https://doi.org/10.2139/ssrn.3749643

Talukdar IH (2019) Obstacles to political participation of the elected female representatives in the local governance of Bangladesh. Bangladesh J Public Adm 25(2). https://doi.org/10.36609/bjpa.v25i2.30

Terry A, Gómez R (2010) Gender and public access computing: an international perspective. Electron J Inf Syst Dev Ctries 43(1):1. https://doi.org/10.1002/j.1681-4835.2010.tb00309.x

Trigueros IMG, de Aldecoa CY (2021) The digital gender gap in teacher education: the TPACK framework for the 21st century. Eur J Investig Health Psychol Educ 11(4):1333. https://doi.org/10.3390/ejihpe11040097

Tzavara D, Wilczek B (2019) Chapter 13 Online: a new 'geography' of learning that supports female access to higher education. In: Innovations in higher education teaching and learning. Emerald Publishing Limited, p 213. https://doi.org/10.1108/s2055-364120190000016017

UN Women (2019) Digital empowerment of women: strategies and good practices. United Nations Entity for Gender Equality

UNESCO (2020) Global education monitoring report 2020: inclusion and education. UNESCO Publishing

UNESCO (2025) Closing the digital divide for women and girls in Africa through education. UNESCO Publishing. https://www.unesco.org/en/gender-equality/education/digital-divide

van Dijk J (2020) Closing the digital divide: a critical perspective on seven kinds of interventions. Digit Policy, Regul GovAnce 22(4):373–388. https://doi.org/10.1108/DPRG-02-2020-0011

Wickens CM, Miller T (2020) Gender, digital literacies, and higher education, p 53. https://doi.org/10.1002/9781119257639.ch3

World Bank (2018) World development report 2018: learning to realize education's promise. World Bank Publications

World Bank Document (2024). https://documents1.worldbank.org/curated/en/099631003072338051/pdf/IDU1116c98a914ebc14dc31a47a1495a00553bae.pdf

Yañez AGB, Alonso-Fernández C, Fernández-Manjón B (2023) Systematic literature review of digital resources to educate on gender equality. Educ Inf Technol 28(8):10639. https://doi.org/10.1007/s10639-022-11574-8

Women Empowerment Through Financial Inclusion: Opportunities and Challenges

Kumud Singh Rajput and Aashka Thakkar

Abstract Women's empowerment is a critical driver of socio-economic development, and financial inclusion plays a pivotal role in fostering gender equality. Financial inclusion refers to access to affordable financial services such as banking, credit, insurance, and digital payment systems, which empower women by enabling financial independence and economic participation. On a global scale financial inclusion has been recognised as an essential parameter for assessing the advancement and economic well-being of society as a whole. In several countries, particularly India, offering inclusive financial products that is, banking services that are accessible and affordable to everyone has taken precedence. The G-20 countries are placing a strong emphasis on promoting financial inclusion as a means of accomplishing other sustainable development objectives, including gender equality (Bhatia et al. in Vikalpa 44(4):182–197, 2020). The groundbreaking strategy of gender equality and women empowerment seeks to shift power dynamics towards the advantage of women and is seen as crucial to the advancement of the world. Therefore, in order to accomplish basic to influential goals, emerging economies are adopting an all-encompassing financial architecture. Despite various governmental and institutional initiatives, a significant gender gap in financial access persists, particularly in developing economies. This study explores the impact of financial inclusion on women's empowerment, examining key enablers, barriers, and policy implications. All published articles between 2000 and 2025 that included the topics Financial Inclusion, Financial Literacy and women's economic empowerment have been retrieved for bibliometric analysis. A total of 1802 English-language papers were extracted from the Scopus database for the purpose of Co—Authorship Analysis, Keyword Analysis, Co—Citation analysis by using VOS Viewer.

K. S. Rajput · A. Thakkar (✉)
Faculty of Management Studies—MBA Program, Parul University, Vadodara, India
e-mail: aashka.thakkar@paruluniversity.ac.in

K. S. Rajput
e-mail: kumudsingh.rajput34219@paruluniversity.ac.in

A. Singh et al. (eds.), *Human-Centered Sustainability: Bridging Well-Being, Gender Equity, and Technology*, World Sustainability Series,
https://doi.org/10.1007/978-3-032-22020-2_7

Keywords Women's empowerment · Financial inclusion · Digital banking · Microfinance · Gender gap · Financial literacy · Economic independence · Policy measures

1 Introduction

Women's empowerment is a critical driver of socio-economic development, and financial inclusion plays a pivotal role in fostering gender equality. Financial inclusion refers to access to affordable financial services such as banking, credit, insurance, and digital payment systems, which empower women by enabling financial independence and economic participation. Despite various governmental and institutional initiatives, a significant gender gap in financial access persists, particularly in developing economies. This study explores the impact of financial inclusion on women's empowerment, examining key enablers, barriers, and policy implications. Achieving women's economic empowerment, one of its objectives under the fifth Sustainable Development Goal on gender equality, depends on financial inclusion. A fifth of Indian women are missing access to a bank account. While our country's efforts to foster financial inclusion have improved the amount of women with bank account access, there are nevertheless substantial differences in account application, accessibility to credit, and savings.

The fundamental principle of economically challenged nations' paths to development is the pursuit of inclusion and equal treatment (Lerner et al. 2010). Women face extreme poverty on a global scale due to cultural and structural impediments (Seguino 2010). Given their lesser degree of education, poorer incomes, and less assets, women have much less disposable income than men (Gonzales et al. 2015). It is anticipated that increasing involvement of women will boost the economies of developing nations (Bharti 2021). Implementing gender-focused financing strategies and services, such as collateral-free financial services, financial education initiatives, and the development of complementary networks, is essential to enabling them to reach their full economic potential (Parwez et al. 2022). India offers a unique and abundant context for studying microfinance-supported women-led transformation.

For plenty of reasons, notably the reality that they are more likely to live separated from a bank branch, lack documentation or mobile phones, and demand guidance when they want to establish and oversee a bank account, women nevertheless encounter challenges whenever seeking out financial services. Financial services availability gives citizens opportunities to generate money, accumulate assets, or participate more fully in capitalism, numerous factors which help promote their economic and social empowerment. Furthermore, financial inclusion provides resilience against shock events such as the COVID-19 epidemic, which focused spotlight to the requirement of delivering formal financial services to poorest populations (Ditta et al. 2025). Financial services accessible offers individuals the opportunity of making money, own assets, or participate more fully in capitalism, numerous factors them help promote financial and social empowerment. Furthermore, financial inclusion promotes resilience versus shock events which include the COVID-19 epidemic, which attracted focus to the necessity of supplying formal financial services to the least fortunate populations. Nevertheless, restrictions persist in the way of women's access to financial services internationally. In accordance with data from the Global Findex Database 2021, women and the poor are more likely to live separated from a bank branch, need support with opening and managing a bank account, and lack documentation or a mobile device.

2 What is Women Empowerment

The process by which women establish ownership of their own lives and gain the ability to make rational choices is commonly referred to as women's empowerment. It involves expanding women's access to financial resources, political representation, are employed, healthcare, and education. Women can challenge traditional gender rules and regulations, eliminate dependence and take part fully and equally in every area of society whenever they are empowered. Basically, women's empowerment is a process with empowerment as an aim. It aims to remove the constraints that hinder women from attaining their full potential and achieve gender equality. As a process, it involves allowing women in social, economic, political, legal, and educational domains in making decisions and to sense assured of their abilities. Empowerment of Women in Reference to Financial Inclusion:

3 Women Empowerment with Reference to Financial Inclusion

The procedure to provide women, particularly those from deprived or excluded populations, equitable access to affordably priced appropriate financial services such digital payment systems, credit, insurance, pensions, and savings spaces has been referred to as "women's empowerment through financial inclusion." Women's total of socioeconomic status has improved as the consequence of this access, and this allows them to develop into financially independent, perform economic endeavours, and make educated financial choices of the people. The implications of women's empowerment on advancement in society and the economy are extensive. Stronger families, more resilient communities, and inclusive economic growth all have been promoted by empowered women (Thakkar et al. 2025).

Women are more likely to participate in civic life, increase family health and nutrition, and support their children's education when they sense empowered. Along with being an issue of justice, the empowerment of women is an essential for sustainable development. In order to promote financial inclusion between all Indian households, the Pradhan Mantri Jan Dhan Yojana was launched in 2014. The initiative's objective is to provide the "financially excluded" access to financial services such as insurance, retirement plans, remittance amenities, need-based credit, and basic savings bank accounts. Over 460 million bank accounts have been opened since then, including 56% of them being held by women and 67% of them located in semi-urban or rural regions, according to data from the Ministry of Finance. From INR 1279 in 2015 to INR 3761 in 2022, the average deposit amount in PMJDY accounts has more than doubled. However, nearly 20% of Indian women are without access to a bank account. There stay gaps in the effective use of bank accounts among those who do have them, and they generally are missing access to credit or savings. Considering financial inclusion has been acknowledged as an important variable that influences economic growth and the elimination of poverty, it is important the gaps be eliminated. It provides that individuals have access to feasible cost-effective financial services that are delivered in a sustainable and ethical manner (Hendriks 2019).

"Financial inclusion" is "the process of ensuring access to financial services and timely and adequate credit where needed by vulnerable groups such as weaker sections and low-income groups at an affordable cost," corresponding to the Reserve Bank of India's National Strategy for Financial Inclusion (2019–2024). Indeed, at least eight of the 17 Sustainable Development Goals (SDGs) have been reported that they are made feasible by financial inclusion that has been recognized as a significant component of development. The affordability of bank accounts, loans, insurance, and other financial services significantly enhances employment, education, and health outcomes. These innovations in turn lead to the mutual goals of eradicating poverty, boosting inclusive growth, and diminishing inequality.

4 Co-Occurrence Network Analysis

The Below Figure provides a list of Keywords related to the field of the terminology "Women Empowerment", as well as their occurrences and total strength of the links which refers to the strong connections between the keywords in the selected articles. The Keyword Network shows that Financial Inclusion, Financial Literacy, Gender, Sustainable Development, Technology, and Financial Education have the highest occurrences, which shows their stronger association. Financial Inclusion is the most significant area with 84 occurrences and a link strength of 412. It has also been observed that Financial Literacy and Gender are also the centrally focused keywords used in the articles analysed with a total link strength of 346 and 237. Keywords such as Domestic Violence, Inequality, Unemployment, Mental Health, and Discrimination also need to be addressed eventually (Fig. 1).

According to the statistics, articles on financial inclusion and women's empowerment demonstrate an extensive degree of global relationship. The creation of a collaborative relationship among these countries is seen in the figure below. When it comes to evaluating the strength of these partnerships, USA tops the list, India comes second, followed by UK, Australia, Canada and China. The analysis also

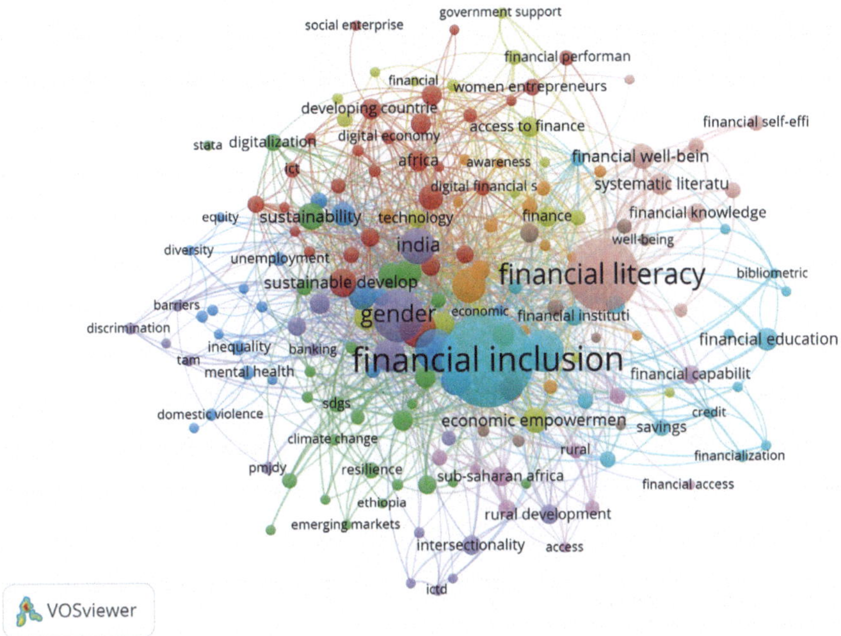

Fig. 1 Co-occurrence of keywords

reflects that countries such as Switzerland, Kenya, Lebanon, Namibia, Uganda, Sri Lanka, Ireland, and New Zealand have contributed very few documents for women's empowerment and Financial Inclusion at the global level (Fig. 2).

5 Introduction of Financial Inclusion for Women Empowerment in India

For economic expansion and a decreasing level of poverty, access to finance and capital is essential. For the goal to accomplish gender equality and economic prosperity in India, women have to have access to financial services and borrowing possibilities. Despite many attempts made to encourage financial inclusion, women continue experience significant barriers while trying getting credit. Better household income, better health and academic performance, and enhanced autonomy and empowerment all have been associated with women obtaining access to credit. (Swains and Wallentin 2017; Hansen 2015). In India, socioeconomic, cultural, and institutional constraints continue that restrict women's access to formal credit. The greater focus on financial inclusion, particularly for women, has given rise to an economic turnaround in India in recent years. Women also actively improving the nation's prosperity as more of the population have access to digital payment systems,

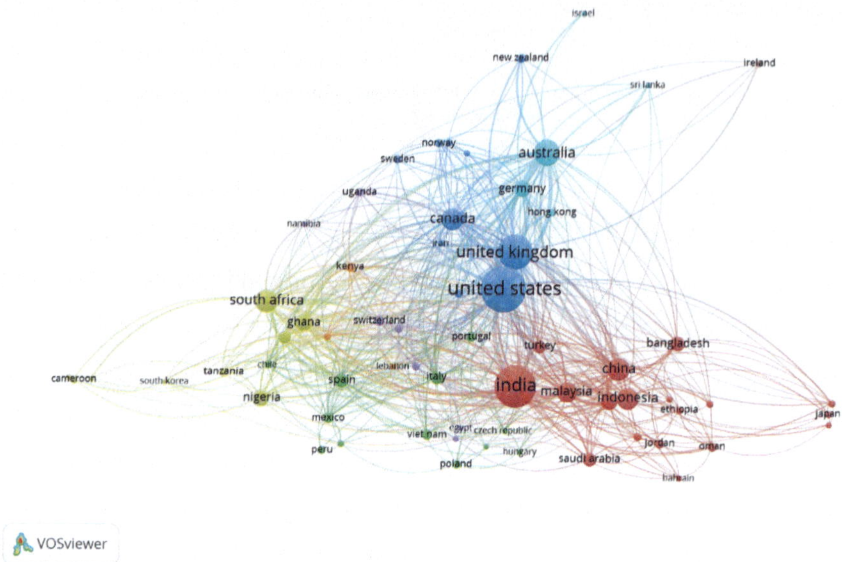

Fig. 2 Co-citation of countries

credit, and financial institutions. Access to cost-effective financial products and services for everybody can be referred to as financial inclusion, and it continues to be an effective means for empowering women, fostering economic growth, and strengthening entire communities. The practice is beginning to gain traction in India, especially since mobile technology and digital banking are helping close the disparities that women traditionally historically faced. Beyond empowered people in general, financial inclusion for women enhances the economy of this nation and encourages community growth (Banna et al. 2022). Based on the World Bank's report from 2024, governments with a higher level of financial inclusion for women suffer from faster GDP growth, reduced income inequality, and better resistance to shocks to their economy. Down to the past 10 years, over 90% of Indian women who resided in rural areas lacked access to recognized financial services; but, because to concentrated initiatives and financial literacy initiatives that the proportion has been steadily increasing. Offering women access to financial resources boosts their probability to invest investments in business ventures, education, and family health, all of which are helping India's growth in economy.

With the objective to foster financial inclusion in every household around the entire country, India established the Pradhan Mantri Jan Dhan Yojana (PMJDY) in 2014. The objective of this program is to offer the "financially excluded" access to financial services such as insurance, retirement plans, remittance facilities, need-based credit, and basic savings bank accounts. Over 460 million bank accounts have been opened since then, via 56% of them possessed by women and 67% situated in semi-rural and rural regions, according to data from the Ministry of Finance. Nearly 54% of

Indian women still have difficulties collecting fundamental financial services owing to community and cultural obstacles, according to the World Economic Forum's 2024 Global Gender Gap Report. However, the Pradhan Mantri Jan Dhan Yojana (PMJDY) alongside additional focused programs in India have proven significant of improving financial inclusion. 250 million women are presently able to open bank accounts by virtue of PMJDY, a zero-balance savings account program that the was implemented in 2014. As per the data from 2024, the number of women who own bank accounts has increased by 36% especially compared to men. Although financial inclusion has been identified as an important component in both economic growth and the decrease of poverty, it is crucial that these gaps be eliminated. It assures that individuals have access to feasible cost-effective financial services that are delivered in a manner that is environmentally friendly and ethical. "Financial inclusion" has become "the process of ensuring access to financial services and timely and adequate credit where needed by vulnerable groups such as weaker sections and low-income groups at an affordable cost," in relation to the Reserve Bank of India's National Strategy for Financial Inclusion (2019–2024).

Gender disparities in financial inclusion in India will be investigated in the remaining part of this brief. For the purpose of to present current evidence on the gender divide, the study makes use of a review of the literature, expert consultations, and data sets from the Global Findex Database 2021, the most relevant rounds of the National Family Health Survey 2019–21, and the All-India Debt and Investment Survey 2019. The World Bank's Global Findex Database compiles data concerning financial services availability worldwide, such as borrowing, savings, and payments. During the COVID-19 spreading, 128,000 adults in 123 economies volunteered in the 2021 edition, encompassing information on digital payments and access to recognized and informal banking services.

According to the Findex reveals, women account for 13% of the globe's unbanked population. According to this index, "account ownership" is defined as possessing an account within a bank, credit union, microfinance organization as well, post office, mobile money service provider, or comparable regulated business. Members of an account are able to employ it to send and receive remittances, store money, manage money, pay bills, acquire credit, and make purchases. In developing economies, the gender gap in account ownership declined from 9%age points to 6% points between 2011 and 2021. In developing economies, 68% of women and 74% of men possess a bank account in 2021. Only seven economies, including China and India, provide home to more than half of the world's unbanked adults. Because of their sheer size, China and India contribute to an important proportion of the worldwide unbanked population (130 million and 230 million, respectively), whilst having high rates of bank account ownership. Latest reports from NITI AAYOG Women's Role in India's Financial Growth Story: The article titled "From Borrowers to Builders: Women's Role in India's Financial Growth Story" was published by NITI Aayog on third March, 2025. In line with an analysis released by Shri B.V.R. Subrahmanyam, CEO of NITI Aayog, more Indian women are actively reviewing their credit scores and asking for credit. Growing financial awareness has been demonstrated by the 27 million women who were maintaining a check on their credit as of December 2024, a 42% increase from the year previously. **Micro Save** Consulting (MSC), NITI Aayog's Women Entrepreneurship Platform (WEP), and TransUnion CIBIL have jointly released the report. The Women Entrepreneurship Platform (WEP) continues to **endeavour** on establishing an atmosphere of hospitality to encourage market associations, mentorship, access to financing, and financial expertise. It involves an integrated effort to give people fair financial access. Developing this momentum involves legislative measures that address structural challenges as well as the role of financial institutions in establishing inclusive products that are particularly suited to women's needs (Bhatia et al. 2024).

The Financing Women Collaborative has been created under the WEP to meet this objective. We have opportunities for additional stakeholders throughout the financial sector to join FWC to promote this mission. One way of guaranteeing employment opportunities for women entering the workforce in India are to promote women to embark on privately held corporations," noted Anna Roy, Principal Economic Advisor for NITI Aayog and Mission Director for WEP. Furthermore, it's an achievable method to boost fair economic growth. Allowing women to embark on independently owned businesses would result in more women accessing the labour force while generating jobs for 150–170 million people in general. In line with the report, women's representation of the global self-monitoring base went from 17.89% in 2023 to 19.43% in December 2024. Compared to women in metro areas, more women from non-metro areas frequently inspect their credit, with an increase in 30% in metro areas and 48% in non-metro areas. In accordance with the report, women's representation of the entire self-monitoring base climbed from 17.89% in 2023 to 19.43% in December 2024. Compared to women in metro areas, more women from non-metro areas regularly examine their credit, with an improvement of 30% in metro areas and 48% in non-metro areas. In 2024, 49% of all self-monitoring women were located in Maharashtra, Tamil Nadu, Karnataka, Uttar Pradesh, and Telangana, with the southern region possessing the largest number at 10.2 million. Compound annual growth rates (CAGR) with active women borrowers over the five years before that were highest in northern and central states, which includes Madhya Pradesh, Uttar Pradesh, and Rajasthan.

6 Expanding the Financial Sector's Reach and Impact

CGAP is focusing on expanding the financial sector's competence and availability for helping women across different incomes, life stages, and risk tolerances in order to hasten progress. For this explanation, we continue to fund ground-breaking studies to understand concerning young women fewer than 25 and women Nano-entrepreneurs. We give them vital information on the obstacles preventing them from obtaining maximum advantage of financial services or from accessing the governing financial system. Although many development professionals acknowledge the need of supporting rural women whose work in agriculture, there is a need for practical business models to reach this audience. Keeping records of how women as agents can benefit themselves, their households, their communities, and digital financial service (DFS) providers has been one of the most significant components of CGAP's work. Nevertheless, they frequently feel extremely restricted by their immediate environment and the business models that employ them. We lack the capacity for improvement what we cannot measure. However, there is still a shortage of market-level knowledge regarding women's financial services exchanges. A game-changer is supply-side gender-disaggregated data (SGDD), which examines how financial institutions approach women. This data might help legislators in modifying rules that will assist women and financial institutions while enhancing the design and delivery of their offerings if it has been made broadly available and used appropriately. As a means to further accomplish this, CGAP is bringing together supervisors and financial regulators from emerging and developing economies to exchange best practices when developing SGDD frameworks. Our objective is to speed up the adoption of policies that foster inclusiveness in financial systems through supporting cross-country learning. In an attempt to develop a financial system that is gender-equitable, CGAP is investigating how governments could gather along public, business, and civil society collaborators in the financial sector to develop ambitious, focused on objectives coalitions. The purpose we have in preserving and highlighting these group initiatives is to motivate and inspire others to do something similar. By doing this, we could set up the conditions needed by women to flourish in their lives, increase their companies and means of survival, and improve the resilience of their households.

The objective of women's financial equality necessitates stronger collaboration deeper engagement, and creative thought. CGAP has committed itself to making sure that financial inclusion results in authentic, persistent empowerment for women internationally through promoting access, upgrading data, confronting gender norms, and forming networks. Women are increasing establishing financial independence through self-monitoring and credit management, building on achievements in financial access. This illustrates their developing financial expertise. The numbers provide an accurate representation of women assuming control of their financial destinies. About 27 million women were actively monitoring their credit by the end of 2024, a staggering 42% increase from 2023. The evolution is reflective of a bigger cultural and economic a revolution, as women increasingly perceive credit health as an effective instrument for impacting their **financial** futures. With the main objective of preserving a girl child's future, the Sukanya Samriddhi Yojana (SSY) scheme came into effect as part of the Beti Bachao Beti Padhao campaign. On January 22, 2015, the Indian government launched a social campaign to address the root cause of its steadily dropping child sex ratio. "Educate the girl child, save girls" is the message of the Beti Bachao Beti Padhao (BBBP) system. The Ministry of Women and Child Development, the Ministry of Health and Family Welfare, and the Ministry of Human Resource Development have been working together on supervising this comprehensive project. This program also aims to give financial access to empower women of the society. The objective of SSY is to tackle an essential problem associated with girls, especially the expenses of marriage and education. The organization's primary objective is to make sure that girls in India experience a bright future by helping their parents prepare money for their child's proper schooling and carefree marriage. For this explanation, SSY launched the Sukanya Samriddhi Account. SSY's maturity period is 21 years from the account opening dates or, in the event she marries after attaining 18, after 18 years. Contributions can only be made for the primary 15 years, unfortunately. The SSY account will thereafter continue to accumulate interest until its maturity.

The Ministry of Women and Child Development launched an online marketing platform for helping female entrepreneurs. Increases women's involvement in e-commerce markets and digital payment mechanisms. By presenting women entrepreneurs and makers to digital the financial sector, their opinions on financial inclusion are enlarged. Women who operate in the unorganized sector are given the opportunity to make investments to pension funds to be part of the Atal Pension Yojana (APY) so they can to plan for retirement. The Women Entrepreneurship Platform (WEP) is an NITI Aayog platform designed to assist women entrepreneurs financially through the provision of credit, mentorship, and business support. The active measures taken by the Indian government to encourage women's financial inclusion were an important step in toward the objectives of economic empowerment and gender equality. These policies establish a foundation for a more inclusive economy through providing access to digital platforms, skill development, entrepreneurial credit, and financial institutions. Financial empowerment for women is a development need of assistance, rather than simply a women's issue. India may accomplish its ambition of inclusive and sustainable growth if women are adequately integrated into the financial system.

7 Challenges of Women Empowerment through Financial Inclusion in India

Patriarchy structures: A socio-ideological concept termed patriarchy maintains the men fathers, siblings, husbands, kids, and others considered superior to women in the family. A further approach to define it is as an institutional framework where men, or patriarchs, command, oppress, and exploit the benefit of women. Scarcity and poverty aren't the only causes of women's discrimination; even wealthy households take benefit of their daughters by denying them their property and land rights. In an attempt to remedy this disparity, the Indian government created a gender-progressive inheritance law; yet, parents consistently rejected their daughters their rights on the assumption of emotion and in compensation for increased dowries and academic achievement (Roy et al. 2023). The abusive treatment of women, which begin in the household of her parents, extends over into her husband's back there, where the arranged unequal distribution of power that developed from patriarchy further weakened her status. Men take over her sexuality, reproduction, and production.

Psychological Factors: Many women lose confidence and self-esteem and approach opportunities with a nervousness of failure for the obvious reasons indicated under the prior section According to an experimental study, women in the lower income group are more inclined than men to be risk mindful as they consider the adverse effects of not being able to settle their financial obligations.

Low Income Wages: Although gender and income inequality have been addressed separately in the literature, they cannot be separated because they continue to interact through disparities in opportunities and outcomes, which are primarily caused by differences in social structures, education, financial access, and personal viewpoints. A woman's disadvantaged compared navigating and bargaining durability, in combination with patriarchy's bias and her own diminishing self-esteem, forces her to enter into social contracts where she may receive fewer rewards and perks than males to complete the same work (Neguse et al. 2022).

Low Financial Accessibility: Women that have access to bank accounts, savings objectives, and other financial resources may be more equipped to control their earnings, personal consumption, and business expenditures (Siddik 2017); individuals who lack these resources are pushed toward obscurity. This was demonstrated by an experimental study performed in India, where demonstrated that women were incapable to establish a business simply because of financial constraints and was incapable to continue operating it due to savings constraints (Brudevold et al. 2025). A woman is subject to prejudicial views that undermine her self-esteem as she endeavours to advance in the male-dominated society, catching her in a cycle of low education, low employment, and low pay. She is additionally denied access to formal finance, including credit, deposits, insurance, payments, and other risk management services (Chakravarty et al. 2025).

Ethnicity: Recent study has demonstrated that ethnicity is an important consideration that must be taken into account while boosting FI in women. Nevertheless, the study found that women utilized benefit from microcredit, it additionally determined that there was some cultural prejudice since Indian women were twice as probably to be obtained for a loan as indigenous women, but indigenous women were only 1.5 times as probably to receive a permit as men. A subsequent study (Kaur and Kapuria 2025) showed support for this the sense that suggesting that households headed by women from lower socioeconomic the beginning were less inclined to be unable to get institutionalized financing. It also suggests that ethnic studies could provide significant knowledge on FI for women.

Road Ahead for Financial Inclusion of Women in India: A report released by the World Bank presents recommendations to eradicate obstacles that restrict women from accessing formal financial services. Following are some notable potential areas for financial sector authorities, together with example of where economies have made significant progress in these areas:

Moving beyond gender-neutral to gender intentional policies: Gender-intentional policies, as in contrast to gender-neutral the ones: Specific challenges that women experience in acquiring and using DFS could potentially be tackled through gender-intelligent policy and regulatory frameworks. To bridge that gap, it is necessary that we recognize alternative identification documents, implement tiered customer due diligence specifications based on risk profiles, and include a vulnerability lens into laws protecting consumers. The negative experiences that women have with DFS can be proactively acknowledged and diminished by legislative bodies incorporating a vulnerability lens into rules and regulations.

Applying sex-disaggregated data (SDD) to shed light on the coming years: More data delivers additional insights into the financial behaviours of different demographics. Disaggregated supply-side and demand-side statistics are necessary for financial sector stakeholders to inform rules and regulations.

Bringing about change via empowerment and education: It is necessary to develop productive initiatives towards digital financial literacy. Setting aside DFL could be aided by interaction amongst important stakeholders with a national financial inclusion strategy. For bettering the ability of women for employing DFS, gender-specific objectives, multidimensional methods of instruction, and context-sensitive approaches are necessary. Initiatives that originate in the community may give tools and possibilities to provide digital literacy to women along with other communities that are underrepresented (Dahlum et al. 2022). Through trained ambassadors and women's cooperatives, Indonesia's female digital ambassadors take advantage of neighbourhood ties. They encourage women in obtaining and acquiring access to DFS furnished by branchless financial services agents.

Building Infrastructure for all: Developing infrastructure for everybody: Regulators should to establish links which provide women access to the financial assistance they demand. Women may participate with greater convenience because of readily accessible infrastructures include collateral registries and interoperable digital payment systems. For improved interoperability, policymakers need to ensure gender-inclusive infrastructure design and promote engagement among financial service providers. Senegal's ID reforms, which dropped additional proof of residency requirements for married women applying for ID cards, ought to demonstrate authorities' certain things. Such ID legislation enables it to be easier for women to secure registration. India's experience with mobile money interoperability minimized expenses, increased competition, and enhanced functionality by eliminating procedural requirements.

Increase women presence in Leadership position in Financial Institutions: An industry which strives at assisting women but lacked female executives would be missing out on complementary opinions. It had been demonstrated that gender diversity in leadership encourages sustainability and brings up new avenues for innovation. It is suggested that the presence of women in the sector and in leadership roles in regulation be regularly and publicly reported. This information could be a significant contribution to raising public awareness of the inadequate representation of women in these institutions, measuring improvements in the financial services industry, and enable the general public to communicate their preference for more diverse business entities, if they choose to do so. The percentage of women in positions of authority could rise if incentives are introduced to promote organizations with female executives.

Ensure digitization of Government Payments: Submitting payments (P2G) or accepting payments from governmental offices (G2P) is a prevalent engagement for women without accounts. The majority of public utility payments and fees for access to public services are still made in cash in developing nations; requiring electronic payment might provide a first experience with a digital payment system. Furthermore, governments frequently deliver a number of public financial subsidies, such as the increasing utilization of cash transfers, to solely financially disadvantaged citizens.

8 Conclusion

Empowering women remains one of the most critical imperatives in the drive toward equitable and sustainable economic development, and financial inclusion has emerged as a central pillar in achieving this goal. This research examined the nexus between financial inclusion and women's empowerment, with a strong focus on the Indian context, where a confluence of governmental initiatives, digital innovation, and grassroots movements has begun to narrow long-standing gender gaps in access to finance. The findings illustrate that financial inclusion goes far beyond access to banking infrastructure it is about equipping women with the autonomy, knowledge, and tools to make independent financial decisions, invest in their futures, and contribute meaningfully to the economy. While policy instruments such as the **Jan Dhan Yojana**, **Unified Payments Interface (UPI)**, and other government-backed initiatives have markedly increased the number of women holding bank accounts, there remains a substantial gap in actual usage, access to formal credit, and financial literacy.

A recurring theme in this study is that the existence of financial infrastructure alone is insufficient. Many women still face multiple layers of exclusion due to socio-cultural norms, limited digital access, and psychological hurdles such as low self-confidence and fear of financial failure. These barriers are especially pronounced in rural and low-income communities, where mobility constraints, lack of collateral and limited financial education impede their participation in the formal financial system. Encouragingly, a growing number of Indian women are beginning to take greater control of their financial trajectories. The surge in credit score monitoring and increased uptake of microcredit and digital wallets signals a shift toward financial self-reliance. However, this progress must be reinforced through targeted, gender-sensitive policies. Addressing systemic issues such as property rights, gender-based financial discrimination, and limited female representation in financial leadership roles is essential for deeper and lasting empowerment. Programs like the **Women Entrepreneurship Platform (WEP)**, **Atal Pension Yojana (APY)**, and **Sukanya Samriddhi Yojana (SSY)** are examples of how state-led financial mechanisms can support women across various stages of life—from education to entrepreneurship and retirement planning. Yet, the impact of these efforts is contingent upon widespread financial literacy, robust digital infrastructure, and institutional willingness to design inclusive financial products. Financial inclusion plays a pivotal role in advancing the **Sustainable Development Goals (SDGs)**, particularly those related to poverty alleviation, education, health, and gender equality. By granting women access to tools for savings, investment, and insurance, nations can foster stronger household economies, boost community resilience, and promote inclusive economic growth. India's progress over the past decade is promising, but closing the remaining gaps requires a coordinated, multi-stakeholder approach.

Elevating female leadership within financial institutions to ensure more equitable policy design and service delivery. Furthermore, leveraging technology for government-to-person (G2P) transfers and digital public utilities can help integrate women into the formal economy, especially those currently excluded due to geographic or socio-economic factors (Esmaeilpour et al. 2023). Strategic policy innovations, when backed by data, community engagement, and inclusive design, have the potential to make financial systems more responsive and accessible to all women, regardless of their background. In essence, this study confirms that financial inclusion is both a means and an end: a means to uplift women economically and socially, and an end in itself as a marker of inclusive progress. For India, embedding financial inclusion in its broader development narrative offers not only a pathway to women's empowerment but also to a more resilient, equitable economy. Achieving this vision demands on-going collaboration between government, financial institutions, civil society, and the women themselves, who are increasingly stepping into their roles as informed, empowered participants in the financial landscape.

References

Arshad A (2023) Nexus between financial inclusion and women empowerment: evidence from developing countries. Gender Manage: Int J 38(4):561–580

Banna H, Mia MA, Nourani M, Yarovaya L (2022) Fintech-based financial inclusion and risk-taking of microfinance institutions (MFIs): evidence from sub-Saharan Africa. Financ Res Lett 45:102149. https://doi.org/10.1016/j.frl.2021.102149

Bharti N (2021) Role of cooperatives in economic empowerment of women: a review of Indian experiences. World J Entrepreneurship Manage Sustain Dev 17(4):617–631

Bhatia S, Dawar G (2024) The impact of financial inclusion on social and political empowerment: mediating role of economic empowerment. J Knowl Econ 15(3):13727–13744

Bhatia S, Singh S (2020) Empowering women through financial inclusion: a study of urban slum. Vikalpa 44(4):182–197

Chakravarty I, Keserwani H, Singh P (2025) The role of Fintech in empowering women: a catalyst for financial inclusion and economic growth. In: 4th Annual international research conference and doctoral workshop. IIM Lucknow, p 91

Dahlum S, Knutsen CH, Mechkova V (2022) Women's political empowerment and economic growth. World Dev 156:105822

Ditta H, Thakkar A, Asmelash A (2025) The prospects of traditional branch banking in the era of digital dominance. In: Strategic workforce reskilling in service marketing, p 25

Esmaeilpour Moghadam H, Karami A (2023) Financial inclusion through FinTech and women's financial empowerment. Int J Soc Econ 50(8):1038–1059

Gonzales MC, Jain-Chandra MS, Kochhar MK, Newiak MM, Zeinullayev MT (2015) Catalyst for change: empowering women and tackling income inequality. Int Monetary Fund 15:1

Hendriks S (2019) The role of financial inclusion in driving women's economic empowerment. Dev Pract 29(8):1029–1038

Lerner M, Pines AM, Schwartz D (2010) Gender differences in entrepreneurship: equality, diversity, and inclusion in times of global crisis. Equality Divers Incl Int J 29(2):186–198

Nguse T, Desalegn G, Oshora B, Tangl A, Nathan RJ, Fekete-Farkasne M (2022) Enhancing women economic empowerment through financial inclusion: evidence from SMEs in Ethiopia. Polish J Manage Stud 25(1):270–291

Parwez S, Patel R (2022) Augmenting women empowerment: a systematic literature review on microfinance-led developmental interventions. J Glob Responsib 13(3):338–360

Seguino S (2010) The global economic crisis, its gender and ethnic implications, and policy responses. Gend Dev 18(2):179–199

Siddik MNA (2017) Does financial inclusion promote women empowerment? Evidence from Bangladesh. Appl Econ Finan 4(4):169–177

Thakkar A, Fentaw AA, Hirpo HD (2025) ChatGPT in academia and research: a comprehensive review of integrating AI in higher education. Multimodal Generative AI 25:233–251

Verma N, Bag A, Pant LM, Kumari N (2024) Analysis of economic development of financial inclusion and women empowerment. In: The economics of financial inclusion. Routledge, pp 224–242

A Qualitative Study of Digital Education's Contribution to the Reduction of Gender Inequalities in Developing Nations and Underserved Communities

Arvind Nain and Archana Singh

Abstract This qualitative paper looks into the importance of digital education in bridging gender gaps among developing countries and marginalized communities. Driven by an increasing global interest in using digital platforms to eliminate historical educational inequalities, this study investigates how both traditional and informal digital learning efforts address educational issues faced by women and girls. The study uses an IPA (interpretive phenomenological analysis) to interpret participants' lived experiences and insights from 612 educational participants from rural and urban institutions in three developing nations (Guatemala in Latin America, Indonesia in Asia, and Nigeria in Africa). The study assesses how socioeconomic factors affect accessibility to digital education and focuses on the way digital platforms are used to alleviate systematic gender inequality. Furthermore, a comparison of pre- and post-digital education implementations sheds light on progress toward closing the gender gap, highlights remaining limits, and makes policy recommendations to stakeholders. All the interviewees were willing to participate in the study, as they were motivated by the purpose of the study, and the research processes of the informed consents, as well as the protection of the data, guarantee that the interview data were collected and conducted in an ethical and proper way. The implications of this study reveal that digital education can be a strong instrument for democratizing learning and promoting an emancipatory environment for women, but also illustrating the impossibility to make it a totally successful and common process for women. This article offers some practical knowledge on best practice in digital education deployment in underdeveloped communities by integrating lessons learned from recent as well as previous studies (Anderson and Burnett In Int J Dig Learn 8(1):5–24 2020; Chen et al in Int J Gender Stud Educ 12(1):55–72 2019) and insights from practitioners. There: Advice for educators and officials who seek to build healthier educational

A. Nain (✉)
Graphic Era deemed to be University, Dehradun, India
e-mail: arvindsnain@gmail.com

A. Singh
School of Management, Graphic Era Hill University, Dehradun, India

A. Singh et al. (eds.), *Human-Centered Sustainability: Bridging Well-Being, Gender Equity, and Technology*, World Sustainability Series,
https://doi.org/10.1007/978-3-032-22020-2_8

ecosystems. This work is a part of the ongoing debate regarding fair education and offers a research-based approach to how digital technology should be used to solve the gender difference.

Keywords Digital education · Gender disparities · Underserved communities · Developing countries · Educational policy · Socio-economic factors

1 Introduction

The increasing acceptance of digital technologies in education shows a promising opportunity to overcome persistent gender inequities. Many developing countries have highly confirmed gender-based educational hindrances as a result of socioeconomic constraints, cultural biases, and limited access to regular education. However, recent shifts toward digital education, which this study defines as the formal learning platforms (such as online courses in higher education) and informal platforms (such as community-based educational applications and mobile-assisted learning initiatives), have begun to transform traditional educational paradigms. This research investigates the complex connection that remains between digital education initiatives and gender equity, focusing on locations with remarkable socioeconomic limitations and past gender disparities.

The rationale of conducting this investigation is diverse. First, there is a fundamental need to comprehend how digital educational platforms can be effectively used to eliminate gender inequities in education systems. Previous research has primarily focused on quantitative measurements of accessibility and rates of enrolment (Johnson 2021; Kumar and Patel 2019), but there is a lack of qualitative accounts that disclose personal narratives along with real-life experiences, which could influence better student-centered policy frameworks. This study bridges the gap by using a qualitative methodology to gather views from educators, administrators, and community members about the implementation and effect of digital education.

The study also looks into the cultural and socioeconomic constraints that specifically impact female students in developing countries. According to scholars like Lee, Rahman, and Gomez (2018), depending on the local context, digital tools have the ability to either worsen or lessen educational disparities. Therefore, by involving voices from three different geographic locations—Nigeria, Indonesia, and Guatemala representing a range of cultural and economic backgrounds, this study seeks to shed light on the ways in which digital education addresses the gender inequities. This paper is organized into multiple sections. Recent developments and research findings on gender equity and digital education are summarized in the literature review. The qualitative approach used, including the recruitment of participants, interview details, and the use of an interpretative phenomenological analysis framework, is described in full in the research methodology. With regard to gender disparities, the data collection and analysis parts provide a methodical comparison of the implementation of digital education before and after. The research findings' implications for the practice and policy of education are further examined in the discussion section. Lastly, suggestions are made for different stakeholders with the goal of optimizing the potential for revolutionary change brought about by digital education. The study incorporates ethical concerns to guarantee that the participant's confidentiality and security of data procedures are strictly upheld. This paper aims to provide a complete examination of how digital education portals are addressing gender obstacles in various cultural contexts. This technology examines the interconnection between local socio-economic conditions and traditional gender roles, and how these factors unlock or prevent the effective use of digital solutions in marginalized societies. In addition, the results of this study can serve as a base for other studies, as well as contribute to educational practices and politics globally.

2 Literature Review

The rapid expansion of digital education over the past decade has long been inspired by warm debate in educational circles about its role in removing educational inequalities. Reviewed based on a series of studies, the results of this research try to offer a nice summary of research on 'online' education and gender.

2.1 Digital Education: Definitions and Scope

The term digital education refers to the use of digital technologies to support learning and teaching (Smith 2019; Thomas and Nguyen 2021). The material uses some online tools, such as virtual learning platforms and smartphone apps for instructional purposes. Digital education is not only supplanting the conventional forms of classroom instruction, but also provides an opportunity to completely transform the landscape of education. It is a catalyst for systemic change, boosting system-wide reach and quality of education, according to Brown and Lee (2020). They infer that online education can be a means to the democratization of education and that it brings such notable learners together in ways previously not possible.

Digital tools can help educators create a more engaging and personalized platform for learning that works best for students. It is more than the embrace of technology; it's indicative of a broader assault for educational justice and transformation." Hence, digital learning is a vital intervention in addressing contemporary issues of education (Smith 2019; Thomas and Nguyen 2021; Brown and Lee 2020). In poor countries, digital education represents an exciting opportunity to expand access to learning. Yet significant issues persist: outdated hardware infrastructure, problematic online connectivity and lack of technology skills (Garcia et al. 2022). Research shows that the socioeconomic context in which digital education is used, greatly influences its performance. But unless these basic issues are addressed, a lot of people will continue to find themselves on the wrong side of the digital divide. In the women's case in particular, the digital divide is usually due to historical gender disparities, making access to digital education significantly difficult. The combination of these constraints highlights the need for targeted interventions that consider technological and social factors. Finally, though digital education is a great area of potential to explore, in underdeveloped countries, it is meant to depend on facing larger structural issues (Garcia et al. 2022; Fernandez and Alkire 2020).

2.2 *Gender Disparities in Education: Global Trends and Context*

Although efforts have been made to eliminate gender gaps in education, inequalities persist worldwide, especially in developing countries (Khan et al. 2020). Most of the existing research on these issues has focused on quantitative indicators such as enrolment ratios, graduation rates and test scores results (Oliveira 2018). Yet, qualitative studies have started to uncover the complexity of these gaps by addressing cultural, economic, and political issues (Davies and Miller 2019).

Rodriguez et al. (2018) contend that gender inequality in education is exaggerated by the convergence of issues such as poverty, remoteness and deeply entrenched cultural norms that prevent girls from attending school. On this point, digital education offers promising evidence by offering flexible, adaptive learning environments tailored to non-dominant individuals (Wang and Li 2021). One of the key opportunities available is to circulate content in local languages on mobile platforms, overcoming linguistic and access barriers that often prevent women in rural areas from attending conventional schools and colleges.

Digital resources may be used to nurture inclusive opportunities for learning in the sense that they can dismantle environmental and cultural barriers to access. But it is not a simple issue and it is based on proper management considering the uniqueness among these groups. As Wang and Li (2021) proposed, when organized properly digital education can be used as an option to counterbalance the persistent disadvantageous educational opportunities of women in resource-poor settings.

Some recent research (Patel and Sharma 2020) suggests that digital education programmes have had some success in increasing literacy and numeracy rates amongst females in certain locations. But their greater impacts are often locally limited early on by socio-economic factors and many projects plod along without realizing their full potential. This contradiction between the promises of digital education and the challenges of making its implementation operational is central to the provocation of this paper. The current literature highlights the importance to conduct in-depth qualitative research to explore how digital tools are being used to meet the specific educational needs of women, especially in culturally unresponsive, socially constraining environments. Understanding these factors is vital if educational practice, which is more effective and more inclusive, is to move beyond superficial change (Patel and Sharma 2020).

2.3 *Digital Education in Developing Countries: Successes and Challenges*

The use of e-learning in underprivileged countries has led to both successes and persistent problems. Studies have indicated that digital learning platforms have certainly opened up access to education in rural areas and were significantly instrumental in empowering girl learners who would have been otherwise marginalized (Adeyemi et al. 2019; Nain et al. 2024; Rahman and Jones 2021). Elsewhere, community-led media work has plugged major gaps by delivering vital educational services where local infrastructure has failed (Singh and Verma 2022). However, there are still challenges to overcome. Some recurrent obstacles this sector faces, such as the weak infrastructure, the unreliable access to the internet and the absence of deep training for the teacher, persist to keep it from evolving (Mendez and Costa 2020; Qureshi et al. 2019). Far more formidable barriers are posed by deep-seated socio-cultural notions of female education that are usually neglected in policy deliberations. These cultural factors significantly influence how individuals consume and utilize digital educational resources. A comparison of rural Nigeria and urban India (Okoro and Patel 2020) suggests these disparities: urbanized areas may have the advantage of technological infrastructures as rural ones grapple with to physical access to technologies and current societal norms that discourage female participation in education. Findings such as these demonstrate the importance of addressing not only technical and logistical gaps, but also deeply embedded cultural issues, to ensure digital education's broad, equitable, and meaningful reach. Unofficial education initiatives have taken on a domineering role in areas where there are no official infrastructures. Self-paced learning which can be driven by peer-to-peer networks and community-owned E-libraries, offers options of learning that are inclusive and adaptive (El-Gamal and Osei 2021). The grassroots initiatives do often cover large educational disparities, especially in deprived communities. Even though these are promising, there are still doubts regarding the long-term sustainability and impact of social media. Their absence of formal certification and standardised curriculum downplays their long term recognisability and effectiveness in addressing structural educational disparities. These platforms need to be high-quality, and they need to be reliable in order to be sustainable long-term solutions to educational inequities (El-Gamal and Osei 2021).

2.4 The Socio-Economic Factors Affecting Digital Education Access

It's often seen that the effectiveness of digital education programs is influenced by socioeconomic issues. The same is shown consistently in the earlier studies too that, there is an impact on students performance in education due to expenditure on technology, cultural expectations, and educational infrastructure made available to them (Bennett and Kumar 2018; Lopez et al. 2022). Its seen in many developing nations that the low-income groups find it so unaffordable to spend on digital gadgets and internet accessibility due to its expensive service cost. As a result these cost restrictions often restrict the accessibility and also effect the opportunities for basic digital learning. In most societies, traditional stereotypes mean that home and caregiving duties fall largely on girls and women, restricting the time they can invest in their studies. This inequitable distribution is not only detrimental to students' participation in school, but also to closing education gaps. Recognizing and addressing these interconnected social barriers will be crucial to developing inclusive digital education strategies that would reach all children, irrespective of their background, their gender.

Female students from poor communities may have to deal with a "double burden" of combining domestic responsibilities with their desire to study and learn (Hasan and Rahim 2021). Digital education provides flexible learning opportunities, as long as it caters to the cultural and social experiences of these students. Without some contextually sensitive dressing up of this interface in programs for women, many women are going to be relegated to the digital platform lower echelons and their potential benefits may not fully be realized. Part of this could be to help support networks and community development projects and to develop locally relevant material to help alleviate some of the burden that current priorities on economic factors create. These aspects not only increase accessibility but also promote long-term engagement for female students (Zhang et al. 2020). Linking digital learning with the needs of poor and marginalized populations, particularly women, in terms of their own unique lived realities, can ensure more equitable educational outcomes and lifelong empowerment.

2.5 Comparative Analyses of Pre and Post Digital Education Implementation

There are a series of case studies in the literature that illustrate the contrast in performances from educational systems before and after the introduction of digital technology. Morales et al. (2021) found that the adoption of digital education platforms in Guatemala has led to marked increases in student engagement and greater flexibility in the way the curriculum is delivered. Similarly, a study in Indonesia noted a dramatic increase in the level of involvement of female students under an educational program fuelled by technology (Putri and Santoso 2019). Such positive results illustrate the potential revolution of digital education in expanding access and diversity.

But these accomplishments are not universally acknowledged. Disparities in location, facilities and infrastructure infrastructure persistently undermine educational outcomes, often to the detriment of residents of rural areas. In some respects, digitalization has been beneficial, but it has also inadvertently exacerbated the digital divide by centralizing digital resources in urban locations and leaving behind rural learners (Ahmed and Rahman 2022). These regional variations showcase how digital education can be an enabler or a disabler. Against this background, the current work draws upon a comparative analysis of digital education in Nigeria, Indonesia and Guatemala to explore more nuanced gendered-impact. This article explores four scenarios in which digital interventions could either challenge or reinforce digital inequality. It is evident through this lens that having access to digital learning is dependent on not only digital infrastructure, but also the extent to which digital technology is adapted to the socio-cultural and infrastructural exigencies of the home.

The available literature offers a strong foundation for exploring the potential and pitfalls of digital education in addressing gender inequality. Despite this progress, continuing obstacles exist which hinder access, particularly the modification of online resources to meet the differing requirements of female students in developing nations. Many of these still do not adequately cover the social, cultural and economic factors which impact on engagement in education. Extending from these findings, this proposal aims to address this deficit by engaging directly with key education stakeholders. It uses qualitative methods to describe the complex contextual issues of digital education, focusing on both the potential for empowerment that digitization brings and the obstacles that persist to prevent its effective implementation. By combining the first-hand views, this research contributes to how digital tools can be applied to support gender equality in education, while also recoginising the crucial gaps which are required to be focused on for sustainable impact.

3 Research Methodology

The present study is a qualitative research endeavour to explore the real-life experiences of the educational community participating in digital education projects in three developing countries—Nigeria, Indonesia and Guatemala. At the heart of this method is the application of the Interpretative Phenomenological Analysis (IPA) framework, which is centred on the examination of how people make sense of personal experiences. Through the use of IPA, the research aims to "show" the nuanced, lived-experiences of teachers, principals, and policymakers by allowing them to tell us about their own experiences of the way they view the impact of digital education in their own location environment. This approach provides an insight not only for opportunities, but also these stakeholders in the challenges that come outside the surface.

3.1 Research Design and Rationale

Using the Discovery Qualitative Design, the study examines insight into questions to 612 members of educational institutions from both rural and urban areas. This is a type of analysis that is ideal for complex social events such as gender inequalities in digital education, which does not accept simple numerical measurements. Unlike quantitative methods, which normalize and reduce vivid experiences for numerical tendencies, qualitative research provides complete praise of ideas, feelings and meanings (Creswell and Poth 2018; Smith and Osborne 2020). This is especially important when dealing with the consequences of cultural, social and economic elements. This study is reported by analysis related to explanatory phenomena (IPA), which is related to how people create the importance of life experiences. Through IPA, it is "involved in ways that stories of individuals are careful or widespread discourses about the benefits and deficiencies of digital education in gender capital. This process provides significant literature in relation to both digital learning and limited factors with a gender perspective.

In addition, the study's qualitative nature allows intensive comparative analysis between three different sociocultural references (Nigeria, Indonesia and Guatemala). "All of these countries face different opportunities (and) challenges in the context of digital education, based on cultural norms, infrastructure and socio-economic situation. Through the participation of a wide range of educational interests such as administrators, teachers, decision-makers and social leaders, studies have a comprehensive approach to achieve a comprehensive attitude for the impact of digital. It provides value to analysis by finding equality and differences between users of the majority of digital initiatives. Such depth can be remembered by quantitative methods (Crosswell and Poth 2018; Smith and Osborne 2020), which is important for capturing intruders of subtle shades of inequality in gender in education.

3.2 Participant Recruitment and Sampling Strategy

By adopting a purposive sampling approach, we were able to invite 612 participants who have a clear understanding of digital education projects, which was conducive to enhancing the relevance and context of the data obtained. Sampling was done across different rural and urban educational settings within Nigeria, Indonesia and Guatemala to obtain a wide array of perspectives. Recruitment involved contacting educational and community-based organizations within each site. The study was also assisted through a snowball sampling method, which enabled the researchers to access a wider representative of the population by participants referring the researchers to other key informants.

The sample design was astutely crafted to ensure a representative gender distribution and to take into account those stakeholders involved in the planning, execution or evaluation of digital education. It included school principals, teachers, government officials, and community leaders. A priority throughout the research journey was the maintenance of ethical integrity. All subjects consented to participate in the study, and their anonymity and personal data were protected. All methods were in accordance with relevant guidelines and regulations as per the institutional review board and international ethical standards for research (Ethics in Education Review, 2019; Johnson et al.

This study intends to provide a holistic picture of digital education across various socio-cultural and geographical contexts by including voices from both female and male teachers, and various educational roles. This strategy increases the trustworthiness and richness of the results, particularly in the examination of gender-related issues in online learning environment.

3.3 Data Collection Procedures

The data for this study was largely collected in semi-structured interviews, which is an optimal means to build up a rich multi-faceted description of the participants' lived experiences. Interviews were 45–75 min in duration and were carried out either face-to-face or through encrypted digital means when face-to-face interviews were not feasible. All interviews were audio taped (with participants' permission) and transcribed (verbatim) for data analysis, to ensure the originality of responses is maintained.

The interview guide was designed to address the following key themes in depth including: reflections on the extent to which digital education has reduced gender-based barriers, observable shifts in student enrolment and participation across female learners, barriers in the implementation of Education technology tools, and recommendations for policy and programming moving forwards. The use of open-ended questions ultimately led to detailed and personal storytelling experiences from participants and therefore a deeper understanding of formal and informal digital learning spaces.

Researchers also collected observational information and field notes from site visits to schools in addition to the interview data. These additional data sources provided important context about the physical, social and cultural environments in which digital education occurs. For example, contextual data informed the physical arrangement of classroom, available equipment and interaction between the students and staff and enriched the analysis.

All data were filed in a secure place and were de-identified to protect the confidentiality of the participants. The research was conducted adhering to ethical standards for ethics (data protection) by following guidelines (Ethics in Education Review 2019; Johnson et al. 2021).

3.4 Data Analysis

The qualitative data in the present research were analyzed using the Interpretative Phenomenological Analysis (IPA), a method best suited for examining the way individuals make sense of first-person, subjective, and lived experiences. This involves an in-depth but flexible sequence of analysts' activities beginning with line by line reading of transcripts to familiarise the researchers with the data and facilitate empathy with participants' accounts. This first engagement helps create a context, or baseline, from which to understand various points of view before deeper analysis ensues (Smith and Osborn 2020).

Following reading, the transcripts were coded inductively according to representative quotations and organised by emerging themes. These themes reflect common ideas and issues regarding digital education effects on gender inequalities. As themes emerged, patterned comparisons were made within and across national contexts Nigeria, Indonesia, and Guatemala to discern both similarities and differences across experiences influenced by sociocultural and infrastructural distinctions.

Qualitative data analysis software was used by the researchers to assist in the processing and analysis of the large amount of textual material. This software supported the sorting of codes, as well as the visual grouping of themes, and encouraged consistent and transparent coding. Special attention was paid to grounding the interpretations in the women's terms, minimizing the possibility of imposing researcher bias, and adhering to the interpretative nature of IPA (Creswell and Poth 2018).

The constant comparative technique was also utilized in the entire analysis. This implied a process of continuously comparing and contrasting individual narratives to identify both converging and diverging experiences with digital education in particular in terms of gendered access, participation and empowerment.

This analytical process helped generate a granular, rich picture of how digital education in multiple socio-economic and cultural contexts is interpreted and lived, as well as supported the study's objective of researching gender equity in more humanistic, participant-influenced ways.

Additionally, all researchers kept a reflective log at every stage of the analytical process in order to increase transparency and lower the risk of bias. The diaries were a key mechanism used to capture the researchers' developing understandings, feelings and analysis, and supported the paramount principle of reflexivity in qualitative research. This technique was consistent with the interpretative phenomenological analysis (IPA) framework of Smith and Osborn (2020), as it recognised researcher as active in making sense of lived experiences presented by participants. The inclusion of reflective diaries increased methodological rigour ensuring that the study findings were credible, reliable and valid. Further, this methodology allowed the study to frame interpretations within participants' voices and contribute usefully to the overall scholarly conversation on gender equity in digital education (Creswell and Poth 2018).

3.5 *Ethical Considerations*

Ethical integrity formed the foundation stone of this study and followed each stage from the recruitment of participants to the presentation of data. Before participating in the study, the participants received a detailed consent form that described the purpose of the study, their participation was voluntary and their oblivion and data privacy would be specifically addressed. Such open legal environments allowed participants to be fully aware of their rights to quit at any time without any unfortunate effects.

Strong data security measures were implemented in order to protect privacy measures. Access to the transcript of digital files and interviews was banned for the group research steam members to be on a securely encrypted server. In addition, false names were awarded to all participants to ensure oblivion in both data analysis and reporting of results. These moral aspects assured that participants' privacy and dignity were seen during the study. Research followed the moral principles approved by an international organization and confirmed through the relevant moral board, Committee de la Investigaci6n Eductiva (2019). These actions, which corresponded to qualitative best practice for research, also strengthened the reliability and reliability of the findings from the study (Creswell and Poth 2018; Smith and Osborn 2020), ensuring that all votes and presentations of the participating experience were handled morality and response.

4 Data Collection and Analysis

Methods for data collection and analysis were important to highlight the complex methods that form inequalities in gender in digital schooling initiatives with low - insival land. The study revealed various approaches from teachers, administrators and social leaders in a semi-composed interview of 612 participants in Nigeria, Indonesia and Guatemala. The explanatory phenomenon analysis (IPA) method was used by researchers to read repeated ties to establish key subjects and patterns with national references. This technique ensured someone's experiences and a deep understanding of socio -cultural determinants affecting them.

This chapter discovered that educational technology has the opportunity to increase learning opportunities, but if it is not implemented inclusive, there is also a risk of further opening the gender difference. For example, in many of the world's least developed countries, teenage girls and young women are significantly less likely to have internet access than are their male peers, limiting their participation in online and digital learning environments. In addition, gender diversity is impeded by cultural norms and security concerns which often restrict girls' access to technology, widening the digital gender gap (Kuteesa and Nsubuga 2024).

Field-level observations complemented interviews as they revealed infrastructure constraints, including limited broadband capabilities and insufficient digital tools in remote areas. These barriers are particularly challenging for girl students, who may already face socio-cultural constraints on schooling. Context-sensitive policies for community engagement are seen as being important for the design of gender-equitable digital education interventions, as proposed in the study. Through centering on women's everyday lives, the research offers valuable perspectives on the complex relationships between technology, education and gender, across various socio-economic locales.

4.1 Participant Demographics and Institutional Backgrounds

Six hundred and twelve respondents including school administrators, officials, teachers, and community members from Nigeria, Indonesia, and Guatemala participated in this study from a balanced sample of rural and urban communities. Nigeria The majority of participants were from government and community schools in rural areas of Nigeria where digital education is in its early phase development. These respondents provided valuable insight into the basic challenges of introducing technology to schools with poor infrastructure.

In comparison, Indonesian participants were predominantly from urban schools in which digital modules were readily taken up so that they could talk of increasing teaching practices as well as the benefits of a more technologically advanced environment. In the meantime, those in Guatemala reported from community education centers that they had direct experience with the implementation of digital learning in under-resourced contexts, underscoring the value of local and homegrown educational approaches.

This multi-site sampling method was very important for the comprehensive comparison study, which explored how specific local conditions affect the influence of digital education and the future development. Digital education, the participants from all three countries said, also had greatly increased female student enrollment in urban areas, where young women are seizing on the new accessibility and convenience. Nonetheless, certain physical/social obstacles have remained and act to block such universal acceptance and utility, especially in rural areas. Common themes that emerged were device cost, limiting gender standards and digital literacy disparities.

The institutionalizing influence of the diverse institutional contexts ranging from well-resourced urban schools to under-resourced rural ones underpinned the broad spectrum of narratives. These perspectives, derived from semi-structured interviews and interrogated through the Analysis of Interpretative Phenomenology (IPA) framework (Smith and Osborn (2020), Creswell and Poth (2018)), reveal that the connection between digital innovation and gender in education is complex.

4.2 Pre-Digital Education Implementation: Baseline Challenges

Before the widespread adoption of digital education, participants identified several constraints that had contributed to gender differentials, with rural areas most likely to experience these. Conventional gender role expectations were a big obstacle, as extensive family responsibilities were laid on most of the female students who were only to spend their time in school, spending no time in their studies. Aside from cultural norms, operational challenges including lack of transportation, and poor infrastructure would have made it impossible for female students to have attended physical education lessons on a regular basis (Okoro and Patel 2020). In interviews, the lack of female teachers in many schools increased gender inequalities through a knowledge of education that supported social norms that viewed education as not for females. And limited community support systems often gave girls little reason and opportunity to pursue education.

Various persistent difficulties plagued female education prior to the digital age, including poor literacy rates, high dropout rates, and limited job growth options. These factors established an intractable cycle of educational exclusion. As one school administrator explained, "For many years, our community assumed that education was only for particular people. The lack of digital technologies further excluded girls from accessing learning resources outside of their immediate environment" (Female Principal, Nigeria, personal communication, May 2023). This remark demonstrates how, previous to digital education, a lack of technology access exacerbated existing gender inequities. Such type of narrations emphasize on the leading issues, which were intended to overcome the efforts in fields of imparting digital education, explaining the crucial role of technology in making available the educational possibilities for underserved communities in different regions.

4.3 Impact of Digital Education: Emergent Themes

The advent of digital education brought a lot of new players and dynamics into the arena. The themes generated, using an IPA approach, presented a number of important themes including increased access, flexibility, culturally appropriate information, and empowered women through self-directed learning. Respondents in Indonesia believe that the shift to online learning platforms facilitated female students to access educational materials in the most convenient time for them and that this in turn reduced the conflict between domestic tasks and academic aspirations (Putri and Santoso 2019). "Digital education has allowed us to break stereotypes on gender that have been deeply rooted in society for millennia, because we can provide children with different perspectives," said leaders in Guatemala, where many parents and educators believe it is taboo to let girls use digital tools. For instance, a teacher in Guatemala reported on a quality digital course that included strong female role models in diverse professional fields, with the programme having a significant positive influence on the goals of female students.

Although they were pleased with the results, participants mentioned concerns about continued problems. Low digital literacy, unreliable access to the internet and the supply of local content are still major issues. Because of inequities in socioeconomics, the advantages of virtual learning did not spread evenly across regions and often left rural areas even further behind urban ones. In Nigeria, said one teacher, "Digital tools have totally revamped our lesson but the high cost of data and gadgets continue to place barriers in the path of many families." (Educator, Nigeria, personal communication, June 2023).

4.4 Comparative Analysis across National Contexts

A cross-country look at feedback from Nigeria, Indonesia, and Guatemala, however, has unearthed some similar and different experiences of what working for digital education initiatives actually entails. he introduction of digital platforms in Indonesia was greatly facilitated by strong communication infrastructure as well as by government programs that provided subsidies for digital devices to female students. The initiatives facilitated smoother and more widespread adoption of digital learning tools, which led to positive outcomes (like fewer female students dropping out). By contrast, in Nigeria and Guatemala numerous challenges, particularly related to infrastructure and cultural resistance, hindered the rollout of digital education.

Results Cut across in Nigeria, there was a varying level of success among participants, with the success often hinging on the existence of technology accessibility and regional differences. This lopsided and inadequate funding led to a lack of reliable access, especially in rural areas. Likewise in Guatemala, community education initiatives have been commended for their ingenuity and flexibility in responding to local needs. But these approaches face barriers to scaling and sustainability, such as lack of government support and inadequate funding.

These differences were further shaped by the socioeconomic context. Although education performance in Indonesia and particularly among girls improved, in Nigeria and Guatemala, deficiencies in infrastructure and cultural obstacles led to slow progress. Notwithstanding these differences, the theme coding process revealed an overall common perspective among participants namely, that digital education has great potential to mitigate gender gaps, but that its actual potentiality relies on surmounting local infrastructure, cultural, and socioeconomic obstacles. This conclusion underlines the need for region-specific policies tailored to the specific needs of each area.

4.5 Data Synthesis and Interpretation

Interpretation from phenomenological approach also identified varied emerging themes, which can add to the understanding about the role of digital education in closing gender gaps. The key topics include:

- Digital education platforms offer flexible and self-driven learning opportunities for women with multiple responsibilities.
- Homegrown programming: localised programming looks at addressing cultural misrepresentations and provides rational role models.
- Infrastructure and socioeconomic challenges like technological access, cost and digital literacy, hinder growth.
- Digital platforms may have the advantage of enab- ling women to access a range of education al materia ls, and insofar as education is associated with socioeconomic empowerment w omen might be expected to benefit from the use of these resources.

This is considered in the context of both past teaching methods, and current movements for inclusivity that are digital based. The comparison findings show that, while digital education efforts have the potential to dramatically eliminate gender-based obstacles, the magnitude of these advantages is heavily influenced by the interaction of technology, culture, and socioeconomic status. The debate that follows delves deeper into its implications on policy and future study.

5 Findings

The study's conclusions, based on data collected from 612 educational instutions in three countries, are divided into four primary themes that indicate the current state and promise of digital education for eliminating gender gaps.

5.1 Increased Accessibility and Participation

Participants in all three national settings agreed that digital education increased educational accessibility. Female students, especially those in cities, have benefited from unrestricted access to information on the internet, decreased travel needs, and the opportunity to customize study schedules around personal responsibilities. According to one Indonesian participant, "online platforms have enabled us to reach a lot of learners who might have dropped due to scheduling conflicts at home" (Teacher, Indonesia, personal communication, April 2022). These narratives are consistent with earlier research (Chen et al. 2019; Wang and Li 2021), which contends that digital education may be an important equalizer in situations with insufficient physical infrastructure. However, in Nigeria and Guatemala, these benefits are offset by disparities in digital infrastructure quality. Although participants acknowledged the improved learning potential, they also noted that uneven internet availability and device pricing still remains significant hurdle, particularly in rural areas. Nonetheless, the general trend shows that increased accessibility is associated with higher rates of enrollment and retention among female students.

5.2 *Shifts in Cultural Attitudes and Empowerment Narratives*

A second important finding of this study was the change that occurred in societal attitudes towards women in education. Respondents said that digital schooling curricula don't just teach young people academic facts but also promote empowering stories that challenge harmful gender beliefs. Such community-based cyber projects in Guatemala, for example, may present instances of accomplishment of local women leaders and thus have also contributed to raising acceptance of women's education in orthodox communities (Morales and Cruz 2021). These changes, while incremental, exemplify a growing awareness of the need to empower women to play a role not only in civil life but also in educational matters. Cultural information added to an online curriculum in Indonesia has also been effective in diminishing stereotypes and advancing gender equity. These are seen as important steps, in overcoming negative attitudes and creating an environment where female students are feeling safer to engage in educational opportunities.

5.3 *Persistent Structural and Infrastructural Barriers*

Even as the big projects are being completed, the study shows, significant structural and infrastructural obstacles remain in place. Participants from Nigeria stated that while they were excited about digital learning owing to the wave of digital education, some communities in Nigeria still encountered obstacles that were too cumbersome, including inadequate power supply, low internet penetration and expensive digital devices (Okoro and Patel 2020). The same issues were found in rural Guatemala where support for digital education is often limited by government funding for the necessary infrastructure. These insights are consistent with the previous literature that a transformative potential of bIDA can only be harnessed if underpinning infrastructure shortcomings are overcome (Ahmed and Rahman 2022; Nwankwo and Musa 2020). That's important for policy makers to know so they can ensure technical improvements really support larger social and economic development goals.

5.4 Comparative Impact Analysis: Pre and Post Digital Education Implementation

Comparing the results with a digital transition in education reveals differences in impact in all three settings. In urban Indonesia digital platform use has led to massive improvement in female student participation and academic success. In contrast, despite some improvements in rural Nigeria and Guatemala, the divide between schools with and without access to the internet is broad. As such these differences emphasize the need for targeted, sub-regional policies that take account of both the progress that has been achieved and the deep structural challenges. In general, the study reveals that the potential of digital education in redress of the markers of gender inequalities is however contingent on the good infrastructure, socioeconomic and regulatory environment. These varied results underpin the subsequent analysis of implications and recommendations.

6 Discussion

The research findings emphasize on digital education's distinct potentiality for decreasing gender inequality in developing nations. This research explores how gender obstacles are tackled by efforts through digital education, the function of socioeconomic variables in assisting or retarding these efforts, and the broader involvement for education policies and practices.

6.1 Digital Education as a Catalyst for Change

It is evident that digital education is already starting to take shape as a positive influence in terms of driving change in the context of long standing gender inequalities. With the opportunity to access learning resources in a timely and flexible manner, digital platforms have enabled women and girls to learn regardless of their geographical location and socio-cultural context. This is in unison with the experimental results in Chen et al. (2019), who proposed that digital platforms can sometimes serve as intellectually enabling presences in the academy. The respondents said that digital learning has not only expanded, but has highlighted the gender standards. While traditional thinking has stopped women's education, successful digital examples began to send their message to cultural traditions. Stories of empowerment in this research reveal the importance of digital education in the creation of a new educational model for girls. These stories are important for efforts to promote a policy related to equal gender education, which focuses on the forces of successes and the forces to offer technology -based education.

6.2 Socio-Economic Constraints and the Digital Divide

Although the promise of digital education is attractive, socio -economic digital dividing mass implementation prevents mass implementation in development areas. The results clearly show that the lack of infrastructure including limited network access and high costs for digital devices, especially in Nigeria and Guatemala, are still major challenges. These findings support the allegations from Nwankwo and Musa (2020) and Ahmed and Rahman (2022) on the need for full investment on digital infrastructure to learn to achieve educational equation. In addition, the role of socio -economic status in relation to digital literacy appeared as a main factor. Girls from poor homes often lack digital literacy required to be served by learning online. Digital distinction not only masks our individual educational progress, but also reinforces this systemic inequality in gender. In order to ensure that the benefits of digital education have been properly distributed, it is also necessary for specific programs such as subsidized access, social awareness campaigns and regulatory changes to ensure.

6.3 Implications for Educational Policy

This research highlights the revolutionary capacity of digital education, which has different political implications.

First, the authorities should prioritize infrastructure expenses to reduce digital differences. This not only increases broadband connection, but also makes the equipment more cheaper and provides training in digital literature programs. Conclusions suggest that such efforts are particularly important in rural areas where technical intervals increase inequality in gender.

Secondly, there is an urgent need to create culturally relevant digital courses. Participants in Guatemala and Indonesia agreed that local and culturally appropriate information is important to increase the commitment and identity of female students. Politicians should work with local teachers and social leaders to ensure that digital materials represent special needs and cultural environments in each society.

Third, the institutional auxiliary mechanism should be constructed, including public-private partnerships that can accelerate digital platforms. This requires cross -sectoral cooperation between public agencies, colleges and universities and technical companies. According to Rahman and Jones (2021), the construction of frameworks for such cooperation can contribute to introducing digital education more efficiently in resources.

6.4 The Role of Informal Digital Education

The results of the study also highlight the role of informal digital education platforms. Community operated internet libraries and social media driven learning networks, among such systems, are gaining traction in ethnographic settings where traditional forms of digital education are at a very early stage of growth. Informal learning networks provide a low-cost supplement that can supplement existing forms of education. However, these tools face the sustainability and recognition problems, and the long-lasting impact of gender gaps reduction has not been well investigated. The distinction between formal and informal online learning suggests the need for integrative policies and structures to nurture both of these developments. Policy makers could investigate mechanisms for valuing and possibly making part of mainstream education, activities surrounding informal learning, thus giving them increased legitimacy and circulation.

6.5 Comparative Perspectives and Cross-Cultural Lessons

The cross-national comparative study reveals useful insights into contextual factors related to the introduction of digital ways of learning. Indonesia has reported the most significant gain in female participation thanks to the more favorable infrastructure, and the opposite is true for Nigeria and Guatemala who are still plagued with serious infrastructural and socio-economic issues. These finding suggest that the "one size fits all" policy is not effective in digital education. Instead, successful programs need to be adapted at the local level, considering local folklore, economic realities, and community needs. Cross case comparisons in this study reveal the importance of technical and culturally-specific approaches to digital education in reducing, or magnifying, gender gaps. Flexibility, reactivity, and adaptation must be guiding principles for policymakers in developing digital education programs.

7 Policy Implications and Recommendations

From a synthesis of the evidence and previous deliberations, we present a detailed package of practical policy recommendations. These approaches are tailored to educational institutions, governments, and community organizations interested in closing gender gaps. It is all about using digital education for gender equity in all aspects of the society.

7.1 Invest in Infrastructure and Digital Literacy

The first recommendation is the need to improve internet connections in underserved areas. We should provide public funding for broadband infrastructure and affordable digital devices. Moreover, setting up community-based digital literacy initiatives will give the female students tools that help them access and use digital teaching resources, contributing to gender equality in learning.

7.2 Develop Locally Relevant Curricula

Governments and schools should partner to develop culturally specific and locally attuned digital content. Including local exceptions and female empowerment examples in online educational materials can be a successful way to challenge stereotypes and role models in traditional gender norms to motivate learners and drive the societal transformation towards more gender equality on the community level.

7.3 Institutional Partnerships and Funding Models

Government, commercial technology companies and educational institutions could work together to popularize digital platforms. PPPs form one modality of funding which can help ensure the sustainability of digital education initiatives in the long-run and pave the way for long-term strategies to combat gender disparities.

7.4 Policy Advocacy and Regulatory Frameworks

Policymakers need to make sure that strong legal and regulatory environments are in place to ensure that all have access to digital education. Such frameworks should involve strict provisions for data privacy and security, and require inclusive, diverse, secure online learning platforms. Safe digital spaces that facilitate and empower female students to take online courses should be specifically supported.

7.5 Monitoring and Evaluation Mechanisms

Finally, rigorous monitoring and evaluation are needed to continually track the impact of digital education interventions. These frameworks need to marry quantitative metrics (enrollment and student performance data) with qualitative data that tell a human story and capture the real-life experiences, barriers and triumphs of girls in school, so they also capture the conversations that drive continued improvements.

8 Conclusion

This qualitative analysis has shown the liberating potential of e-learning in reducing gender inequalities for developing countries and marginalised groups. The 612 respondents from Nigeria, Indonesia and Guatemala, who made up the survey, highlight the promise of online platforms within schools and the barriers that continue to stand in the way.

The study reveals that the digital education offers flexible, accessible, locally meaningful learning environments where women life-long learners, who challenge embedded gender bias. While the study detected significant Infrastructural and socioeconomic limitations that still hamper full enjoyment of the digital education. Hence, targeted policy initiatives which include infrastructure spending, relevant curriculum development, and effective public-private partnerships are needed to bridge the chasm between promise and praxis. Distributed Symmetric Key Establishment (DSKE) is not a panacea for all gender-related barriers to education, but it is a critical element of the broader strategy to advance education equality and increased economic opportunity for girls in the developing world. Further research is essential to understand the long-term impacts and to improve policy interventions using digital technology to promote gender inclusivity in education.

9 Limitations and Future Research

Like any qualitative exploration, there are limitations that would benefit from cautious consideration of these results. Although a representative sample of 612, the sample size is limited and the views expressed may not fully represent the experiences of all stakeholders throughout all countries. In addition, the challenges of sociocultural context demand the interpretation of broad conclusions in light of local circumstances.

Future work should seek to incorporate larger, more diverse samples, combined-methods approaches that link qualitative perspectives with quantitative observations.

A longitudinal study would allow us to better ascertain the long-term consequences of online education programs on gender inequality. Moreover, more evidence form informal digital educational institutions and potential practices of bringing it into the formal education could serve as useful guiding principles for general policy-making.

References

Adeyemi BO, Johnson K, Adekunle T (2019) Digital pedagogy in Nigeria: bridging educational gaps in rural communities. African J Inf Syst 11(3):123–142

Ahmed S, Rahman P (2022) Examining the digital divide in rural education: challenges and prospects in developing countries. J Edu Technol Soc 25(2):45–63

Anderson T, Burnett M (2020) Digital education pathways: empowering marginalized communities through innovation. Int J Dig Learn 8(1):5–24

Bennett R, Kumar S (2018) Socio-economic influences on digital literacy in low-income populations. J Educ Res 46(4):367–385

Brown L, Lee A (2020) Reforming education through technology: digital platforms in developing countries. Education Today 34(2):89–108

Chen Y, Martinez R, Wu X (2019) Gender and digital education: an empirical study of policy impacts in Asia. Int J Gender Stud Educ 12(1):55–72

Creswell JW, Poth CN (2018) Qualitative inquiry and research design: choosing among five approaches, 4th edn. Sage Publications

Davies P, Miller C (2019) Gender inequality in education: traditional obstacles and modern digital opportunities. J Modern Educ 15(3):203–221

El-Gamal A, Osei R (2021) Informal learning and digital platforms: analyzing community-based initiatives in Africa. African Rev Educ 19(1):77–95

Ethics in Education Review (2019) Guidelines for ethical research in educational settings. International Ethics. Publications

Fernandez M, Alkire S (2020) Overcoming educational inequities: the role of technology in developing countries. J Dev Stud 28(2):134–152

Garcia H, Silva R, Torres P (2022) Digital education: bridging infrastructure gaps in remote communities. J Rural Dig Initiat 10(1):41–59

Hasan R, Rahim N (2021) Digital empowerment: facilitating female education through online learning in South Asia. South Asian J Educ 11(3):211–229

Johnson PR (2021) Quantifying educational disparities: new insights into digital learning impacts. J Educ Assess 39(4):310–327

Johnson L, Matthews R, Lee D (2021) Ethical conduct in educational research. Res Ethics Q 17(2):89–106

Khan A, Suzuki H, Mehta S (2020) Gender disparities in education: a comparative study of South Asia and Africa. Comp Educ Rev 64(1):53–74

Kumar S, Patel R (2019) Digital revolution in education: impact on gender equity in South Asia. J Educ Chang 20(3):275–293

Kuteesa K, Akpuokwe C, Udeh C (2024) Gender equity in education: addressing challenges and promoting opportunities for social empowerment. Int J Appl Res Soc Sci:6.631-641.10.51594/ijarss.v6i4.1034

Lee J, Rahman M, Gomez L (2018) Cultural barriers in education: overcoming gender biases with technology. J Glob Educ 14(2):145–163

Lopez M, Nguyen T, Ramirez E (2022) Socio-economic challenges to digital learning adoption in Latin America. Latin Am J Educ Technol 17(4):189–207

Mendes R, Costa A (2020) Virtual bridges: the impact of digital education on remote communities. J Virtual Learn 12(2):99–119

Morales E, Cruz F (2021) Online learning and gender empowerment: a case study from Guatemala. Central Am J Educ 9(1):67–85

Nain, A., Bohra, N. S., Tewari, D., Grewal, H., & Ahmad, V. (2024, November). AI for teacher Well-being: enhancing job satisfaction and reducing stress in Indian secondary schools. In: 2024 4th international conference on advancement in electronics and communication engineering (AECE). IEEE, pp 668–673

Nwankwo J, Musa A (2020) Digital divide in Nigeria: challenges and prospects for female education. Niger J Educ Technol 5(3):211–230

Okoro C, Patel R (2020) Educational barriers in rural Nigeria: gender, culture, and digital conversion. African Educ Rev 13(1):47–66

Oliveira P (2018) Gender disparities in education: trends and challenges. J Soc Educ 3(2):101–119

Patel S, Sharma N (2020) Digital learning platforms and gender equity: remote education in the age of technology. Int J Dig Pedagogy 7(1):33–55

Putri D, Santoso B (2019) The effect of digital learning tools on female education in urban Indonesia. Indonesian J Educ Technol 11(4):211–228

Qureshi I, Ahmed Y, Siddiqui F (2019) Challenges in implementing digital education in underdeveloped regions. J Global Learn 6(2):135–152

Rahman A, Jones M (2021) Policy interventions for digital education: a critical review of Southeast Asia's experience. J Technol Pol 8(3):179–197

Rodriguez L, Perez G, Santos M (2018) Gendered challenges in education: the role of technology in bridging educational divides. J Feminist Educ Stud 5(2):65–84

Singh R, Verma S (2022) The rise of informal digital education in rural India: implications for gender equity. J Commun Educ 14(1):55–73

Smith J (2019) Defining digital education in the 21st century. Digital Learn J 4(1):21–38

Smith J, Osborn M (2020) Interpretative phenomenological analysis as a qualitative research method. Qual Res Educ 15(3):105–123

Thomas R, Nguyen L (2021) Digital technology and economic empowerment in education: a study of emerging economies. J Glob Digital Stud 10(2):89–107

Wang Y, Li F (2021) Digital inclusion in education: addressing gender disparities through technology. Asian J Digital Educ 9(1):45–63

Zhang P, Chen J, Lu K (2020) Localizing digital content: strategies for enhancing gender equity in online education. J Local Learn 7(2):101–119

Implications of Digital Technologies on Gender Inequality in G20 Countries

Ajay Kumar Singh and S. Srinivasan

Abstract Existing theoretical and empirical literature emphasized that digital technologies (DTs) are conducive to make change in social, economic, environmental, health, business, and financial sectors. Application of DTs is also creating barriers and challenges in the social sector. DTs are helpful to increase communication of people and help to increase social development. DTs are providing multiple opportunities for increasing women empowerment. Consequently, use of DTs and digital infrastructure would be conducive to reduce gender inequalities. Although, limited research could pay significant attention to examine the impact of DTs on gender equality. While gender equality is a prime of SDGs. Hence, it cannot be achieved without gender equality. Thus, this research assesses the role of DTs in gender inequality in G20 countries using a log-linear regression model. Gender inequality index is used as a representative indicator for gender inequality in the model. While six crucial variables in the domain of digitalization are applied as core explanatory variables. Thereupon, human development index, health and education expenditure, female employment participation rate, female self-employed, inflation, environmental degradation, trade openness, sex ratio and women population are considered as core explanatory variables in the statistical and empirical investigation. Gender inequality has declined due to increase in human development, education and health expenditure, cereal yield, per capita energy consumption, renewable energy consumption, trade openness, sex ratio, female self-employed and women business. While environmental degradation, female population and female unemployment produce a positive impact on gender inequality. Mobile cellular and ICT goods exports appear effective to reduce gender inequality.

Keywords Gender equality · Human development · Sustainable development · Women empowerment · Social development · Digital technologies · Education

A. K. Singh (✉) · S. Srinivasan
Department of Humanities and Social Sciences, Graphic Era Deemed to be University, Dehradun, Uttarakhand, India
e-mail: a.k.seeku@gmail.com

S. Srinivasan
e-mail: s.srinivasan.hss@geu.ac.in

A. Singh et al. (eds.), *Human-Centered Sustainability: Bridging Well-Being, Gender Equity, and Technology*, World Sustainability Series,
https://doi.org/10.1007/978-3-032-22020-2_9

1 Introduction

Digital technologies (DTs) are effective to increase dissemination of information and knowledge among large users within and across countries (Prabha et al., 2023). There are many DTs like mobile technologies, mobile smartphone, blockchain, artificial intelligence (AI), internet, broadband, etc. are creating a favourable ecosystem for people to attain their different goals in social, economic, marketing, education, technological, environmental sectors, and labour market (James, 2013; Asongu. 2015; Grau-Sarabia & Fuster-Morell, 2021; Kouladoum, 2023; Lu et al., 2023; Fatima et al., 2025). Numerous studies have found positive implications of DTs and ICTs on economic development. Social inclusion is increasing due to use of DTs and ICTs. DTs and ICTs have already revolutionized the marketing sector. The quality of the education sector is improved as adoption of digital tools and infrastructure in the academic sector. ICTs are worthy and useful to increase technological change and advancement. Application of DTs and ICTs are providing scientific processes and alternatives to increase the environmental quality. Use of ICTs and mobile phones are found supportive to increase the income and profits of the farmers in China (Zhang et al. 2025). In other developing countries, the ICTs and DTs are helping to make the social and economic changes of the people. ICTs and DTs are increasing the mobilization of the labour market. Accordingly, the labour market is creating ample job opportunities for skilled and unskilled workers. In the financial sector, ICTs and DTs are helping to increase online transactions of money with minimum prices. DTs and ICTs are minimizing the operating cost in many sectors including financial, services and online marketing.

DTs and AI are providing new opportunities for people to increase their relationships (Grau-Sarabia and Fuster-Morell 2021). Thereupon, DTs are creating new income generation and employment opportunities in the labour market (Prabha et al. 2023). DTs are increasing the labour productivity and efficiency in most sectors (Wajcman et al. 2020; Rani et al. 2022). Technological innovation is getting significant benefits from DTs. Hence, DTs are increasing technological change and innovation. ICT is improving technological advancement and total factor productivity of the system (Chen 2004). DTs are a crucial driver to increase the network ability of economic agents in the market (Prabha et al. 2023). Business communities are developing new business and ventures through application of DTs. Capital-intensive business gets benefits from the internet (Panichsombat 2016). ICTs drive entrepreneurship, business and education development (Tchamyou et al. 2019; Fatima et al. 2025). ICTs are promoting the efficiency of the government (Adam and Alhassan 2021). DTs are helping to increase trade and business (Prabha et al. 2023). Subsequently, ICTs have created several new opportunities for improving the quality of life across countries (Gurumurthy 2006; Adam and Alhassan 2021). ICTs appeared positive to increase income, wage and equal distribution of income at micro level in China (Zhang et al. 2025). Furthermore, ICTs have the potential to increase accessibility of information and enhance social justice (O'Donnell and Sweetman 2018). Accordingly, digital infrastructure creating indicators are supportive to increase inclusive growth (Kouladoum 2023).

Application of DTs develop the platform for digital economy. Use of DTs are increasing the significant opportunities for women to increase their participation in the workforce (Prabha et al. 2023). ICTs are supportive to increase the skills of women and enhance their participation in the economic and social system (Chen 2004). ICT is fruitful to increase the equitable accessibility of resources for male and females. Application of ICTs is becoming useful for women to increase their economic empowerment (Gurumurthy 2006; Olayemi and Olayemi 2024). ICTs help to increase market and resource accessibility for women (Fatima et al. 2025). Digital economy and digital transformation are significant promoters for female employment (Lu et al. 2023; Ye and Cai 2024). Grau-Sarabia and Fuster-Morell (2021) recognized that application of ICTs creates a conducive ecosystem to increase gender equality as it increases the accessibility of women in social, economic and environmental sectors. DTs develop multiple ways to increase the participation of women in the society. ICTs may be helpful for women to increase their skills and employability (Grau-Sarabia and Fuster-Morell 2021). Thus, ICT is working as a valuable driver for increasing gender equality and SDGs (Chen 2004; O'Donnell and Sweetman 2018; Shah and Krishnan 2024). Nowadays, ICT is a significant driver for economic growth and sustainable development (Chen 2004). ICT is beneficial to increase business, entrepreneurship, and living standards of women (Asongu et al. 2019). ICTs is also creating a driving force to increase educational entrepreneurship, technology literacy and social engagement and new innovative skills of the people (Olayemi and Olayemi 2024). Agricultural growth is increasing due to use of ICT (Asongu et al. 2019). Agricultural development is also an important driver to increase women empowerment.

Moreover, DTs provide ample possibilities for women to increase their social and economic outcome (Wajcman et al. 2020). DTs are found suitable to increase gender equality (Wajcman et al. 2020; Shikongo et al. 2023; Shah and Krishnan 2024; Wu et al. 2024). Past literature has been examined the impact of DTs, ICTs and digital economy on gender diversity, gender inequality, women education, women's workforce participation, and women empowerment in different countries (Chen 2004; Sorgner 2019; Grau-Sarabia and Fuster-Morell 2021; Prabha et al. 2023; Shikongo et al. 2023; Olayemi and Olayemi 2024; Shah and Krishnan 2024; Fatima et al. 2025). These studies could assess the role of single or specific variables related to ICT on gender equality or inequality. Thus, limited studies could develop the interconnection of gender equality with different DTs and other determinants. Therefore, this chapter is achieved the answer of following research questions:

- What are the main determinants of gender equality?
- Can DTs help to promote gender equality?
- What may be the negative consequences of gender inequality on SDGs?
- How can DTs be favourable for increasing gender equality in G20 countries?

This article is attained following objectives:

- To observe the crucial determinants of gender equality as per the theoretical and empirical literature.
- To examine the impact of DTs on gender equality in G20 countries.
- To suggest policy proposals to increase gender equality and SDGs through application of DTs.

2 Review of Literature

Gender equality means to create equal opportunities for females and male in all sectors like jobs, business, wage, education, justice, etc. Gender equality is a situation in which male and female are treated equally and there is no existence of discrimination based on gender (Olayemi and Olayemi 2024). It is a significant pillar for improving prosperity of the society and sustainable development. Furthermore, many SDGs cannot be promoted without improving gender equality. Thereupon, gender equality is supportive to increase economic and human development, and holistic development of society (Chen 2004). For instance, family planning is not possible without the participation of women in the society. Thus, women play a kay role to maintain the social-cultural and economic activities (Olayemi and Olayemi 2024). Other social issues like poverty, well-being, and physical and mental health are significantly associated with gender equality (Olayemi and Olayemi 2024). Zhang et al. (2025) also determined that health status, educational level, family size and social capital reflects the overall income inequality.

Moreover, women are an essential component of the workforce in a nation (Lu et al. 2023). Most specifically, gender equality is an important SDGs of UNDP (Prabha et al. 2023; Fatima et al. 2025). Gender inequality is caused to increase poor nutritional, maternal mortality rate, poverty, illiteracy and infant mortality rate (Olayemi and Olayemi 2024). Further, gender inequality showed a negative impact on gross enrolment rate (Mbodji 2024). Gender inequality creates more vulnerability for females than male. Women receive relatively low opportunities in social, economic, business, education, health and other sectors due to gender inequality. Women also face mental and psychological stress due to gender inequality. Gender inequality, thus, is a crucial challenge for global communities to achieve SDGs. Moreover, gender inequality in employment is a critical issue at global level (Anyanwu and Augustine 2013; Lu et al. 2023). Further, high gender inequality is caused to increase diversity in other components of the society. Thus, gender equality is essential to increase women empowerment. Further, women empowerment creates a better platform for social development.

Gender inequality increases due to limited career development, low wage rate, poor employment structure, and low employment opportunities for women in the labour market. Labour market is unable to provide the ample scope for further growth of women. For example, the ICT sector does not have ample scope for women employment (Ye and Cai 2024; Grau-Sarabia and Fuster-Morell 2021).

Existing researchers explained the impact of gender inequality and equality on social and economic prosperity. For example, Chen (2004) found the positive impact of gender equality on economic growth. Mbodji (2024) accessed the impact of gender inequality on education in sub-Saharan countries. Olayemi and Olayemi (2024) examined the association of gender equality with sustainable development in Nigeria.

Further, existing researchers pursue different indicators for explaining the progress of gender equality in different countries. For instance, sex ratio, fertility rate and infant mortality rate of females signify the progress of gender diversity in a nation (Anyanwu and Augustine 2013). Gender equality depends on micro and macro level indicators and activities (Potrafke and Ursprung 2012). Potrafke and Ursprung (2012) examined the impact of globalization on gender equality in 100 developing countries. Family characteristics like income, family size, cultural and religion, occupations etc. have a significant impact on gender equality. Most indicators in the domain of social, economic, technology, environment and cultural sectors help to increase gender equality. Gender equality depends on multiple factors. The factors can be broadly divided in following categories:

Social: Education is an essential component to increase gender equality (Chen 2004). Female education is highly effective to reduce gender inequality (Chen 2004). Thus, equal education for male and females works as a catalyst for gender equality (Grau-Sarabia and Fuster-Morell 2021). Education can be effective to increase skill development and human capital (Chen 2004; Grau-Sarabia and Fuster-Morell 2021). Equality in employment may also increase as education increases (Chen 2004). Therefore, education may be valuable to reduce income inequality as it provides the accessibility of health care-related information (Olayemi and Olayemi 2024; Ndoya and Asongu 2024). Thus, public expenditure in the education sector is necessary to increase inclusive education (Mbodji 2024). Public expenditure on education also increases the female enrolments in the academic sector (Chen 2004). Anyanwu and Augustine (2013) reported that gender equality could be improved due to the increase in primary education in African countries.

Gender based violence is caused to increase gender inequalities at greater extent. Chronic poverty hampers the livelihood of women. Thus, poverty leads to increased gender inequality (Anyanwu and Augustine 2013). There are many social obligations like children and family responsibility which further increase gender inequalities.

Economic Development: Lack of equal economic opportunities for females create more gender diversity. Equal wage rate for male and female is a key driver to increase gender equality. Creation of equal jobs for male and female maintains gender equality. It improves the participation of women in social and economic sectors. Accordingly, it ensures the responsibilities of women in the society. Low economic opportunities for women also determine gender inequality. Gender wage gap is a crucial driver for gender inequality. Gender wage gap is caused to increase income inequality and gender disparities. While employment diversity for male and female creates more gender inequalities (Anyanwu and Augustine 2013). Chen (2004) revealed that economic development is crucial to increase gender equality. Inappropriate working culture for women also drives gender inequality. Thus, inclusive growth is highly imperative to increase gender equality.

Resource Availability and Accessibility of Finance: Limited accessibility of resources like finance, education, infrastructure, etc. are creating obstacles to increase gender equality. While financial accessibility for women helps for them to be socially and politically viable.

Health and Culture: Women's health is deprived due to multiple reasons like low dietary pattern, low awareness of women towards their health and additional responsibilities towards children and old age people. Hence, growth of the agricultural sector may be positive to increase gender equality as it meets food security and creates jobs for women. Moreover, traditional culture in society does not allow women to be involved in the social and economic system.

Psychological Behavior: Psychological behaviour of the people may also refute the equal contribution of females in the social and economic system.

Business and Entrepreneurship: Business and entrepreneurship opportunities for women are essential drivers to reduce gender inequality. Involvement of women in micro-enterprises helps to increase their income and gender equality to a greater extent.

Government Policies: Government policies are promotive to increase gender equality. For example, micro finance programs help women to increase their financial capability to be engaged in micro enterprises. Further, micro finance programs enhance education and participation of women in social and economic prosperity.

Technology and Innovation: Innovation is a key driver to increase social and economic development as it creates extensive jobs and increases growth patterns (Grau-Sarabia and Fuster-Morell 2021). Technological innovation could be helpful for women to acquire knowledge and communicate worldwide (Fatima et al. 2025). Gender inequality exists due to variation in analytical skills like reading, writing, analytical numeracy, thinking and learning of women (Sorgner 2019). While the variation is, the mentioned variables exist due to higher technological illiteracy of women than men (Grau-Sarabia and Fuster-Morell 2021). Therefore, technological innovation may be an important driver to increase gender equality. While inequality in technological accessibility is caused to reduce employability and skills of the women (Grau-Sarabia and Fuster-Morell 2021). Women have a low participation in the field of technologies (Grau-Sarabia and Fuster-Morell 2021). Thus, digital women illiteracy is caused to increase gender disparities.

Digital Technologies: Applications of DTs are crucial to increase the role of women in economic and social activities. Thus, advancement of DTs can be favorable to reduce gender inequality (Shikongo et al. 2023). Existing researchers have been accessing the impact of different components associated with digital infrastructure and ICT on various indicators like gender equality, education and employment inequality among male and female, business opportunities for women, quality of life, digital inequality, poverty and income inequality, social network across countries (Panichsombat 2016; Marler 2018; Tchamyou et al. 2019; Adam and Alhassan 2021; Zhang et al. 2025). However, most research is carried out by researchers in the context of developing countries. ICT infrastructure helps to increase gender equality in education and employment (Chen 2004). ICTs could be useful to increase gender equality and create a platform for SDGs (Olayemi and Olayemi 2024).

James (2013) examined the implications of mobile phone use on inequality in 11 African countries. Wakunuma (2013) also claimed the mobile phone has potential for social and economic development. Asongu (2015) perceived that mobile penetration has a positive impact on income equality in 52 African countries. Adam and Alhassan (2021) noted that smartphones are cultivating a new opportunity for businessmen, government and common citizens. Panichsombat (2016) assessed the association of internet penetration and income inequality in the Asia Pacific region. It noted that internet penetration rate shows a higher impact on income equality in developed countries than developing countries. Marler (2018) demonstrated the review to examine the contribution of smartphones in inequality. It claimed that the impact of mobile on digital inequality and social networks varies. Sorgner (2019) examined the role of digital technologies to maintain gender equality in the developing countries. It found a gender difference due to DTs across countries and industrial sectors.

Asongu et al. (2019) observed the significance of ICT and inclusive education in inequality in 42 sub-Saharan African countries. Cooray (2012) explored the role of gender equality in the education system. Tchamyou et al. (2019) reported that mobile phone and internet decrease income inequality through increasing education in 48 African countries. Furthermore, fixed broadband helps to increase lifelong learning and economic growth. Mobile phones can be applied to increase social equality in society (Burrell 2010). Grau-Sarabia and Fuster-Morell (2021) explored the consequences of digital divide in gender approach. Adam and Alhassan (2021) observed the implications of mobile phone penetration on quality of life in 114 countries. Rani et al. (2022) examined the labour process in the era of digitalization. Prabha et al. (2023) observed the participation of women in the agricultural sector due to adoption of DTs.

Wu et al. (2024) observed the impact of digitalization in female employment and gender wage gap in the manufacturing sector of China. Digitalization could be effective to create jobs for women and reduce wage gap. Digitalization is helpful to increase output and productivity (Wajcman et al. 2020; Wu et al. 2024). It is noted an improvement in gender equality in the labour market due to application of digitization (Sorgner et al. 2017). Lu et al. (2023) also realized that female employment could be improved due to the increase in the digital economy in China. Shah and Krishnan (2024) examined the impact of ICT on gender and income inequality in 86 countries. It found a crucial impact of ICT on gender inequality in these countries. Ndoya and Asongu (2024) captured the effect of digital divide on inequality in sub-Saharan Africa. It claimed that the impact of digital divide on income inequality varies across countries in Africa. Despite that digital divide produces a negative impact on income inequality. Bouanza et al. (2024) examined the role of ICT on business constraints and its moderating effect on income inequality in 48 sub-Saharan African countries.

Table 1 Time period of respective G20 countries

Country	Time period
Austria, Belgium, Brazil, Canada, China, Czechia, Denmark, Estonia, Finland, France, Germany, Hungary, Italy, Japan, Korea (Republic of), Latvia, Malta, Mexico, Netherlands, Portugal, Spain, Sweden, United Kingdom, United States	2000–2022
Argentina, Australia, Cyprus, India, Lithuania, Poland, Romania, Slovenia, Turkey	2001–2021 2001–2022
Ireland, Luxembourg, Russian Federation, Slovakia, South Africa	2002–2022
Croatia, Greece, Indonesia, Saudi Arabia	2003–2022
Bulgaria	2004–2022

Source Author's compilation as per the availability of data

It revealed that ICT helps to moderate the effect of business constraints on income inequality. Ye and Cai (2024) examined the role of digital transformation for creation of female employment in China. It claimed the digital transformation could be helpful to increase employment for women. Using the primary data, Zhang et al. (2025) noticed that income inequality is significantly influenced by ICTs in China.

3 Research Methods and Materials

3.1 Study Area

The G20 countries have a significant share of GDP in the world. Thereupon, G20 countries are giving more priority to adopt digitalization in the different sectors. Many countries in this group could reduce gender inequality in the last decade. However, few countries are unable to improve gender equality due to certain reasons like low digital adoption capacity, low digital literacy, low education, low per capita income, low digital infrastructure, low digital skills etc. Therefore, this chapter explores the progress of gender inequality and assesses its affecting factors in G20 countries. The statistics of most variables are available during 2001–2022 for G20 countries. Hence, imbalanced country-wise panel data during the mentioned time period is considered for further statistical and empirical investigations in this research (Table 1).

3.2 Sources of Data

This article is based on secondary data that is taken from the website of United Nations Development Programme (UNDP), World Development Indicators (the World Bank), Organization for Economic Cooperation and Development (OECD).

Moreover, useful data related to digitalization, women empowerment and gender inequality, and other variables are taken from the website of International Labour Organization (ILO) and published research articles and government reports.

3.3 Explanations of the Variables

Gender inequality can be measured in various forms like ratio of male to female enrolment in education, ratio of female employment with male employment, sex ratio, female contribution in total employment, and female labour force participation rate (Chen 2004). Social institutions and gender index is also used to examine the progress of gender equality (Potrafke and Ursprung 2012). Gender inequality index value (*GIIV*), thus, is taken as a proxy indicator for gender inequality and dependent variable (*IV*) in this chapter. While number of internet users (*IUI*) (Chen 2004; Asongu et al. 2019; Kouladoum 2023; Prabha et al. 2023), fixed broadband subscribers (*FBS*) (Asongu et al. 2019; Tchamyou et al. 2019; Kouladoum 2023; Prabha et al. 2023), fixed telephone subscription (*FTS*), and Mobile cellular subscriptions (*MCS*) are considered as proxy variables for DTs. Moreover, ICT goods exports and imports are supportive to create digital infrastructure. Accordingly, these variables are considered as core explanatory variables for measuring their impact on gender inequality in G20 countries.

Thereupon, human development index value (*HDIV*) (Grau-Sarabia and Fuster-Morell 2021), education expenditure (*EE*) and health expenditure (*HE*) also enhance gender equality. Employment opportunities for females also helps to minimize gender inequality. Hence, female labor force participation rate (*FLFPR*) and ratio of female to male labor force participation rate (*RFMLFPR*) are included as control variables to assess their impact on gender inequality. While female unemployment (*FU*) has an adverse impact on gender equality. Food security and energy consumption are essential components to increase the health of women. Hence, cereal yield (*CY*) and per capita energy consumption (*PCEU*) may be useful to increase gender equality. Consequently, female life expectancy (*FLEB*) is useful to increase gender equality.

Inflation GDP deflator (*IGDPD*) reduces real income of people. Hence, it is predicted that *IGDPD* has a negative impact on women empowerment and gender equality (Kouladoum 2023). Female population (*FP*) may increase the pressure on resources. Thus, gender inequality may increase due to increase in *FP*. While sex ratio (*SR*) signifies the equality in male and female. Thus, *SR* also produces a significant impact on gender equality (Anyanwu and Augustine 2013). For instance, sex population ratio produces a negative impact on gender equality in African countries (Anyanwu and Augustine 2013). While in some other countries, *SR* could be helpful to increase gender equality.

The impact of environmental degradation on human health would be negative. Hence, environmental degradation would be more accountable for increasing gender inequality. Accordingly, per capita CO_2 emissions (*PCCOE*) are used to capture the impact of environmental degradation on gender inequality. While, renewable energy consumption (*REC*) helps to reduce environmental degradation and enhance gender equality.

GDP per person employed (*GDPPE*) is also considered as an explanatory variable to examine its impact on gender equality. Extensive business opportunities for women would lead to increased women empowerment and gender equality. Thus, female self-employed (*FSE*) and Women Business and the Law Index Score (*WBLIS*) are also applied as mediator variables in this research. Furthermore, exports of goods and services is used as a representative variable for trade openness (*TO*) in regression analysis.

3.4 Theoretical Framework and Formulation of Empirical Model

The existing researchers have formulated various forms of empirical models to explore the interconnection of gender inequality with ICT related variables, and social-economic and technology promoting and other indicators (Asongu 2015). Chen (2004) examined the role of ICT in gender equality and economic development using a linear regression model. Panichsombat (2016) developed a linear regression equation to assess the impact of mobile penetration in income equality. Potrafke and Ursprung (2012) observed the impact of globalization on gender equality using a linear regression model. Adam and Alhassan (2021) applied the PLS-SEM technique to examine the impact of ICT tools on quality of life. It observed that ICT regulation produces a significant impact on quality of life. Mbodji (2024) applied a fixed-effect panel model to examine the impact of gender inequality on inclusive education in African countries using panel data investigation. Kouladoum (2023) applied a Driscoll-Kraay technique to examine the impact of inclusive growth in African countries.

The descriptive analysis is used to observe the variation in different variables across G20 countries in this chapter. Karl-Pearson correlation analysis is used to examine the association among the variables. While the log-linear regression model is formulated to examine the impact of DTs and other variables on gender inequality. Following form of regression equation is adopted in this chapter:

$$
\begin{aligned}
(GIIV)_{ct} = {} & \beta_0 + \beta_1 \ln(HDIV)_{ct} + \beta_2 \ln(EE)_{ct} + \beta_3 \ln(PCCOE)_{ct} \\
& + \beta_4 \ln(CY)_{ct} + \beta_5 \ln(HE)_{ct} + \beta_6 \\
& \ln(FLFPR)_{ct} + \beta_7 \ln(FSE)_{ct} + \beta_8 \ln(RFMLFPR)_{ct} + \beta_9 \ln(FU)_{ct} \\
& + \beta_{10} \ln(PCEU)_{ct} + \beta_{11} \ln(TO)_{ct} + \beta_{12}
\end{aligned}
$$

$$\ln(FBS)_{ct} + \beta_{13}\ln(FTS)_{ct} + \beta_{14}\ln(ICTGE)_{ct} + \beta_{15}\ln(ICTGI)_{ct}$$
$$+ \beta_{16}\ln(IUI)_{ct} + \beta_{17}\ln(MCS)_{ct} + \beta_{18}$$
$$\ln(GDPPE)_{ct} + \beta_{19}\ln(IGDPD)_{ct} + \beta_{20}\ln(FLEB)_{ct} + \beta_{21}\ln(FP)_{ct}$$
$$+ \beta_{22}\ln(REC)_{ct} + \beta_{23}\ln(SR)_{ct} + \beta_{24}$$
$$\ln(WBLIS)_{ct} + \lambda_{ct} \quad (1)$$

Here, β_0 is constant coefficient; $\beta_1, \ldots, \beta_{24}$ are coefficients of corresponding variables; ln is natural logarithm of associated variables; t time; c is cross-sectional country; λ_{ct} is error-term in Eq. (1). The brief explanation of other variables is presented in Table 2.

Furthermore, the elasticities of gender inequality with respect to individual variables is also estimated. The following formula is applied to estimate the elasticity of *GIIV* with respective explanatory variables:

$$\text{Elasticity of GIIV with respect to HDIV}$$
$$= \text{Regression coefficient of HDIV} \times [\text{Mean value of HDIV/GIIV}] \quad (2)$$

Similarly, the elasticities of *GIIV* with respect to other variables are also observed.

3.5 *Finalization Model, Validity of Coefficients and Application of Statistical Software*

The regression coefficients are estimated through Driscoll-Kraay technique (Kouladoum, 2023). While it also examines the correlation coefficients of various lags of error-term (i.e., estimated through log-linear regression equation) to check the robustness of empirical results. SPSS statistical software is used for descriptive analysis of the secondary data. STATA statistical software is applied to complete the country-wise panel data analysis.

4 Results and Discussion

4.1 *Descriptive Results*

The statistical summary of the variables is given in Table 3. It is noted a high variation in all variables (except *HDIV*) overall, between and within G20 countries. Also, the minimum and maximum values of respective variables infer that G20 countries have significant variation. Therefore, the impact of explanatory variables on *GIIV* would also vary in G20 countries. Moreover, log forms of undertaken variables (except,

Table 2 Explanation of explained and explanatory variables

Variables	Code	Unit	Variable's category
Gender Inequality Index (value)	*GIIV*	Number	*DV*
Human Development Index (value)	*HDIV*	Number	
Education expenditure (% of GNI)	*EE*	%	
Carbon dioxide (CO_2) emissions per capita (t CO_2e/capita)	*PCCOE*	t CO_2e/capita	Mediator
Cereal yield (kg per hectare)	*CY*	Kg./Ha	Control
Current health expenditure (% of GDP)	*HE*	%	Control
Female labor force participation rate (ages 15–24, % of female)	*FLFPR*	%	Control
Female self-employed (% of female employment)	*FSE*	%	Control
Ratio of female to male labor force participation rate (%)	*RFMLFPR*	%	Control
Female unemployment (% of female labor force)	*FU*	%	Control
Energy use (kg of oil equivalent per capita)	*PCEU*	Kg./Capita	Mediator
Exports of goods and services (% of GDP) (Trade Openness)	*TO*	%	Mediator
Fixed broadband subscriptions (per 100 people)	*FBS*	Number	Core
Fixed telephone subscriptions (per 100 people)	*FTS*	Number	Core
ICT goods exports (% of total goods exports)	*ICTGE*	%	Core
ICT goods imports (% total goods imports)	*ICTGI*	%	Core
Individuals using the Internet (% of population)	*IUI*	%	Core
Mobile cellular subscriptions (per 100 people)	*MCS*	Number	Core
GDP per person employed (constant 2021 PPP $)	*GDPPE*	PPP $	Control
Inflation GDP deflator (annual %)	*IGDPD*	%	Control
Female life expectancy at birth (years)	*FLEB*	Number	Control
Female population (% of total population)	*FP*	%	Control
Renewable energy consumption (% of total final energy consumption)	*REC*	%	Mediator
Sex ratio at birth (male births per female births)	*SR*	Number	Control
Women Business and the Law Index Score (scale 1–100)	*WBLIS*	Number	Mediator

Source Author's compilation as per the rationality of the variables

EE, *HE*, *PCCOE*, *FLFPR*, *FU*, *PCEU*, *TO*, *ICTGE*, *ICTGI*, and *FSE*) are not in normal forms. Thus, double-logging of these variables are included in the regression analysis.

GIIV is negatively correlated with *HDIV*, *EE*, *CY*, *HE*, *FLFPR*, *RFMLFPR*, *PCEU*, *TO*, *FBS*, *FTS*, *ICTGE*, *IUI*, *MCS*, *GDPPE*, *FLEB*, *FP*, *WBLI* (Table 4). The estimates suggested that human development, education expenditure, health

Table 3 Statistical summary of the variables

Variable		Mean	SD	Min	Max	Observations	Skewness
ln(*GIIV*)	Overall	− 1.8521	0.7276	− 4.7105	− 0.2523	N = 952	− 0.2896
	Between		0.6798	− 3.3018	− 0.6012	n = 43	
	Within		0.2766	− 3.2608	− 1.0892	T-bar = 22.1395	
ln(*HDIV*)	Overall	− 0.1720	0.1066	− 0.7032	− 0.0492	N = 952	− 1.7767
	Between		0.1017	− 0.5442	− 0.0737	n = 43	
	Within		0.0365	− 0.3451	− 0.0489	T-bar = 22.1395	
ln(*EE*)	Overall	1.4877	0.2730	0.5822	2.0882	N = 952	− 0.7775
	Between		0.2582	0.5825	1.9765	n = 43	
	Within		0.0962	0.9513	1.8942	T-bar = 22.1395	
ln(HE)	Overall	2.0053	0.3427	0.7514	2.9346	N = 952	− 0.6315
	Between		0.3278	1.0305	2.7495	n = 43	
	Within		0.1167	1.5639	2.4587	T-bar = 22.1395	
ln(*PCCOE*)	Overall	1.9516	0.5930	− 0.0633	3.2597	N = 952	− 0.4856
	Between		0.5814	0.3430	2.9341	n = 43	
	Within		0.1597	1.2033	2.3367	T-bar = 22.1395	
ln(*CY*)	Overall	8.4095	0.4298	5.1722	9.1944	N = 952	− 1.3138
	Between		0.3899	7.2967	9.0687	n = 43	
	Within		0.1880	6.2850	9.0075	T-bar = 22.1395	
ln(*FLFPR*)	Overall	3.5895	0.4200	1.8658	4.4169	N = 952	− 0.6842
	Between		0.4140	2.1678	4.3988	n = 43	
	Within		0.1109	3.1917	4.2456	T-bar = 22.1395	

(continued)

Table 3 (continued)

Variable		Mean	SD	Min	Max	Observations	Skewness
ln(*RFMLFPR*)	Overall	4.2814	0.2360	3.1479	4.5045	N = 952	− 2.5566
	Between		0.2346	3.3208	4.4856	n = 43	
	Within		0.0616	3.9617	4.7169	T-bar = 22.1395	
ln(*FU*)	Overall	2.0030	0.5408	0.7738	3.5908	N = 952	0.4183
	Between		0.4680	1.1027	3.3256	n = 43	
	Within		0.2864	0.8975	2.9063	T-bar = 22.1395	
ln(*PCEU*)	Overall	7.9897	0.5860	5.9749	9.1516	N = 952	− 0.6113
	Between		0.5851	6.2838	8.9809	n = 43	
	Within		0.1127	7.3237	8.4216	T-bar = 22.1395	
ln(*TO*)	Overall	3.6809	0.5995	2.2012	5.3623	N = 952	0.0578
	Between		0.5829	2.4299	5.1576	n = 43	
	Within		0.1669	3.1161	4.2444	T-bar = 22.1395	
ln(*FBS*)	Overall	2.3603	1.6982	− 6.3284	3.8756	N = 952	− 1.9894
	Between		0.9456	− 0.9566	3.4493	n = 43	
	Within		1.4236	− 5.5641	4.4869	T-bar = 22.1395	
ln(*FTS*)	Overall	3.3561	0.7015	0.3577	4.2781	N = 952	− 1.6226
	Between		0.6468	0.9134	4.0688	n = 43	
	Within		0.2994	1.7181	4.5468	T-bar = 22.1395	
ln(*ICTGE*)	Overall	1.4128	1.2690	− 3.5066	4.1532	N = 952	− 0.8021
	Between		1.2122	− 2.0141	3.3266	n = 43	
	Within		0.4537	− 0.2336	3.4269	T-bar = 22.1395	

(continued)

Table 3 (continued)

Variable		Mean	SD	Min	Max	Observations	Skewness
ln(*ICTGI*)	Overall	2.1311	0.4706	1.0886	3.7464	N = 952	0.2783
	Between		0.4152	1.4101	3.1048	n = 43	
	Within		0.2276	1.2555	3.2372	T-bar = 22.1395	
ln(*IUI*)	Overall	3.9434	0.7422	− 0.4155	4.6052	N = 952	− 2.1318
	Between		0.4828	2.0886	4.4362	n = 43	
	Within		0.5708	1.3788	5.8784	T-bar = 22.1395	
ln(*MCS*)	Overall	4.5635	0.5219	− 0.5007	5.3038	N = 952	− 3.5431
	Between		0.2757	3.3407	4.9086	n = 43	
	Within		0.4448	0.7221	5.6808	T-bar = 22.1395	
ln(*GDPPE*)	Overall	11.2823	0.5651	8.9046	12.6658	N = 952	− 1.1749
	Between		0.5556	9.6184	12.5633	n = 43	
	Within		0.1470	10.3483	12.0539	T-bar = 22.1395	
ln(*IGDPD*)	Overall	0.9317	1.1623	− 8.3453	4.5647	N = 952	− 1.2061
	Between		0.7413	− 0.5956	3.0502	n = 43	
	Within		0.9070	− 6.8181	3.1001	T-bar = 22.1395	
ln(*FLEB*)	Overall	4.3842	0.0621	4.0158	4.4740	N = 952	− 2.3361
	Between		0.0601	4.1372	4.4584	n = 43	
	Within		0.0207	4.2627	4.4842	T-bar = 22.1395	
ln(*FP*)	Overall	3.9257	0.0449	3.6158	3.9945	N = 952	− 4.0792
	Between		0.0465	3.6681	3.9902	n = 43	
	Within		0.0073	3.8735	4.0130	T-bar = 22.1395	

(continued)

Table 3 (continued)

Variable		Mean	SD	Min	Max	Observations	Skewness
ln(*REC*)	Overall	2.4256	1.0516	− 3.7134	4.0587	N = 952	− 1.4329
	Between		0.9486	− 0.4159	3.8167	n = 43	
	Within		0.4824	− 0.8719	4.3995	T-bar = 22.1395	
ln(*SR*)	Overall	0.0569	0.0166	0.0373	0.1638	N = 952	4.2896
	Between		0.0157	0.0382	0.1454	n = 43	
	Within		0.0051	0.0105	0.0875	T-bar = 22.1395	
ln(*WBLI*)	Overall	4.4344	0.1929	3.1676	4.6052	N = 952	− 3.3187
	Between		0.1734	3.5379	4.5843	n = 43	
	Within		0.0986	3.9797	5.2786	T-bar = 22.1395	
ln(*FSE*)	Overall	2.5692	0.7418	0.1705	4.5059	N = 952	0.2661
	Between		0.7407	0.6050	4.4383	n = 43	
	Within		0.1611	2.1346	3.9369	T-bar = 22.1395	

Source Estimated by authors

expenditure, energy use, trade openness, fixed broadband subscriptions, fixed telephone subscriptions, internet use, mobile cellular subscription, GDP per person employed, female life expectancy, female population and women business would be positive to reduce gender inequality. Female unemployment and inflation are positively correlated with gender inequality. Hence, these variables increase gender inequality.

4.2 Empirical Results

The coefficients of core, control and mediator variables with *GIIV* are presented in Table 5. The *F*-value is found statistically significant at 1% significance level. Thus, it is evidence that these variables have an independent impact on gender inequality. The negative association of *HDIV* with *GIIV* infer that human development helps to reduce gender inequality. Gender inequality is expected to be decreased by 3.32% as human development increases by 1%. *HDI* is an integration of per capita real income, life expectancy rate and literacy rate. Hence, these variables are effective to reduce gender inequality. Previous researchers like Chen (2004) also reported that education and economic development are supportive to increase gender equality and women empowerment. Moreover, the above-mentioned results are also supported by the negative coefficient of *FLER* with *GIIV*. Gender equality is predicted to be decreased by 1.499% due to a 1% increase in female life expectancy rate.

Education expenditure helps to increase equal accessibility of education for males and females, quality of education and human skills. Thus, the education expenditure must be effective to reduce gender equality. The education expenditure showed a negative impact on *GIIV*. The estimate implies that gender inequality is predicted to be declined by 0.19% as education expenditure increases by 1%. Past literature also signifies that education expenditure is an essential driver to increase gender equality (Grau-Sarabia and Fuster-Morell 2021; Mbodji 2024).

Health expenditure is useful to enhance the health security of people. Thus, health expenditure may enhance gender equality. The negative coefficient of *HE* with *GIIV* infers that health expenditure is found positive to increase gender equality. The result indicates gender inequality is likely to be decreased by 0.03% as health expenditure increases by 1%.

Environmental degradation has a negative impact on physical and mental health of people. While women's health is adversely affected due to environmental degradation. Thus, the positive coefficient of *PCCOE* with *GIIV*, reveals that environmental degradation is caused to increase gender inequality. Gender inequality is expected to be increased by 0.31% as environmental degradation increases by 1%. While renewable energy consumption is a significant driver to reduce CO_2 emissions (Ganda & Ruza, 2025). Hence, renewable energy consumption would enhance gender equality. It is noted as a negative coefficient of *REC* with *GIIV* in G20 countries. Thus, it is apparent that renewable energy consumption would be positive to increase gender

Table 4 Correlation coefficients among the variables

Variables	GIIV	HDIV	EE	PCCOE	CY	HE	FLFPR	FSE	RFMLFPR	FU	PCEU	TO	FBS
GIIV	1	− 0.834**	− 0.171**	− 0.256**	− 0.377**	− 0.520**	− 0.370**	0.584**	− 0.706**	0.248**	− 0.390**	− 0.315**	− 0.653**
HDIV	− 0.834**	1	0.383**	0.445**	0.429**	0.639**	0.358**	− 0.767**	0.532**	− 0.236**	0.574**	0.329**	0.706**
EE	− 0.171**	0.383**	1	0.196**	0.113**	0.346**	0.178**	− 0.524**	0.148**	0.148**	0.334**	0.192**	0.219**
PCCOE	− 0.256**	0.445**	0.196**	1	0.107**	0.220**	0.197**	− 0.473**	0.108**	− 0.076**	0.883**	0.193**	0.138**
CY	− 0.377**	0.429**	0.113**	0.107**	1	0.463**	0.205**	− 0.244**	0.212**	− 0.238**	0.224**	0.249**	0.419**
HE	− 0.520**	0.639**	0.346**	0.220**	0.463**	1	0.438**	− 0.532**	0.495**	− 0.063*	0.378**	− 0.087**	0.517**
FLFPR	− 0.370**	0.358**	0.178**	0.197**	0.205**	0.438**	1	− 0.200**	0.448**	− 0.356**	0.252**	− 0.055*	0.209**
FSE	0.584**	− 0.767**	− 0.524**	− 0.473**	− 0.244**	− 0.532**	− 0.200**	1	− 0.515**	− 0.032	− 0.552**	− 0.344**	− 0.418**
RFMLFPR	− 0.706**	0.532**	0.148**	0.108**	0.212**	0.495**	0.448**	− 0.515**	1	− 0.198**	0.241**	0.190**	0.461**
FU	0.248**	− 0.236**	0.148**	− 0.076**	− 0.238**	− 0.063*	− 0.356**	− 0.032	− 0.198**	1	− 0.132**	− 0.154**	− 0.246**
PCEU	− 0.390**	0.574**	0.334**	0.883**	0.224**	0.378**	0.252**	− 0.552**	0.241**	− 0.132**	1	0.194**	0.283**
TO	− 0.315**	0.329**	0.192**	0.193**	0.249**	− 0.087**	− 0.055*	− 0.344**	0.190**	− 0.154**	0.194**	1	0.324**
FBS	− 0.653**	0.706**	0.219**	0.138**	0.419**	0.517**	0.209**	− 0.418**	0.461**	− 0.246**	0.283**	0.324**	1
FTS	− 0.543**	0.585**	0.175**	0.362**	0.322**	0.468**	0.378**	− 0.448**	0.313**	− 0.192**	0.394**	0.128**	0.223**
ICTGE	− 0.054*	− 0.019	− 0.177**	0.022	0.157**	− 0.092**	0.205**	0.021	− 0.049	− 0.329**	− 0.021	0.146**	− 0.039
ICTGI	0.070*	− 0.110**	− 0.180**	0.097**	0.136**	− 0.036	0.324**	0.048	− 0.052	− 0.295**	0.029	− 0.055*	− 0.159**
IUI	− 0.679**	0.778**	0.317**	0.276**	0.406**	0.510**	0.243**	− 0.539**	0.478**	− 0.198**	0.412**	0.340**	0.893**
MCS	− 0.372**	0.473**	0.240**	0.130**	0.179**	0.198**	− 0.126**	− 0.423**	0.253**	0.010	0.148**	0.283**	0.563**
GDPPE	− 0.504**	0.679**	0.349**	0.558**	0.448**	0.395**	0.149**	− 0.539**	0.203**	− 0.104**	0.644**	0.580**	0.453**
IGDPD	0.416**	− 0.291**	− 0.076**	− 0.127**	− 0.229**	− 0.235**	− 0.195**	0.205**	− 0.308**	0.078**	− 0.176**	− 0.152**	− 0.252**
FLEB	− 0.775**	0.849**	0.172**	0.264**	0.346**	0.496**	0.297**	− 0.512**	0.361**	− 0.292**	0.385**	0.247**	0.645**
FP	− 0.309**	0.090**	− 0.238**	− 0.279**	− 0.127**	0.131**	0.080**	− 0.150**	0.550**	− 0.138**	− 0.256**	0.025	0.018
REC	0.039	− 0.187**	0.063*	− 0.460**	− 0.200**	− 0.027	0.072*	0.265**	0.172**	− 0.011	− 0.237**	− 0.119**	0.058*
SR	0.085**	− 0.369**	− 0.457**	− 0.154**	− 0.069*	− 0.332**	0.056*	0.475**	− 0.032	− 0.168**	− 0.253**	− 0.085**	− 0.135**
WBLI	− 0.706**	0.588**	0.140**	− 0.049	0.224**	0.501**	0.294**	− 0.384**	0.693**	− 0.156**	0.106**	0.285**	0.627**

Table 4a Correlation coefficients among the variables conti…

Variables	FTS	ICTGE	ICTGI	IUI	MCS	GDPPE	IGDPD	FLEB	FP	REC	SR	WBLI
GIIV	– 0.543**	– 0.054*	0.070*	– 0.679**	– 0.372**	– 0.504**	0.416**	– 0.775**	– 0.309**	0.039	0.085**	– 0.706**
HDIV	0.585**	– 0.019	– 0.110**	0.778**	0.473**	0.679**	– 0.291**	0.849**	0.090**	– 0.187**	– 0.369**	0.588**
EE	0.175**	– 0.177**	– 0.180**	0.317**	0.240**	0.349**	– 0.076**	0.172**	– 0.238**	0.063*	– 0.457**	0.140**
PCCOE	0.362**	0.022	0.097**	0.276**	0.130**	0.558**	– 0.127**	0.264**	– 0.279**	– 0.460**	– 0.154**	– 0.049
CY	0.322**	0.157**	0.136**	0.406**	0.179**	0.448**	– 0.229**	0.346**	– 0.127**	– 0.200**	– 0.069*	0.224**
HE	0.468**	– 0.092**	– 0.036	0.510**	0.198**	0.395**	– 0.235**	0.496**	0.131**	– 0.027	– 0.332**	0.501**
FLFPR	0.378**	0.205**	0.324**	0.243**	– 0.126**	0.149**	– 0.195**	0.297**	0.080**	0.072*	0.056*	0.294**
FSE	– 0.448**	0.021	0.048	– 0.539**	– 0.423**	– 0.539**	0.205**	– 0.512**	– 0.150**	0.265**	0.475**	– 0.384**
RFMLFPR	0.313**	– 0.049	– 0.052	0.478**	0.253**	0.203**	– 0.308**	0.361**	0.550**	0.172**	– 0.032	0.693**
FU	– 0.192**	– 0.329**	– 0.295**	– 0.198**	0.010	– 0.104**	0.078**	– 0.292**	– 0.138**	– 0.011	– 0.168**	– 0.156**
PCEU	0.394**	– 0.021	0.029	0.412**	0.148**	0.644**	– 0.176**	0.385**	– 0.256**	– 0.237**	– 0.253**	0.106**
TO	0.128**	0.146**	– 0.055*	0.340**	0.283**	0.580**	– 0.152**	0.247**	0.025	– 0.119**	– 0.085**	0.285**
FBS	0.223**	– 0.039	– 0.159**	0.893**	0.563**	0.453**	– 0.252**	0.645**	0.018	0.058*	– 0.135**	0.627**
FTS	1	0.176**	0.044	0.217**	– 0.047	0.450**	– 0.318**	0.605**	0.091**	– 0.350**	– 0.129**	0.303**
ICTGE	0.176**	1	0.859**	– 0.075**	– 0.208**	– 0.126**	– 0.175**	0.086**	0.001	– 0.268**	0.331**	– 0.128**
ICTGI	0.044	0.859**	1	– 0.139**	– 0.262**	– 0.178**	– 0.091**	– 0.064*	– 0.052	– 0.261**	0.301**	– 0.212**
IUI	0.217**	– 0.075**	– 0.139**	1	0.654**	0.511**	– 0.221**	0.636**	0.015	0.029	– 0.255**	0.614**
MCS	– 0.047	– 0.208**	– 0.262**	0.654**	1	0.327**	– 0.096**	0.341**	0.036	– 0.038	– 0.231**	0.344**
GDPPE	0.450**	– 0.126**	– 0.178**	0.511**	0.327**	1	– 0.194**	0.558**	– 0.158**	– 0.216**	– 0.312**	0.351**
IGDPD	– 0.318**	– 0.175**	– 0.091**	– 0.221**	– 0.096**	– 0.194**	1	– 0.317**	– 0.092**	– 0.03	– 0.025	– 0.330**
FLEB	0.605**	0.086**	– 0.064*	0.636**	0.341**	0.558**	– 0.317**	1	– 0.002	– 0.125**	– 0.093**	0.486**
FP	0.091**	0.001	– 0.052	0.015	0.036	– 0.158**	– 0.092**	– 0.002	1	0.176**	– 0.159**	0.516**
REC	– 0.350**	– 0.268**	– 0.261**	0.029	– 0.038	– 0.216**	– 0.03	– 0.125**	0.176**	1	0.060*	0.220**
SR	– 0.129**	0.331**	0.301**	– 0.255**	– 0.231**	– 0.312**	– 0.025	– 0.093**	– 0.159**	0.060*	1	– 0.232**
WBLI	0.303**	– 0.128**	– 0.212**	0.614**	0.344**	0.351**	– 0.330**	0.486**	0.516**	0.220**	– 0.232**	1

Source Estimated by Authors. **. Correlation is significant at the 0.01 level (1-tailed). *. Correlation is significant at the 0.05 level (1-tailed)

Table 5 Regression coefficients of explanatory variables with GIIV

Number of obs	952		*R-squared*		0.7953	
Number of groups	43		*Mean VIF*		6.21	
F(24, 22)	108,343.1		*Prob > F*		0.000	
ln(*GIIV*) = [*DV*]	*Coef*	*Std. Err*	*t*	*P* > |*t*|	[95%Conf. Interval]	
ln(*HDIV*)	− 3.3206	0.4294	− 7.73	0.000	− 4.2112	− 2.4300
ln(*EE*)	− 0.1902	0.0712	− 2.67	0.014	− 0.3377	− 0.0426
ln(*HE*)	− 0.0343	0.0490	− 0.7	0.491	− 0.1360	0.0674
ln(*PCCOE*)	0.3131	0.0567	5.52	0.000	0.1955	0.4308
ln(*CY*)	− 0.1495	0.0425	− 3.52	0.002	− 0.2377	− 0.0613
ln(*FLFPR*)	− 0.2632	0.0272	− 9.69	0.000	− 0.3195	− 0.2068
ln(*RFMLFPR*)	− 0.3132	0.0648	− 4.83	0.000	− 0.4476	− 0.1788
ln(*FU*)	0.0169	0.0290	0.58	0.567	− 0.0433	0.0771
ln(*PCEU*)	− 0.5301	0.0571	− 9.29	0.000	− 0.6484	− 0.4117
ln(*TO*)	− 0.1602	0.0142	− 11.31	0.000	− 0.1896	− 0.1308
ln(*FBS*)	0.0202	0.0125	1.62	0.120	− 0.0057	0.0461
ln(*FTS*)	0.2150	0.0484	4.44	0.000	0.1147	0.3153
ln(*ICTGE*)	− 0.0100	0.0182	− 0.55	0.587	− 0.0477	0.0276
ln(*ICTGI*)	0.1578	0.0272	5.81	0.000	0.1015	0.2141
ln(*IUI*)	0.1399	0.1092	1.28	0.214	− 0.0866	0.3665
ln(*MCS*)	− 0.0367	0.0509	− 0.72	0.479	− 0.1422	0.0689
ln(*GDPPE*)	− 0.1678	0.0451	− 3.72	0.001	− 0.2614	− 0.0742
ln(*IGDPD*)	0.0661	0.0191	3.45	0.002	0.0264	0.1057
ln(*FLEB*)	− 1.4992	1.2872	− 1.16	0.257	− 4.1687	1.1703
ln(*FP*)	1.0522	0.3150	3.34	0.003	0.3989	1.7055
ln(*REC*)	− 0.0583	0.0133	− 4.37	0.000	− 0.0859	− 0.0306
ln(*SR*)	− 7.9382	1.6271	− 4.88	0.000	− 11.3125	− 4.5638
ln(*WBLI*)	− 0.5126	0.1221	− 4.2	0.000	− 0.7658	− 0.2595
ln(*FSE*)	− 0.1318	0.0200	− 6.6	0.000	− 0.1732	− 0.0904
Con. Coef	11.6533	6.4025	1.82	0.082	− 1.6247	24.9314

Source Estimated by Authors

equality. Gender inequality is projected to be decreased by 0.058% as renewable energy consumption increases by 1%.

Conversely, cereal yield and energy consumption would maintain the health of women. Hence, the coefficients of *CY* and *PCEU* with *GIIV* are found negative and statistically significant at 1% significance level. Hence, the estimates simplified that cereal yield and per capita energy consumption enhance gender equality. Gender equality is predicted to be increased by 0.15% and 0.53% due to 1% increase in cereal yield and per capita energy consumption, respectively. Furthermore, cereal yield also explores the overall agricultural development. Further, agricultural development may nurture jobs for women. Thus, agricultural development helps to increase gender equality and women empowerment.

Extensive employment opportunities for females would enhance women empowerment and gender equality. Thus, the coefficients of *FLFPR* and *RFMLFER* with *GIIV* are noted negative. The estimates signify that gender equality improves as female labour force participation rate and ratio of female labour in total labour force increase. Further, it is perceived that gender inequality is likely to decrease by 0.26% and 0.31% due to 1% increase in *FLFPR* and *RFMLFPR*, respectively. The findings of earlier scholars like Anyanwu and Augustine (2013) also proposed that employment would be valuable to increase gender equality. Conversely, gender equality is negatively associated with female unemployment. Gender inequality, thus, increases as the female unemployment rate increases. The positive coefficient of FU with GIIV, reveals that female unemployment is caused to increase in gender inequality.

Exports of goods and services is an important indicator to explore the performance of trade openness (*TO*). While *TO* improved after globalization. Thus, it produces a vital impact on gender equality. The negative coefficient of *TO* with *GIIV*, specifies that trade openness showed a negative impact on gender inequality. Thereupon, the estimate infers that gender inequality is projected to decrease by 0.16% as trade openness increases by 1%.

FBS, *FTS*, *ICTGE*, *ICTGI*, *IUI* and MCS are favourable variables to nurture an effective ecosystem for digitalization and digital infrastructure. These are the important digital technologies (DTs) which provide ample opportunities to the economic agents. However, the positive coefficients of *FBS*, *FTS*, *ICTGI* and *IUI* with *GIIV*, demonstrate that gender inequality could be increased due to an increase in these variables. While *ICTGE* and *MCS* are observed to reduce gender inequality. Previous researchers like Asongu (2015); Tchamyou et al. (2019) also recognized that mobile phone technologies could be useful to reduce gender and income inequality.

The coefficient of *GDPPE* with *GIIV* is observed negatively. The estimate, therefore, reveals that GDP per employment leads to decreased gender inequality. Gender inequality is predicted to decrease by 0.1678% due to 1% increase in GDP per person employed. While inflation reduces the purchasing power, saving and investments. Inflation also produces a negative impact on the level of employment and consumption, and demand and supply of products in the market. Hence, inflation may cause increased gender inequality. The positive coefficient of *IGDPD* with *GIIV* implies that gender inequality is expected to be increased by 0.0661% due to 1% increases in inflation.

The female population also has a positive impact on gender inequality. Gender inequality is projected to increase by 1.0522% due to 1% increase in the female population. While gender equality is expected to increase as sex ratio increases. The negative coefficient of *SR* with *GIIV*, suggested that gender inequality is predicted to increase by 7.9382% as sex ratio increases by 1%.

A business ecosystem for women is highly imperative to increase gender equality. The negative coefficients of *WBLI*, and *FSE* with *GIIV* suggest that gender inequality decreases as the women business and law index, and the female self-employed

Table 6 Validity of the regression coefficients

No. of lags	Lag (1)	Lag (2)	Lag (3)	Lag (4)	Lag (5)	Lag (6)	Lag (7)	Lag (8)
Lag (1)	1.0000							
Lag (2)	0.9889*	1.0000						
Lag (3)	0.9832*	0.9889*	1.0000					
Lag (4)	0.9774*	0.9836*	0.9894*	1.0000				
Lag (5)	0.9724*	0.9783*	0.9848*	0.9902*	1.0000			
Lag (6)	0.9670*	0.9729*	0.9788*	0.9851*	0.9900*	1.0000		
Lag (7)	0.9616*	0.9675*	0.9735*	0.9787*	0.9853*	0.9900*	1.0000	
Lag (8)	0.9556*	0.9640*	0.9693*	0.9736*	0.9794*	0.9858*	0.9906*	1.0000

Source Estimated by authors

increase. The statistical values suggested that gender inequality may decline by 0.5126% and 0.1318% due to 1% increase in *WBLI* and *FSE*.

4.3 Validity Verification of Regression Coefficients

Table 6 provides the results that were used to check the validity of the regression coefficients. The validity is verified as estimating the correlation coefficients among the lags of error-terms. A statistically significant correlation among the various lags of error-terms. The estimates, thus, claim that regression coefficients of core, mediator and control variables with *GIIV* have validity.

4.4 Elasticities of Gender Inequality

Elasticity measures the % variation in a specific variable due to 1 unit or 1% increase in its determinants. The results imply that it is an elastic relationship between *GIIV* with most variables (except, *FU*, and *ICTGE*). *CY*, *PCEU*, *GDPPE*, *FLEB*, *FP*, and *WBLI* are perfectly elastic with *GIIV* (Table 7). Thus, cereal yield, per capita energy use, GDP per person employed, female life expectancy, female population, and women business has a greater impact on gender inequality. Further, the results emphasized that variation in gender inequality also depends on *HDI*, *FLFPR*, *RFMLFPR*, *PCCOE*, *FSE*, *TO*, *FTS*, *IUI*, *MCS*, and *SR*. Remaining variables like *EE*, *HE*, *FBS*, *ICTGI*, *IGDPD* and *REC* have low impact on gender inequality.

Table 7 Elasticities of gender inequality with EVs

Variables	Mean	Reg. Coef	Elasticity
GIIV	0.2006		
HDIV	0.8465	– 3.3206	– 14.0110
EE	4.5842	– 0.1902	– 4.3462
PCCOE	8.2675	0.3131	12.9031
CY	4862.2463	– 0.1495	– 3623.3963
HE	7.8465	– 0.0343	– 1.3416
FLFPR	39.2828	– 0.2632	– 51.5378
FSE	17.5807	– 0.1318	– 11.5502
RFMLFPR	74.0258	– 0.3132	– 115.5693
FU	8.6636	0.0169	0.7298
PCEU	3441.8131	– 0.5301	– 9094.5828
TO	47.5385	– 0.1602	– 37.9617
FBS	20.1913	0.0202	2.0331
FTS	34.2732	0.215	36.7308
ICTGE	7.6205	– 0.01	– 0.3799
ICTGI	9.4491	0.1578	7.4325
IUI	61.6208	0.1399	42.9717
MCS	104.7376	– 0.0367	– 19.1605
GDPPE	90,828.3408	– 0.1678	– 75,971.5569
IGDPD	4.5118	0.0661	1.4866
FLEB	80.3200	– 1.4992	– 600.2343
FP	50.7378	1.0522	266.1138
REC	16.6224	– 0.0583	– 4.8306
SR	1.0587	– 7.9382	– 41.8936
WBLI	85.6099	– 0.5126	– 218.7461

Source Estimated by authors

5 Conclusion and Policy Implications

The main objective of this chapter is to examine the crucial determinants of gender equality. Thereupon, it evaluates the impact of DTs and other variables on gender inequality in G20 countries using a log-linear regression model. Country-wise imbalanced panel data during 2002–2022 is implemented to attain the mentioned objective. Spearman's rank correlation coefficient technique is applied to examine the association of gender inequality with specified 24 variables. These 24 variables are considered as core, mediator and control variables. The elasticities of gender inequality with these variables are also calculated in this research.

Human development, education expenditure, health expenditure, energy use, trade openness, fixed broadband subscriptions, fixed telephone subscriptions, internet use, mobile cellular subscription, GDP per person employed, female life expectancy, female population and women business are negatively associated with gender inequality. While gender inequality is positively correlated with female unemployment and inflation. The regression results reveal that human development, education and health expenditure, cereal yield, participation of women in employment, per capita energy use, trade openness, GDP per person employed, female life expectancy, renewable energy consumption, sex ratio, self-employed are found as important determinants to reduce gender inequality in G20 countries. Hence, these variables would be supportive to SDGs in G20 countries. Female unemployment, inflation and female population are leading to increased gender inequality. Moreover, mobile cellular and ICT goods exports have a negative impact on gender inequality. The results also suggested that gender inequality is perfectly elastic with most variables (excluding, female unemployment, ICT goods exports, and inflation). While cereal yield, per capita energy use, GDP per person employed, female life expectancy, female population, and women business have a greater impact on gender inequality as compared to other explanatory variables.

Digital skills of women are essential for increasing their participation in the labour market (Grau-Sarabia and Fuster-Morell 2021). Thus, the G20 countries should increase the digital skills of women to reduce gender inequality in all dimensions. G20 countries should create business opportunities for women to reduce gender inequality. Equal rights for women in the education, business and financial sector would ensure gender equality in G20 countries. It is necessary to increase huge investment in the digital sector to increase the ICT accessibility of women (Grau-Sarabia and Fuster-Morell 2021). The government should ensure internet accessibility for all to reduce inequalities. Extensive employment opportunities for women would enhance gender equality (Anyanwu and Augustine 2013). Initiation of household welfare programmes would be promotive to increase gender equality (Anyanwu and Augustine 2013). Agricultural development, per capita energy use, renewable energy and women business opportunities would also enhance gender equality.

References

Adam IO, Alhassan MD (2021) The effect of mobile phone penetration on the quality of life. Telecommun Policy 45(4):1–10. https://doi.org/10.1016/j.telpol.2021.102109

Anyanwu JC, Augustine D (2013) Gender equality in employment in Africa: empirical analysis and policy implications. African Development Rev 25(4):400–420. https://doi.org/10.1111/1467-8268.12038

Asongu S (2015) The impact of mobile phone penetration on African inequality. Int J Soc Econom 42(1):706–716. https://doi.org/10.1108/IJSE-11-2012-0228

Asongu SA, Orim SMI, Nting RT (2019) Inequality, information technology and inclusive education in sub-Saharan Africa. Technol Forecast Soc Change 146:380–389. https://doi.org/10.1016/j.techfore.2019.06.006

Bouanza JRFK, Asongu SA, Nnanna J (2024) Business constraints, information technology and income inequality in Sub-Saharan Africa. J Entrepreneurship 33(4):725–760. https://doi.org/10.1177/09713557241306884

Burrell J (2010) Evaluating shared access: social equality and the circulation of mobile phones in rural Uganda. J Comput-Med Commun 15(2):230–250. https://doi.org/10.1073/pnas.1909326117

Chen D (2004) Gender equality and economic development: the role for information and communication technologies. *World Bank*. https://documents1.worldbank.org/curated/en/288621468778204692/pdf/WPS3285.pdf

Cooray A (2012) Suffrage, democracy and gender equality in education. Oxf Dev Stud 40(1):21–47. https://doi.org/10.1080/13600818.2011.646976

Fatima U, Ahmad N, Angeliki Menegaki N, Wang H (2025) How do information and communication technologies (ICTs) empower women? Women's Stud Int Forum, 109(1):1–13. https://doi.org/10.1016/j.wsif.2025.103057

Ganda F, Ruza C (2025) Unveiling the non-linear dynamics: quantile regression of financial development's impact on environmental degradation in BRICS and G7 nations. Discover Sustainability 6(60):1–26. https://doi.org/10.1007/s43621-025-00792-y

Grau-Sarabia M, Fuster-Morell M (2021) Gender approaches in the study of the digital economy: a systematic literature review. Human Soc Sci Commun 8(1):1–21. https://doi.org/10.1057/s41599-021-00875-x

Gurumurthy A (2006) Promoting gender equality? Some development-related uses of ICTs by women. Dev Pract 16(6):611–616. https://doi.org/10.1080/09614520600958298

James J (2013) Mobile phone use in Africa: implications for inequality and the digital divide. Soc Sci Comput Rev 32(1):113–116. https://doi.org/10.1177/0894439313503766

Kouladoum JC (2023) Digital infrastructural development and inclusive growth in Sub-Saharan Africa. J Soc Econom Dev 25(1):403–427. https://doi.org/10.1007/s40847-023-00240-5

Lu J, Xiao Q, Wang T (2023) Does the digital economy generate a gender dividend for female employment: evidence from China. Telecommun Pol 47(6):1–10. https://doi.org/10.1016/j.telpol.2023.102545

Marler W (2018) Mobile phones and inequality: findings, trends, and future directions. New Media Soc 20(9):3498–3520. https://doi.org/10.1177/1461444818765154

Mbodji YC (2024) Inclusive education, gender inequality and public expenditure in sub-Saharan Africa. Rev Educ 12(1):1–10. https://doi.org/10.1002/rev3.3452

Ndoya H, Asongu SA (2024) Digital divide, globalization and income inequality in sub-Saharan African countries: analysing cross-country heterogeneity. Soc Respons J 20(1):1–19. https://doi.org/10.1108/SRJ-07-2022-0277

O'Donnell A, Sweetman C (2018) Introduction: gender, development and ICTs. Gender Dev 26(2):217–229. https://doi.org/10.1080/13552074.2018.1489952

Olayemi OM, Olayemi KJ (2024) Promoting gender equality through ICT: a pathway to achieving sustainable goals in Nigeria. J Susta Dev 1(2):1–12

Panichsombat R (2016) Impact of internet penetration on income inequality in developing Asia: an econometric analysis. ASR: CMU J Soc Sci Human 3(2):151–167. https://www.thaiscience.info/Journals/Article/CMUS/10989410.pdf

Potrafke N, Ursprung HW (2012) Globalization and gender equality in the course of development. Eur J Pol Econ 28(4):399–413. https://doi.org/10.1016/j.ejpoleco.2012.04.001

Prabha TP, Alok S, Kumar R, Singh S (2023) The digital revolution—implications of digital technologies on women's workforce participation. In: Sharma R, Shishodia A, Gupta A (ed) Fostering sustainable development in the age of technologies. Emerald Publishing Limited, Leeds, pp 303–318. https://doi.org/10.1108/978-1-83753-060-120231021

Rani U, Castel-Branco R, Satija S, Nayar M (2022) Women, work, and the digital economy. Gend Dev 30(3):421–435. https://doi.org/10.1080/13552074.2022.2151729

Shah CS, Krishnan S (2024) ICT, gender inequality, and income inequality: a panel data analysis across countries. Inf Syst Front 26(1):709–727. https://doi.org/10.1007/s10796-023-10396-4

Shikongo S, Iyawa G, Hamunyela S (2023) Can digital technologies tackle gender inequality? a review. In: Proceedings of the international conference on information systems and emerging technologies (ICISET). https://doi.org/10.2139/ssrn.4663419
Sorgner A, Bode E, Krieger-Boden C (2017) The effects of digitalization on gender equality in the G20 economies. Kiel Institute for the World Economy, Kiel, Germany
Sorgner A (2019) The impact of new digital technologies on gender equality in developing countries. Department of policy, research and statistics working paper 20/2019, United Nations Industrial Development Organization, Vienna. https://downloads.unido.org/ot/16/76/16760725/WP_20_FINAL.pdf
Tchamyou VS, Asongu SA, Odhiambo NM (2019) The role of ICT in modulating the effect of education and lifelong learning on income inequality and economic growth in Africa. African Dev Rev 31(2):261–274. https://doi.org/10.1111/1467-8268.12388
Wajcman J, Young E, Fitzmaurice A (2020) The digital revolution: implications for gender equality and women's rights 25 years. https://www.unwomen.org/sites/default/files/Headquarters/Attachments/Sections/Library/Publications/2020/The-digital-revolution-Implications-for-gender-equality-and-womens-rights-25-years-after-Beijing-en.pdf
Wakunuma KJ (2013) Mobile for development in Africa. Feminist Africa 18(1):131–139
Wu Y, Li L, Zheng Y (2024) The impact of digitization in manufacturing on female employment and gender wage. J Asian Econom 95(1):1–12. https://doi.org/10.1016/j.asieco.2024.101821
Ye R, Cai X (2024) Digital transformation, gender discrimination, and female employment. System 12(5):1–17. https://doi.org/10.3390/systems12050162
Zhang X, Chang M, Zhang C, Zhang S, Lin Q (2025) The impact of internet use on income inequality from different sources among farmers: evidence from China. Agriculture 15(8):8–18. https://doi.org/10.3390/agriculture15080818

The Impact of Artificial Intelligence (AI) and Technology on Gender Dynamics in the Workplace

Arpita Nayak, Ipseeta Satpathy, and Vishal Jain

Abstract AI, in conjunction with digital technology, causes profound shifts in employment relations, the organization of work, and the way business norms are handled at work. The advancement of technology that enhances output increases complexity, erasing the gains made in the pursuit of gender equality within professions. This paper explores how AI implementations, both regarding gender bias, entail the maximum exacerbation of inequality measures existing when technical development teams are not diverse or algorithms and datasets are opaque, while regarding equitable employment potential, there is at the same time likelihood-maximisation of transformative potential. This research combines fundamental principles of intelligent systems development along with gender studies and labor economics to promote workforce digitization which places humans at its core. Our research demonstrates the necessity of finding an equilibrium between technological advancement and gender equality protection to ensure digital work transformation brings the same value to every workforce member. Our study presents tangible action items which policymakers along with organizational leaders and technology developers, should implement to establish inclusive and equal artificial intelligence-driven work environments.

Keywords AI bias · Workplace technology · Gender equality · Algorithmic discrimination · Digital transformation · Inclusive innovation · Workplace dynamics · Diversity and inclusion

A. Nayak (✉)
KIIT School of Liberal Studies, KIIT DU, Bhubaneswar, India
e-mail: 2181158@ksom.ac.in

I. Satpathy
KIIT School of Management, KIIT DU, Bhubaneswar, India
e-mail: ipseeta@ksom.ac.in

V. Jain
Department of Computer Science and Engineering, School of Engineering and Technology, Sharda University, Greater Noida, India

A. Singh et al. (eds.), *Human-Centered Sustainability: Bridging Well-Being, Gender Equity, and Technology*, World Sustainability Series,
https://doi.org/10.1007/978-3-032-22020-2_10

1 Introduction

Work and organizational processes can be considered having gone through a significant historical revolution through the integration of artificial intelligence (AI) and the latest digital technologies. Over the past ten years various advancements in the categories of machine learning automation and data analysis revolutionized sectors and industries as well as changed job duties and the training of current and prospective employee forces. Modern technological tools ranging from predictive analytics, robotics, and process automation in business operations improves business operations and productivity besides resulting in the generation of enormous economic value. While the novelties in technology brings operational and productivity improvements to industries, it also results in transformation of labor market complexities. It becomes easier to classify the current changes in employment paradigms, reframing work forms and reshaping working environments of today's world into the context of the digital age. The systems of AI are changing the nature of the done work and the methods of work process organization as well as estimating and appraising the contemporary organizational work. For technological changes to be effective in such a process, it needs careful direction for successful technological advancement since such improvements should be able to bring about a transformation in the workplace as well as the impact on employment and the organization in the technological era must be managed since AI technology coupled with today's advancement demands humans to engage in different working relations between human and machines and this warrants new forms of skills development across all the organizational levels. Thus, increased productivity in conjunction with germination of new problems of workforce development presents a challenging paradox which requires development of a proper strategy. The workplace digitalization process has sped up because of remote collaboration tools and AI-powered recruitment systems and performance monitoring tools that make human–machine partnership an indispensable requirement. The technological transformation of workplaces impacts different workers differently because gender inequalities continue to worsen in the automated society (Mirbabaie et al. 2022).

The advancement of artificial intelligence has shifted from basic computational operations to include complex decision-making in employee selection and business project management as well as creative tasks. Cloud platforms provide real-time intercontinental teamwork capabilities yet IoT technologies generate large-scale productivity measurements of employees. Such advances in technology result in dual effects that destroy manual and cognitive positions especially in administrative and manufacturing but generate new combined roles that require technical expertise together with emotional competencies and ethical decision-making capabilities. Organizations were compelled by the COVID-19 pandemic to implement digital workflows at breakneck speed which led many employees to fail in their adaptive efforts. The adjustment of the workforce takes place while keeping in place existing workplace inequalities that limit upskilling accessibility and technical resources distribution. Women who encountered systemic barriers before digital workplaces now deal with both new digital versions of previous obstacles and totally unique forms of discrimination in algorithm-based professional settings.

The various obstacles that hinder gender equity continue to endure even though surface-level evidence suggests increased progress. Female labor participation has risen to near half the worldwide workforce but they still working in lower-valued care sector positions that cannot be replaced by automation technologies. The "glass ceiling" barrier has developed new forms in Fortune 500 companies during 2021–2023 since female CEO numbers fell from 7.4% to 6.6% even though middle-management diversity grew. Gender disparities in the workforce experience increased complexity due to digital workplaces because technology tools that use historic data for training risk maintaining gender biases against women and because endless availability of virtual work worsens the situation for woman who must maintain household management. The technologies that contribute to advancement of gender inequality possess tools that could potentially solve these problems by using blockchain credentialing for transparent career paths and VR training for leadership development. The main problem requires engineers to build digital solutions with focused gender considerations instead of assuming technical neutrality (Howard 2019).

The efforts to achieve gender equality in the last hundred years have yielded substantial improvements yet women remain impeded by fundamental labor system inequalities in their occupational paths and high positions of power. Women are consistently fewer in technical fields that pay well but appear more frequently in risky positions which automatically qualify for termination. Women experience substantial wage discrepancies compared to men during similar work and leadership positions overwhelmingly belong to men particularly in STEM and corporate boardrooms. Structural barriers together with societal rules and workplace discriminators block women from obtaining mentorship programs and networking opportunities while preventing their advancement in their careers. AI technology presents organizations with new hiring possibilities by utilizing algorithms that must be carefully designed to avoid bias creation but allows flexibility in digital training methods which help women succeed in the workforce. The effective comprehension of these dynamics stands essential to establish technological progress which suppresses pre-existing inequalities and serves as an equitable driver for human growth (Tiwari 2024). Through market trend and statistical information with industry-specific examples this analysis shows how technological automation dismantles gender traditions and enables new worker empowerment through job training and flexible employment systems and unbiased AI platforms. Research-oriented primary target seeks to transform workplaces using technology so all workforce members gain equivalent support regardless of gender while establishing diverse leadership frameworks. Best practices will thereafter be provided to guide both policy makers and business and advocacy stakeholders in their endeavour to promote digital workplace equity.

I. **Research Questions**

- In what ways are AI and digital technologies reshaping gender roles and participation in the modern workplace across sectors?

- How do algorithmic decision-making systems impact gender equality in recruitment, evaluation, and leadership representation, and what safeguards can promote fairness?

II. **Research Objectives**

- To explore how AI and emerging workplace technologies are influencing gender dynamics, with a focus on both opportunities and challenges.
- To critically examine the role of algorithmic bias and ethical considerations in shaping equitable and inclusive digital work environments.

III. **Understanding Gender Dynamics at Work**

Hailed as the most revolutionizing process in work and Management of organizations' advancement in digital technologies and artificial intelligence. The last ten years saw the improvements in automation and growth in the field of machine learning as well as in data analytics that revolutionized the jobs and created shifts in the work capabilities of employees in various industries. Business processes witness revolutionary changes with help of AI and intelligent solutions based on predictive analytics and robotic process automation to improve operational efficiency and productivity, as well as to support better decisions for all kinds of enterprises. The economic saving benefits and operation improvements derived from these IT applications bring about complex tasks that redefine work setting. Digital transformation challenges the set work practices since it eliminates some occupations and requires new ways of working. Advanced Administration in contemporary organizations undergo drastic change in work activity facilitated through AI systems concerning the accomplishment of tasks and their appraisal as well as employee value estimation. Fragile change management now requires managing technological advancement in parallel with the transformation of employees leads to achieving positive outcomes of digital transformation that maintain employment systems and structures. The contemporary revolution brings a new human–machine partnership that supplements human abilities in a way that demands repeated updates of skills and change of attitude in many tiers of an organization. Digital transformation offers an opportunity to achieve productivity improvement as never before but at the same time raises considerable workforce development issues and job security threats thus requiring sound policies. (Turesky 2020).

Technological innovations particularly, the use of enhanced ICTs and the integration of artificial intelligence have been strongly recognized as the most revolutionary innovation applicable to work and organizational processes. The recent years have seen major advancements in automation, concurrent to artificial and data analytic intelligence, and these disrupted industries by reconstructing job functions and evolving skills and competencies. Targeted business operations reap unthinkable changes through the application of artificial intelligence prominently marked by integrated data analysis and robotic mechanisms to increase functionality and productivity on business processes as well as the capability to make decision based solely on analytical findings across all business sectors. The many benefits that ensue from business and economic returns and operational improvements resulting from such technology adoption can be vast in terms of work transformations. Digitalization erases some traditional work structures, forcing new management system while on the other hand, it calls for new ways of working. Essentially, the required change requires the concurrent leadership of technology advances together with employee adaptation to guarantee the achievement of digital transformation outcomes that secure employment systems and structure. The current wave of automation and artificial intelligence heralds a new form of symbiosis between man and machine, enhancing the human potential but demanding constant skill enhancement and shift in attitudes at individual, group and organizational levels. The new opportunities created by digitalisation present an opportunity to achieve productivity growth rates that are almost unprecedented but this comes with corresponding workforce development challenges and threats to job security which have to be addressed with proper policies (Oguns n.d.; Chowdaiah et al. 2024).

There is need for an intersectional analysis in organizational gender studies in terms of how diverse factors that include race and social status impact on women's workplace barriers.Black women while comparable in educational standards as their white counterparts are only 3% of tech sector employees but they experiences both racial and gender discrimination.. Working-class women experience different digital exclusion barriers than middle-class women since they lack home broadband access and do not have enough money to acquire required certifications needed for tech industry employment (Ceia et al. 2021). The job selection process begins by using filtration algorithms to exclude HBCU institutions so that it continues with performance evaluation tools designed from predominantly white male behavioral data. The Covid-19 pandemic sharpened the growing divide between sexes because women of color held frontline essential roles outside home but carried the most caregiver duties during school shutdowns. These effective solutions combine separate types of discrimination into a single all-inclusive package that includes digital skills instruction along with child care help and hardware loan programs. Research shows that advanced workplace technology achieves better results through intersectional design methods that build platforms for multi-dimensional communication capabilities across cultural demographics. Multiple identities require systematic integration during inclusive design since human diversity's complex nature makes single-axis diversity programs struggle to succeed (Campbell 2020).

During industrialization work-from-home gained institutional independence from workplace roles which established norms for men to earn outside their homes while women performed household duties. The historic valuation of jobs persists in current pay differences because jobs historically designated female such as nursing and teaching receive lower pay than comparable positions monitored by males. Men hold higher authority roles than women at all career levels even when carrying out the same programming tasks because front-end positions attract more women than men who choose back-end systems roles. Experts developed these patterns when computer programming moved from female-dominated clerical work to genius-level labor status yet preserved its male cultural roots in the middle of the twentieth century. History reveals the importance of understanding biases because it shows how gender stereotypical beliefs from the past continue to exist in apparently neutral categories of work which new advanced technologies will automate (Hartmann 1976).

2 AI and Emerging Technologies in the Workplace

Modern workplace transformation results from artificial intelligence (AI) and automation alongside advanced analytics because these technologies implement powerful changes to organizational processes and employee resource allocation and decision-making systems. Due to their ability to perform candidate screening and administrative tasks using robotic process automation (RPA) machine learning algorithms deliver exceptional operational efficiency but create significant privacy issues and work system stability questions. The current AI technology system controls complete employee domains through its authority over both hiring decisions and rating evaluations along with working language analysis and workforce sentiment assessments. The widespread adoption of remote work due to COVID-19 has made digital collaboration tools into vital systems which analyze employee communication activities. Modern technology implementations produce variable employment results that depend on each worker's specific situation. Studies show that artificial intelligence can worsen income disparities between genders which results in discrimination against women and underrepresented groups during job recruitment and career advancement cycles. Organizations must understand all technological potentials to design workplaces which maximize innovation and minimize secondary consequences (Mellul 2018).

Employment structures are undergoing transformation due to multiple disruptive technology convergences. Modern AI technology coupled with machine learning systems performs resume assessments alongside employee turnover predictions through enormous datasets that reveal patterns which traditional managers would normally miss. Analytical tools that predict employee behavior analyze multiple metrics which include productivity measures and device-based metrics as well as biometric data for performance evaluation and wellness assessment. Robotic process automation has taken over standard cognitive and manual duties throughout accounting and legal document evaluation and customer service which threatens to shift women out of their traditional administrative work positions. Manufacturing and healthcare sectors use collaborative robots (cobots) to work together with humans thus transforming physical labor operation. These technological systems together form a paradoxical situation. New hybrid roles which blend technical capabilities with emotional intelligence and creativity and ethical judgment emerge from technology-driven job elimination processes despite these technical advancements often developing through normative patterns of gender socialization (Zirar 2023).

Human resource departments leverage AI implementations which showcase the optimistic possibilities and urgent concerns associated with workplace technological adoption. Algorithmic systems show potential to decrease human biases by eliminating prejudice from the process but they tend to maintain previous discriminatory attitudes in their operational algorithms. Amazon scrapped its AI recruiting platform because it demonstrated bias against applications using terms such as "women's" and female-only educational information. Productivity software for remote work follows keystrokes and tracks video calls and monitors computer usage including mouse movement but fails to measure team-based assignments where women excel. AI-driven leadership assessment tools monitor email content as well as attendance at meetings while studying network interactions which might produce disincentives based on the language patterns typical to female and non-native English-speaking candidates. AI tools designed for career pathing show women alternative professional paths though their models do not reflect current insights about gender differences in interests and abilities. Workplace technologies demand gender-aware design features while ongoing audits for novel algorithmic biases form the core requirement for modern workplaces (Cheon et al. 2021).

Remote work transformations together with hybrid work models created fresh connections between workplace technology systems and gender-based equality patterns. Employee business continuity improves by using video conferencing tools such as Microsoft Teams and Zoom and Slack alongside new types of visibility biases show women face video observation 20% more often. When caregivers rely on asynchronous communication platforms the technology should create flexibility however these platforms often force caregivers to maintain constant availability due to unwritten expectations. Productivity monitoring systems operating remotely fail to recognize the dual role working mothers handle which results in performance consequences for offline caregiving activities in standard work hours. Working under hybrid systems threatens to generate two work-levels because regular in-person meeting participants who are mostly male non-caregivers gain superior mentorship access and career advancement possibilities. Remote work technologies possess the potential to improve equitable engagement yet existing deployments end up sustaining or intensifying conventional workplace gender structures during online communication (Fan et al. 2025).

Multiple gender-specific challenges emerge from the work process integration of advanced technologies which need immediate solutions. Algorithmic management focuses on perpetual productivity measurements which harm staff members who need care responsibilities yet emotion recognition technology within customer service forces employees to adhere to specific gendered emotional protocols. Future virtual workspaces and metaverse platforms will absorb real-world discrimination unless they are purposefully built to be inclusive. Technology provides both beneficial opportunities while simultaneously containing unique possibilities for transformation. The combination of blockchain credential systems with AI mentor matching stands to provide returning female workers a way to demonstrate acquired skills and break through traditional business networking restrictions. Through digital twin simulations organizations can evaluate flexible scheduling policies ahead of implementation to discover any unintended gender bias. Proactive interventions through gender impact assessment of new workplace technology alongside diverse team participation and algorithm transparency frameworks will help uncover positive technological potential (Laat 2023).

Resolving gender-based impacts from workplace technologies requires united efforts throughout various sectors. The implementation of auditing requirements for hiring algorithms and workplace surveillance tools needs standardization from legislators who draw their models from the EU's proposed AI Act for high-risk systems. Organizations should establish prize-based programs that motivate workers to discover discriminatory operation patterns within their company systems. Companies that run HR departments must gain firsthand understanding of technical systems to assess vendor declarations regarding algorithm neutrality effectively. Organizations that lack ongoing cultural transformation alongside their digital initiatives will make existing inequalities worse—digital transformation succeeds when organizations invest in leadership development for inclusivity and caregiving support. Organizations should implement three emerging best practices starting with employee-led technology councils to test new systems while performing gender-based impact analyses before implementation and developing standardized protocols to bypass algorithm recommendations in cases where human judgment reveals bias (Bacchi 2017).

The progression of workplace technologies requires more research on multiple important concerns. The introduction of generative AI tools such as ChatGPT presents questions about how their implementation will impact the value of writing and communication abilities that women disproportionately master yet receive minimal professional appreciation. Information of women's advancement in occupations in AI saturated systems as well as the conventional employment in connection with International overviews of AI work might be useful. More research is required to understand the impact of multiple group memberships on the use of technologies such as how speech recognition services work differently between genders and ethnicities and how AI productivity tools work for employees with neurodiverse workplaces.

Choosing AI integration with the workplace technologies today directs systems to two very different paths: There are thus two possibilities: the systems that replicate the discrimination of the past or the new working models. When it comes to making decisions about the design of the system, to the ways of implementing it, and all the policies and regulations that are to be put in place to support this process, they build two different futures. These technological developments inevitably become seen as manifestations of development team philosophy where engineering specializations collaborate on and contribute towards design instead of merely appended to the process. Those organizations that undertake effort to analyze the technological influence and maintain consistent pattern of resource distribution tend to attain higher market standings. These approaches offer such advantages as higher employee retention and a considerably broader range of possibilities to choose talents, better compliance with the societal tendencies. This is because ethical AI systems allow organizations to reap the risk reduction advantages and at the same time allow the organization to garner a competitive advantage than the others in value-oriented customer markets. Theoretical and empirical evidence can guide business organizations exactly on how to adopt technology for gender in a way that organizational and ethical constraints can allow. operation guidelines for real-world case explorations that lays down functional strategies to setup workplace technologies for furthering fair treatment rather than for perpetuating biased work culture. Our modern epoch is the decisive era of workplace design because now delineating the form that organizational technologies of work will take in the future often means proscribing unfair treatment of employees. The following frameworks will create ways in which workplaces can promote equity rather than preserve inequity.

3 AI and Gender Bias: Risks and Realities

Risky and unbounded, the day-to-day decision-making at the workplace mediated through artificial intelligence entails a certain set of paradoxes. As has been put forward, being bias free, artificial intelligence systems maintain gender-based discrimination both at its current level and escalate it. AI systems learn employment discriminations during the training phase because of the historical data processed using methods that may appear to be statistical. The belief in unprejudiced AI stems from misconceptions about statistical bias and the notion that algorithms are capable of outcompeting human rationality while at the same time they do not consider the role that historical data does not capture future possibilities. The case therefore underlines the urgent need to develop stringent bias evaluation measures alongside continued algorithm oversight and active people control in order to prevent development of the digital gender inequality problem. Current solutions that manage to 'remove bias' maintain a troubling mirage of neutrality, while perpetuating tones of inequality that need changes during technology design and application at the workplace. To proceed, we must recognize that AI operates as a tool for expressing organizational biases that require conscious effort in algorithm design, so as not to embed existing discrimination into our automated systems today. New technologies use past employment record for gender bias and employ statistical analysis techniques to embed it in systems and make it less apparent. In employment cycle mechanisms women's issues arise in that when they leave their professions, they experience discrimination thus, the lack of proper understanding of team leadership redefines such behavior as failure. There are two conflicting questions about technology that researchers have to answer here. it can either perpetuate existing biases in places of work or drive them out. This phenomenon needs immediate examination of automated discrimination protections and including bias points into the algorithmic systems (Manasi 2022; Franzoni 2023).

Modern AI recruitment systems help companies with their hiring needs but their core technical components often embed gender discrimination within the system processes. An analysis of major employer hiring systems reveals that screening algorithms used by 75% of such organizations reduce candidate rankings when female attributes appear in resumes such as attendance at women's colleges or involvement in women's sports or professional obligations. Female candidates experience disadvantage during video interview evaluation because processing systems incorporate male-centric standard responses to measure verbal skills and emotional along with vocal attributes. Some recruitment systems create more problems by interpreting pauses between parental responsibilities in professionals' résumés as marks of incompetence since these algorithms do not understand the life transitions women commonly experience. These biases extend beyond hiring: Promotion recommendation systems favor workers who display continuous linear job progression patterns rather than candidates who have diverse professional experience. Wage discovery algorithms demonstrate poor understanding of intersectional influences so they fail to uncover the increased salary discrimination experienced by female minority workers. Regrettably both examples demonstrate a critical error in data-driven methodology while displaying a lack of insight into historic biases which lead AI systems to blindly reproduce unexamined biases (Pathak et al. 2023).

The implementation of algorithms results in observed gender bias patterns that create real-world hiring inequalities in employment selection processes. Amazon's experimental recruitment algorithm started to select male candidates over females by learning gender stereotypes after training with a decade of engineering hires. Terms such as "women's" and colleges admitting only women caused the model to distrust data preventing gender stereotypes from breaking. A consulting organization's artificial intelligence leader selection process demonstrates how deeply gender biases infiltrate the automated systems used for candidate evaluation. The algorithm left out female candidates on par with their male counterparts based on leadership skills although the usage of gendered term in the assessment records show no discrimination against the female candidates. The workforce analytics platforms direct the ladies to non-tech careers by using traits like agreeableness as personality indicators, because these are perceived as feminine but paradoxically are associated with non-technical incompetence. From the examples it becomes apparent of how bias finds its way into the AI systems, thus becoming the basis of discrimination of women and minority groups. Distorted definition at the beginning (masculinized leadership definition), chosen training data (promotion records) and the designed feature schema (psychological profiles as traits) served as the basis for bias formation. The errors of these systems perpetuate the discrimination of women and those in the minority, as well as maintaining the current ratios of power. It thus serves the purpose of an almighty feedback loop which imputes differential input data sets into each subsequent iteration in order to produce computerised discrimination on a large scale. Algorithmic bias needs remedies that affect the technical aspect because it is a part of a system that has inherent historical assumptions of organization and discrimination from the time of its design (Nadeem et al. 2022).

Artificial Intelligence business decisions are inclusive of sexism that gets amplified due to inherent flaws in data inputs and problems in designing the systems. Machine learning algorithms show the GIGO problem by mimicking the datasets fed to it which means that it will continue to uphold male leaders as a result of male-dominated hiring practices in the past. Several related biases are evident in these systems on account of their operation. Training data imbalances contribute to representation bias; measurement bias in leadership assessments includes masculine traits, and aggregation bias leads to generalized gender classification. Sentiment analysis tools show an example of this through their ironic expectations, which deem certain women as too forceful or, contrarily, not assertive enough to deserve and be worthy of winning; which are mentalities humans set when assessing women. As a result, the training on historical outcomes in positive male experiences leads these systems to equate success with male-selective characteristics while the training data is numerically equal. The use of self-reinforcing feedback is another way through which discrimination is entrenched into the larger society as modern systems feedback biased means as the new system means. AI functions to perpetuate human biases through standardized procedures that transform gender-based stereotypes into automated data conclusions yet generate obstacle-to-discovery for existing discrimination challenges. Technical solutions alone will not solve these problems because we need to fundamentally redefine the way algorithms determine merit and success and potential. Proficient AI systems maintain discriminatory behaviour through multiple operation cycles by receiving biased employment decisions which generate discriminatory training output that perpetuates bias and discrimination. The ongoing destruction of this pattern requires automated methodologies (adversarial de-biasing) and human analysts to study AI pattern acquisition and discrimination targets (Collet and Dillon 2019).

We need fundamental updates both during AI system development and post-deployment to address gender bias in workplace artificial intelligence systems. Ethical robotics development requires diverse expert teams because gender inequality continues to exist throughout technological sectors thus enabling teams to predict numerous foreseeable risks. Once organizations conclude their practice of algorithmic process secrecy decision-makers should subject their systems to independent audits for transparency purposes especially for employment-related systems. AI technology bias reduction efforts benefit from two proven methods which operate simultaneously. TheAIfield benefits from "algorithmic impact assessments" for predicting future data inequality levels during development cycles alongside "bias bounties" for researchers who discover biases in AI systems. New York City pioneers the mandatory monitoring of automated employment assessment devices yet current guidelines typically overlook gender-based protections. AI debiasing operations need sustained partnerships among data scientists and HR professionals and ethical experts and employee representatives to maintain ongoing improvements. AI systems require transparent processes for job challenges because automated systems which generate biased recommendations need efficient resolution methods. Seeking neutral results remains unattainable thus systems fighting historical inequities need to replace modern automation methods (Buslón et al. 2023).

Artificial intelligence programs in workplaces fight against gender bias through isolated methods that operate independently from one another. Counterfactual testing helps operators detect how protected variables such as gender affect results alongside synthetic augmentation providing balanced representation. AI-based process systems need future implementation to integrate human involvement through both artificial intelligence suggestion reviews by panels and a decision-determining elements display system. AI ethics departments within corporate structures must recruit gender experts who enable automated explanations to appear when algorithms modify employee professional prospects. Overall, analyzing unpredictable career paths in particular, the "Feminist AI" approach appears to have potential because it does endeavour to act intentionally against male-dominating systems. Each proposed strategy implements the same basic principle as their core foundation. Today's technology functions inside established power systems where human operators must constantly manage its interactions with these frameworks. The organizations that will be successful in the future must actively implement equal systems instead of depending completely on automated precision.

The existing gender bias in workplace interactions with AI technology creates both clear warnings about possible consequences and demands actionable solutions. The examples show how blind acceptance of algorithmic systems creates fast and large-scale implementation of discriminatory practices. The same technology has the potential to identify and fix human biases that have persisted for numerous generations as long as it receives careful design and rigorous monitoring. For positive stocks to ensue, the move beyond lifecycle-focused AI solutions is imperative since they remain fundamental to data programs and algorithmic systems as well as organizational criteria. Those responsible for deciding the future of organizations require transparency and accountability due to the dispersed work required to redesign systems with gender equality for algorithmic work. Engaging with this issue on the part of organizations serve two purposes, namely preserving the integrity of the organization while at the same time fostering the diversity of the workforce during the digital transition.

4 Opportunities for Advancing Gender Equality

AI solutions present the mechanisms of bias retention as well as the ways of advancing women's rights and making gender equality at work quicker. Currently, users stand to gain from artificial intelligence detection of inequality, thus enhancing the tools promoting equality advancement and social inclusion of women. Innovative audit techniques are supported with a complementary digital solution made up of two modules for revealing undisclosed gender pay bias and AI-based mentorship for all employees. New technologies enable effective solutions by dismantling organizational defensive positions that hinder traditional diversity initiatives thus enabling organizations to expand current solutions beyond traditional barriers. Superior performing AI systems acknowledge that real gender equality needs AI support beyond simple system changes modified for female users. Organizations need artificial intelligence help to find and restructure existing infrastructure systems. Organizations have begun adopting AI-driven diversity management approaches because they understand gender equality solution serves business needs while respecting morality thus these approaches are advancing from pilot testing to becoming fundamental elements of talent management strategies (Carroll 2024).

AI-powered advanced analytics help organizations to both identify and resolve the gender disparities that exist in compensation levels and career advancement. Current pay equity tools integrate traditional regression techniques with machine learning models to identify nuanced discrimination patterns which normal analysis methods often overlook. These tools track how differences in entry-level wages tend to grow as organizations distribute raises and bonuses inconsistently. The systems provide ongoing measurement unlike once-a-year reporting which shows only static information through example trends such as the point where female promotion data demonstrates decline across career periods. Modern platforms use natural language processing to scan performance reviews for gender biases through their analysis of equivalent achievements which receive different descriptions of "aggressive" for women and "assertive" for men. Predictive equity tools represent an innovative breakthrough because they demonstrate through simulation how existing pay decisions influence workforce evolution thus allowing organizations to intervene appropriately. These applications show how AI uses unique capabilities to disclose hidden discriminatory patterns that organizations can act upon but success requires openness from within the organization to respond to uncovered evidence (Volodzkiene 2025).

The deployment of artificial intelligence systems now helps create fairer work environments through advanced detection and elimination of workplace exclusionary practices. Current sentiment analysis systems use internal communication data in email threads and meeting transcripts to search for indications of interrupting behavior along with credit-taking abuse and gender-biased "office housework" distribution. By using network analysis tools organizations can uncover exclusionary events that prevent women from accessing informal communication networks and essential mentor relationships. Artificial intelligence software tracks meetings using virtual assistants to measure speech durations which support participation from every team member. Workplace assessment tools provide remarkable value by resolving common problems with measuring inclusion in online work settings. New applications detect workplace layouts using virtual reality technologies to identify built environments that signal unplanned exclusion zones. High-quality inclusive platforms use team operational discoveries to guide team dynamics improvements while avoiding individual-focused identification. The effectiveness of employee-developed technological work environments that recognize women's worth in the workplace grows stronger when privacy protection standards are strong (Araujo et al. 2024).

Multiple workplace equity issues emerge because educational platforms that implement artificial intelligence technology can successfully reduce the technical abilities gender gap. Adaptive learning systems personalize their training materials through machine learning algorithms which determine how students should enter their subjects and adjust their training speeds to maximize STEM practitioner success. Virtual reality training spaces supported by AI tutors create safe learning environments that ensure both stereotype-free training conditions and instant errorless feedback for teaching technical abilities. Promising educational frameworks emerge when skill development methods merge with contemporary credential systems. The digital badge infrastructure from blockchain technology provides validation of nonstandard professional achievements to eliminate hidden gender assumptions which affect female professionals with atypical career paths. AI technology tools help organizations evaluate job skills then identify gender-based skill differences to suggest personalized training programs. These tools reach their peak effectiveness alongside institutional support through child-care provision and post-training work support structures since access limitations alone cannot fix institutional problems. The swift digital transformation needs AI-based learning tools properly implemented to develop essential frameworks that enable female participation in future workforces (Patel 2024).

Artificial intelligence creates smarter mentor matching systems using algorithms that address traditional networking difficulties women encounter. Today's analytics platforms create mentor–mentee connections through skills assessment and goal and personality assessment profiles that allow employees at all levels to foster relationships beyond traditional networking limits. Mentorship applications use NLP for two reasons: providing tips for discussions during the mentorship meetings as well as understanding that leadership may require guidance as well. Organizational technology follows the productivity of the employees' work experience, which attracts the promotion one requires from senior executives. AI trainers and virtual reality systems shall be the future career tools as they offer dynamic support through AI workplace coaching and spice up the practice of professional advocacy through simulated scenarios provided by VR. Technological solutions are greatest when they foster structured cross-gender mentoring which educates men about the barriers women face at workplace and give women expanded opportunities to develop better professional networks. The latest advanced solutions complement talent management systems to design career development mentors from the programs to the actual advancement of programs. These tools make sure that equality in relation to relationships that shall facilitate career growth is achieved while eradicating techniques that have been designed to provide entry to producers based on the demographic tendencies of personal populations (Rasool 2017).

These technologies require several key implementation issues to be addressed in order to achieve their full interoperability. Careful management of privacy issues remains important since tools which evaluate employee communications or behavior exist. Organizations must develop strict rules on how to respond to AI-generated information to prevent unrealistically high expectations that do not result in meaningful organizational change. These systems function optimally when used to amplify rather than diminish human decision-making power along with organizations' active pursuit of equity goals. In the future, multiple emerging concepts will expand AI capabilities to accelerate gender equality progress. the development of intersectional algorithms that account for how gender combines with race, disability, and other factors. Improved intersectional algorithms will combine gender analysis with studies of race and disability through research clusters and enhanced use of generative AI to generate bias-aware training data, along with shared de-identified diversity data by consortium organizations for improved algorithmic fairness. Organizations will benefit most from these maturing technologies by transforming their thinking about gender equity from an inspirational objective to a measurable element of continuous business performance improvement.

AI enables rapid advancement of workplace gender equality through strategic deployment; however, organizations need to use these systems while actively monitoring their implementation to drive systemic transformation. Some of the tools include those that audit pay equity, promote inclusive practice and skill building, and others that support mentorship services go beyond the merely technical. These technologies offer ways and means of altering the culture and balance of power in organizations. However, through demonstrations of its ability to identify social bias, it is quite apparent that AI has so much more to offer in the way of equality for women than the sum total of untaken human decisions. Organizations that bring and harness AI potentials for fair workplaces with the management of risks create favourable organisational environments in place for all talents within the organisation. It means that gender equality in the algorithmic period is still in the sphere of human choices rather than in the development of the software itself.

5 Policy, Governance, and Best Practices

Gender is kept in equality from societal bias while making these options apparent by real AI during options beyond basic human options. When organizations align these brave workplace development capabilities with risk management strategies, it is stated that organizations develop strong organizational environments that all talent members in the organization would reap. There are indications that the claim of gender equality during the algorithmic period is still a function of human decision-making over the functioning of the technology. It is only through the coordinated use of legal measures and technical standards and best practices to align the aims and objectives of stakeholders that viable solutions may be found. AI policy on gender protection implies that gender would need to be factored in at the design and in the training data collected right through to the implementation phases as well as have strong accountability systems established in the technology at every stage that it is being used All businesses must build bias prevention methods throughout their complete process development cycle because early prevention cannot be delayed until the final stage of development. Organizations that fail to establish governance structures in their AI systems must face competitive losses as well as legal actions and employee and reputation damage. This represents a critical business requirement. Premier organizations stand at the forefront through their commitment to gender-equal AI systems because these systems represent both ethical responsibilities and strategic benefits that attract diverse talent and serve multiple markets.

Today's gender-inclusive AI policy standards build their foundation from three core principles - transparency alongside accountability and redress systems. According to the AI Act of the European Union employment decision tools qualify as high-risk systems which demand thorough bias auditing in addition to human oversight with explicit gender impact assessment requirements. Canada adapted the Algorithmic Impact Assessment tool for its government requiring gender-based analysis during AI system deployments and the technique has evolved to become routine across private sector AI implementations. As part of its Recommendation on the Ethics of Artificial Intelligence UNESCO establishes a requirement which forces member states to maintain gender equality through AI system protocols preventing gender bias amplification. Policies get to their targets, hence enforce hard policy coupled with more flexible soft laws. Organizations in Singapore can voluntarily use the AI Verify framework to test human resources technology systems while following benchmark standards for detecting gender biases and Iceland's equal pay certification program now forces auditors to include AI decision system checks during assessments. The regulatory methods demonstrate that achieving gender equity within AI requires both process standards that enforce diverse team participation and technical specifications that represent all demographic populations. The main frameworks demonstrate real progress through their intersectional bias testing protocols, including California's upcoming regulations for testing along race-gender subgroups, because gender bias consists of multiple dimensions (Björkdahl 2020; Sang 2022).

Leadership organizations deploy multiple governance processes to successfully develop and deploy gender-responsive artificial intelligence solutions. Managing workplaces with AI requires gender-expert AI ethics boards combined with algorithmic accountability officers and employee oversight committees tracking AI systems to ensure best performance. The AI governance framework at Microsoft mandates gender impact assessment testing across each development phase and IBM provides Fairness 360 as a tool for persistent bias monitoring during system deployment. Technical implementations do not eliminate biases, which led Intel to mandate that female representatives make up at least 30 percent of AI development teams globally. Governance researchers currently investigate two new approaches: the use of employee rewards for finding discriminatory algorithms and the implementation of performance disclosure systems through algorithmic nutrition labeling for gender-related influences. A comprehensive AI governance framework enhances existing DEI measures through Unilever's method, which detects sudden changes to gender outcomes from automated choices throughout different company levels and operational units. The development of powerful governance frameworks requires both strong execution of disciplinary methods and comparative oversight mechanisms. Executives at Salesforce receive compensation packages featuring AI ethics performance criteria which verify organizational gender equity improvements through verified leader accountability for real change (Sauer 2018; Nyiransabimana 2018).

The organizational dynamics and the distribution of power within organizations should change in order to strive for equal opportunities in the places where AI is utilized in workplaces. Good governance frameworks articulate policies that point towards the need to debias a system for the real purpose of fairness, but address how society creates paradigms for the development of technology and the automated manner in which benefits and rewards are distributed. The integration of an AI system in business ensures success in performance and gender-sensitive design principles in business operations enhance the creativity and output of normal business processes. Due to the continuous advancement in AI, there must be a corresponding improvement in the mechanisms that oversee and coordinate governance. Culturally sound leadership reinforced by sound governance structures, together with policy-compliant applications of 'best practices will turn the emerging technologies into places of work that are free from discrimination, with gender parity entrenched at each ladder of the career progression.

6 Conclusion

Achieving gender parity within computer-mediated workplaces cannot be achieved with the help of technology alone; power dynamics and decision-making in organizations need to be changed. Governing strategies that reduce bias in algorithms are effective, although we must disband conventions that influence the design of technologies while interrogating the reward and success system. This means that organizational workplace protocols of common business need to incorporate the AI technology, to come up with inclusive design solutions that can help them progress sustainably. Society has a problem in regulating developments in AI in the same pace as technology to exert better control on the governance systems. With strategic policy initiatives, sound governance structures and enviable leadership, these formidable technologies can change into workplaces with gender equality rights incorporated in their system.

References

Anurag VN, Kumar Lilhore U (2023) Transforming work: the impact of artificial intelligence (AI) on modern workplace. In: 2023 3rd international conference on technological advancements in computational sciences (ICTACS)

Araujo MC, Elizabeth BB, Bauhoff S, Berlinski S, Bosch M, Busso M, Brenes (2024) Expanding opportunities: policies for gender equality and inclusion

Bacchi C (2017) Policies as gendering practices: re-viewing categorical distinctions. J Women, Polit Pol 38(1):20–41. https://doi.org/10.1080/1554477x.2016.1198207

Björkdahl A, Somun-Krupalija L (2020) Governance for urban services: access, participation, accountability, and transparency, pp 107–126

Buslón N, Cortés A, Catuara-Solarz S, Cirillo D, Rementeria MJ (2023) Raising awareness of sex and gender bias in artificial intelligence and health. Frontiers in Global Women's Health, 4

Campbell K (2020) Gender and technology: social context and intersectionality. Handbook of research in educational communications and technology: Learning design, pp 115–204

Carroll P, Esposito Amideo A (2024) Gender equality: opportunities and challenges for the OR community. J Oper Res Soc 75(6):1011–1029. https://doi.org/10.1080/01605682.2024.2343343

Ceia V, Nothwehr B, Wagner L (2021) Gender and technology: a rights-based and intersectional analysis of key trends

Cheon E, Zaga C, Lee HR, Lupetti ML, Dombrowski L, Jung MF (2021) Human-machine partnerships in the future of work: exploring the role of emerging technologies in future workplaces. In Companion publication of the 2021 conference on computer supported cooperative work and social computing

Chowdaiah D, Mathew S (2024) Technology-driven economy: policy suggestions for increasing women's involvement with artificial intelligence, big data, and cloud infrastructures. library of progress-library science. Inf Technol Comput 3

Collett C, Dillon S (2019) AI and gender: four proposals for future research

de Laat K (2023) Remote work and post-bureaucracy: Unintended consequences of work design for gender inequality. Ind Labor Relat Rev 76(1):135–159. https://doi.org/10.1177/00197939221076134

Fan W, Moen P (2025) The future (s) of work? Disparities around changing job conditions when remote/hybrid or returning to working at work. Work Occup 52(1):91–129

Franzoni V (2023) Gender differences and bias in artificial intelligence. In: Gender in AI and robotics. Springer International Publishing, pp 27–43

Hartmann H (1976) Historical roots of occupational segregation-capitalism, patriarchy, and job segregation by sex. Signs 1(3):137–169

Hervías Parejo, V., & Radulović, B. (2023). Public policies on gender equality. In: Gender-competent legal education. Springer International Publishing, pp 405–428

Hervías Parejo V, Radulović B (2023). Public policies on gender equality. In: Gender-competent legal education. Springer International Publishing, pp 405–428

Howard J (2019) Artificial intelligence: implications for the future of work. Am J Ind Med 62(11):917–926. https://doi.org/10.1002/ajim.23037

Manasi A, Panchanadeswaran S, Sours E, Lee SJ (2022) Mirroring the bias: gender and artificial intelligence. Gend Technol Dev 26(3):295–305. https://doi.org/10.1080/09718524.2022.2128254

Mellul C (2018) Emerging technologies in higher education and the workplace: an assessment. The Future

Mirbabaie M, Brünker F, Möllmann NR, Stieglitz S (2022) The rise of artificial intelligence–understanding the AI identity threat at the workplace. Electron Markets 1–27

Nadeem A, Marjanovic O, Abedin B (2022) Gender bias in AI-based decision-making systems: a systematic literature review. Australian J Inf Syst 26. https://doi.org/10.3127/ajis.v26i0.3835

Nyiransabimana V (2018) Gender equality-key challenges and practical solutions to women participation in local governance in Rwanda. African J Publ Affairs 10(3):1–22

Oguns EO (nd) Tech-driven gender equality: Empowering women in the digital age

Patel J (2024) Gender equality: a promise to attain sustainable development. Int J Sustainab Build Technol 7(1):48–53

Pathak S, Solanki VK, Linh NTD (2023) Gender biasness: a victim of artificial intelligence-based development. In: Communication technology and gender violence. Springer International Publishing, pp 81–98

Pyburn R, Van Eerdewijk A (2021) Advancing gender equality through agricultural and environmental research: past, present, and future. Intl Food Policy Res Inst

Rasool S (2017) Gender equality and social development. Future directions in social development, pp 119–139

Sauer AT (2018) Equality governance via policy analysis?: the implementation of gender impact assessment in the European union and gender-based analysis in Canada

Shang B (2022) Tackling gender inequality: definitions, trends, and policy designs. International Monetary Fund

Tiwari R (2024) The impact of artificial intelligence in the workplace and its effect on the digital wellbeing of employees. Int J Progress Res Eng Manage Sci (IJPREMS) 4(6):2422–2427

Turesky M, Warner ME (2020) Gender dynamics in the planning workplace: the importance of women in management. J Am Plann Assoc 86(2):157–170

Volodzkiene L (2025) 4 gender equality. Societal Challenges and Opportunities of Low-Carbon Energy Transformations

Zirar A, Ali SI, Islam N (2023) Worker and workplace artificial intelligence (AI) coexistence: emerging themes and research agenda. Technovation 124(102747):102747. https://doi.org/10.1016/j.technovation.2023.102747

Gender Equity in India's Five-Year Plans: Policy Perspectives on Women's Empowerment

Ankur Nandi, Tupai Barman, Tapash Das, and Tarini Halder

Abstract This research contributes to the global agenda of sustainable and inclusive development by advocating for gender-responsive policies, enhanced women empowerment. The Government of India has initiated several measures under different Five-Year Plans towards the development of women. This study analyzes policy measures and planning strategies that aim to eliminate gender-based disparities in India. Objectives of this chapter is to highlight different constitutional Articles and the initiatives by the Government of India for women's development. This paper also brings to light the empowerment status of women through first to twelfth Five-Year Plans in India. The Indian government had acknowledged the importance of women's role in development from the first to the twelfth five-year plan (1951–1956). During the process of development, the position of women has been raised by numerous programs, acts, and initiatives. Governments and the voluntary sectors should make empowerment policies for women and implement strategies in actions that eliminate disparities in access to and control of resources, economic opportunities, power, rights, and political voice.

Keywords Women's empowerment · India's five-year plans · Gender equity · Policy perspectives · Sustainable development

1 Introduction

Women comprise almost 48.4% of the total population, as stated in the Census Report (Report of Census 2011). Women play an important role in the process of family development, society, and the national stage. The movement of 'empowerment of

A. Nandi (✉) · T. Barman · T. Halder
Department of Education, Kalyani University, Nadia, West Bengal, India
e-mail: ankurpalashdanga@gmail.com

T. Das
Kazi Nazrul University, Asansol, West Bengal, India
e-mail: tapash.das@knu.ac.in

A. Singh et al. (eds.), *Human-Centered Sustainability: Bridging Well-Being, Gender Equity, and Technology*, World Sustainability Series,
https://doi.org/10.1007/978-3-032-22020-2_11

women' has emerged as a global phenomenon. Women are the most marginalized section of people in the Indian democracy. Women are socially exploited, economically dependent, and politically powerless. Notably, women's participation has been measured in national development in this period and used as a gauge for inclusive democracy. Post-1950, the Planning Commission launched a five-year plan to address the different dimensions of advancing our democratic nation (Pal 2013). In the Five-Year Plan, women were weaker members of society and the focal point of welfare (Gaganamali 2020). In consideration and furtherance of the Constitutional mandate in the interest, benefit, security, and welfare of women, the Government of India through its Planning Commission established provisions in the first and all later Five-Year Plans (Barman 2019). Post-independence, our planners designed various constitutional clauses protecting women's interests and established a range of socioeconomic development programs meant to promote women's standing. Also, several voluntary organizations are putting a lot of effort into the emancipation of women (Gaganamali 2020). In line with the provisions in the Indian Constitution, it is a statutory requirement that women are treated equally to men in every facet of society. In this context, the Ministry of Women and Child Welfare stands tall, providing for the right development of women and children in India (Sharman and Rahaman 2021). The term "welfare" in the 1970s changed to "development" in the 1980s and "empowerment" in the 1990s in terms of policy approaches (Barman 2019). Women have been accorded particular significance since the implementation of the sixth Five-Year Plan. Significantly, the seventh and eighth five-year years have placed a strong emphasis on 'development' and 'empowerment', respectively. As time went on, the 11th and 12th five-year plans emphasized inclusive growth (Pal 2013). Since 1951, the Indian government has implemented five-year programs to achieve economic development. Every Five-Year Plan includes a specific part devoted to schemes and projects created with women in mind. The Indian government has implemented many schemes and projects aimed at promoting the advancement of women in each of its Five-Year Plans. The paper intends to focus on these initiatives undertaken by Govt. of India for the development of women.

Gender Equity, Five-Year Plans, and Sustainable Development Goals.

The Sustainable Development Goals were set by the United Nations. Mainly, it helps to fulfill the purpose of SDG 5 (gender equality), where the evolution of the government's policy has been analyzed for women's empowerment through the various five-year plans. This study has realized the continuity of the development of women's welfare and subsequently to empowerment, which reflects a deep commitment to gender equality in the governance system. Furthermore, it is compatible with SDG 4 (qualitative education), where various steps have been analyzed, including education, vocational training, and adult literacy programs, to promote women's access to education.

In addition, the research is important from the point of view of SDG 3 (well -being and welfare), where the programs related to maternal health, nutrition, reproductive health care, and family welfare are highlighted. This study strengthens SDG 10 (asymmetry reduction) through the analysis of conservation principles and Women's

Material Plan (WCP), where attempts to eliminate gender discrimination in the political, social, and economic fields have been shown. More importantly, it is helpful to SDG 16 (peace, justice, and strong institutions), where the role of institutions such as constitutional provisions, legal protection, and the National Women's Commission has been highlighted in the formation of an inclusive and just society. Overall, this study questioned in favor of a sustainable and inclusive development, where gender-sensitive plans, equality-based access to resources, and women's participation in nation formation were encouraged.

2 Objectives

1. To study the Constitutional basis of Women's Empowerment
2. To study the different strategies as reflected in various Five-Year Plans regarding Women's Empowerment in India.

3 Methodology

The approach of the study is qualitative, and the study is documentary in nature. The content analysis method was used. The study is based on Secondary sources, and the data have been collected from journals, Government websites, books, reports, and magazines by different databases such as Google Scholar, JSTOR, ERIC, etc. Search terms included "Women's Empowerment and SDGs," "Women's Empowerment in Five-Year Plans," and "Gender Equity and SDGs." The selected literature was categorized according to the study's objectives and analyzed thematically.

4 Discussion

4.1 Concept of Empowerment

Social, political, and economic facets are all included in the multifaceted concept of empowerment (Baruah 2013). The process of acquiring power over resources and decision-making is known as empowerment. It is a process of increasing awareness and developing ability, which increases involvement and gives power to make decisions. Thus, empowerment is about giving individuals greater freedoms and choices of action. It is significant both for its intrinsic value and for its potential to help the impoverished achieve better development outcomes. Power is 'the ability to control people or things-right or authority of a person or group to do something,' the Oxford Dictionary (2000) states, and empowerment is 'to give somebody the power or authority to do something; to authorize; to give somebody more control over their own life or the situation they are in'. Panda (2007) asserts that the term "empowerment" has been in use since the seventeenth century and, despite its previous meaning of "to invest with power and authority," it is currently seen as an enabling component. The International Women's Year and the Women's Decade, which ran from 1975 to 1985, were proclaimed by the UN (Naqvi 2008). The concept of empowerment was first applied to women's development and the study of women's social subordination in India in the 1980s.

4.2 Concept of Women's Empowerment

According to the Programme of Action 1992, societal change must start with the empowerment of women. Women's empowerment refers to their right to possess and enjoy authority over their life as both individuals and members of society. It is how women increase the range of options available to them and so take charge of their life. It is a process by which women learn to organize themselves to become more independent, to stand up for their autonomous rights to control resources and make decisions, and to help confront and end their own subjugation. It supports women in improving their dignity and sense of self-worth. It helps the women develop confident and positive self-images. It promotes decision-making and activity as well as group bonding. It guarantees equitable involvement in the process of enacting social change. To transform society, it promotes collective action. It gives women financial independence and self-sufficiency. It enables women controlling their own lives.

4.3 The Constitutional Basis of Women's Empowerment

See Table 1.

4.4 Approaches to Women Upliftment in Various Plans

The development of women was seen as a "Welfare" issue in the First Five-Year Plan (1951–1956) and combined with the welfare of underprivileged groups (such as criminal tribes, ST, SC, and OBC). Thus, the First through Fifth Five-Year Plans regarded the advancement of excluded and vulnerable populations as "Welfare." The sixth Plan included a new chapter titled "Women and Development," which shifted the focus from "welfare" to "development." The eighth plan attempted to move the "Development" approach by including a sub-theme called "Development of Women" in the chapter on "Social Welfare." The ninth Plan changed the focus from "Development" to "Empowerment of Women," treating women as a distinct

Table 1 Constitutional basis of women empowerment

Article	Objectives	Focussed on
Article 10	Women's equality as well as their legal protection.	Equality and protection
Article 14	Women have equal legal protection that guarantees a life of dignity without discrimination based on race, religion, sex, creed, faith, or place of worship.	No discriminate
15 (1)	Gender discrimination is forbidden (it is forbidden to treat any citizen unfairly based merely on their religion, race, caste, sex, or place of birth). Discriminating against a citizen based on their gender, race, religion, caste, or any combination of these is prohibited.	No discriminate
Article 15(3)	Relaxes this prohibition, enabling the government to favor women by creating favorable policies and special measures that will raise their social standing and guarantee social, political, and economic justice. Where women and children are involved, the state is required to make certain provisions.	Provision for women
Article 16	All individuals have to be treated equally with respect to employment or appointments to offices of the government. No individual should be excluded from or discriminated against in any state employment or office based on sex, religion, place of birth, race, caste, residence, or a combination of these. Since the rights are available to all individuals, irrespective of gender, the constitution hence indirectly provides equal rights for women.	Equal rights for women
Article 16 (1) & (2)	All citizens will stand on an equal platform regarding jobs as well as appointments to any government post.	Equality of opportunity
Article 38	To ensure the welfare of their population, states need social, economic, and political equity. Reservation of 33% in parliament and state assemblies was also implemented by the 81st amendment bill.	Reservation for women
Article 39(a)	According to the directive principles of state policy, states must focus their policies on ensuring that both women and men have an equal right to a sufficient standard of living.	Securing for women and men equally
Article 39 (d)	Equal pay for equal labour performed by women and men.	Pay for equal work for both women and men
Article 39(e)	Ensure that workers' health and strength, both male and female, and children's tender age are protected, and that citizens are not compelled to pursue occupations that are inappropriate for their age or strength out of a lack of resources.	Secure the strength and Well-being of labourers, whether male and female,
Article 45	All children receive free and required schooling up until the age of 14.	Education for all

(continued)

Table 1 (continued)

Article	Objectives	Focussed on
Article 51 A (e)	Encourage unity and a sense of fraternity among all Indian women and refrain from mistreating them.	Promote unity and a spirit of brotherhood among all women
Article 243 D	Women have been allocated at least one-third of the seats designated for chairpersons of Zila Parishads, panchayat Samiti, and gram panchayats (including SCs and STs).	1/3 of the seats have been reserved for women (including SCs/STs).

Source: Authors work

Fig. 1 Approaches to women empowerment. *Source* Authors work

target group and as agents of societal change and advancement. The Tenth Five-Year Plan's development program gave women's empowerment top priority as we adopted human development as the goal of all our developmental endeavours (Fig. 1).

4.5 Detail Elaboration of Five Years Plans

The national economic policies included in Five-Year Plans are coordinated and centralized. Since gaining its independence, India has adopted Five-Year Plans to achieve economic development since 1951. In this time frame, Indian has finished twelve plans. Setting clear goals and targets and implementing appropriate strategies to reach them are the key components of economic planning. Every Five-Year Plan has a dedicated section on programs and initiatives created specifically with women in mind. The Indian government has launched several scheme and projects aimed at advancing women's development in each of its five-year plans (Barman 2019). An important objective of Five-year plan is Women Development. The Planning Commission identified three key areas where they have focused specifically on the empowerment of women:

1. Education
2. Social welfare and
3. Health

4.5.1 First Five Ears Plan

The "Central Social Welfare Board (CSWB)" was established in 1953 to advance social welfare initiatives and carry out welfare programs via nonprofit organizations for women, children, and the disabled. Its goal is to empower women and children by raising understanding of the law, fighting social injustices, and developing capacity. (www.sociologyguide.com/women-and-society/central-social-welfare.php). Mahila Mandals are traditional associations that integrate gender-equality or resistance movements that enable women to more successfully enter the public domain. "Community Development Program and National Family Planning Program" in 1951 aims to empower individuals and communities to make choices and translate them into action. The purpose of the Community Development Program, initiated in many countries in the mid-twentieth century, was to involve local communities in socio-economic development. This program aimed at Empowerment and Participation, Capacity Building, Infrastructure Development, Poverty Reduction etc. (Mukerjee 1952). The "National Family Planning Program" was launched to control population growth, particularly in countries facing rapid population expansion. Its objectives included: Population Control, Maternal and Child Health, Education and Awareness, Sustainable Development etc., (Seltzer 2002) (Table 2; Fig. 2).

The Central Government established a "Central Social Welfare Board", primarily to assist voluntary organizations in planning welfare programs for women, children and persons with disabilities (Konwar 2019; Kumari 2009; Sewa 2023; Poonam and Singh 2023; Barman 2019). Specifically designed to monitor women's issues, in 1953, the "Central Social Welfare Board" was established (Sewa 2023; Singh 2021; Golpan: Lakshmisha 2017). The Primary objective of this board is to promote the welfare of women by encouraging girls' participation in "primary, university, and

Table 2 Initiatives for women's empowerment under first five year plan

Plan	Initiatives for women empowerment
First five year plan (1951–1956)	Establishment of the "central social welfare board" to support women, children, and people with disabilities.
	Focus on organizing women into "women's clubs or Mahila Mandals" for community support and empowerment.
	Launch of the "community development program and National Family Planning Program" in 1951.
	Objective to promote the welfare of women by encouraging girls' participation in "primary, university, and vocational education", and preparing them for employment.
	Emphasis on "educating women" to help them become better mothers and manage households effectively.
	Promotion of training for female midwives, physicians, and dais to improve women's healthcare and childbirth practices.
	Aims to enhance women's employability through "technical training."

Source: Authors work

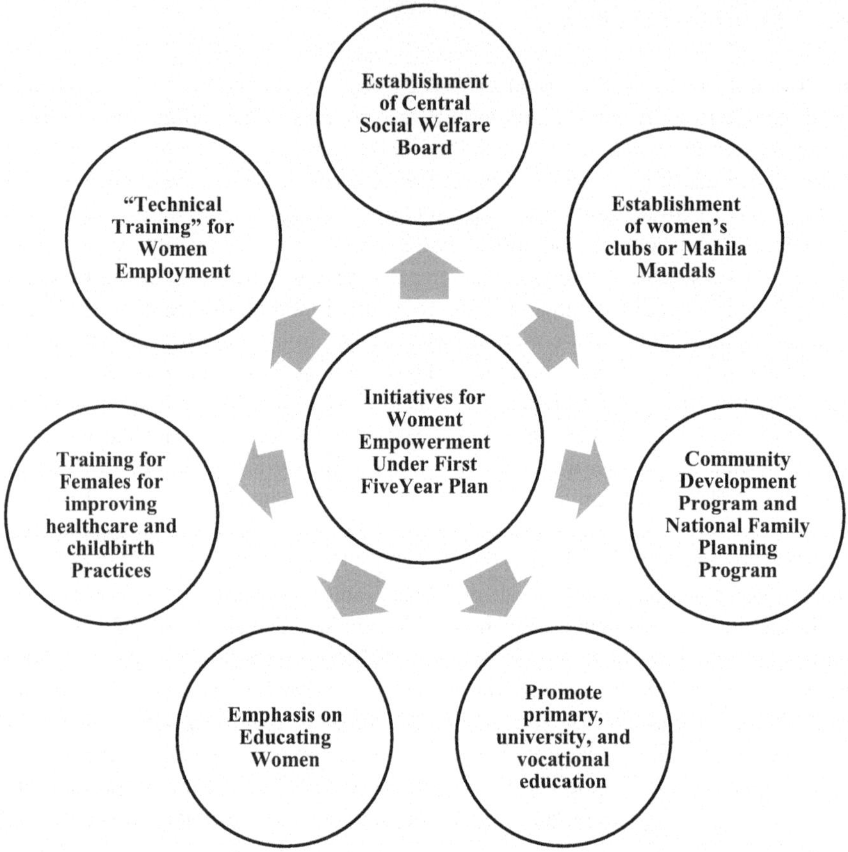

Figure 2 Initiatives for women's empowerment under first five year plan. *Source* Authors work

vocational education", and preparing them for employment. Emphasis on "educating women" to help them become better mothers and manage households effectively. Promotion of training for female midwives, physicians, and dais to improve women's healthcare and childbirth practices. Aims to enhance women's employability through "technical training." (Barman 2019). The necessity of forming women's clubs or mahila mandal has been shown by this board (Sewa 2023; Singh 2021). In this Plan the Community Development Program (1952) and the National Family Planning Program (1951) were launched for women's development (Singh 2021). Helped women to fulfil their rightful place in the family and community to "promote the welfare of women" (Sewa 2023).

Table 3 Initiatives for women empowerment under second five year plan

Plan	Initiatives for women empowerment
Second five year plan (1956–1961)	To encourage "girls' education" through their parents, and to provide education greater pertinent to the requirements of girls and the employment of female educators.
	Aims for "welfare initiatives for women."
	To allocate various forms of labour in the field of labour while attending to the physical and biological impairments of women at work.
	Emphasis on activities of "Mahila Mandal"
	Focussed on decent work safe working condition for women workers and also takes maternity benefits.

Source Authors work

4.5.2 Second Five Years Plan

Welfare initiatives for women are designed to address gender inequality and promote the empowerment and well-being of women in various sectors of society. The primary purposes of welfare initiatives for women are Economic Empowerment, Health and Well-being, Education and Literacy, Legal Rights and Protection, Social Inclusion and Gender Equality etc. (Table 3; Fig. 3)

The second five-year strategy encouraged the creation of Mahila Mandals to be used in grassroots initiatives. Mahila Mandal is the key focal area for women's development at the primary level in this five-year plan. The strategy suggested that women employees ought to fight against hazardous jobs and the maternity benefits should be considered. The government also recommends the equal pay for equal labour principle (Singh 2021; Sewa 2023). To promote girls' education by engaging parents and providing education tailored to their needs, with a focus on employing female educators. Efforts should also address gender-specific challenges in the workplace and allocate suitable forms of labour for women (Barman 2019).

4.5.3 Third Five Years Plan

Health-related programs are initiatives aimed at improving the physical, mental and social well-being of individuals and communities. These programs are specifically designed for women nd to address the challenges that women face throughout their lives. The program aims to promote Maternal Health and Safe Motherhood, Reproductive Health and Family Planning, Prevention and Management of Gynaecological Health Issues, Combating Gender-Based Violence (GBV), Sexual and Reproductive Health Education, Breastfeeding and Infant Health, Mental Health Support for Women, Nutrition and Anemia Prevention, Cancer Prevention and Early Detection, HIV/AIDS and STI Prevention etc. (WHO 2021) (Table 4; Fig. 4).

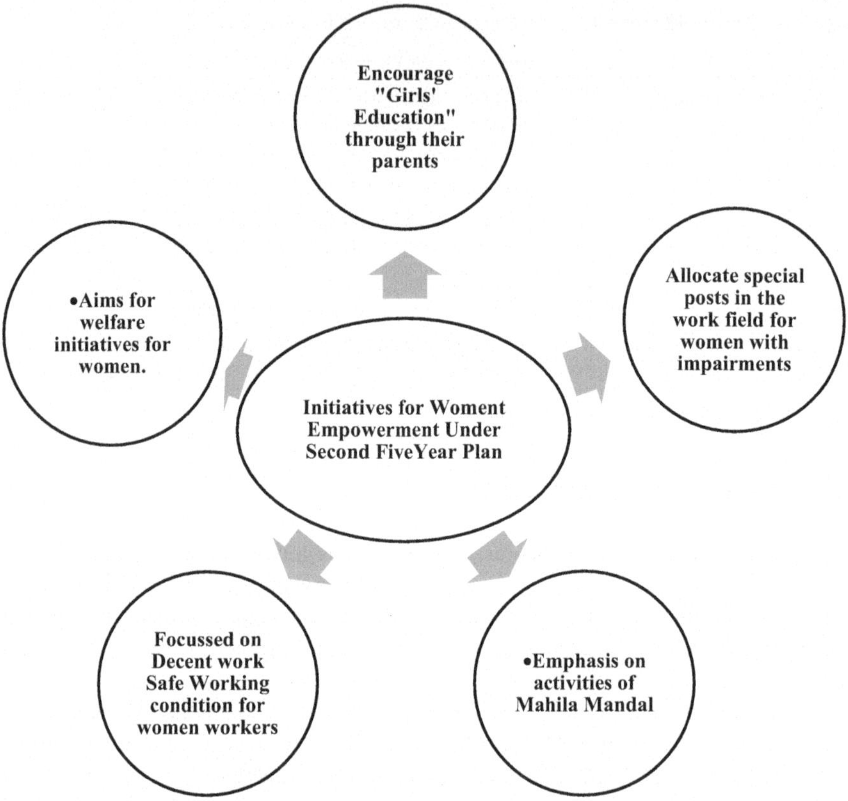

Fig. 3 Initiatives for women empowerment under second five year plan. *Source* Authors work

Table 4 Tabular presentation of Initiatives for women empowerment under third five year plan

Plan	Initiatives for women empowerment
Third five year plan (1961–1966)	Focuses on "female education."
	Launching the "health-related program."
	The provision of services for family planning, nutrition, maternal health, and the welfare of children and women.

Source Authors work

The government's five-year plan emphasizes female education above all else. It also includes a health program that focuses on providing services for child welfare, addressing maternal health issues, providing sustenance, and family planning (Singh 2021; Sewa 2023; Poonam and Singh 2023). The third five-year plan maintained its focus on "women's welfare" (Patel 2017). Included provisions for "nursing, prenatal and maternal health services, additional feeding for children, women's education, and other health care programs" (Lakshmisha 2017; Barman 2019).

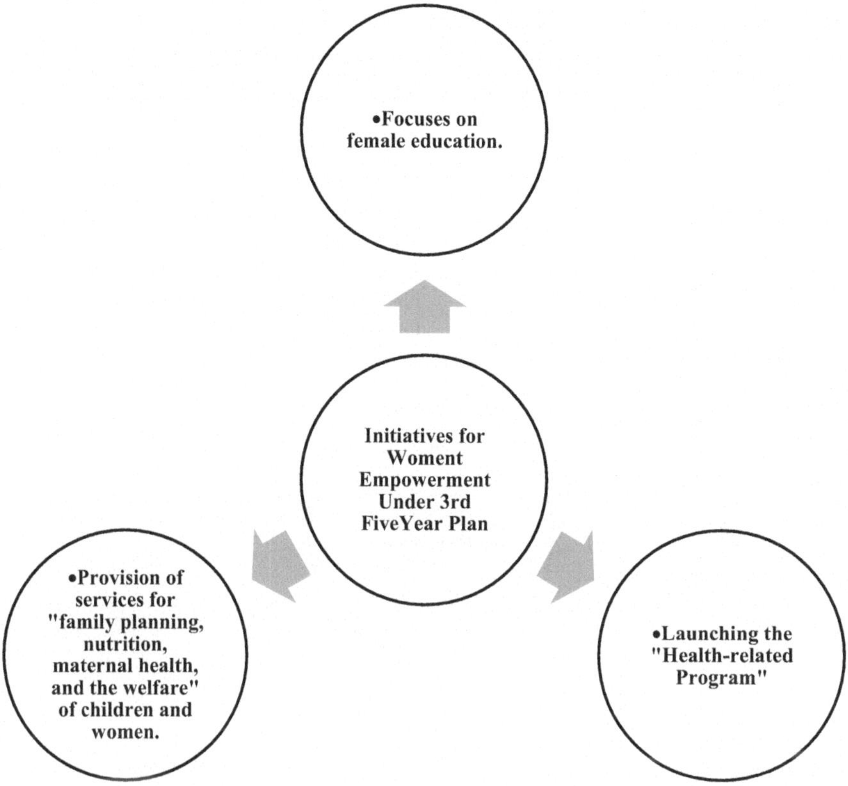

Fig. 4 Tabular presentation of Initiatives for women empowerment under third five year plan. *Source* Authors work

4.5.4 Fourth Five Years Plan

"Socio-economic programs" and "Courses of education for adult women" are essential tools for women's welfare, aiming to uplift their social and economic status by providing opportunities for skill development, education, and financial independence. These programs are designed to address the unique challenges that adult women face, particularly in developing and underserved communities, and contribute to long-term gender equality and empowerment. The main purposes of these programs are Economic Empowerment and Financial Independence, Educational Advancement and Lifelong Learning, Social Inclusion and Gender Equality, Skill Development for Employment, Entrepreneurship and Small Business Support, Health and Well-being Awareness, Legal Awareness and Rights Education, Support for Marginalized and Vulnerable Women, Digital Literacy and Technological Inclusion etc. (Kabeer 2001) (Table 5; Fig. 5).

The fundamental strategy of this plan is to prioritize the wellbeing of women within the family (Singh 2021; Barman 2019). The Central Social Welfare Board

Table 5 Initiatives for women empowerment under fourth five year plan

Plan	Initiatives for women empowerment
Fourth five year plan (1966–69) and Interim plans (1969–74)	"Central social welfare board" adopted two programs - "Socio-Economic Programs and Courses of education for adult Women for women welfare".
	To promote women's welfare within the home as the cornerstone of the system.
	Laid emphasis on "women's education" and the advancement of women's welfare, particularly family planning, within the family.

Source Authors work

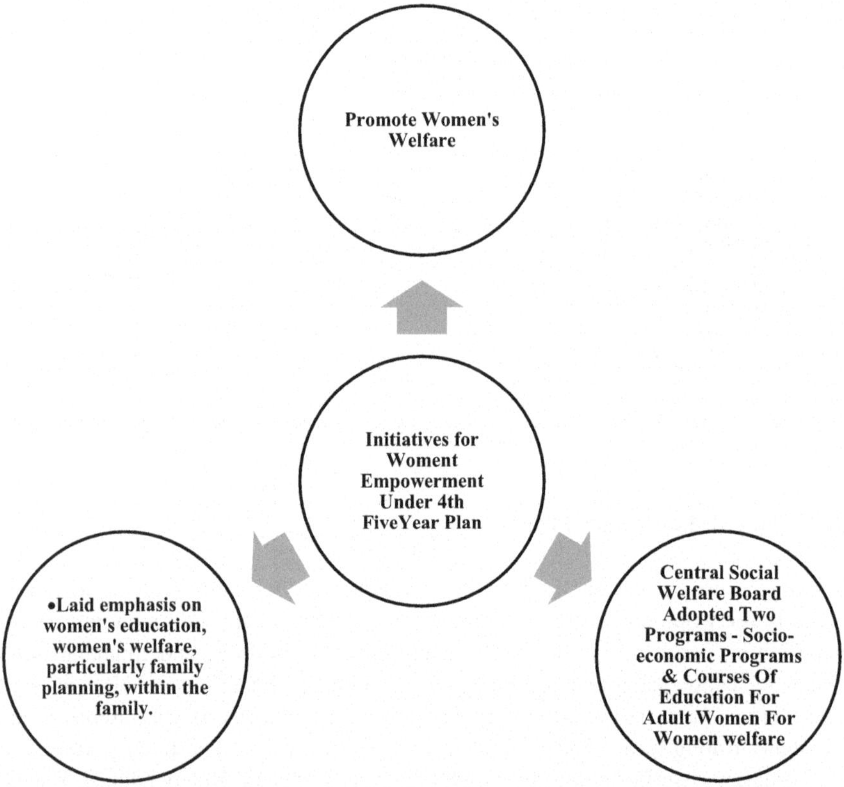

Fig. 5 Initiatives for women empowerment under fourth five year plan. *Source* Authors work

adopted two women's welfare programs throughout this five-year plan (1969–74): Socio-Economic Programs and Education Programs for Adult Women (Singh 2021). Laid emphasis on promoting women's welfare within the family, including women's education and family planning (Sewa 2023; Barman 2019).

4.5.5 Fifth Five Years Plan

The purpose of the "Women's Welfare and Development Bureau" is to empower women and promote their well-being by implementing programs and policies that address their socio-economic, legal, and health needs. It plays a critical role in advancing gender equality and ensuring that women can fully participate in all aspects of life. The specific purposes of the bureau are Promoting Gender Equality: To advocate for equal opportunities for women in education, employment, and leadership, ensuring that women have the same access to resources and decision-making roles as men (United Nations (2015). Enhancing Economic Empowerment: To provide resources, skills training, and financial support to women, helping them achieve economic independence and reduce poverty (International Labour Organization 2017). Improving Health and Reproductive Rights: To ensure women have access to healthcare services, including reproductive health, family planning, and maternal care, promoting better health outcomes for women and children (WHO 2020). Raising Legal Awareness: To educate women about their legal rights and provide them with the knowledge and tools to access justice, protect their rights, and challenge discrimination (International Development Law Organization 2019). Supporting Vulnerable and Marginalized Women: To assist women in vulnerable situations, such as single mothers, widows, elderly women, and those with disabilities, through targeted social welfare programs that improve their living conditions and social integration (UNICEF 2021). Encouraging Women's Political Participation: To promote women's involvement in governance and decision-making processes by supporting leadership training, advocating for gender quotas, and fostering political engagement (UN Women 2020) (Table 6; Fig. 6).

Table 6 Initiatives for women empowerment under fifth five year plan

Plan	Initiatives for women empowerment
Fifth five year plan (1974–1979)	A strong emphasis on "Women's training" for meet all their needs, including safety and money for survival.
	The Ministry of Social Welfare established the "Women's welfare and development bureau."
	Changing the focus from "welfare" to "development."
	Submitted in conjunction with the international Women's decade and the filing of the Indian Women's status committee report.

Source Authors work

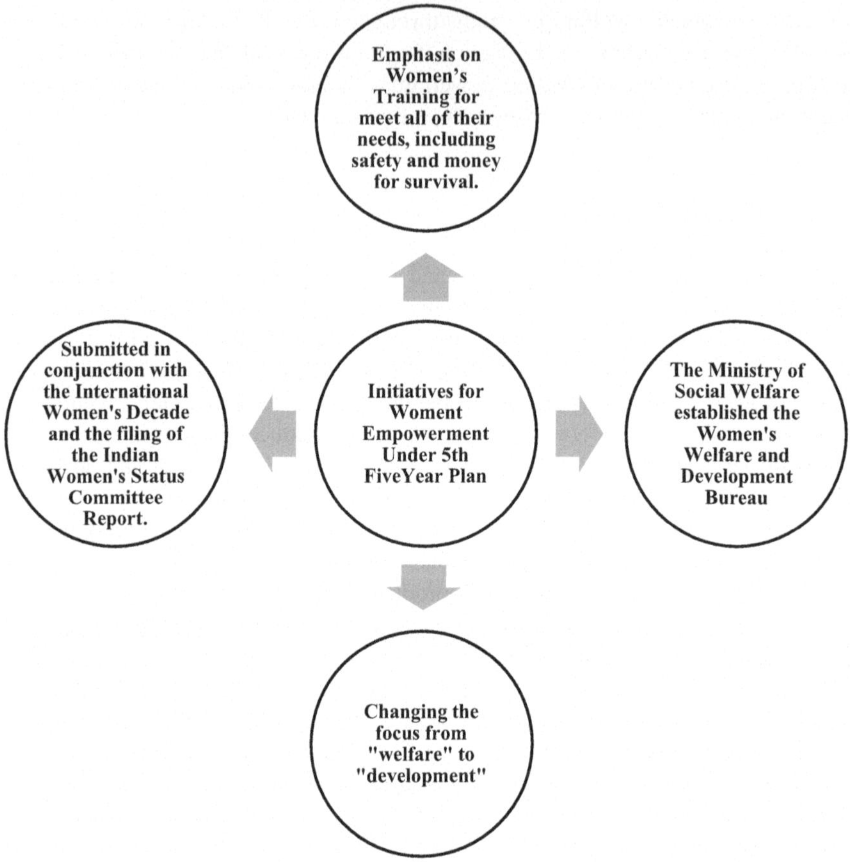

Fig. 6 Initiatives for women empowerment under fifth five year plan. *Source* Authors work

In this five-year plan Welfare organizations must provide training to women to fully meet their needs, which include protection from undesirable people and an income to survive (Singh 2021; Barman 2019). It was a new approach of integrated welfare with development services laying emphases on training of women (Lakshmisha 2017). The "Committee on the Status of Women" in India produced a report at the same time as this plan, which was in line with the International Women's Decade. Subordinate to the Ministry of Social Welfare was established the "Women's Welfare and Development Bureau". Has shift on focus from "welfare" to "development" of women with emphasis on training (Lakshmisha 2017; Sewa 2023; Konwar 2019). Aiming to serve as a focal point within the Indian government and to coordinate policies, programs, and initiatives for women's development, the Ministry of Social Welfare established the "Women's Welfare and Development Bureau" in 1976. (Konwar 2019). This plan was submitted in conjunction with the International Women's Decade and the filing of the Indian Women's Status Committee Report.

4.5.6 Sixth Five Years Plan

Purpose of Constitutional and Legal Guarantees for women is to protect their rights, ensure equality, and provide a legal framework that promotes their welfare and empowerment. These guarantees aim to eliminate discrimination, address gender-based inequalities, and create an environment where women can fully participate in social, economic, and political life (Table 7; Fig. 7).

A program called the Development of Women and Children in Rural Areas (DWCRA) was unveiled in September 1982 as an IRDP (Integrated Rural Development Programme) sub-plan. Providing chances for self-employment and drawing attention to women living in poverty are the main goals (Shaikh 2021). The main goals of DWCRA are to improve the abilities of rural women and give them access to finance and assets that can generate a profitable income (Mishra, nd).

The "Integrated Rural Development Programme (IRDP)" aims for women to uplift their socio-economic status by providing access to credit, subsidies, and skill development, enabling them to contribute to household income and participate more actively in rural economies.

This Plan was considered a turning point in the advancement of women (Konwar 2019). This five-year plan reviews the status and circumstances of women and provides them with constitutional and legal guarantees. To promote women's empowerment, the Indian government implemented three phased strategies: employment, healthcare, and education. In addition, the government took several more steps to

Table 7 Initiatives for women empowerment under sixth five year plan

Plan	Initiatives for women empowerment
Sixth five year plan (1980–1985)	This five-year plan reviews the status and circumstances of women and provides them with "constitutional and legal guarantees".
	To encourage "women's empowerment" through work, education, and health care.
	The focus was laid on "providing support to women to overcome poverty", emphasized the importance of increasing their access to resources, and owe joint titles to land and property.
	A program called "development of Women and children in rural areas (DWACRA)" was initiated in 1982, primarily to help women take up and manage income-generating activities.
	To encourage the promotion of functional literacy, boost the enrolment of girls in elementary school, and support the advancement of women's education in underdeveloped areas.
	Women's nutrition needs were taken into consideration, along with maternal and child welfare programs, to enhance health and family welfare.
	In this five-year plan "integrated rural development programme (IRDP)" was introduced, but no new initiatives for the employment of women in agriculture or industry were developed.

Source Authors work

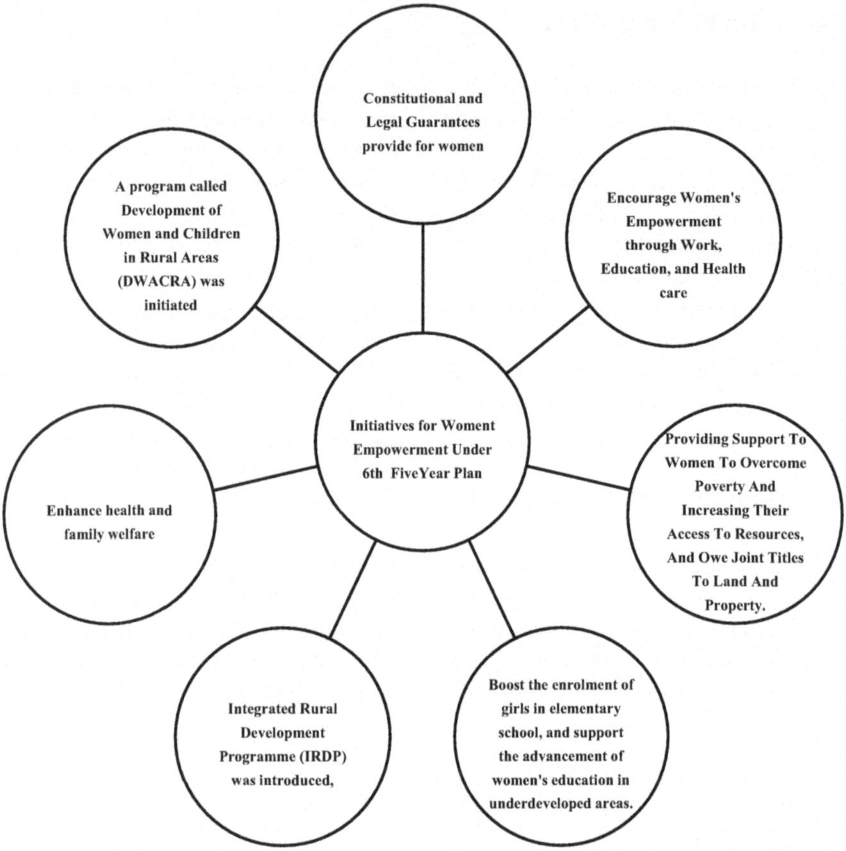

Fig. 7 Initiatives for women empowerment under sixth five year plan. *Source* Authors work

address women's issues in a variety of fields (Singh 2021; Sewa 2023; Lakshmisha 2017). The focus was laid on providing support to women to overcome poverty, emphasized the importance of increasing their access to resources, and owe joint titles to land and property (Sewa 2023; Lakshmisha 2017). A program called "Development of Women and Children in Rural Areas (DWACRA)" was initiated in 1982, primarily to help women take up and manage income-generating activities (Poonam and Singh 2023). To encourage functional literacy, boost girls' enrolment in elementary schools, and advance women's education in underdeveloped areas. Women's nutrition and maternal health were prioritized through welfare programs to improve family well-being. Vocational training aimed to increase jobs, though the Integrated Rural Development Programme lacked new initiatives for women's employment in agriculture or industry (Barman 2019).

4.5.7 Seventh Five Years Plan

Department of Women and Child Development: aims to create a supportive environment that promotes the well-being, development, and empowerment of women and children. The "Beneficiary Oriented Schemes (BOS)" (1986) are designed to economically and socially empower women, particularly those from marginalized and underprivileged backgrounds. The "National Perspective Plan for Women (NPPW, 1988–2000)" was developed by the Government of India to provide a comprehensive framework aimed at promoting gender equality and empowering women across the country (Table 8; Fig. 8).

The objective of the seventh five-year plan was to include women in the principal area of governmental development (Konwar 2019). The seventh five-year plan recognized the critical role that women play in agriculture and related fields and placed a strong emphasis on the necessity of empowerment and gender equality. To improve women's economic and social standing and integrate them into the mainstream of national development, a strategy was created (Sewa 2023; Lakshmisha 2017). The "Department of Women and Child Development" was founded by the "Ministry of Human Resource Development" in 1985 with the aim of providing women and children with the necessary impetus for their comprehensive growth

Table 8 Initiatives for women empowerment under seventh five year plan

Plan	Initiatives for women empowerment
Seventh five year plan (1986–1991)	A focus on the necessity of "empowerment" and "gender equality".
	To promote women's and children's holistic development, established in 1985, the "Department of Women and Child Development".
	To provide women with greater chances for their everyday survival, several craft and art centres open for them.
	To encourage activities that provide employment for women's groups and women from the weaker sections, Women development corporations were founded in all of India's states.
	To promote the 27 "beneficiary oriented schemes (BOS)" (1986) that benefit women directly across various ministries/departments.
	The Government of India released the "National Perspective Plan for Women (NPPW, 1988–2000)". The NPPW emphasizes the importance of reservations for women in several professions.
	To increase awareness of women's rights and benefits to empower them.
	To prepare women for more fruitful careers in the workforce.
	To develop support services to lessen the overwhelming everyday stress that women bear from domestic work.
	To encourage women's participation in the building of a sustainable environment by increasing their access to science and technology.
	The aim is to encourage women's corporations in each state, with equal funding contributions from the federal and state governments.

Source Authors work

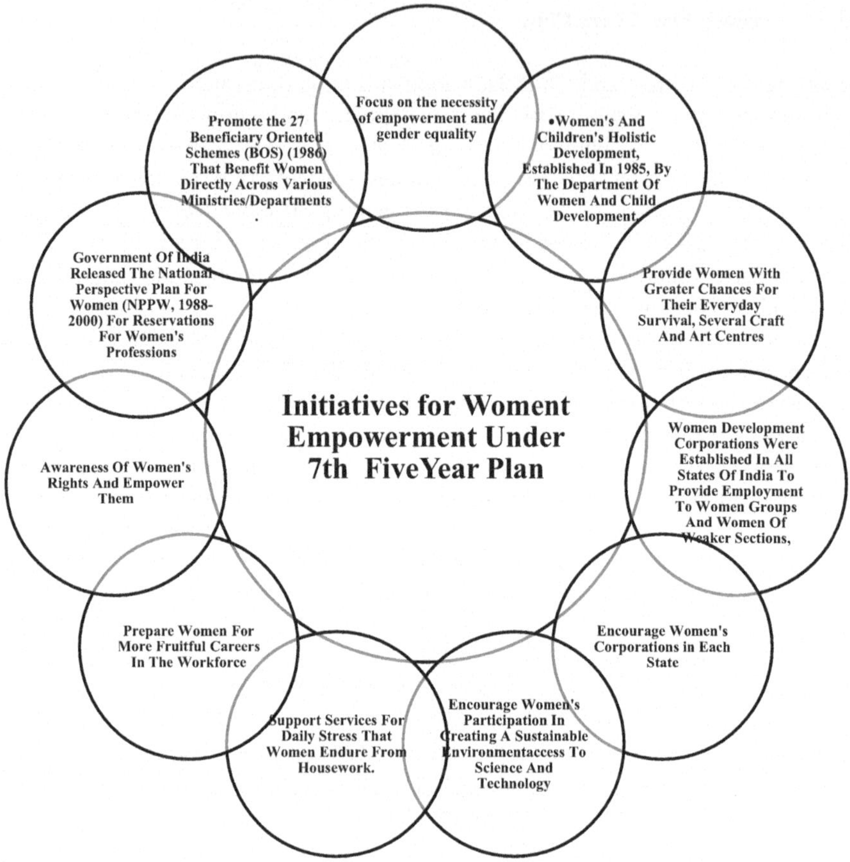

Fig. 8 Initiatives for women empowerment under seventh five year plan. *Source* Authors work

(Konwar 2019). All Indian states established Women Development Corporations with the goal of advancing work-generating initiatives for women's groups and women from underrepresented backgrounds. Promoted beneficiary-focused initiatives for women throughout several development areas (Sewa 2023). According to this plan, the government should work to give all societal segments, particularly the impoverished ones, access to high-quality facilities. Many art and craft centers are offered to women to provide them with greater chances for day-to-day survival (Singh 2021). One of the most important aspects of the seventh plan was the identification and promotion of 27 "Beneficiary Oriented Schemes (BOS)" (1986) that provided direct benefits to women in various Ministries/Departments. The GoI released the "National Perspective Plan for Women (NPPW)" (1988–2000) in 1988. This document's goals are to assess how much women have been included in India's development and offer suggestions for social justice and equity for all women. Reservations for women in a variety of professions are highly valued by the NPPW (Konwar 2019; (Poonam and

Singh 2023). Empower women by raising awareness of their rights, preparing them for successful careers, and reducing domestic work stress through support services. Promote their involvement in sustainability through access to science and technology. Support women's corporations with equal federal and state funding (Barman 2019).

4.5.8 Eighth Five Years Plan

"The Kasturba Gandhi Education Scheme (KGBV)", initiated in 2004 by the Government of India, aims to enhance the educational opportunities for girls, especially those from marginalized and disadvantaged backgrounds. "The National Commission for Women (NCW)" was established in 1992 by the Government of India to address the concerns of women and ensure their rights and empowerment (Table 9; Fig. 9).

To strive toward the goal of enabling women to attain equality and equal participation in all areas of life by securing her rights and entitlements through appropriate policy formulation, legislative action, efficient law enforcement, the execution of programs and policies, and the creation of solutions for issues or circumstances resulting from discrimination and crimes against women (National Commission for Women http://ncw.nic.in/mission-and-vision/mission).

Women's development was greatly aided by the eighth Plan (1992–1997), with human development as its main goal (Konwar 2019). The eighth plan's approach indicated a shift in focus from "development" to "empowerment" of women (Konwar 2019; Sewa 2023; Barman 2019; Gaganamali 2020; Poonam and Singh 2023). It promised to make sure women did not lose out on the advantages of progress coming

Table 9 Initiatives for women empowerment under eighth five year plan

Plan	Initiatives for women empowerment
Eighth five year plan (1992–1997)	The eighth plan's strategy indicated a shift from 'development' to 'empowerment' of women.
	Encourage women to participate in the development process as equal partners and participants.
	"Kasturba Gandhi education Scheme" established on august 15, 1997, girls' schools were established in districts with low female education rates.
	This plan specifically focuses on empowering women at the local level through "panchayat raj institutions".
	The department of "Women and child development" also prioritized the development of women in its 1989–2000 action plan for women.
	The "National Commission for Women" was constituted as the first statutory body to deal with women's problems.
	The ultimate goal of all these initiatives is to empower and enable women to support themselves financially.

Source Authors work

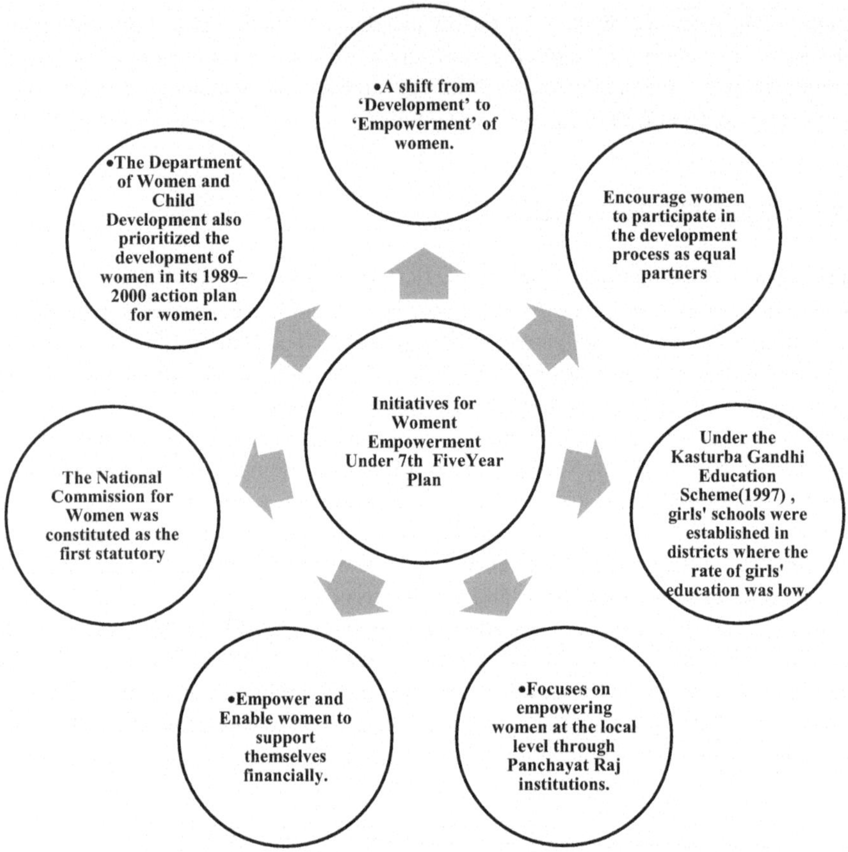

Fig. 9 Initiatives for women empowerment under eighth five year plan. *Source* Authors work

from various sectors (Singh 2021). Encourage women in the development process so they can take part and contribute equally by implementing programs that support the general development initiatives and monitor the benefits that women receive from other development sectors (tenth FYP, Plan Comm. GOI, p. 218; Sewa 2023; Lakshmisha 2017; Gaganamali 2020; Singh 2021). Under Kasturba Gandhi Education Scheme established on August 15, 1997, girls' schools were established in districts with low female education rates (Konwar 2019). This initiative leveraged panchayat Raj institutions to empower women, particularly at the local level. A plan of action for women was also established by The Department of Women and Child Development also with a focus on women's development during the years 1989–2000 (Konwar 2019; Sewa 2023; Barman 2019). Mahila Samridhi Yojana and Indira Mahila Yojana, were launched in this Plan (Poonam and Singh 2023). Emphasized the empowerment of women and their participation in the reservation system-based local self-government decision-making process in both urban and rural areas. As the first governmental body focused on women's empowerment, the National Commission for Women was established (Sewa 2023; Singh 2021). Making women economically independent and self-sufficient is the goal of all these initiatives (Konwar 2019).

4.5.9 Nineth Five Years Plan

The Women Component Plan (WCP) was introduced by the Government of India to ensure the integration of gender perspectives in various development programs and policies. The primary purposes of the WCP to Empowerment of Women.

The Special Nutrition Programme (SNP) launched in 1970–71 by the Ministry of Social Welfare, Government of India. It was launched in India to address the issues of malnutrition among vulnerable groups, particularly women and children. The primary purposes of the SNP are Combat Malnutrition, Improve Nutritional Status, Support Maternal and Child Health etc. (Table 10; Fig. 10).

The Mid-Day Meal Scheme (MDMS)," launched by the Government of India in 1995, is designed to provide free meals to schoolchildren. While the "Mid-Day Meal Scheme (MDMS)" is primarily aimed at improving children's nutrition and education, it also has significant indirect benefits for women. The purpose of the

Table 10 Initiatives for women empowerment under ninth five year plan

Plan	Initiatives for women empowerment
Ninth five year plan	Empowerment of every woman" was the primary aim of this initiative.
	Reservation of one-third seats in state assemblies and parliaments for women.
	Convergence of currently offered services, the "Women component plan (WCP)", infrastructure, resources, and labour in sections pertaining to and specifically for women.
	Implementing the "Women component plan (WCP)" specific strategy to guarantee that women from other development sectors receive at least 30% of cash and benefits.
	Setting up women SHGs are a key mechanism for empowering women.
	Child health care related to reproduction is highly prioritized.
	Making the "special nutrition programme" (SNP) and "mid-day meals" (MDM) a universal supplementary feeding program.
	Ensuring certain that women and girls have simple and equal access to education through the 1998 special action plan.
	Gender bias must be eliminated from all educational initiatives.
	Girls have free education up to the college level, including professional courses.
	Encouraging women to acquire skills in contemporary, emerging trades, etc.
	Adoption of "National Policy for empowerment of Women".
	The government implemented the "Rajrajeswari Mahila Kalyan Yojana" and the "Bhagya Shree Bal Kalyan policy" on October 19, 1998, to improve the situation of girls and to give them insurance protection.

Source Authors work

scheme as it relates to women are Encouraging Girls' Education, Reducing Child Labor and Gender Disparities, Promoting Health Awareness etc.

The "National Policy for the Empowerment of Women (2001)" was introduced by the Government of India with the primary aim of advancing gender equality, improving the status of women, and promoting their participation in all aspects of national development. The key purposes of the policy for women are Enhancing Women's Participation in Decision-Making, Protection of Women from Violence, Gender Sensitization and Awareness etc.

The "Rajrajeswari Mahila Kalyan Yojana" was designed to empower women, particularly those from marginalized and economically disadvantaged backgrounds, by providing them with financial, social, and health support.

The "Bhagya Shree Bal Kalyan Policy" is primarily designed to improve the status of girl children, but it also has significant benefits for women, particularly mothers and families. The key purposes for women under this policy are Encouraging the Birth and Care of Girl Children, Financial Security for Girls, Promoting Girls' Education, Empowering Mothers etc.

The ninth Plan (1997–2002) significantly changed the traditional approach to planning for women in two ways (Konwar 2019; Sewa 2023). This plan becomes effect on April first, 1997 (Singh 2021). became one of the nine main goals and initiatives to establish a supportive atmosphere where women can freely express their rights as equal partners with men both within and outside the house (Konwar 2019; Sewa 2023; Poonam and Singh 2023). This plan's primary goal was to empower each woman (Barman 2019). The strategy aims to create a supportive environment for women in all settings. It includes vocational training to meet their needs and calls for childcare, hostels, and affordable housing to enable women's broader societal participation (Gaganamali 2020; Singh 2021; Barman 2019). Adopted a strategy of allocating at least 30% of funds to programs targeted specifically at women (Sewa 2023). Women have been key drivers of societal progress, seen as empowering forces. The "Women Component Plan" (tenth FYP) aimed to streamline services specific to women. On October 19, 1998, the government implemented the Rajrajeswari Mahila Kalyan Yojana and Bhagya Shree Bal Kalyan Policy to improve the status of girls and provide them with insurance protection (Konwar 2019). The 73rd and 74th Constitutional Amendment Acts of 1993 reserve 1/3 of local body seats for women in rural and urban areas, strengthening their role in democracy. This has elevated women's status in rural governance (Kumari 2009).

4.5.10 Tenth Five Years Plan

Swayamsidha The term "swayamsidha" literally means "to help oneself." It is a government project that was introduced in 2001 by the Indian government's Ministry of Women and Child Development (*Yadav, nd)*. This Scheme primarily targets women who participate in Self-Help Groups (SHGs), thereby enhancing their level of empowerment. Women who participate in these programs learn to be self-sufficient and solve difficulties on their own. Promoting women's empowerment was the government's only goal (*arthapedia.in*) (Table 11; Fig. 11).

The Swashakti Scheme, launched by the Government of India, is designed to empower women by promoting self-reliance, financial independence, and social development through the creation of Self-Help Groups (SHGs).

The Indira Mahila Yojana (IMY) was launched by the Government of India with the goal of empowering women, particularly those from rural and underprivileged communities, by enhancing their socio-economic status and enabling them to participate actively in community development. National Common Minimum Generation Activities Program (NCMP)".

Self-Help Groups serve as a powerful tool for women's empowerment and community development, providing a platform for individuals to improve their socio-economic status while fostering solidarity and collective growth. Self-Help Groups are made up of small, mostly female-only groups of people that join in rural areas to save money and lend money to one another. Poor individuals in rural areas, particularly women, are empowered by self-help groups (Next IAS Team 2023).

The Support for Training and Employment Programme (STEP) (1987) The Ministry has been administering 'Support to Training and Employment Programme

Table 11 Initiatives for women empowerment under tenth five year plan

Plan	Initiatives for women empowerment
Tenth five year plan (2002–2007)	Continuing the main strategy of "empowerment of Women" as agents of social change and development.
	Adopted "Swayamsidha" scheme.
	Adopted "Swashakti" scheme.
	Adopted "Indra Mahila Yojana" scheme.
	Adopted Women's component plan (WCP) scheme.
	National Common Minimum Generation Activities Program (NCMP)" scheme.
	Self help groups (SHGS).
	Training cum centres for Women "(Swawlamban previously known as NORAD'S women economic programme)" (1982–85)
	Support for training and employment programme (STEP)" (1987)
	"Rashtriya Mahila Kosh (RMK)" etc.
	The following schemes were implemented to provide support service hostels for working Women-
	"Short stay homes" (1969–70)
	"Swadhar" for Women under Various Circumstances (2001–02)

Source Authors work

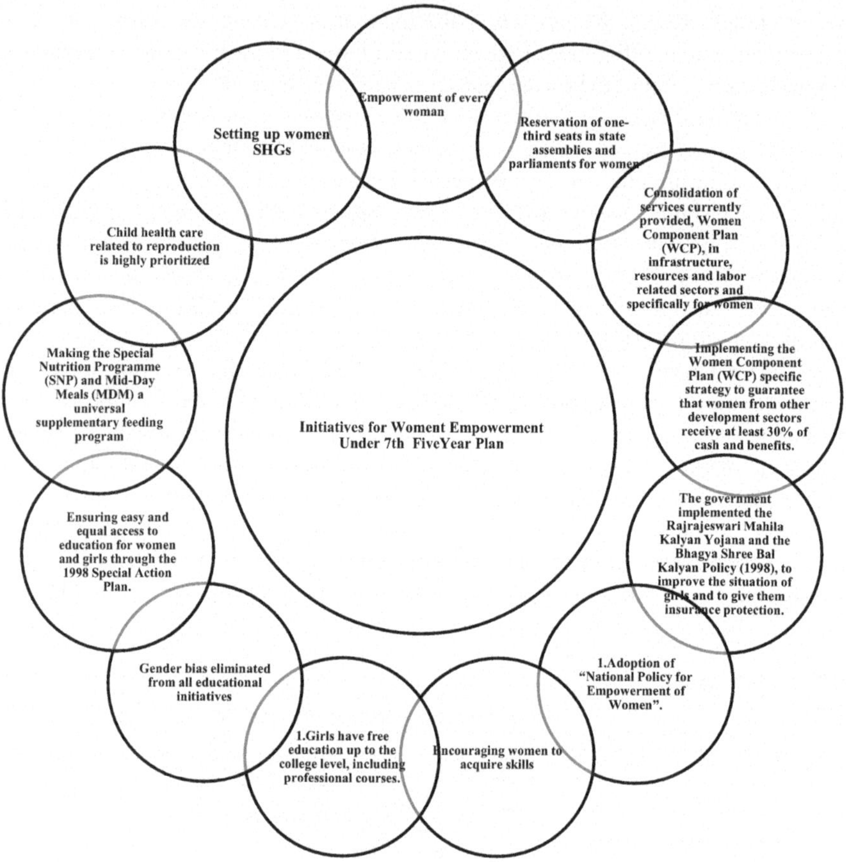

Fig. 10 Initiatives for women empowerment under ninth five year plan. *Source* Authors work

for Women (STEP) Scheme' since 1986–87 as a 'Central Sector Scheme'. The guidelines of STEP Scheme have been revised in December, 2014 by Ministry of Women and Child Development. The STEP Scheme aims to provide women with employability skills as well as the abilities and skills necessary for them to pursue self-employment or entrepreneurship. The program is meant to help all women in the nation who are 16 years of age or older, especially those who live in rural areas (Ministry of Women and Child Development, 2015).

In 1993, the Indian government founded Rastriya Mahila Kosh (RMK), a national organization under the Ministry of Women and Child Development, as a society under the Society Registration Act, 1860, with the goal of uplifting the socioeconomic status of underprivileged women. Along with microfinancing, credit and thrift, capacity training, marketing connections through the Self-Help Groups (SHGs) model, and enterprise development for low-income women, RMK has implemented

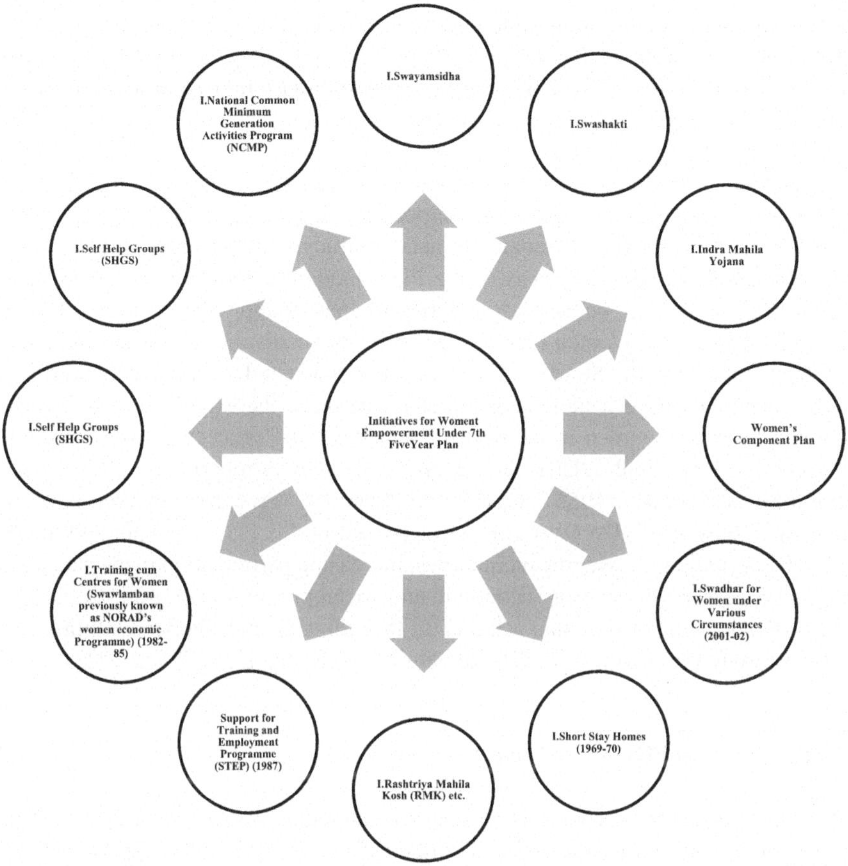

Fig. 11 Initiatives for women empowerment under tenth five year plan. *Source* Authors work

several promotional initiatives to popularize the idea of women's empowerment (Ministry of Women and Child Development, 2023).

Short Stay Homes" (1969–70) Short Stay Homes for Women and Girls was a program started by the Indian government in 1969 in the Central Sector to protect and rehabilitate women and girls who are in danger on a social and moral level because of family issues, mental health issues, social rejection, exploitation, or other circumstances (Sociology Guide 2023).

Swadhar The Swadhar Greh Scheme, which is being carried out by the Ministry of Women and Child Development, is aimed at assisting women who have been subjected to adversity and require institutional help for rehabilitation to live with dignity. For these women, the scheme aims to provide economic and social security along with housing, food, clothes, and health care (Ministry of Women and Child Development 2024).

Women have specific, monitorable targets defined under the Tenth Plan (2002–2007). They are as follows: Women are grouped into Self-Help Groups (SHGs) under the Swyamsidha initiative, which is a recast of the Indira Mahila Yojna and is used for generating income (Gaganamali 2020). The Tenth Plan seeks to operationalize the 2001 National Policy for Women's Empowerment, focusing on protecting, empowering, and advancing the rights of women and children. It emphasizes a rights-based approach to make women agents of socioeconomic change (Sewa 2023; Kumari 2019). Programs to expand female farmers' employment prospects and production were another major emphasis of the Plan. made a significant commitment to empowering women by putting into practice the newly approved "National Policy for Empowerment of Women (2001)", ensuring women's and children's survival, protection, and growth (Sewa 2023). To ensure that gender equity concerns and ensure that plans have a beneficial effect on women, there is a need to expand the engagement of women in the development planning process at every stage of the process (Gaganamali 2020). The "Married Women's Property Act (1874)"; the "Hindu Succession Act (1956)" was amended and the "Protection of Women from Domestic Abuse Act" (PWDVA 2005) was passed in this Plan (Poonam and Singh 2023). It also includes lessening inequalities, increasing participation in the decision-making process, and increasing economic and social possibilities for all people and groups (Barman 2019). Swadhar scheme for Providing shelter, shelter, clothing, and medical care to blind women in difficult situations (Poonam and Singh 2023).

4.5.11 Eleventh Five Years Plan

The Gender Budgeting scheme is an initiative aimed at ensuring that government budgets and policies are analyzed and structured to address gender disparities. The purpose of the scheme is to promote gender equality and women's empowerment by integrating gender perspectives into the planning, allocation, and implementation of government budgets (Table 12; Fig. 12).

The Eleventh Plan considers women as agents of social and economic advancement in addition to being equal citizens. The government prioritizes economic empowerment and ensures an environment free from physical, financial, social, psychological, and other types of violence against women. It also gives women fundamental rights to confront the effects of globalization on women and their realities (Barman 2019). Also, along with concentrating on gender-related issues, we also work to preserve the conditions that allow women to advance economically, socially, and politically. The three areas of violence against women, women's health, and economic empowerment are the primary targets of the government's approach to eradicate the issues (Singh 2021). Aimed to guarantee that every woman and child can reach her full potential and enjoy the rewards of economic growth and prosperity; to put an end to the various forms of exclusion and prejudice that affect them. The strategy acknowledged women as equal members of society and as drivers of social and economic advancement. Implemented specific initiatives for equity and female empowerment. Additionally, a gender budgeting strategy was unveiled (Sewa 2023).

Table 12 Initiatives for women empowerment under eleventh five year plan

Plan	Initiatives for women empowerment
Eleventh five year plan (2007–2012)	Make sure there is no physical, economic, social, psychological, or other type of violence against women in the environment.
	To maintain the circumstances that enable women to achieve political, social, and economic empowerment.
	The three main areas of focus are "economic empowerment", "women's health", and "violence against women".
	To ensure that every woman and child may reach their full potential and take part in the advantages of wealth and economic growth.
	A scheme on "gender budgeting" was also introduced.

Source Authors work

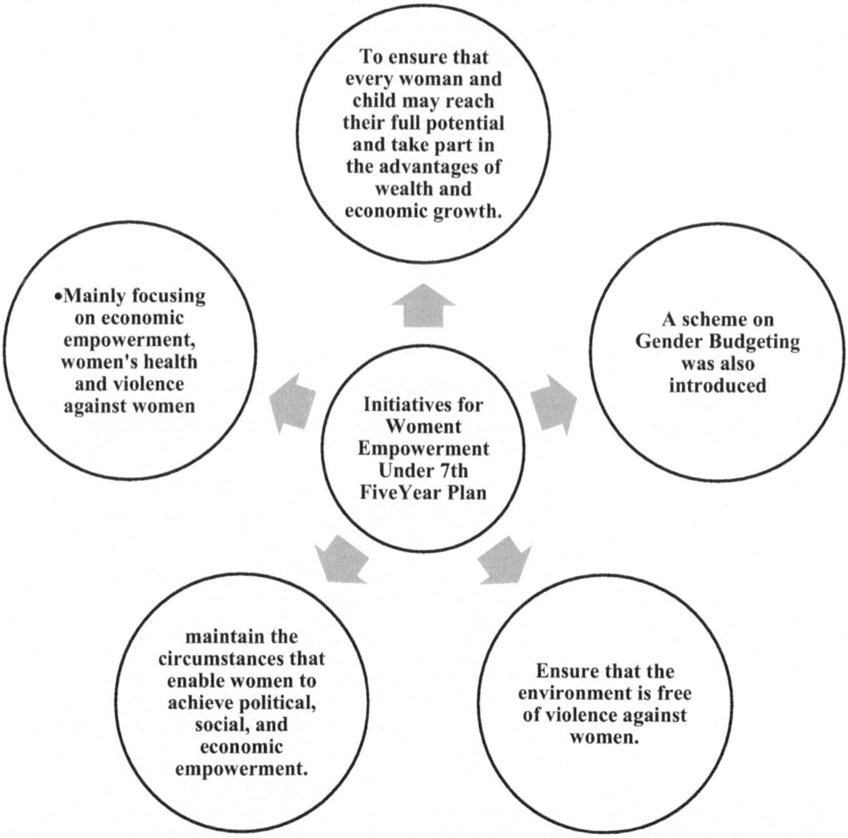

Fig. 12 Initiatives for women empowerment under eleventh five year plan. *Source* Authors work

4.5.12 Twelfth Five Years Plan

In this plan, women were viewed as inferior human beings and second-rate individuals while men enjoyed all positions of authority and prestige in both the public and private spheres. "Faster Sustainable and More Inclusive Growth" is the name of this plan. This strategy has a decided growth rate of −8.2% (Singh 2021). The plan seeks to improve women's status by removing institutional and structural barriers and strengthening gender mainstreaming. It aims for inclusive, sustainable growth, ensuring women from all societal segments participate equally in development. The plan promotes "freedom" and "choice" for women through awareness-raising, creating frameworks to challenge patriarchal norms, and addressing the impacts of economic reforms and globalization on women and children. By tackling structural issues, it ensures progress in women's conditions and the realization of their rights and ambitions (Sewa 2023). The plan aims to empower adolescent girls (11–18) through healthcare, nutrition, and life skills education, while enabling marginalized women to achieve economic empowerment. It provides relief and rehabilitation to vulnerable women lacking family support or income, combats child and women trafficking, and protects women from discrimination and all forms of violence (Barman 2019) (Table 13; Fig. 13).

Table 13 Initiatives for women empowerment under twelfth year plan

Plan	Initiatives for women empowerment
Twelfth five year plan (2012–2017)	This plan entitled as "Faster Sustainable and More Inclusive Growth."
	It also emphasizes building knowledge and institutional frameworks to support women to give them more "freedom" and "choice."
	Empowering adolescent girls (ages 11 to 18) on health, nutrition, and life skills.
	Encouraging marginalized women's economic empowerment.
	Providing help and rehabilitation to disadvantaged and vulnerable women who do not have independent sources of income or the backing of their families or the community.
	The prevention and suppression of child and women trafficking.
	Defending women against prejudice and all types of violence.

Source Authors work

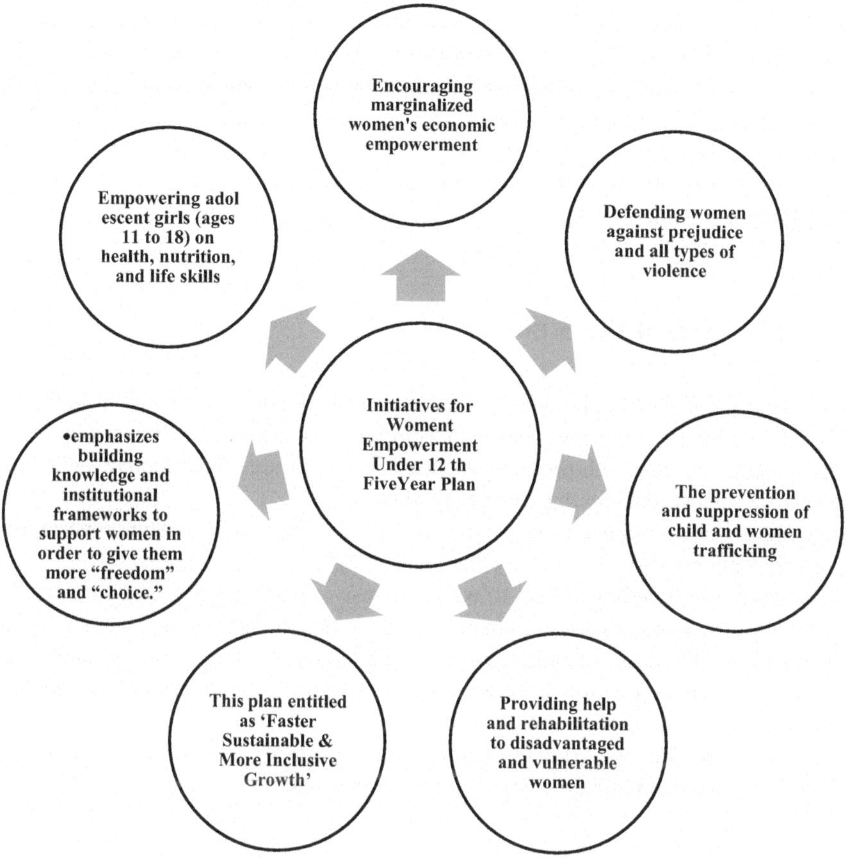

Fig. 13 Initiatives for women empowerment under twelfth five year plan. *Source* Authors work

5 Conclusion

Finally, we can conclude that women's empowerment has been given special emphasis since independence. Thus, several efforts have been made for the empowerment of women in various parts of the Indian Constitution and in the Indian Plans. As per Nayak and Mahanta (2017), the Indian government has made efforts to empower women in various ways since independence. The issues of women's empowerment have been given attention during various planning periods. But they have not been put into practice, so women's empowerment has not reached its peak. Singh (2021) states that all Five-Year Plans (1951–2017) had both positive and negative points. Plans at times don't work due to a lack of resources and income in the health sector and women's empowerment. There remains a vast difference between the stated goals in the Constitution, legislation, policies, plans, programs, and related mechanisms despite enough provisions and endeavors for women's protection and advancement in many sectors (Sewa 2023). The government has been unable to do as much as women have been subjected to more insecurity, poverty, illiteracy, casteism, traditionalism, backwardness, corruption, criminalization, and male dominance (Sewa 2023). But the government and voluntary organizations are still working to empower women, so girls are now leading in many respects. Thus, women's empowerment is possible only by the concerted efforts of the family, society, people, government, and voluntary organizations.

6 Implication of the Study

- The study helps to shaping gender-sensitive policies and governance in India.
- By analyzing the evolution of women's empowerment initiatives in various Five-Year Plans, the research provides a critical views to identify effective steps and persisting gaps in achieving gender equality.
- It emphasizes the need to move from welfare-based approaches to empowerment-centric frameworks.
- The study urges policymakers to adopt inclusive, participatory, and sustainable strategies for women's development.
- It highlights the necessity for continuous implementation, monitoring, and evaluation of gender-specific programs across sectors such as education, health, employment, and administration.
- It underscores the importance of integrating constitutional provisions, international commitments like the SDGs, and grassroots participation.

References

Barman BC (2019) Five year plans and women development in India. Res Rev Int J Multidisc 4(1):290–292. https://old.rrjournals.com/past-issue/five-year-plans-and-women-development-in-india/

Baruah B (2013) Role of electronic media in empowering rural women education of north East India. Abhivayakti 1:23–26

Five Year Plans of India (1951–2017) Planning commission government agency in New Delhi, India. planningcommission.gov.in

Five-Year Strategic Plan (2011–2016) Ministry of women and child development. Government of India

Gaganamali SB (2020) A study of Women empowerment through five year plans in India. In J Creat Res Thoughts IJCRT) 8(3):2557–2563. https://ijcrt.org/papers/IJCRT2003356.pdf

Girl Child in the Eleventh Five Year Plan (2007–2012) Ministry of women and child development government of India Shastri Bhawan New Delhi. https://www.aicte-india.org/downloads/girl_child_XI_five_year_plan.pdf

International Labour Organization (2017) Women's economic empowerment in the changing world of work. ILO. https://www.ilo.org/sites/default/files/wcmsp5/groups/public/@ed_dialogue/@act_emp/documents/publication/wcms_619803.pdf

International Development Law Organization (2019) Access to justice for women: legal empowerment and inclusion. IDLO. https://www.idlo.int/news/highlights/justice-women-access-empowerment-and-representation

Kabeer N (2002) Resources, agency, achievements: reflections on the measurement of Women's empowerment. Int Inst Soc Stud 30(3):435–464. https://doi.org/10.1111/1467-7660.00125

Konwar P (2019) Women empowerment in the five-year plans of India. Int J Res Cult Soc 3(4):64–69. https://www.researchgate.net/publication/332780543_

Kumari MP (2009) Empowerment of women and rural development: a study of self-help groups in west Godavari district. [Acharya Nagarjuna university]. http://hdl.handle.net/10603/9845

Naqvi H (2008) The path of women's empowerment in India. Kurukshetra 56(11):26–28

Nayak P, Mahanta B (2017) Women empowerment in India. Bull Polit Econ 5(2):155–183. https://doi.org/10.2139/ssrn.1320071

Next IAS Team (2023) Self-help groups (SHGs) in India—functions. Advantages Probl. https://www.nextias.com/blog/self-help-groups-shgs/

Mishra N (nd) DWRCA: development of women and children in rural areas. https://in.search.yahoo.com/search?fr=mcafee&type=E210IN826G0&p=Development+of+Women+and+Children+in+Rural+Areas+(DWACRA)+for+the+purpose+for

Ministry of Women and Child Development (2015) Support to training and employment programme. Government of India. https://pib.gov.in/newsite/PrintRelease.aspx?relid=118658

Ministry of Women and Child Development. Rashtriya Mahila Kosh. https://pib.gov.in/PressReleasePage.aspx?PRID=1795476

Ministry of Women and Child Development (2024) SWADHAR Greh (a Scheme for Women in difficult circumstances). https://www.myscheme.gov.in/schemes/sg

Mukerjee B (1952) Community development in India. Community Development Bulletin, United Nations

Pal S (2013) Women in twelfth five-year plan in India: an analysis. J Humanit Soc Sci 2(1):121–127. https://www.academia.edu/4155519

Panda S (2007) Political empowerment of tribal women in Orissa. Kurukshetra 55(5):29–32

Patel A (2017) Unit-4 Women in planning and social security measures policy. In: IGNOU self learning materials, master of arts in women's and gender studies, Block-5 Lagislation. Policy, Social Protection and http://hdl.handle.net/123456789/6185

Patel UK (2019) Study of different dimensions of women empowerment in India: a sociological perspective. Int J Res Soc Sci 9(4):1730–1735. https://www.ijmra.us/project%20doc/2019/IJRSS_APRIL2019/IJRSSApril19uTPALM.pdf

Prasad N, Arshad md, Kumar V, Jindal P (2016) Education for empowerment of women. Directorate distance education Tripura University. Vikas Publishing House Pvt. Ltd.. E28, sector 8, Noida-201301 (UP). www.vikashpublishing.com

Poonam, Singh Y (2023) Development of Indian Women by five year plan and various Indian government schemes. South India J Soc Sci 22(1):177–183. https://www.researchgate.net/publication/371761145

Sewa N (2023) The empowerment of Women through planning and budgeting in India with special reference to five year plans. Int J Multidisc Res (IJFMR) 5(4):1–6. https://www.ijfmr.com/papers/2023/4/4972.pdf

Seltzer JR (2002) The origins and evolution of family planning programs in developing countries. RAND Corporation. https://www.rand.org/pubs/monograph_reports/MR1276.html

Singh K (2016) Importance of education in empowerment of Women in India. Motherhood Int J Multidisc Res Dev 1(1):39–48. https://www.motherhooduniversity.edu.in/pdf/Publications/2016/Khushboo%20Singh.pdf

Singh SP (2021) An analytical study of the five-year plans: a special reference to health and women empowerment. Int J Multidisc Edu Res 10(2):66–70. https://s3-ap-southeast-1.amazonaws.com/ijmer/pdf/volume10/volume10-issue2(3)/15.pdf

Shaikh I (2021) Development of Women and children in rural areas. https://thedailybrunch.com/2016/05/05/development-of-women-and-children-in-rural-

Sociology Guide-A student guide to Sociology (2023) https://www.sociologyguide.com/women-and-society/short-stay-homes.php
SwayamsidhaScheme,arthapedia.in http://arthapedia.in/index.php?title=Swayam_Sidha_Scheme
The Oxford Advanced Learner's Dictionary (2000) Sixth Edition, OUP, New Delhi
United Nations (2015). Gender equality and Women's empowerment: transforming the lives of Women and girls through the SDGs. United Nations. https://www.un.org/sustainabledevelopment/gender-equality/
UNICEF (2021) Social protection and Women's economic empowerment. UNICEF. https://www.unicef.org/media/112591/file/Social-Protection-Gender-Equality-Outcomes-Across-the-Lifecycle-Full-Report.pdf
Vishal Yadav, Swayamsidha Scheme, govinfo.me. https://govinfo.me/swayamsiddha-scheme/
World Health Organization (2021). Improving maternal health. WHO. https://www.who.int/health-topics/maternal-health#tab=tab
World Health Organization (2020) Women's health and rights: a critical area for sustainable development. WHO. https://iris.who.int/bitstream/handle/10665/332070/9789240005105-eng.pdf
Women UN (2020) Women's political participation and leadership. UN Women. https://www.unwomen.org/en/what-we-do/leadership-and-political-participation/facts-and-figures

Global Perspectives on Empowering Women Through Financial Literacy: A Pathway to Inclusive and Sustainable Development

Vikas Garg and Vandana Ahuja

Abstract This chapter examines how financial literacy can be a powerful tool to empower women worldwide and contribute to inclusive, sustainable development. Using theoretical frameworks, including feminist economics, the capability approach, and human capital theory, it emphasizes the importance of financial literacy as individual knowledge advanced, but as a means of encouraging independence, improving economic participation, and enhancing social outcomes. While advancements in financial inclusion have been observed, substantial gender gaps remain driven by structural barriers such as socio-cultural norms, digital exclusion, and institutional inefficiencies. Drawing on global examples, the chapter features best practices and case studies from various parts of the world, mobile money innovations in Africa, Self-Help Group models in Asia, conditional cash transfer programs in Latin America, and policy-driven strategies in Europe and North America. This chapter highlights continuing challenges and proposes a multidimensional strategy that involves curriculum development, digital tools, inclusive policies, and cross-sector partnerships. Finally, it highlights future avenues, including calls for greater intersectional, longitudinal, and technological equity research. At the global level, it is imperative to connect financial literacy with the Sustainable Development Goals, like SDG 1, SDG 4, SDG 5, and SDG 8, helping women globally to have the knowledge, confidence, and resources to contribute to resilient families, inclusive economies, and sustainable communities. Thus, literacy in private finance does not become just a set of skills but a fundamental and critical enabling component of global gender equity and development.

V. Garg (✉) · V. Ahuja
Symbiosis Institute of Business Management, Noida, India
e-mail: vikasgargsir@gmail.com

Symbiosis International (Deemed University), Pune, India

V. Ahuja
e-mail: director@sibmnoida.siu.edu.in

A. Singh et al. (eds.), *Human-Centered Sustainability: Bridging Well-Being, Gender Equity, and Technology*, World Sustainability Series,
https://doi.org/10.1007/978-3-032-22020-2_12

Keywords Women's empowerment · Financial literacy · Inclusive development · Gender equality · Sustainable Development Goals (SDGs) · Digital financial inclusion

1 Introduction

1.1 Background

Financial literacy is now seen as an essential catalyst for economic inclusion and individual agency, especially for women, in the twenty-first century. As such, and given the advancement of economies worldwide towards a more digital and interdependent ground, making informed financial choices becomes not just an individual benefit but a necessity for society as a whole. Despite promising progress toward gender equality, women around the world, and in particular, those in developing economies, are still disproportionately excluded from financial systems, preventing inclusive and sustainable development (OECD 2022a, b, c, d, e). It enhances women's ability to make strategic life choices, supports economic independence, and contributes to gender-inclusive growth trajectories.

Broader development goals, notably the United Nations Sustainable Development Goals (SDGs), underlie the intersection between financial literacy and women's empowerment. Goals 5 (Gender Equality) and 8 (Decent Work and Economic Growth) directly connect to the need for financial literacy, including equitable access to financial resources and economic participation. Women with good financial literacy are more likely to start businesses, invest in their children's education, and invest in the well-being of their community, sustaining a virtuous cycle of socio-economic development (UN Women 2023a, b).

1.2 Rationale for the Study

Although financial inclusion has gained ground worldwide, a formidable gender gap exists. According to the Global Findex Database (World Bank 2022a, b), women are still 7% less likely than men to have a bank account, and in poor economies, the gap is even wider. In addition, access does not mean understanding. Access itself is only part of the solution; many women with access to financial services do not have the knowledge and confidence to use these tools effectively. However, the issue becomes even more acute due to the combined effects of poor education infrastructure, regional socio-cultural conditions, and digital illiteracy, especially on the margins, such as rural and older women and women of varied physical ability.

The study of financial literacy as a weapon of change for women's empowerment makes this chapter vital in reducing these inequalities. It aims to add to the increasing literature that acknowledges financial literacy as a human right and a fundamental skill for attaining equitable development. Moreover, the chapter is also timely in light of the global movement toward improving after COVID-19 specifically undermined women's financial and social conditions (ILO 2023a, b, c).

1.3 Objectives of the Chapter

This chapter will discuss the global approach to empowering women through financial literacy, the results of such efforts, and the challenges experienced. The specific objectives are:

- To learn about global initiatives and policymaking efforts to promote women's financial literacy and inclusion;
- To determine the socio-economic effect of financial literacy on women's empowerment and sustainable development;
- To locate effective best practices and scalable frameworks that have worked in different regional contexts;
- To provide practical recommendations for governments, NGOs, and the private sector to advance the development of gender-responsive financial literacy programs.

The chapter aims to link the scholarly world with the policy world, providing directions for further thinking and debate among future researchers and policy practitioners.

1.4 Methodological Approach

This chapter applies a qualitative and analytical methodology based on secondary research. Furthermore, information was obtained from a narrative review of global and regional reports, peer-reviewed journal articles, and policy briefs. Key data sources include:

- The World Bank's Global Findex Database (2022): Cross-National Data on Financial Inclusion;
- OECD's Financial Literacy Surveys and Gender Reports (2021–2023), comparative perspective;
- Development indices by UN Women and UNDP, with a focus on empowerment and capability enhancement;
- Case studies from disparate economies, including mobile banking initiatives in Kenya, SHG models in India, and fintech interventions in Latin America.

The analytical framework draws on the capability approach (Sen 1999), which highlights freedom and opportunity, and feminist economic theories, which criticize the gendered aspects of financial systems. Using a gender-transformative lens, we assess the intersectionality of financial literacy with other socio-economic and cultural factors.

2 Theoretical Framework and Literature Review

2.1 Theoretical Underpinnings

A multidimensional theoretical framework is needed to comprehend the mechanisms of women's financial empowerment via literacy. **Theories from feminist economics, the capability approach, and human capital theory** help us understand how financial literacy impacts the economic position and presence of women in society.

Feminist economics provides a critical approach to economic models that frequently ignore unpaid care work, gendered divisions of labor, and systemic marginalization of women from financial structures (Folbre 2021). It highlights the structural inequalities that prevent women's participation in the economy and critiques the gender-neutral assumptions that underlie conventional economic policies.

The **capability approach** was developed by Amartya Sen (1999) and focuses on the fundamental freedoms people have to live lives they value. Empowered women are full of potential, and it all starts with financial literacy, which improves their choices, determination, and control over their finances. Such an approach also understands that empowerment is not just based on access to capital but on opportunity and the need to consider the quality and context of financial engagement.

This aligns with the **human capital theory**, highlighting the importance of obtaining education and skills to be economically productive (Becker 1993). Financial literacy, a type of education of an economic nature, is a massive investment in human capital. The impact of financial literacy on women goes further than personal, as when women become financially literate, they positively improve their economic situation as well as that of the household and the national economy.

2.2 Understanding Financial Literacy

Financial literacy means more than just managing money. This includes a set of knowledge, skills, attitudes, and behaviors that equip individuals to make informed, quality decisions with their financial resources (OECD 2022a, b, c, d, e). The OECD/INFE framework divides financial literacy into three constituent parts:

Financial literacy: Understanding interest rates, inflation, risk diversification, and budgeting.
Financial behavior (e.g., saving, future planning, comparing financial products).
Financial mindsets: Attitudes towards financial planning, debt, and long-term goal setting.

Socio-economic constraints, restricted education, and lack of specific financial literacy initiatives affect women in all three areas (Lusardi and Mitchell 2020). Thus, financial literacy has to be context-sensitive and inclusive and respond to women's particular barriers.

2.3 Women and Financial Literacy: Global Trends

Developed and developing countries alike show a persistent gender gap in financial literacy, as global data reveals. As per the World Bank (2022a, b) Global Financial Literacy Survey, 35% of women are generally financially literate, against 44% of men. The gap is even wider in low-income countries, often reflecting structural barriers such as gender-based discrimination, limited digital access, and little control over household finances.

Meanwhile, women perform lower than men on financial literacy tests in OECD Countries, although they frequently run household budgets (OECD 2022a, b, c, d, e). In the Asia–Pacific region, financial literacy programs for rural women have had some success, but coverage is uneven owing to linguistic and cultural divides (Asian Development Bank 2021). While mobile banking services have opened the door for greater financial inclusion among women in Africa (e.g., M-Pesa), they haven't necessarily enhanced financial literacy (Demirgüç-Kunt et al. 2022).

Importantly, these trends illustrate that access is not empowerment. Lack of understanding and ill-informed contextual knowledge often leads to underutilization or misuse of financial services. Thus, financial literacy must be coupled with tools, hands-on support, and conducive environments.

2.4 Review of Existing Literature

Much research shows a positive relationship between women's financial literacy and empowerment. Lusardi and Mitchell (2020) note that significantly more savings, investing in children's education, and entrepreneurship positively correlate with financial education among women. The study, which is based on evidence from more than 20 countries, shows that financial literacy among women results in better long-term financial planning and less risky borrowing behavior.

In their 2022 study, Klapper and Singer show that women with higher financial literacy are more resilient to economic crises, such as the recent COVID-19 pandemic. Women understanding risk management and savings are more equipped to weather income shocks and are more likely to sustain stability in the household. In India, the Self-Help Group Bank Linkage Programme (SHG-BLP) demonstrated that joint financial training and microcredit offered during exclusive financial services improve economic well-being and increase household bargaining power (NABARD 2021). Likewise, in the case of Kenya, the combination of access to mobile financial services and basic financial education leads to higher success rates for female micro-entrepreneurs (Suri and Jack 2021). However, the literature shows that gaps exist (2023). UNDP further states that many existing financial literacy programs do not cater to the socio-cultural factors restricting women's financial autonomy. There are no mechanisms to evaluate long-term impact, but partly because there are no mechanisms to assess impact.

Recent studies underscore the importance of intersectional approaches, taking caste, disability, age, and so on into account, to build inclusive financial literacy programs. Some groups, like adolescents or older women, require different approaches to pedagogy and other support systems (Plan International 2023a, b).

3 Global Landscape and Best Practices

3.1 Policy Frameworks and International Initiatives

Financial literacy has emerged as an important area of focus on the global development agenda in recent decades. Numerous international frameworks and initiatives reinforce its significance in pursuing gender equality and sustainable development. At the forefront of these is the United Nations 2030 Agenda for Sustainable Development, under which Sustainable Development Goal (SDG) 5 stresses the importance of gender equality and SDG 8 calls for inclusive economic growth and decent work. The principles of financial literacy undergird these aims by assisting women in accessing and commanding economic resources to develop autonomy and financial resilience (UN Women 2023a, b).

The G20 Global Partnership for Financial Inclusion (GPFI) has also been a key driver of women's financial inclusion. The 2023 policy compass focuses on collecting gender-disaggregated data, tailoring digital financial services to the needs of women, and addressing the necessity for a supporting regulatory context (GPFI 2023). The World Bank's Global Findex Dataset remains an essential source of information for understanding women's engagement with financial institutions, highlighting both improvement and ongoing gaps (Demirgüç-Kunt et al. 2022).

Furthermore, the OECD/INFE has released policy guidance on gender-sensitive national financial literacy strategies, promoting inclusive education programs from the school system to the workplace (OECD 2022a, b, c, d, e).

3.2 Case Studies by Region

A. **Africa: Mobile Finance and Informal Savings Groups**

Mobile financial inclusion has always been a front-runner for Africa. So far, the best-known example is M-Pesa in Kenya, which enables users to transfer, receive, and save money using mobile phones. Research shows that with access to M-Pesa, women's financial independence and household savings significantly increased. Between 2008 and 2014, the service lifted an estimated 194,000 Kenyan households out of poverty, with female-headed households benefiting the most (Suri and Jack 2021).

Also, a different type of innovation is being implemented in Africa through Village Savings and Loan Associations (VSLAs). These are self-managed groups of women who put money into a common fund and access micro-loans for small-scale businesses. In Uganda and Tanzania, increased participation in VSLAs has improved women's confidence, decision-making power, and income-generating capacity (CARE 2021). However, challenges remain. Despite using digital platforms, most women, particularly in rural areas, are financially illiterate. This has encouraged organizations like BRAC and FINCA to combine financial literacy modules with mobile banking products for sustainable use (BRAC 2023).

B. **Asia: Self-help Groups and Institutional Models**

The **Self-Help Group (SHG)–Bank Linkage Program in India** is one of the world's most significant financial inclusion initiatives. This initiative has linked more than 10 million women to the doorstep of formal financial institutions and is managed by NABARD. Financial literacy is ingrained in the SHG model, which continues through regular group meetings, training sessions, and peer learning. Studies find that members of self-help groups save more, engage in entrepreneurial activity, and have more autonomy in budgeting (NABARD 2021).

That's particularly true in the case of **Bangladesh**, which is offering the world an astonishingly powerful example, thanks in part to the Grameen Bank and the more recent Joyeeeta Foundation, both dedicated to microfinance and enterprise development exclusively for women. In the Grameen model, financial literacy is integrated into the loan distribution process so that women know the terms, cash flow management, and productive investment (Yunus & Moingeon 2020).

For example, in one Asian country, **Indonesia**, the government had targeted conditional cash transfers for health and education through Program Keluarga Harapan, complemented by financial education workshops (ADB 2021).

C. **Latin America: Cash Transfers and Microfinance Innovations**

Latin America has made great strides through conditional cash transfer (CCT) programs like Bolsa Família in Brazil and JUNTOS in Peru. These programs target women who are the heads of households and seek to interrupt the cycle of poverty across generations. CCTs have increased women's consumption smoothing/insurance behavior and improved savings and budgeting, especially when combined with financial literacy training (World Bank 2022a, b).

The national strategy for financial education (ENIF) in Mexico was developed with the involvement of NGOs and financial institutions and is presented through workshops to women entrepreneurs in informal sectors. These initiatives frequently employ gamification and storytelling to navigate literacy obstacles (OECD 2022a, b, c, d, e).

Microcredit institutions, like BancoSol in Bolivia and Compartamos Banco in Mexico, have developed simplified financial products with inbuilt education modules. This calls for a combination of access and knowledge to ensure sustainable empowerment (J-PAL 2021a, b).

D. **Europe and North America: Digital Tools and Policy Innovations**

Financial literacy is regarded as a vital element of lifelong learning in Europe. The United Kingdom's Money and Pensions Service (MaPS) provides "Talk Money Week" every year and funds programs aimed at single mothers, elderly women, and survivors of domestic abuse (MaPS 2023).

In the U.S., the FDIC's Money Smart for Adults program offers budgeting, credit management, and fraud protection components. Particular outreach is geared toward low-income women, immigrants, and survivors of financial abuse (FDIC 2022). In addition to providing digital investment platforms, fintech companies that serve women, such as Ellevest, offer financial coaching and education.

The Canadian government's Financial Literacy Strategy focuses on giving special access to underserved groups, including Indigenous women, to community-based workshops that combine cultural relevance and practical financial skills (Government of Canada 2023).

Such efforts are evidence of a continuing struggle against gender inequities, even in high-income countries and among marginalized populations. However, innovation in digital tools and policy assistance is tackling these gaps.

3.3 Role of NGOs, Private Sector, and Public–Private Partnerships

NGOs have played a crucial role in addressing the gaps left by government efforts. Women's World Banking, CARE International, and Plan International have implemented gender-transformative financial literacy programs on several continents. These programs do not only give you knowledge but also develop your soft skills like negotiation, goal-setting, and leadership.

The private sector, especially fintech companies, has begun taking the lead. Apps like Tala and Kiva, available on mobile devices, provide micro-loans and embedded financial education. In India, Airtel Payments Bank wrapped its mobile banking offering with a layer of a digital literacy program in a partnership with SEWA (Self-Employed Women's Association) (GSMA 2022).

Public–private partnerships (PPPs) are following suit. In Rwanda, for example, the government partnered with the MasterCard Foundation to implement the "Hanga

Ahazaza" initiative, which integrates entrepreneurship training and financial literacy for young women (MasterCard Foundation 2023).

Such collaborations are most effective when anchored in local conditions with explicit language, visual aids, and culturally sensitive materials. Additionally, implementing monitoring and evaluation frameworks ensures outcomes are not limited to access but result in authentic empowerment.

4 Challenges and Barriers

Despite financial literacy's potential to empower women and pave the way for inclusive development, the path to mass and meaningful adoption is fraught with challenges. These include cultural, economic, technological, and institutional barriers that unduly substitute systemic inequalities and restrict the transformative influence of financial education programs. Knowing these obstacles is key to providing inclusive and sustainable solutions.

4.1 Socio-Cultural Constraints

Socio-cultural norms remain one of the most enduring barriers to women's financial literacy. In large parts of the world, deeply rooted patriarchal values limit women's access to education, mobility, and decision-making power. These customs are reflected in family positions that render women unexposed to the economy too(l)s, placing them as passive receivers instead of active agents in financial transactions (UNESCO 2022).

In many countries in South Asia, the Middle East, and parts of Sub-Saharan Africa, women's financial choices are influenced or controlled by male family members. Women may even have financial literacy but be put off using their autonomy because of social stigma or fear of backlash (OECD 2022a, b, c, d, e). In conservative communities, some women are also restricted in their mobility and, therefore, are unable to attend financial education workshops or open bank accounts.

Another intersecting barrier is gender-based violence (GBV). GBV survivors are generally poor and are not in a position to access financial services or to go into training programs. Without an effort to bridge these sociocultural barriers, financial literacy initiatives may not only reach the women who need them the most but, over time, can also increase the gap (UN Women 2023a, b).

4.2 Economic and Structural Challenges

Economic vulnerability is a key barrier to financial empowerment. Women, especially in the Global South, work in the informal economy where earnings are irregular and social protection is weak. Lacking consistent revenue or official employment, women may not meet the criteria for financial products, such as loans, insurance, or savings accounts (World Bank 2022a, b).

In addition, poverty narrows the scope of what is considered beneficial about financial literacy; for women struggling to make ends meet, long-term planning, investing, or retirement saving might feel irrelevant or unrealistic. In these contexts, financial education should be tailored to meet women where they are focusing on immediate, practical skills such as household budgetary oversight, consumer debt management, and short-term savings.

Moreover, the absence of childcare and time poverty caused by unpaid domestic labor leaves very little room for women to take part in training sessions or make financial decisions. Women are performing 76% of the world's unpaid care work (ILO 2023a, b, c), encouraging less time and energy that they can spend on financial literacy initiatives.

Structural inequities in property rights and inheritance laws further aggravate these challenges. In some countries, women are not allowed to own or inherit land and other assets (such as houses), which decreases their collateral and creditworthiness, which are crucial for financial access (FAO 2022a, b). Even where policies are in place to promote gender equality, implementing these policies is often weak or inconsistent.

4.3 Digital Divide and Technological Barriers

Whilst digital financial services have been a game-changer for access to banking and financial education, they have also created a further exclusion for women lacking digital literacy. The gender digital divide, in owning a mobile phone, accessing the internet, and feeling confident online, is a stubborn barrier, especially for women among low-income and rural populations.

Women are 19% less likely than men to use mobile internet globally, and in some regions (e.g., South Asia), the gap is getting larger (GSMA 2023). Most financial literacy programs today are offered via apps, online modules, or SMS-based tools, which, albeit unintentionally, leave behind digitally illiterate women and those lacking ownership of their own devices. In addition, when women do have access to digital tools, they may share devices with family members or have little control over their usage (Plan International 2023a, b).

Privacy and security issues have kept many women away from digital financial platforms. Few in a community like this one, where reports of financial scams are standard, may participate, in part due to unfamiliarity with, or fear of, online banking. To address this barrier, financial inclusion through financial education should be complemented with digital literacy training that is accessible, culturally and linguistically appropriate, and community-based. Technologically, gender-responsive tech design, voice-enabled banking or visual tools can help close this divide.

4.4 Institutional and Policy Gaps

The efficiency of financial literacy programs is often limited by institutional and policy blind spots. In many instances, national strategies of financial education excluded a gender lens; consequently, the programs did not take into account the specific barriers that women face (OECD 2022a, b, c, d, e). Even where there are gender-sensitive policies, the absence of gender-disaggregated data means that it is difficult to measure progress or fine-tune interventions.

One significant drawback has been the fragmentation of initiatives. NGOs, financial institutions, or donor agencies often operate independently, leading to overlapping efforts and outcomes not being reusable. Without centralised monitoring and quality control, low engagement and limited follow-up, as well as a lack of proper evaluation metrics, are commonplace across many programmes (UNDP 2023a, b).

Additionally, the financial institutions themselves may carry implicit biases that make female clients feel unwelcome. Lengthy paperwork, male-dominated atmospheres, and non-inclusive marketing materials can push women away from financial institutions, especially women from marginalized communities (Women's World Banking 2022). There is also an urgent gap in training the trainers. Even fewer facilitators of financial education programs are trained to address gender-specific issues or participatory, inclusive pedagogy. This leads to sessions being too generic or not building trust with their women members.

5 Impact Assessment and Measurable Outcomes

Financial literacy initiatives for women must be evaluated on measured outcomes to ensure their long-term success, which would not only show improved knowledge but also behavioral change, empowerment, and well-being. Assessing these outcomes is critical to improving current programs, supporting resource allocation, and informing policy reform. Here, we delve into significant areas of impact, from empowerment to economics to social and environmental change, and through common tools that assess these dimensions.

5.1 Empowerment Outcomes

These financial literacy programs aim to empower women to make informed decisions to exercise control over their economic resources. Empowerment occurs across a variety of fields:

- **Household decision-making**: According to studies, women who have financial literacy are more often involved in household decision-making, including budgeting, education of children, healthcare, and asset purchasing (UN Women 2023a, b). This change in intra-household power relations increases the women's agency and the sense of self-efficacy.
- **Entrepreneurial Activities**: Financial literacy can improve women's abilities to start and operate small businesses. For instance, in India, 23% more members of SHGs with finance training reported running a profitable enterprise (NABARD 2021). They also showed improved pricing practices, improved record-keeping, and greater use of digital payment platforms.
- **Personal Autonomy and Mobility**: The ability to access financial services is important to a woman's mobility, especially when coupled with knowledge, as it gives her the ability to pay for personal expenses, leave abusive relationships, or invest in self-development (GSMA 2023). This liberation leads to better well-being and assurance in using public spaces and services.
- **Empowerment Effect**: Women with greater community involvement can lead to a better future for their children, especially young girls. Financially savvy mothers are also more likely to open savings accounts for their kids, plan for their higher education, and teach them financial values at home (Plan International 2023a, b).

5.2 Economic Development Outcomes

Enhancing women's participation in formal economies and financial literacy increases economic growth. Micro and macro levels at which the following development outcomes can be noticed:

- **Increasing Growth of Women-Led MSMEs**: Financially literate women are more empowered to access credit, manage their cash flows, and expand their micro, small, or medium enterprises. IFC (2022) found that small firms owned by women with financial training achieved 18% higher revenue growth over 12 months than those without training in Sub-Saharan Africa.
- **Higher Household Savings and Investment**: Plans that offer savings education have higher saving frequency and volume. In Peru, participants in a financial literacy program increased their use of informal savings by 38% over the year (World Bank 2022a, b).
- **Lower Debt Strains and Economic Risks**: By facilitating responsible behavior in choosing loans and credit, financial learning enables women to steer clear of predatory lending and over-indebtedness. For example, in Bangladesh, women

who received training from BRAC's Microfinance Plus program were more likely to repay loans on time and less likely to use exploitative lenders (BRAC 2023).
- **Better Labor Market Outcomes**: Educated women are more adept at scrutinizing employment contracts, negotiating wages, and investing in upskilling. They are especially evident among young women joining the workforce or moving from informal to formal employment.

5.3 Social and Environmental Impact

Apart from economic metrics, financial literacy also plays a role in advancing broader social and environmental challenges and thus contributes towards the United Nations Sustainable Development Goals (SDGs).

- **Community Development and Leadership**: Financially empowered women often lead in community associations, SHGs, or cooperatives. These spaces act as spaces for civic engagement and advocacy, further strengthening women's voices in local governance and development planning (CARE 2022).
- **Support for Health and Education**: Exposure to financial literacy promotes preventive healthcare and quality education for children, leading to benefits in human capital in the long run.
- **Environmental Sustainability and Green Actions**: Women with financial literacy are more likely to invest in renewable energy products, green businesses, and organic farming, all of which support environmental sustainability (FAO 2022a, b). In such countries, financial literacy can significantly integrate climate risks into investment decisions and adapt to suitable technologies.

5.4 Tools for Assessment

To measure and keep track of the benefits of financial literacy programs for women, several tools and indicators are being used:

- **Financial Literacy Index (FLI)**: The FLI is a composite index representing individuals' financial concepts, attitudes, and behaviors. Gender-disaggregated FLI scores can help policymakers detect disparities and evaluate their interventions' effectiveness (OECD 2022a, b, c, d, e).
- **Gender Empowerment Measure (GEM)**: GEM measures women's participation in economic and political life, resource control, and decision-making autonomy. Though these scores are not directly linked to financial literacy, every financial education effort could correlate with these score improvements (UNDP 2023a, b).

- **Theory of Change (ToC)**: A Theory of Change (ToC) is widely adopted in NGOs and development programs, which maps inputs from financial education to empowerment outcomes. Therefore, this model is essential for analyzing longer-term, multi-layered consequences (CARE 2022).
- **Impact Evaluations and RCTs**: Randomized control trials (RCTs) have become more prevalent in establishing causality between financial literacy and behavior change. One example is J-PAL's evaluation of a financial education program in India, which showed participants a 17% increase in regular savings behavior six months after the intervention (J-PAL 2021a, b).
- **Qualitative Assessment**: Focus groups, life stories, and participatory evaluations are crucial for capturing non-meaningful outcomes such as changes in confidence, agency, and social status that cannot necessarily be quantified, especially in contexts where non-statistical data may not provide a complete picture.

6 Strategies and Recommendations

Progress towards financial literacy programs for women globally has been significant, and with the right innovative, inclusive, and context-appropriate strategies, equitable and sustainable empowerment is possible. Building on global evidence and best practices, this section provides actionable recommendations in four thematic pillars: education, technology, policy, and stakeholder engagement.

6.1 Enhancing Financial Education

Financial education has to dig deeper than straightforward content to be effective. It must be inclusive, culturally illuminating, and lifecycle-based, spanning the various phases of a woman's life, enticing youth through to the autumn of her life.

Curriculum Development and Pedagogy

It builds the argument for making financial literacy compulsory in schools and colleges, especially for girls. Early preparation helps smooth economic decision-making and builds financial confidence over the longer term (OECD 2022a, b, c, d, e). This should be done by introducing age-appropriate modules in the national curriculum on budgeting, saving, digital banking, and rights as consumers.

Interactive and participatory methods, such as storytelling, role-playing, and visual tools, have led to better retention among adult learners, particularly in low-literacy environments, compared to conventional classroom methods (UNESCO 2022). Programs must use local languages and real-life contexts to make them more relatable.

Targeted Outreach

Content tailored for marginalized women should be created, such as:

- Rural women
- Widows and single mothers
- Migrant and displaced women
- Women with disabilities

Also, target vulnerable transition points when financial knowledge can safeguard resilience, like post-disaster recovery, retirement planning, or divorce.

6.2 Leveraging Technology

There is a powerful opportunity to use digital tools to expand financial literacy efforts, as long as the digital divide is appropriately addressed.

Mobile-Based Learning

The first exposure to formal finance for many women in developing countries is mobile phone-based. Tala, Kiva, and M-Pesa are examples linked with SMS-based financial literacy modules. In low-bandwidth areas, on-demand, self-paced learning content can be provided through such tools (GSMA 2023).

Gamification and the Integration of EdTech

As is evident, games and simulation-based tools have proved effective in teaching abstract financial concepts. Plan International, for example, reported a 32% increase in budgeting confidence among girls in Bangladesh who participated in a financial literacy game (Plan International 2023a, b).

Digital learning platforms (such as Khan Academy, Coursera, and country-specific portals like India's National Center for Financial Education) can be leveraged to increase free, multilingual content.

Bridging the Digital Gap

Infrastructure needs to be invested in, including device access, low-cost data plans, and digital literacy training. Recruiting male family members to join digital literacy campaigns can encourage co-learning and lower household barriers (ITU 2022).

6.3 Inclusive Policy Formulation

Government commitments to financial literacy must be anchored in national financial inclusion efforts and gender equity initiatives through enabling regulations and the political will to enforce them.

Inclusive Planning and Gender Budgeting

Gender budgeting practices should be adopted with a detailed allocation of funds for women's financial literacy and entrepreneurship training. Financial literacy must be part of wider empowerment strategies manifested in national action plans (UN Women 2023a, b).

Regulatory Reforms

Financial regulators should:

- Requiring financial institutions to collect data disaggregated by gender.
- Reduce Know Your Customer (KYC) processes for women.
- Promote the development of financial products sensitive to gender.

Central banks and finance ministries could establish dedicated gender and financial inclusion cells to oversee progress and coordinate stakeholders (World Bank 2022a, b).

Monitoring and Evaluation

Governments and NGOs need to establish strong monitoring frameworks based on outcome-oriented indicators, including:

- Ever-growing financial products used by women.
- Change in behavior (e.g., more saving, lower informal borrowing).
- Shift in the decision-making power in households.

Gender-sensitive tools, such as the Financial Empowerment Scorecard and Theory of Change frameworks, can provide nuanced insights into impact (CARE 2022).

6.4 Multi-stakeholder Engagement

However, there's not a single sector solution to the issue, and empowering women through financial literacy must come from all sectors. It demands active cooperation between governments, civil society, academia, the private sector, and women.

Role of Governments

Public institutions should be at the forefront of providing funding, policy coherence, and integrating financial education into wider gender equality programs. So, government-led programs can act as anchor models for private and NGO participation.

NGOs and Community Organizations

Grassroots organizations have deep community trust and can tailor delivery mechanisms to wide-ranging demographic groups. Community-based training with follow-up mentorship is evidenced by programs such as SEWA in India and FINSCOPE in Africa.

The Role of the Private Sector and Financial Institutions

Banks and the fintech sector can support by:

- Inclusive financial product development (e.g., microloans with loose terms).
- Providing free digital training for future female customers.
- Incorporating workplace financial wellness programs for women employees.

Mastercard's Girls4Tech program provides training in STEM and financial concepts to adolescent girls in more than 50 countries (Mastercard Foundation 2023).

Academia and Research Institutions

Universities can be essential in developing, piloting, and evaluating financial literacy curricula. Working with academic institutions guarantees that the interventions are evidence-based and that the system is updated ceaselessly.

7 Future Directions and Research Agenda

With the ever-changing global landscape, new socio-economic, technological, and environmental challenges have emerged in some cases, which require adopting a responsive approach to their needs in financial literacy for their empowerment. Today's data primarily focuses on publication counts and the number of citations rather than understanding science regarding trends or gaps that must be filled through targeted research and global collaboration. This chapter describes critical areas for innovation, inquiry, and international engagement.

7.1 Emerging Trends

Data-Driven Changes Through Technology

Three trends are reshaping how financial services are delivered: rapid advances in financial technology (fintech), artificial intelligence (AI), and blockchain. Such popular tools have potential as low-cost, scalable, and tailored financial literacy interventions. For instance, AI-powered chatbots can emulate financial scenarios and tailor advice to women in real time, including in low-literacy contexts (World Bank 2022a, b).

Women lacking traditional credit scores can use blockchain technology to create secure and verifiable financial histories, increasing their access to formal credit (IFC 2023). In addition, decentralized finance (DeFi) platforms could create new channels for peer-to-peer saving and investing, though they need digital education to be meaningful and safe.

Gamification and EdTech

The younger population has been seeking out gamified financial education. Mobile applications and online platforms that use storytelling, rewards, and simulations can better introduce the abstract concept of finance to children and make it fun. This approach is well-suited for young people and adults with low levels of formal education (Plan International 2023a, b).

Financial Literacy Linked to Climate

As the effects of climate change increasingly affect livelihoods in agriculture, that's particularly true for women working in informal economies as well (FAO 2022a, b): they require financial literacy that enables climate-resilient planning, one that opportune weather-based insurance, green microloans, and sustainable investment options.

7.2 *Areas for Further Research*

Many critical knowledge gaps are still missing that need to be filled to the attention of academics and policy-makers despite such efforts across the world:

Longitudinal Studies

Most of the existing evidence on financial literacy focuses on specific outcomes in the short term. Longitudinal studies are required to investigate how this financial education impact persists dynamically over time, such as through what behavior is maintained, if there are effects across generations, and whether long-term wealth is accumulated.

Access and Outcomes: An Intersectional Perspective

Future research could also investigate how gender interacts with other identities like disability, caste, age, sexual orientation, and refugee status. These intersections should guide the design of programs that are not just inclusive but also equity-oriented (UNDP 2023a, b).

Digital Exclusion and Algorithms' Discriminatory

As fintech assumes a more significant part in financial education, it is imperative to explore how systems from algorithms can entrench or erase gender bias. Further research should help understand how digital financial tools could be designed to meet diverse users' diverse needs, mitigate bias, and contribute to women's privacy and safety (ITU 2022).

Behavioral Economics in Women's Finance

There is also room to increase behavioral economic studies focusing on how emotions, biases, and social norms shape women's financial decisions. This knowledge can help create better programs, ones with an understanding of psychology baked in.

7.3 Global Collaboration

Global partnerships are vital for scaling successful models and facilitating knowledge transfer across regions.

South-South Cooperation

Countries in the Global South often have similar socio-economic contexts. South-South exchanges and peer-learning platforms can drive innovation in financial literacy programming. For example, savings-led models in India's self-help groups could inform East African engagement, and mobile money frameworks in Kenya might influence downscaled efforts in Southeast Asia (UN Women 2023a, b).

Resilience Planning for the Post-pandemic World

The Code highlighted women's financial vulnerabilities, specifically those in informal sectors; the pandemic made those vulnerabilities more glaring. In the future, pandemic preparedness and recovery financing should be part of the financial education agenda so that women can achieve financial resilience in the face of crisis (ILO 2023a, b, c).

Integration, Sustainability, and SDGs

Future action must be more coherent with the Sustainable Development Goals (SDGs). Beyond mere economic development, initiatives promoting financial literacy can also and should be evaluated in terms of their contributions to gender equality (SDG 5), inclusive growth (SDG 8), and climate action (SDG 13) (UNDP 2023a, b).

8 Conclusion

8.1 Recapitulation of Key Points

In this chapter, we have discussed the importance of financial literacy as a transformative tool for empowering women and as a catalyst for inclusive and sustainable development. Anchored in theoretical paradigms like feminist economics, the capability approach, and human capital theory, the conversation has focused on financial literacy being not just about knowledge but expanding freedoms, creating agency, and allowing informed decision-making.

Our global overview of regional efforts in Africa, Asia, Latin America, Europe, and North America has shown that the contextualized nature of financial literacy approaches has enabled women to start businesses, manage household finances, save and invest wisely, and actively participate in community development. Whether SHGs in India, mobile money in Kenya, or conditional cash transfer programs in Latin America, financial education has increased women's economic role and standing in society.

Simultaneously, the chapter has delineated root challenges that persist as obstacles to progress. Deep-rooted socio-cultural norms, economic marginalization, digital exclusion, and institutional inefficiencies are significant barriers to overcome. Women have structural barriers that lead to the inability to have the tools to engage in financial behavior.

However, these barriers can be strategically addressed with evidence-based approaches like introducing financial education in schools, utilizing technology to reach the most vulnerable, embedding financial literacy in gender-sensitive policies, and employing multi-stakeholder partnerships. The evidence shows that well-designed programs improve savings behavior, decision-making power, resilience, and inter-generational benefits for families and communities.

8.2 Final Reflections

In the world, we are trying to achieve the Sustainable Development Goals (SDGs), especially those linked to gender equity (SDG 5), inclusive economic growth (SDG 8), and reducing inequality (SDG 10). Empowering women through financial literacy must be at the center of our efforts. Financial education is not some add-on or nice-to-have initiative; it is, therefore, core to allowing women to engage in and benefit from economic and social systems.

We must rethink financial literacy moving forward, not only as an individual skill set but as a collective achievement, and one that requires cooperation from governments, educators, financial institutions, NGOs, technology providers, and communities. Programs must be intersectional, recognizing that women have different identities and experiences across geographies, abilities, and socio-economic contexts.

Continuous learning and innovation are equally important. With evolving trends such as digital finance, climate risk, and global crises like pandemics, the content and delivery of financial education must change. Newer technologies, including AI, blockchain, and gamification, are exciting prospects for inclusive outreach if conceived through an empathetic, accessible, and equitable lens.

To wrap up, here's a big vision: A world in which every woman, no matter who she is, understands her financial life and has the confidence and resources necessary to control it, a world where empowered women create strong families, inclusive economies, and sustainable societies.

Realizing this vision involves more than commitment; it requires collective action, investment, and innovation. Financial literacy isn't simply teaching women to manage money; it is a catalyst for changing lives, communities, and the trajectory of global development.

References

ADB (2021) Women's empowerment through conditional cash transfers in Indonesia. Asian Development Bank. https://www.adb.org

Asian Development Bank (2021) Gender equality and financial inclusion in Asia and the Pacific. https://www.adb.org/publications/gender-equality-financial-inclusion-asia-pacific

Becker GS (1993) Human capital: a theoretical and empirical analysis, with special reference to education (3rd ed). University of Chicago Press

BRAC (2023) Empowering women through microfinance plus programs. https://www.brac.net

CARE (2021) VSLA programming guidance. https://www.care.org

CARE (2022) Measuring empowerment: toolkit for financial inclusion programs. https://www.care.org

Demirgüç-Kunt A, Klapper L, Singer D (2022) The Global Findex database 2021: financial inclusion, digital payments, and resilience in the age of COVID-19. World Bank. https://globalfindex.worldbank.org/

FAO (2022a) Gender equality and land rights. Food and Agriculture Organization of the United Nations. https://www.fao.org

FAO (2022b) Gender, agriculture and climate change. Food and Agriculture Organization of the United Nations. https://www.fao.org

FDIC (2022) Money smart for adults curriculum. https://www.fdic.gov/resources/consumers/money-smart/adults.html

Folbre N (2021) The rise and decline of patriarchal systems: an intersectional political economy. Verso Books.

Government of Canada (2023) National financial literacy strategy 2021–2026. https://www.canada.ca

GPFI (2023) Policy guide on advancing women's digital financial inclusion. G20 Global Partnership for Financial Inclusion. https://www.gpfi.org

GSMA (2022) The mobile gender gap report. https://www.gsma.com

GSMA (2023) The mobile gender gap report 2023. https://www.gsma.com/mobilefordevelopment/resources

IFC (2022) Women-owned businesses and financial capability in Africa. International Finance Corporation. https://www.ifc.org

IFC (2023) Blockchain and financial inclusion: gender perspectives. International Finance Corporation. https://www.ifc.org

ILO (2023a) Building forward fairer: women's employment and financial resilience post-COVID. International Labour Organization. https://www.ilo.org

ILO (2023b) Care at work: investing in care leave and services for a more gender-equal world of work. International Labour Organization. https://www.ilo.org

ILO (2023c) World employment and social outlook: trends 2023. International Labour Organization. https://www.ilo.org/global/research/global-reports/weso/2023/WCMS_865332/lang--en/index.htm

ITU (2022) Bridging the gender digital divide: guidelines for inclusive tech. International Telecommunication Union. https://www.itu.int

J-PAL (2021a). Financial literacy and economic outcomes. Abdul Latif Jameel Poverty Action Lab. https://www.povertyactionlab.org

J-PAL (2021b). Improving savings behavior through financial education in India. Abdul Latif Jameel Poverty Action Lab. https://www.povertyactionlab.org

Lusardi A, Mitchell OS (2020) Financial literacy and the need for financial education: evidence and implications. Swiss Journal of Economics and Statistics 156(1):1–8. https://doi.org/10.1186/s41937-020-00046-3

MaPS (2023) UK strategy for financial wellbeing. Money and Pensions Service. https://www.maps.org.uk

Mastercard Foundation (2023) Girls4Tech impact report. https://www.mastercardfoundation.org

NABARD (2021) Status of microfinance in India 2020–21. National Bank for Agriculture and Rural Development. https://www.nabard.org

OECD (2022a) Empowering women through financial education and awareness. Organization for Economic Co-operation and Development. https://www.oecd.org/finance/empowering-women-through-financial-education.htm

OECD (2022b) Measuring financial literacy: OECD/INFE survey 2022. https://www.oecd.org

OECD (2022c) Recommendations on financial literacy education. https://www.oecd.org

OECD (2022d) Women and financial education: OECD/INFE toolkit. https://www.oecd.org

OECD (2022e) Women and financial literacy: OECD/INFE evidence, policies and practice. https://www.oecd.org/financial/education/women-financial-education.htm

Plan International (2023a) Empowering girls through financial education. https://www.plan-international.org

Plan International (2023b) Girls and financial literacy: global best practices toolkit. https://www.plan-international.org

Sen A (1999) Development as freedom. Oxford University Press

Suri T, Jack W (2021) The long-run poverty and gender impacts of mobile money. Science 354(6317):1288–1292. https://doi.org/10.1126/science.aah5309

UN Women (2023a) Gender snapshot 2023: accelerating gender-responsive finance. https://www.unwomen.org

UN Women (2023b) Progress on the sustainable development goals: the gender snapshot 2023. https://www.unwomen.org/en/digital-library/publications/2023/09/progress-on-the-sustainable-development-goals-the-gender-snapshot-2023

UNDP (2023a) Gender equality in human development report 2023. United Nations Development Programme. https://hdr.undp.org

UNDP (2023b) Human development report 2023: advancing gender equality in the digital age. United Nations Development Programme. https://hdr.undp.org

UNESCO (2022) Global education monitoring report 2022: gender in education. https://unesdoc.unesco.org

Women's World Banking (2022) Designing for her: a playbook for financial services. https://www.womensworldbanking.org

World Bank (2022a) Global Findex database 2021: financial inclusion and women's empowerment. https://globalfindex.worldbank.org

World Bank (2022b) The Global Findex Database 2021: financial inclusion, digital payments, and resilience in the age of COVID-19. https://www.worldbank.org/en/publication/globalfindex

Yunus M, Moingeon B (2020) Building social business models: lessons from the Grameen experience. Long Range Plan 43(2–3):308–325

Sustainable Business, Governance and Responsible Leadership

From Chappals to Change: How Relaxo Scales Towards a Sustainable Future

Binod Kumar Singh, Ishita Girdhar, and Srishti Mehta

Abstract The case highlights Relaxo's evolving product mix and aggressive advertising efforts aimed at becoming a strong Indian footwear player. Over the past 43 years, Relaxo has adopted both Push and pull strategies to enhance visibility of its product portfolio in the Indian market. The buying behavior of consumers is significantly changing over the years, with consumers becoming increasingly brand-centric, demanding the latest trends and contemporary styles. In today's competitive world, companies are integrating product design with supply chain operations to provide customers with new products and services faster. The Indian Footwear Industry has the potential to reach US$29.2 billion in revenue by 2025. From what was once a basic necessity, footwear has now become a fashion statement and status symbol. However, growth sometimes comes with challenges. Rapidly changing customer preference is one of the biggest challenges faced by the footwear industry. This case highlights changing consumer behavior and brand positioning within the footwear segment. It also explores the retail growth strategies and the supply chain challenges faced by the industry. Through inclusive branding, supply chains rooted in real-world needs, and locally driven impact, the case examines how businesses in emerging economies can approach sustainability not just as a checkbox, but as a way of thinking. The dilemma Relaxo faces, whether to decentralize and how much to invest in inclusion without hurting margins are the very same tensions that define the journey of many Indian enterprises trying to do well and do good.

Keywords Footwear market · India · Sustainability · Growth drivers · Consumer behavior

As Relaxo Footwears scaled its operations across India, one pressing challenge troubled CEO Ramesh Kumar Dua: how to address growing supply chain inefficiencies

B. K. Singh (✉) · I. Girdhar · S. Mehta
UPES, Dehradun, India
e-mail: binodbhu@gmail.com

A. Singh et al. (eds.), *Human-Centered Sustainability: Bridging Well-Being, Gender Equity, and Technology*, World Sustainability Series,
https://doi.org/10.1007/978-3-032-22020-2_13

in the southern and western parts of India without compromising the company's well-established value-for-money positioning.

Relaxo's centralised manufacturing hub in North India, once a strong asset, started showing signs of strain despite the brand's solid reputation and rising demand in fast-growing regions. The company faced challenges like longer lead times, higher logistics costs, and regular stock shortages in distant markets, especially in the South and West. To improve responsiveness and support future growth, expanding the network with additional Regional Distribution Centres (RDCs) seemed like a good move. However, this required an important capital investment and a fundamental shift in how the business operates, which could potentially impact profit margins in a market that's quite sensitive to pricing.

Making this decision was not easy. Should Relaxo prioritise decentralizing its distribution and logistics network to better serve underserved areas, or stick with its existing approach and focus on improving overall system efficiency? Dua knew that the choice could influence costs, customer satisfaction, and the company's ability to stay competitive in the long run.

1 Overview and Background

In a time when businesses are being held accountable not just for profits but for people, purpose, and planet, Relaxo Footwears offers a quietly powerful case of how large-scale manufacturing can intersect with affordability, inclusion, and sustainability. As India's largest volume-based footwear producer, Relaxo has built a legacy not by appealing to luxury markets, but by embedding value and dignity in mass-market products reaching the underserved, underrepresented, and often overlooked consumers across Tier II, III, and rural India.

What sets Relaxo apart and makes this study relevant is its consistent focus on human-centered design and equitable distribution at scale. Founded in 1976 by brothers Ramesh Kumar Dua and Mukand Lal Dua with just ₹10,000 and a vision to create quality footwear for every Indian, Relaxo grew from a modest chappal manufacturer into a multi-brand powerhouse. Despite early challenges such as limited capital, government restrictions, and market skepticism, their unwavering belief in offering Affordable, Beautiful, Comfortable, and Durable (ABCD) products laid the foundation for Relaxo's future success. Today, it operates with over 70,000 retailers, exports to 34 countries, and carries iconic brands like Sparx, Flite, and Bahamas (Stock Daddy n.d.).

Ramesh Kumar Dua's determination to study rubber technology, despite academic rejections, and his eventual acceptance at the Plastics and Rubber Institute in London, proved pivotal. The technical expertise gained there became the backbone of Relaxo's product quality. Officially incorporated as Relaxo Footwears Ltd. in 1984, the company was well-prepared to seize the opportunities presented by India's economic liberalization in the 1990s. A major inflection point came in 1995, when Relaxo invested ₹7.5 crore in a modern facility with a daily capacity of 50,000 pairs, aligning perfectly with the surge in market demand (Stock Daddy n.d.; Your Story 2023).

This chapter highlights how Relaxo has navigated economic liberalization, shifting consumer behavior, and the rise of digital commerce, all while prioritizing functional design, regional inclusivity, and localized sustainability practices. From women- and children-focused product lines to investments in rural education, water conservation, and clean energy, Relaxo's strategy is not driven by flashy CSR, but by practical, rooted interventions that reflect a deep understanding of its consumer base.

Yet, as Relaxo expands further into South and West India regions historically underserved by its North-centric supply chain, a central dilemma emerges: Should Relaxo decentralize its logistics and invest in regional distribution centers (RDCs), risking increased costs, or continue optimizing its centralized operations, even if it means slower deliveries and stock shortages in fast-growing markets? This question challenges the core of Relaxo's value proposition which is balancing affordability with access, efficiency with equity.

By framing this dilemma within broader themes of stakeholder theory, inclusive growth, and sustainable business practices, the case adds to a growing conversation around what responsible scaling looks like in emerging economies. Because Relaxo's story isn't only about shoes it's about building systems that strive to be efficient, empathetic, and fair, even when resources are tight and decisions aren't easy.

2 Indian Footwear Industry: An Overview

India's footwear industry had grown rapidly and is now recognised as one of the fastest expanding sectors in retail and manufacturing, with estimates projecting it is worth to surpass ₹1.4 lakh crore by 2030 (Indus Footwear 2025). This upward trajectory was largely fuelled by both urban and rural consumers increasingly prioritizing style and brand reputation. India remained the second-largest global producer of footwear, manufacturing over 2.58 billion pairs annually and satisfying more than 90% of its domestic demand (Invest India n.d.).

Per capita footwear consumption had risen to 2.3 pairs each year, aligning more closely with global averages (Assomac n.d.). This shift was driven by a youthful, aspirational demographic, a burgeoning middle class, broader digital access, and supportive government initiatives like the Production Linked Incentive (PLI) scheme for leather and footwear production. Major brands such as Relaxo, Bata, Puma, and Metro continued expanding their retail presence, while smaller players faced challenges related to distribution logistics and the price sensitivity of lower-income consumers.

As omni-channel retailing continued to evolve, consumer expectations increased considerably. This shift had driven brands to focus not only on speed and value but also on incorporating style into their offerings. Having a strong aesthetic appeal became an essential part of the industry's competitive edge.

3 Growth Drivers for Indian Footwear Industry

3.1 Growth Opportunities in Women/ Kids Segment

In 2024, women's footwear was projected to capture over 30.2% of the total market share in India, making it one of the fastest-growing segments in the footwear industry (Statista 2024). This surge was driven by a rise in working women, higher disposable incomes, and an increased emphasis on comfort paired with personal style. Consumers sought shoes that blended fashion and practicality, moving beyond basic options. As a result, there was an increase in demand for athleisure footwear, premium offerings, and diverse design choices.

At the same time, the market share for children's footwear had grown to nearly 12% (Indus Footwear 2025). Urban and semi-urban parents were increasingly opting for branded, comfortable, and stylish shoes for their children. The growth of e-commerce platforms, which provided greater convenience, competitive prices, and a wider variety of options, further accelerated this trend.

Both women's and children's footwear segments presented major opportunities for growth as consumer preferences evolved, and the market became more sophisticated. In response, companies launched niche product lines, prioritizing innovative designs, and personalizing their marketing strategies to appeal to these increasingly brand-conscious audiences.

3.2 Evolution of E-Com Channel

Relaxo Footwears puts more emphasis on its online sales channels, recognizing how important e-commerce has become in India's retail scene. This shift was a key part of the company's growth strategy. Relaxo was active on its own direct-to-consumer websites and popular online marketplaces, allowing it to connect with a broader audience. The company was especially focused on reaching younger, tech-savvy shoppers and expanding into Tier 2 and Tier 3 cities. To support this push, Relaxo invested in improving its digital infrastructure, simplifying its supply chain, and ramping up marketing efforts to better target the increasing number of consumers shopping for footwear online (Motilal Oswal Financial Services 2024).

Relaxo boosted its online footprint by investing in its own e-commerce website and expanding its presence on top platforms like Amazon, Flipkart, and Myntra. The company also explored omni-channel strategies that blended offline accessibility with the ease of digital shopping to better connect with younger, tech-savvy consumers.

By 2027, digital retail was projected to represent significant growth out of total footwear sales (IMARC Group 2024). To stay competitive in this increasingly digital-first market, important investments were needed in logistics, last-mile delivery, and personalized shopping experiences.

3.3 *Preference for Branded Footwear*

In 2025, Indian consumers showed an even stronger preference for branded footwear, driven by a young, urban population that's increasingly connected online, along with rising incomes. Several customers saw footwear not just as a necessity but as an important part of personal style and self-expression. This shift reflected a broader cultural move where footwear was valued for its design and brand identity, rather than just utility.

This trend was especially noticeable in Tier II and Tier III cities, where easier access to online and offline stores made branded options more appealing. Consumers in these areas favoured well-known brands, prioritising quality and stylish designs over cheaper, generic alternatives (Ernst and Young 2022). To keep up with this demand, brands like Metro, Bata, Puma, and Relaxo expanded their product ranges and gained a larger share of the market.

Besides, government policies, such as reductions in GST on affordable footwear and other initiatives, had given this movement extra momentum (Press Information Bureau 2021). In the changing environment, branding, building customer trust, and providing a great shopping experience became key differentiators. Companies that focused on these areas were building loyalty and enjoying steady growth.

4 Brands and Product Portfolio

Beginning with Relaxo rubber/EVA slippers, the company moved up the value chain by adding fashionable footwear 'Flite' and 'Sparx' sportswear in 2005 and later stacking up its defence in the organized footwear industry by foraying into Men's formals through 'Boston', launching 'Schoolmate', footwear for school kids and 'Mary Jane and Elena' for women. The company sells more than 16 crore pairs a year across 10 different brands and 6000 SKUs competing both with the unorganized sector and the MNCs (Trendlyne 2019). Relaxo's impressive growth and deep market penetration are largely driven by its rich and varied brand portfolio, each tailored to specific consumer segments and needs.

The Company operates in 3 major categories i.e., Relaxo and Bahamas (rubber slippers), Flite (EVA and PU Slippers) and Sparx (sports and canvas shoes, sandals and sporty slippers). The company sells its products through 50,000 retailers served through distributors, 390 retail outlets, e-commerce, modern trade, etc. It has a portfolio of brands including major brands like Relaxo, Flite, Sparx, Bahamas, and Schoolmate.

5 Relaxo's Business Strategy

Over the years, Relaxo has developed a multi-brand approach that closely aligns with the changing needs of Indian consumers. In a retail environment that was both fast-paced and style-conscious, the company's strength lay in smoothly blending trendy designs with affordability, all while ensuring its products were accessible across the country.

Ramesh Kumar Dua, the Managing Director of Relaxo Footwears, noted that while customers used to prioritize footwear that was simply durable and comfortable, their preferences have evolved. Today, they also seek footwear that reflects their style, helps express their identity, and holds aspirational value (Business Standard 2023). To meet these shifting expectations, Relaxo had adopted a customer-centric approach built on four strategic pillars: Structured Market Research, Product innovation, Portfolio rationalization and Brand Equity.

The shift was especially evident in Tier II and Tier III cities, where consumers' buying patterns started to shift due to easier access to branded footwear both online and in physical stores. For these consumers, emphasis on quality, trendy designs, and trusted brand names became more important than choosing unbranded or generic options (Deloitte n.d.). Leading footwear brands responded by expanding their presence and launching more diverse collections to meet increasing demand.

Government initiatives also played a key role in supporting this trend. Reduced GST rates on affordable footwear and encouragement of local manufacturing sped up the adoption of branded products (Press Information Bureau 2021). In a competitive environment that changed rapidly, brand reputation, customer trust, and the overall shopping experience became critical factors. Companies that consistently focused on these areas were able to sustain long-term growth and build stronger customer loyalty.

Relaxo approached this market with a dual push-and-pull strategy, combining celebrity endorsements to generate aspirational appeal with strong distribution networks to ensure product availability. Bollywood actors like Ranveer Singh (Flite), Akshay Kumar (Sparx), and Salman Khan (Hawai) helped enhance Relaxo's brand image across different consumer segments.

Their distribution approach was also personalised by region. Relaxo developed targeted regional strategies rooted in brand resonance rather than just managing categories. For example, while Relaxo was especially popular in the North and East, Sparx gained major traction in South and West India. As Gaurav Dua, EVP of Marketing, explained:

> Each brand has a different geography where it thrives. Our focus now is to expand Bahamas and Flite in western India and strengthen Sparx in the East. We are growing state by state, city by city. (Sinha 2023)

6 Manufacturing and Distribution

Relaxo Footwears has built one of India's strongest distribution and manufacturing networks. It operates nine advanced manufacturing units across Haryana (Bahadurgarh), Rajasthan (Bhiwadi), and Uttarakhand (Haridwar), with a daily capacity of 10.5 lakh (1.05 million) pairs. These facilities follow lean practices like MOST and yield improvement initiatives, ensuring efficiency and quality. All products are ISI-marked, with ISO 9001:2015 and ISO 14001:2015 certifications. 22 A recent 50,000-pair/day expansion at Bhiwadi reflects its commitment to scale. Relaxo controls 100% of its manufacturing, from design to packaging. Its products reach the market through ~ 650 distributors and over 70,000 retailers/MBOs. It also runs 399 Exclusive Brand Outlets (EBOs), contributing ~ 9% to sales, and is active on major e-commerce and quick-commerce platforms (Amazon, Flipkart, Myntra, Blinkit, Zepto, Swiggy Instamart), which together make up ~ 10% of sales.

Relaxo is India's largest non-leather footwear manufacturer. Capacity grew from 7.5 lakh pairs/day in FY19 to 10.5 lakh in FY24. North and East India contribute ~ 70% of sales, while new geographies are being tapped. It began exports in FY07 and, as of June 2024, exports to 34 countries, accounting for ~ 4–5% of revenue. An overseas office in Dubai supports international growth (n.d. 2025). Its asset-heavy model delivers economies of scale, with the average selling price rising from ₹48 in FY08 to ₹150 in FY24 (7.4% CAGR). Through BIS compliance and premiumization of brands like Sparx and Flite, Relaxo continues to strengthen its position in the organized market.

7 Relaxo Supply Chain Strategies

One of Relaxo's key strengths lay in its supply chain, which allowed the company to deliver consistent value through efficiency, agility, and scale. Maintaining a smooth and cost-effective supply chain had been critical to upholding its leadership position across various channels, including wholesale, retail, exports, e-commerce, and omni-channel platforms.

By 2025, Relaxo operated ten manufacturing plants in North India, producing over a million pairs of shoes every day. Supported by a broad distribution network of more than 60,000 multi-brand outlets (MBOs) and over 450 exclusive brand stores (EBOs), these facilities served more than 600 towns and cities (Relaxo Footwears Limited 2024). In addition, Relaxo has extended its footprint globally, exporting to over 35 countries and establishing a regional office in Dubai to support international growth.

Managing operations at this scale definitely presented some unique challenges. Since most manufacturing facilities in India were mostly located in the north, it had been difficult to ensure fast, cost-effective deliveries to the more remote western and southern regions. To tackle this, Relaxo strategically expanded its network of

Regional Distribution Centers (RDCs) in carefully chosen locations. These RDCs helped cut down lead times, improve regional service levels, and bring down overall logistics costs (Relaxo Footwears Limited 2024).

Relaxo also adopted integrated IT systems and robust ERP solutions to manage inventory, scheduling transportation, and production planning as the business grew. To further boost efficiency, the company simplified its logistics network by partnering only with well-organized and reliable transportation providers, which helped ensure consistent service quality across regions (Relaxo Footwears Limited 2024).

On the growth front, Relaxo has upgraded its warehouse operations by integrating advanced technology to handle larger inventories more efficiently and reduce processing times. Also, the company has simplified its outbound shipping processes through improved logistics solutions, ensuring sustainable growth and long-term success (Relaxo Footwears Limited 2024). This reduced fixed operating costs and allowed Relaxo to focus more on core activities such as boosting demand, strengthening its brand, and innovating new products.

In an interview with Celerity, Vijay Wadhwani, Assistant Vice President—Supply Chain, emphasized that modern supply chains must be agile and insight-driven: "Today, customers expect speed and personalization. Managing complexity means improving visibility, collaboration with suppliers, and having tighter control over operations." (Logistics Insider 2024).

By 2025, multi-brand outlets (MBOs) contributed nearly 80% of Relaxo's domestic revenue, reflecting their strong market presence. Meanwhile, e-commerce sales had steadily grown and made up about 12% of total sales. The company also generated around 5% of its revenue through exports and explored new markets in the Middle East and Southeast Asia (Relaxo Footwears Limited 2024).

Relaxo improved its ability to serve India's vast and varied regions by continuously refining its supply chain and customizing its approach to different market needs. These efforts not only helped meet local demands but also laid a solid foundation for international expansion, balancing affordability with a focus on delivering great customer experiences.

8 Marketing and Branding

Relaxo has built a strong brand identity through celebrity endorsements, regional appeal, and an evolving digital strategy. With sub-brands like Bahamas, Sparx, Flite, and Schoolmate, it caters to diverse urban and semi-urban consumers across income groups. Celebrity endorsements play a key role in brand positioning—Salman Khan represents Bahamas, embodying vibrant, youthful energy, Akshay Kumar endorses Sparx, highlighting performance and ambition, Katrina Kaif and Sonakshi Sinha have fronted Flite, enhancing its style and comfort appeal among women. These associations are aligned with the identity of each sub-brand, increasing visibility and trust in a crowded market. Relaxo's digital strategy, once slow, is gaining pace. It has strong followings on Facebook (160 K+) and LinkedIn (63 K+), but weaker engagement on Instagram and Twitter. To improve this, the company is investing in influencer marketing and video campaigns like Sparx's "It's In Me" featuring Akshay Kumar aimed at Gen Z and millennials. However, there's still a need for more consistent, personalized digital storytelling.

Marketing efforts are integrated with Relaxo's omnichannel distribution: 394 + Exclusive Brand Outlets; 50,000 + multi-brand retail points; Strong presence on Amazon, Flipkart, Myntra, etc. According to Gaurav Dua, Executive Director—Sales and Marketing at Relaxo, the company's growth will continue to be driven by a strong presence in both multi-brand outlets and e-commerce platforms. Rather than building its own app, Relaxo uses a brand-as-seller model on established platforms to retain control over pricing, service, and inventory while scaling efficiently.

By blending traditional retail trust with digital reach, Relaxo has built a hybrid marketing model that supports brand growth and customer acquisition. While there's room to enhance digital agility, its celebrity-driven branding, retail strength, and omnichannel approach make it a leader in India's value- to-mid footwear market (Shastri n.d).

9 Sustainability and CSR Initiatives

Relaxo Footwears has integrated sustainability and corporate social responsibility directly into its core strategy, emphasizing environmental stewardship, community development, and governance. Guided by its CSR and ESG committee and aligned with global standards like TCFD, the company ensures cohesive efforts across all its operations.

9.1 Environmental Sustainability

Zero Liquid Discharge and Water Conservation—Relaxo has implemented Zero Liquid Discharge (ZLD) systems across manufacturing plants and installed rain-water harvesting facilities, reducing external water dependency and conserving vital resources. Clean Energy Transition—The company shifted boilers to PNG (piped natural gas), adopted hybrid CNG-diesel generators, and installed rooftop solar systems and wind turbines, underscoring its commitment to achieving NetZero emissions (Relaxo Footwears Ltd. 2024) Material Resource Management- Relaxo focuses on waste reduction, energy efficiency, and eco-friendly logistics, including electric and CNG vehicle adoption in its supply chain.

9.2 Community Welfare and Social Impact

Education and Skill Development—The Parivartan Model Schools project began in Haridwar (Uttarakhand), upgrading infrastructure in 13 schools with digital classrooms, WASH facilities, learning spaces, and teacher training; later expanded to 32 schools (India CSR 2021). The Student-to-Scholar initiative in Haryana enabled over 6000 students to access scholarships, contributing significantly to educational empowerment (CSR Box n.d). A Remedial Education Program in Delhi's Bawana colony has supported 150 + children since 2018 through remedial learning, leadership training, and life skills, supported by partner Navjyoti India Foundation (INDIA CSR 2022). For, Healthcare and Sanitation, they launched a Nayan Avoidable Blindness Project in Rajasthan, screening tens of thousands across 194 villages, established vision centers, and training local youth in optometry. Operated Mobile Health Units in Bhiwadi, Rajasthan, in partnership with Smile Foundation, providing essential medical services and constructing public toilets (CSR Box n.d). They provided pandemic Relief and Social Support through the Relaxo Foundation, the company donated 25,000 pairs of slippers to North Delhi Municipal Corporation sanitation workers and supplied medical equipment to hospitals in Delhi, Haryana, and Uttarakhand during COVID19 (INDIA CSR 2020). They also contributed to Environment-Based Community Projects and participated in watershed conservation in Alwar, Rajasthan, involving reservoir construction to support agriculture for 1000+ families. Sponsored large-scale afforestation near Sariska Tiger Reserve, planting 50,000 saplings (25–30% fruit-bearing) to restore biodiversity and support livelihoods.

9.3 Governance and ESG Framework

The company follows strong governance and transparency, with its CSR/ESG committee chaired by board members overseeing ethics, supply chain audits, whistleblower policies, and cybersecurity. It partners with the ESG Research Foundation to promote public engagement on sustainability, encouraging customer pledges via in-store QR codes. Relaxo also ensures inclusive workforce practices—diversity, fair pay, and robust health and safety systems integrate with its wider social agenda (n.d, ESG at Relaxo 2025).

10 Financial Performance

Relaxo Footwears Ltd. experienced a moderate financial contraction in FY25, reflecting consumption headwinds and internal restructuring initiatives. The company reported net revenue of ₹2,790 crore, a 4% decline compared to ₹2914 crore in FY24. This decline was primarily due to subdued demand in the mid-range segment and a realignment of its distribution channels. Despite the revenue dip, Relaxo maintained a strong operating profile. Its EBITDA stood at ₹382 crore, slightly lower than ₹407 crore in FY24. The EBITDA margin for the year came in at 13.7%, indicating continued focus on operational efficiency. In Q4FY25, Relaxo reported EBITDA of ₹112 crore with a robust quarterly margin of 16.1%. The company's net profit for FY25 was reported at ₹170 crore, down from ₹197 crore in the previous fiscal, a 13.7% year-on-year decline.

This translated into a net profit margin of 6.1%, which, while reduced, remained healthy for a value-driven consumer business. The Q4FY25 PAT stood at ₹56 crore. Relaxo's management acknowledged the temporary drag on earnings but reiterated that the internal restructuring, particularly of distribution and retail strategy, was necessary groundwork for long-term growth. Chairman and MD Ramesh Kumar Dua emphasized that the company would now shift its strategic focus to margin enhancement and profitability, led by digital initiatives, operational streamlining, and sharper brand-positioning (Madhiraju 2025).

11 Way Forward

Over the past decade, the Indian footwear industry has undergone a revolutionary shift. Consumers expected more than affordability; they want stylish, comfortable, and sustainable products delivered quickly. This developing demand pushed manufacturers to rethink their entire supply chains, integrating market responsiveness, manufacturing agility, and innovative product design to create a more simplified and efficient approach to delivering value.

For a company like Relaxo, aligning design with supply chain operations moved from being a competitive edge to an absolute necessity. This integration made it possible to launch new products faster, improve forecasting accuracy, cut manufacturing costs, and shorten time-to-market. It also enabled quicker adaptation to seasonal trends and regional preferences while minimising inventory risks.

Looking ahead, Relaxo's success depended on three key objectives: maintaining flexibility in decision-making and execution, excelling in last-mile delivery, and offering a balanced value proposition that combines affordability with aspirational appeal. With its strong supply chain, diverse portfolio of brands, and deep consumer insights, Relaxo appeared well-positioned to achieve these goals and sustain its reputation as one of India's most trusted footwear brands.

Bibliography

Assomac (n.d.) India: increase in footwear prices may lead to drop in demand, hurt manufacturers. Assomac News. https://assomac.it/en/news/news-from-world/india-increase-in-footwear-prices-may-lead-to-drop-in-demand-hurt-manufacturers/

Business News Today (2024, May 11) Relaxo Footwears Q4 and FY25 results: Consolidation year sets stage for FY26 recovery. https://business-news-today.com/relaxo-footwears-q4-and-fy25-results-consolidation-year-sets-stage-for-fy26-recovery/

Business Standard (2023, November 13) Relaxo looks to bring a lifestyle quotient to its portfolio. https://www.business-standard.com/companies/news/relaxo-looks-to-bring-a-lifestyle-quotient-to-its-portfolio-123111300898_1.html

CSRBOX (n.d.) Relaxo footwears limited—healthcare and sanitation activities—Haryana. https://csrbox.org/India_CSR_Project_Relaxo-Footwears-Limited-Healthcare-and-Sanitation-Activities-Haryana_4016

CSRBOX (n.d.) Relaxo student to scholar program—Haryana. https://csrbox.org/India_CSR_Project_Relaxo-Footwears-Limited-Relaxo-Student-to-Scholar-Program-Haryana-_14564

Deloitte (n.d.) The Indian consumer at the forefront: Navigating India's dynamic retail landscape. https://www2.deloitte.com/content/dam/Deloitte/in/Documents/consumer-business/in-cb-consumer-at-the-forefront-noexp.pdf

EY (2022) Future of consumption in India: the new consumer equation. https://assets.ey.com/content/dam/ey-sites/ey-com/en_in/topics/consumer-products-retail/ey-future-of-consumption-india-new-consumer-equation.pdf

Grand View Research (2024) India footwear market size and outlook, 2024–2030. https://www.grandviewresearch.com/horizon/outlook/footwear-market/india

IMARC Group (2024) India footwear industry. Size, share and analysis 2025–2033. https://www.imarcgroup.com/india-footwear-market

IndianCompanies.in (2023, June 24) Relaxo footwears Ltd—products and brands list. https://indiancompanies.in/relaxo-footwears-ltd-products-brands/

India CSR Network (2020, July 9) Corporate social responsibility (CSR): Relaxo Foundation contributes 25,000 pair of slippers for MCD sanitation staff. India CSR. https://indiacsr.in/corporate-social-responsibility-csr-relaxo-foundation-contributes-25000-pair-of-slippers-for-mcd-sanitation-staff/

India CSR Network (2021, March 8) CSR: Relaxo Foundation adopts 32 primary schools for redevelopment and concept learning. India CSR. https://indiacsr.in/csr-relaxo-foundation-adopts-32-primaryschools-for-redevelopment-and-concept-learning/

India CSR Network (2022, April 19) CSR: Relaxo supporting the cause of remedial education of 150 children of Bawana resettlement colony, Delhi. India CSR. https://indiacsr.in/csr-relaxo-supporting-the-cause-of-remedial-education-of-150-children-of-bawana-resettlement-colony-delhi/

Indus Footwear (2025, February 19) Overview of the Indian footwear industry: key players and innovations. https://www.indusfootwear.com/Overview-of-the-Indian-Footwear-Industry.php

Invest India (n.d.) Leather and footwear sector in India: overview and opportunities. https://www.investindia.gov.in/sector/leather-footwear

LeadIQ (2024) Relaxo Footwears Limited—company profile. https://leadiq.com/c/relaxo-footwears-limited/5a1d7d08240000240056be70

Livemint (2023, November 7) Bata India banks on casualisation, premiumisation as formal wear fades. https://www.livemint.com/companies/news/bata-india-banks-on-casualisation-premiumisation-as-formal-wear-fades-11699351044439.html

Logistics Insider (2024, May 22) Building a resilient supply chain in the age of disruption. https://www.logisticsinsider.in/building-a-resilient-supply-chain-in-the-age-of-disruption/

MarketScreener (2023, August 7) Relaxo Footwears Limited increases production capacity. https://in.marketscreener.com/quote/stock/RELAXO-FOOTWEARS-LIMITED-46729789/news/Relaxo-Footwears-Limited-Increases-Production-Capacity-44589704/

Motilal Oswal Financial Services (2024, February 9) Relaxo Footwears Limited: Q3FY24 result update. Motilal Oswal Financial Services.

Press Information Bureau (2021, December 31) GST Council recommended change in GST rates for footwear and textiles. Government of India. https://pib.gov.in/PressReleasePage.aspx?PRID=1786697

Relaxo Footwears Ltd. (2024) Directors report. India Infoline. https://www.indiainfoline.com/company/relaxo-footwears-ltd/reports/directors-report

Relaxo Footwears Ltd (2024) Environmental, social and governance (ESG). https://relaxofootwear.com/pages/esg

Relaxo Footwears Ltd (2024) Manufacturing. https://relaxofootwear.com/pages/manufacture

Relaxo Footwears Ltd (2024, July 20) Q1 FY25 results presentation. Investment Guru India. https://investmentguruindia.com/editorial/uploads/news-pdf/7145629e_Relaxo-Footwear_IC_Q1FY25.pdf

Relaxo Footwears Limited (2024, May 29) Investor Presentation Q4 FY24. https://www.relaxofootwear.com/investor-relations/

Shastri A (n.d.) Marketing strategy of Relaxo Footwear. IIDE—Indian Institute of Digital Education. https://iide.co/case-studies/marketing-strategy-of-relaxo-footwear/

Sinha M (2023, July 25) From flip-flops to fashion: how Relaxo footwears is walking the talk. Financial Express. https://www.financialexpress.com/business/brandwagon/from-flip-flops-to-fashion-how-relaxo-footwears-is-walking-the-talk-3180295/

Statista (2024) Revenue share of the footwear market in India in 2024, by segment. https://www.statista.com/statistics/1324747/india-revenue-share-of-footwear-market-by-segment/

StockDaddy (n.d.) Footwear stocks in India: top footwear companies in India. https://www.stockdaddy.in/blog/footwear-stocks-in-india

The Pioneer (2025, February 24) The footwear industry takes a step towards sustainability and social empowerment. https://www.dailypioneer.com/2025/columnists/the-footwear-industry-takes-a-step-towards-sustainability-and-social-empowerment.php

Trendlyne (2019, June 20) Relaxo takes on Bata with celebrity endorsements. https://trendlyne.com/posts/1935691/relaxo-takes-on-bata-with-celebrity-endorsements

YourStory (2023, August 9) The journey of Relaxo: How this footwear brand became a household name. https://yourstory.com/2023/08/relaxo-footwear-journey-success

Beyond Compliance: Assessing Corporate Social Responsibility Activities by Mining Companies in India

Ashna Siddiqui and Shouvik Kumar Guha

Abstract Corporate Social Responsibility (CSR) began as a compass for societal upliftment but has not been more than a cog in the wheel in Corporate Governance, in the last decade, with minimal contribution in overall upliftment of society. Corporates show major interest in projects that gets them visibility and capitalise on the same, without any significant development in the society. The legislative provision relating to CSR in India does outline a framework for a holistic development via possible CSR activities but that remains implemented in only the letter of law and not the spirit. India became a torchbearer in mandating CSR to profitable corporates however, the impact assessment of those CSR activities needs a critical and stricter lens. Sectors digging more from the society, nature especially, need to be at the forefront to have a CSR policy that has impactful CSR activities ensuring its alignment with Sustainable Development. Mining sector remains a sector that has been in the limelight not for the right reason, corporates under this sector have a major role to play in increasing environmental issues. The CSR policy of these mining corporations have little to offer in rebuilding the environment and that needs to be relooked into. The chapter aims to assess the impact of the CSR activities carried out by mining companies in two of the major mining states in India i.e. Odisha and Chhattisgarh, and suggest implementable solutions for a long term impactful CSR policy.

Keywords Corporate Social Responsibility · Companies Act 2013 · Sustainable development · Mining companies

---

A. Siddiqui (✉) · S. K. Guha
WBNUJS, Kolkata, India
e-mail: ashnaphd2020@nujs.edu

S. K. Guha
e-mail: shouvikkumarguha@nujs.edu

A. Siddiqui
Birla Global University, Bhubaneswar, India

A. Singh et al. (eds.), *Human-Centered Sustainability: Bridging Well-Being, Gender Equity, and Technology*, World Sustainability Series,
https://doi.org/10.1007/978-3-032-22020-2_14

1 Introduction

Corporate Social Responsibility as a concept has existed for centuries and is much older than a decade of it coming into existence as a legal provision in India. However, the first usage of the word is traced in Howard R Bowen's work in 1953 'Social responsibilities of the businessman' (Bowen 1953). It emphasized that the decision making of the businessman should be centered around the values of and responsibilities towards, the society. In 1960 Keith Davis in his work 'Can business afford to ignore social responsibilities'. Davis (1960) has highlighted the businessman's ever evolving role in assessing and re-assessing their responsibilities towards the society as the time changes. Social responsibility is an 'indefinable idea' and that's why if one looks at it from a management perspective it is more or less a businessman's decision taken partially for something that goes further than the firms economic interest. Keith outlines social responsibility in two faces, one where the businessman does the responsible work towards the economic development, welfare of the community that holds the businessman's unit, and the second face thrives to nurture the ethos and values of human beings and society. In 1970s, CSR witnessed massive acceptance as debates went berserk, discussions became more intriguing and there came a power shift from businessman's profit interest to community driven work. Carroll in his work in late 1970s mentioned that social responsibility of corporates encompasses economical, ethical, legal and discretionary hopes that the society has from them (Carroll 1979). Further refinement of the concept CSR was done by Carroll in 1999 when he used the pyramid model (Carroll 1999) to explain the four major responsibilities that shall ensure that the corporates are applauded for their work towards society. He also mentioned that "*Social responsibility can only become reality if more managers become moral instead of amoral or immoral.*" It is pertinent to mention that the society has always been in need of profitable stakeholders to morally and ethically contribute in development. The year 2000, also the year when Millennium Development Goals were adopted by the United Nations, with its 8 goals, though none of them strategically linked with social responsibility of the business but had potential for UN and private sector cooperation for achievement of those eight goals. Later in the 2005 CSR witnessed a transitional shift from the idea of 'minimal commitment to strategic necessity' (Chandler 2016), Chandler in their work realized the importance of strategically placing their business in a role where it actively looks into the needs of the society, ensures the work and gains admiration in the eyes of its stakeholders. Amid all these decades of conversations, committee reports, policy and guidelines to try that one's country does maintain balance between private public enterprise and cordial uplifting relation between corporates and society, it still has a long way to see light and the situation of the society largely remains the same i.e. in dire need of positive change.

With the gap in legislative provision's implementation in just pen and paper and not in practice, it becomes extremely crucial to conduct research that highlights this gap. The study is undertaken to emphasise on this research problem and the dire need of policy reforms for CSR activities of companies. Where the companies that damage the environment more, must be made to compensate more in replenishing the depleted resources.

2 Literature Review

2.1 CSR and Sustainable Development Goals:

The word of the decade is Sustainable Development. "Sustainable development is development that meets the needs of the present without compromising the ability of future generations to meet their own needs" (IISD n.d.) but not many care to unfold what lies beneath its definition. The approach to sustainable development can be complete only when it has environmental, economic and social development all combined.

As the paper discussed Millennium Development Goals above, the discussion of an even more ambitious set of goals i.e. Sustainable Development Goals 2030 adopted after Millennium Development Goals is imperative. The shift from eight goals under Millennium Development Goals to 17 goals under Sustainable Development Goals suggest a broader agenda and an overall development desired to be achieved. Sustainable Development Goals specifically focus on some key goals like Climate Action under SDG-13(United Nations n.d.-a). It is even more important that the corporates come together as a stakeholder and help in achieving these goals.

The International Council for Science (ICSU) reviewed these goals and criticised "that the framework as a whole might not be internally consistent and as a result not be sustainable', if interlinkages, complex dynamics and conflicting relations between the goals are ignored". The Sustainable Development Goals Report offers the precise scenario of development. Even the Secretary General of the United Nations has highlighted that

> Unless we act now, the 2030 Agenda will become an epitaph for a world that might have been. -António Guterres, Secretary-General, United Nations"

Corporates via their CSR activity can contribute a lot to actual development and thereby achievement of these goals. Alignment of CSR activities with Sustainable Development Goals 2030 is a strategic suggested move that could help both, the corporates bring a good name, build a more trusted brand, and the country to contribute better in achievement of each of these goals. The Schedule VII of the Act that lists down the activities companies may undertake, aligns well with the seventeen Sustainable Developments Goals (SDGs). The same when mapped with each other (as shown below in Table 1) highlights the interdependency and the role that corporates have to play effectively and efficiently while doing their business to uplift the society. These goals do not work in seclusion but inclusion, activity in one area shall affect the other. This dependent relationship between CSR activities and SDGs is illustrated in Table 1.

Table 1 Mapping Schedule VII of Companies Act 2013 with relevant Sustainable Development Goals 2030

Schedule VII of Companies Act 2013	Relevant Sustainable Development Goals
1. Eradicating hunger, poverty, and malnutrition	– SDG 1: No Poverty
	– SDG 2: Zero Hunger
2. Promoting education, including special education	– SDG 4: Quality Education
3. Promoting gender equality and empowering women	– SDG 5: Gender Equality
4. Ensuring environmental sustainability	– SDG 13: Climate Action
	– SDG 15: Life on Land
5. Health care, including preventive health care	– SDG 3: Good Health and Well-Being
6. Rural development	– SDG 1: No Poverty
	– SDG 11: Sustainable Cities and Communities
7. Sanitation and drinking water	– SDG 6: Clean Water and Sanitation
8. Contribution to the Prime Minister's National Relief Fund	– SDG 17: Partnerships for the Goals
9. Setting up homes and hostels for women and orphans	– SDG 11: Sustainable Cities and Communities
10. Training to promote livelihood for youth	– SDG 8: Decent Work and Economic Growth
11. Promoting sports and cultural activities	– SDG 3: Good Health and Well-Being

Adapted from United Nations SDGs 2030 and Schedule VII, Companies Act, 2013

Every year the United Nations releases its report regarding the progress made towards SDGs. In the year 2023, the report highlights that there has been a mixed progress pertaining to SDG-3, but pertaining to the change in climate i.e. SDG-13 it broke almost every record and became the warmest year (United Nations 2024) and under SDG 15 the decline is world's forest area is still rapid (United Nations n.d.-b.)

A look into the Sustainable Development Goals 2030 and a comparison of that with Schedule VII of Companies Act 2013 helps us understand that both of them intend for one common thing that is protection of the rights of the community. No community protection is possible until and unless the protection of each and every individual of the society or community is ensured. It is pertinent to highlight that human rights comprises of rights relating to self and the environment around you. At all cost the environment must be protected for mankind to survive and for the planet to revive for future generation.

Companies must align their CSR activities with the Sustainable Development Goals, as they are a profitable stakeholder. Attainment of these goals is only possible when all the stakeholders especially the privileged and powerful ones like profitable corporates come forward and align their activities with SDGs. No social responsibility of mining corporates shall make sense if it doesn't align with environment related SDGs especially SDG-13(Climate change) among others. The study intends to highlight the need of mining companies aligning their CSR policy enlisting CSR activities with SDGs, most importantly SDG-13 to ensure that the resources aren't exhausted by these mining giants.

2.2 India's Tryst with Corporate Social Responsibility

With the history of Daan and Zakaat being practiced for many centuries, a country so culturally vibrant and rich, knitted with values and ethics, needed a legislative provision to remind the corporates of their responsibility to give back to the society. Section 135 of the Companies Act 2013 laid down the mandatory provision for profitable corporates to spend two percent of their average net profit of last three years (Ministry of Corporate Affairs 2013a). It further delved into ensuring that these corporates frame a CSR policy alongwith the help of CSR Committee and undertake CSR activities as outlined in Schedule VII of the Companies Act 2013 (Ministry of Corporate Affairs 2013b). The CSR committee in the case of a company which has an Independent director requirement must have the Independent director in their CSR Committee (Ministry of Corporate Affairs 2013c). The legislative provision Section 135 of the Act is historic, as India became the first ever country to make CSR transition from a voluntary act to a mandatory one. Not just does the provision mandates companies to spend in CSR it also lays down the activities, strategies to frame policy, involvement of independent director (if any) in CSR committee for an effective CSR. The provision has been aided with Companies (Corporate Social Responsibility) Rules 2014 that further elaborates and enlists the requirement necessary for the CSR activity (Ministry of Corporate Affairs 2014a). Both the act and rules has gotten various amendments to fine tune the act of CSR. Timely circulars from Ministry of Corporate Affairs relating to CSR also enhances the activities being undertaken by the corporates. But as there is nothing perfect, only if these written letter of law were adopted in spirit by the corporates then we would have seen overflowing case studies of impact and transformation in society. The Schedule VII of the Act (Ministry of Corporate Affairs 2013b) has tried to cover almost all the ambits that could be there, starting from education, alleviation of poverty, zero hunger, environment, rural development, and others, for the holistic development of the society, but still lacks uniformity and transparency in distribution of funds. India despite making it a mandatory act hasn't seen any major achievement in the progress of the society that could be owed to CSR.

Foundational thinkers such as Bowen, Davis et al. have comprehended the ethical and moral compass of CSR, the transition of CSR strategically aligned with SDGs have come in the eye of a global perspective. However, the gap in transition of CSR from a compliance-based activity to a transformative tool is clearly visible. The reviewed literature exposes the lack of empirical results as to how CSR helps in its alignment with SDGs, and especially sectors that causes degradation mitigate the damage caused, by their environmental based CSR activities. This gap calls for a focused inquiry in CSR activities of corporates such as mining companies and their assessment of real contribution towards environment, in mining rich but ecological sensitive Indian states like Chhattisgarh and Odisha.

3 Research Methodology

The limited impact of CSR activities of companies especially mining companies, in environment-based goals represents the critical research problem. Greenwashing and compliance based check boxing aids in the sustenance of this research problem.

Qualitative analytical approach to examine the secondary data drawn from the Annual reports, CSR disclosures and official CSR portals of the mining companies is adopted to study and suggest solutions for the said research problem. This method shall help in critically assessing the disconnect between mere CSR compliance and actual environmental outcomes, and suggest interventions to realign CSR policies with objectives of SDGs. This methodological approach helps in evaluating the qualitative nature of CSR expenditures and the comparative analysis of different states helps in identifying patterns of such expenditures in environmental initiatives.

4 Analysis

4.1 Mining Sector in India and Other Mining Rich Countries

Some sectors play a crucial role in development while some like mining owe their massive profits, to development. Extraction of resources has led to tremendous development, but it is imperative to understand at the cost of what. An entire idea of giving back to the society comes to haul the moment mining sector comes into picture. A sector which takes away from the environment much more than it has to offer in return. For the longest of time this sector has just exploited and exhausted a region and moved on to the next to deplete their resources, without batting an eye. The idea of Sustainability while performing mining operations unfortunately takes a back seat. To curb this practice by the sector, countries have been coming up with legislations and supplementary rules to keep a check and balance.

The Mines and Minerals Rules of 2015, India, that reverbed the act of Mines and Mineral(Development and Regulation) Act 1957 illustrates the need of District Mineral Foundations established by State government but sponsored and funded by Mining companies (Ministry of Mines 2015). The major work of the foundations remains to ensure that the affected parties get compensated, and their interests are protected. The guidelines released by the governments in the state act a roadmap in building the projects in and around the welfare of the affected parties. Minimisation of adverse impact of the industries tremendous growth, and rehabilitation wherever required remains crucial. Another act that mandates activities like mining on non-forestland permissible only if the compensatory funds are allocated for the same under Compensatory Afforestation Fund Management and Planning Authority (CAMPA) (Ministry of Environment, Forest and Climate Change 2016). The foundation has been there for around a decade and it still is unfolding its effectiveness. The 1957 act also mentions under Sec 18 (1), the duty of the Central Government to ensure that there must be systematic development with regard to minerals and protection of environment by prevention of any pollution causing mining operation must take place (Ministry of Law and Justice 1957).

In Zamibia, Section 12 of the Mines and Minerals Act 2015 explains that the chapter of mining and non-mining rights lays down strict provision for any exploration or mining without any environmental impact assessment, failing which imprisonment to individuals and hefty punishment to corporates (Government of Zambia 2015a, b). In South Africa, a mineral rich country, under their Mineral and Petroleum Resources Development Act mandates the consent of the would be affected community, local municipality and empower them to raise objection if any, and only then can a mining company obtain a license (Government of South Africa 2002). Australia that leads in the mining sector has allowed its states to have their own legislation for mining. Despite being a federal state its New South wales has a Mining Act of 1992 (Government of Australia 1992), Queensland has a Mineral Resources Act, 1989 that emphasizes on assessment of mineral resources for practicable land use management. South Australia and Western Australia having their Mining Act of 1971 and 1978 respectively to lay down the regulation relating to mining operations.

Although CSR strikes out as a balanced approach of responsibility than the above compliances and compensations. Mining companies are under the screening considering there needs to be more done than what has been, these companies should be the flagbearer in CSR initiatives and strive to compensate the depletion of mineral and other natural resources. This also means that the idea of Sustainable Development takes a backseat when mining companies come into question. Any CSR activity that doesn't largely resonate with the idea of protection of environment wouldn't be seen in the same light as CSR activity of other companies from different sector. The Polluter pays principle is not clearly in sync with the idea of environmental responsibility of mining companies but it connotes that there is a strong nexus between mining companies and environmental concerns thereby leading to suggestion of damage control activities in form of CSR, by the profit making corporate giants in mining sector.

Mostly CSR activities by the mining companies in India focus on a plethora of activities laid out in Schedule VII of the Companies Act 2013 but very rarely do they focus on environmental upgradation. CSR funds spent for these activities listed in CSR policy recommended by CSR Committee highlights that the idea of CSR is seen as a check box exercise and not a strategic result yielding activity. Since India is among the top ten mining countries of the world, and it has the badge to be the first country to mandate CSR provision, it needs to also be a flagbearer and contribute positively to the cause of climate change and make it biggest player and stakeholder i.e. mining companies more accountable towards the environment. There is no denial that mining is required for the development of the nation and the approach towards mining companies could be less strict, however it is important to realise that the business activities of these mining companies need to strike a balance between their profit, production and social responsibility. Over the years mining companies have tried to identify themselves as the companies who do high level of CSR to wash their otherwise bad image. The enlightened self-interest of these mining companies do not help them wash away and not reciprocate to the society with activities that are important for environmental upgradation.

4.2 CSR Activities in the Indian Mining States of Chhattisgarh and Odisha

The state of Chhattisgarh and Odisha are two of the major mining states in the country (Ministry of Mines n.d.). Chhattisgarh a state in its two decades of formation and ample natural resources is in a race to move towards rapid development and attract more industries, for it has resources to offer. Recently it was declared that the first lithium mines of the country are to be opened in Chhattisgarh, the decision came in at National Mineral Exploration Trust's sixth governing meeting (Economic Times 2024). With the opening of a mine of its own importance the state of Chhattisgarh shall strive to become even more developed in the next few decades. It is also to be noted that the state has seen tremendous increase in mineral revenue in the year 2023–2024, if compared with the year 2000, the year of formation of the state, mineral revenue has increased for over 30% (Ministry of Mines n.d.).

The state is exponentially growing due to its visibility as a mineral rich state and the state makes significant economic development through minerals. However, the dissatisfying part of the economic development of the state is its unsustainable nature. The significant increase in mineral revenue is directly proportional to the felling of trees, deforestation and that is directly proportional to environmental degradation. Another issue pertaining to the fast paced industrialization is displacement of the community, the dependents of the forests, without any replacement.

Odisha, a state that has an ecology to boast about, pristine waterbodies, lush green forests, ample mineral resources but yet remains one of the poorest state with a high number of people falling in the category of below poverty line, the growth of the state is also very disproportionate as the development is seen only in and around the capital district, many other districts remain below the average of the state (The New Indian Express 2023). Despite the poverty stricken condition, Odisha has offered an extremely welcoming approach to the projects relating to infrastructure by companies and industries. Odisha has strategically put itself at the forefront by calling itself the hub of mining and metals, however it is important to observe that this approach may go against the very welfare of its own people (Down to Earth n.d.). As more than the help, projecting the state's citizen onto environment depleting activities in a calamity consistent state is more extracting than uplifting.

An analysis of CSR in Chhattisgarh would help in understanding the situation clearly, in a minerally rich state, with many projects by corporate giants, where a total of hundred crores is spent on CSR (see Fig. 1) but the sector or the activity relating to environment, climate action (see Fig. 2) in various districts of Chhattisgarh combined, received only few crores, if the data of 2021–2023 is analysed.

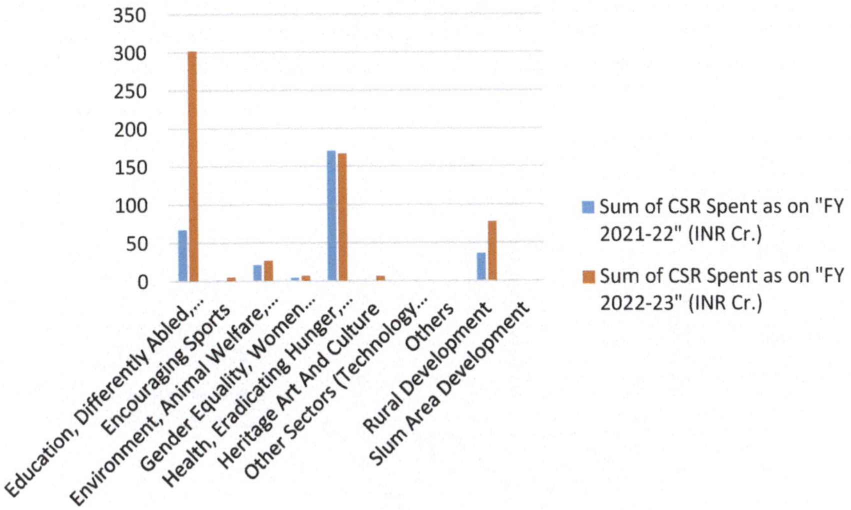

Fig. 1 Total CSR spending in Chhattisgarh (2021–2023). *Source* csr.gov.in

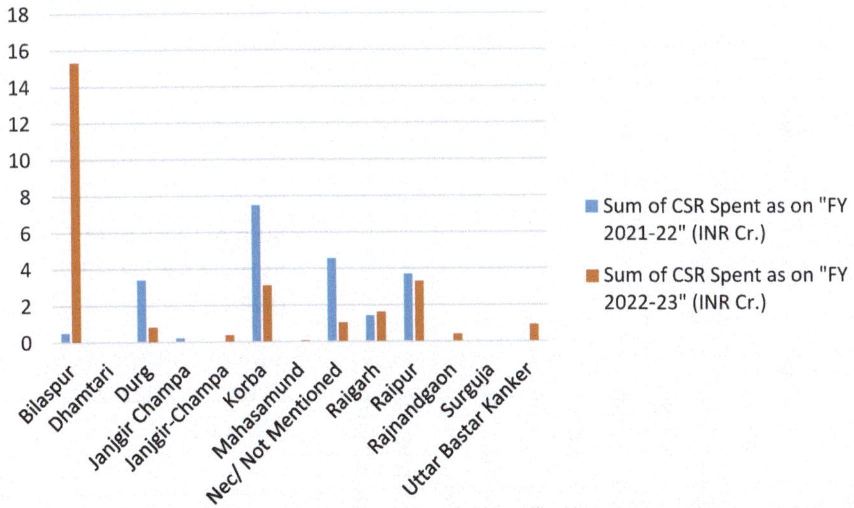

Fig. 2 Total CSR spending on Environment in districts of Chhattisgarh (2021–2023). *Source* csr.gov.in

Similar analysis of CSR in Odisha would help in understanding the situation clearly, in another minerally rich state, with many projects by corporate giants, where a total of more than four hundred crores is spent on CSR (Fig. 3) but the sector or the activity relating to environment, climate action in various districts of Odisha combined, received less than few crores in each of its district (Fig. 4), if the data from 2021 to 2023 is analysed.

It is imperative to highlight that both Odisha and Chhattisgarh have CSR funds being poured in by giant corporates working in all sectors, but as can be seen needs more spending of fund specifically in the goal of environment betterment and climate

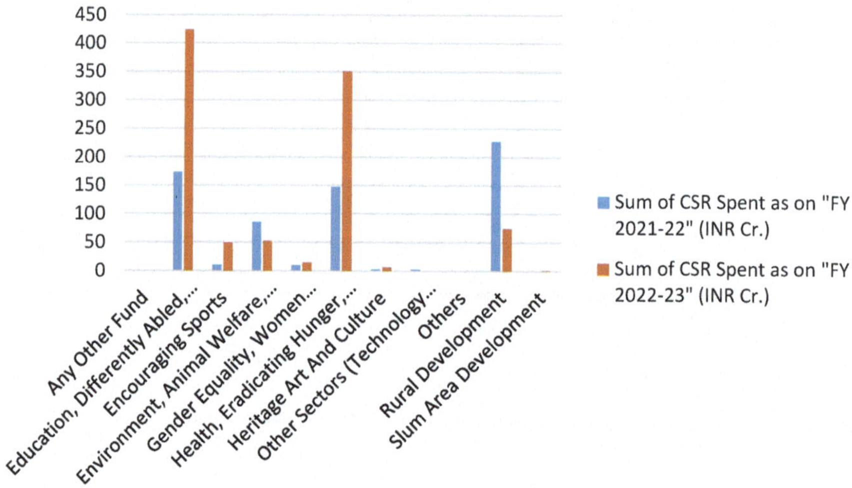

Fig. 3 Total CSR spending in Odisha (2021–2023). *Source* csr.gov.in

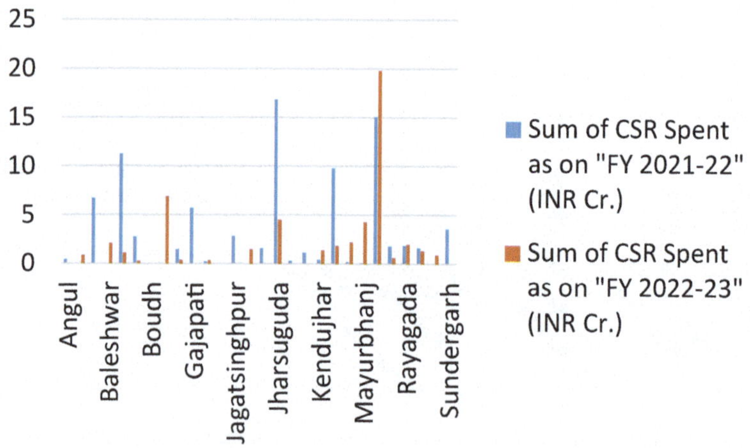

Fig. 4 Total CSR spending on Environment in districts of Odisha (2021–2023). *Source* csr.gov.in

action, given the rich heritage and culture as well as the protection of the thriving indigenous population of these states.

5 Discussion

Mining corporates and their social responsibility activities in Chhattisgarh and Odisha:

One must highlight the activities undertaken by Mining companies present in Chhattisgarh and Odisha to assess their CSR activities and its impact. It is of utmost significance to note that these companies drain the natural resources present in the state for the betterment of their companies but they need to give it back to the state, and the environment most importantly. Other companies engaged in different sector and businesses may not have as much as the responsibility as these companies engaged in mining sector to work for the betterment of environment.

An analysis of the CSR activities of few corporate giants present in the state of Odisha and Chhattisgarh shall help in understanding the need to go beyond mere compliance and actually work and aim for better impact of the CSR activities enshrined in the CSR policies of the companies.

Mahanadi Coalfields Limited (MCL), is a Public Sector Undertaking in India, and now a Miniratna. In 1992 it was carved out of South Eastern Coalfields Limited. The company has its headquarter at Sambalpur, Odisha. MCL spent in the year 2022–2023 only around 12 crores in Schedule VII, entry 4 concerning environment, and also animal welfare etc., out of the total 152 crores CSR fund spent by them in that year. From 12 crores around 7.5 crores was spent in one activity that is the 'Construction of compound wall around Rani park, Talcher' (Mahanadi Coalfields Limited 2023).

> MCL shall undertake select CSR activities out of the themes listed in Schedule-VII of the Companies Act as amended from time to time. The themes in the scope of CSR policy must be interpreted liberally so as to capture the essence of the subjects enumerated in Schedule-VII of Companies Act. Any modifications in Schedule VII of Companies Act or directions from DPE or MCA shall also deemed to have been incorporated in the scope of CSR policy of MCL from the date of such changes being notified by the Government. (Mahanadi Coalfields Limited n.d.-a).

National Aluminium Company Limited (NALCO), a 'Navratna' company under the Schedule 'A' category of CPSE's, was founded on 7th January, 1981 with its registered office in Bhubaneswar, Odisha. It is one of the country's largest bauxite, alumina, aluminium and power complex. In the year 2022–23 NALCO has spent in Environmental Sustainability and conservation of natural resources, a total of 25.87 lakhs, in one activity i.e. 'Support for plantation in and around Chandaka Damapara Wildlife Sanctuary' from Rs. 3954.44 lakhs, the overall total CSR fund spent by NALCO in 2022–23 (National Aluminium Company Limited n.d.-b).

Tata Steel is one of the world's most geographically diversified steel producers. They are one of the few steel operations that are fully integrated from mining to the manufacturing and marketing of finished products. They set up their second greenfield steel plant of 3 MnTPA in the eastern state of Odisha in 2016, the expansion to 8 MnTPA in currently underway.

Despite the philanthropic approach since decades, many voluntary charitable work, and overspending than the due CSR fund, it is to be noted that when it comes specifically to work for the environmental protection the company has a long way to go. Tata steel has allocated for approved CSR programs for the year 2022–23, in item iv of Schedule VII only about 14 crores to activity under animal welfare, protection of flora and fauna, and maintenance of water quality from a total of around Rs 400 crores allocated as CSR fund approved for CSR programs for that year (Tata Steel n.d.-a, n.d.-b).

South Eastern Coalfields Limited (SECL) is an eminent mineral company which produces coal and is located at Bilaspur, Chhattisgarh. It controls a total of sixty-seven mines, 41 mines located in Chhattisgarh and the remaining in Madhya Pradesh. SECL in the year 2022–23 has undertaken many CSR activities in the state of Chhattisgarh and Madhya Pradesh, however out of the total 12.08 crores spent across four projects in Chhattisgarh, only 2.89 crore was denoted in the category of environment and that too in a faraway land from the actual mining location, for the development of Bilasa Udyan and protection work of River Arpa Ghat at Bilaspur (South Eastern Coalfields Limited 2023).

National Mineral Development Corporation (NMDC), a Navratna Public Sector Enterprise under the Ministry of Steel, Government of India is the single largest producer of iron ore in India. It owns and operates highly mechanized iron ore mines in Chhattisgarh and Karnataka. NMDC in the year 2022–23 has spent a mere 11.28 crores in Chhattisgarh out of the total 87 crores spent across the states, for a river water front development and plantation works in Dankini and Shankini Rivers at Dantewada district of Chhattisgarh (National Mineral Development Corporation n.d.-a).

Bharat Aluminium Company Limited (BALCO0 has Mor Jal Mor Maati, Vedanta Skill School, Unnati, Nayi Kiran, Arogya, Mobile Health Van and Balco Medical Centre and Connect, initiatives that had the highest attention from the pool of Rs 24.53 crores in the year 2022–23. But unfortunately a mining company that is degrading the environment didn't find any flagship initiative under the environment category (Bharat Aluminium Company Limited 2023).

It is important that one must try to compensate at the place where the damage is done, therefore the CSR activities of mining companies present in the state of Chhattisgarh and Odisha must focus on and be made in a manner, that it uplifts the environment in the state especially in the locality/districts they are based at, ensure less to no degradation and comply with the environmental requirements and elevates the affected and displaced local community.

6 Conclusion

The legislative actions undertaken ever since the introduction of provision for CSR can be understood as either seeing corporates as partners for the development of society or strengthening the responsibility of business towards society. There have been various guidelines from the Government time to time to remain relevant with respect to dynamic changes in the global business environment. Another regulatory body in India, Securities and Exchange Board of India (SEBI) has also laid down a framework in form of Business Responsibility Reporting framework for top listed companies. This is still at an infant stage and companies are getting acquainted with reporting from a perspective of Environmental Social Governance (ESG). Policies and guidelines are formed and released time to time, the year 2019 witnessed revision in the earlier guideline to form National Guideline for Responsible Business Conduct (NGRBC) adhering to Companies Act 2013 ensuring it is in sync with globally adopted and followed Sustainable Development Goals (SDGs) and United Nations Principles on Business and Human Rights.

> The building of human rights will be one of the foundation stones on which we would build in the world an atmosphere in which peace could rule- Eleanor Roosevelt

No business could be built by harming the peace and damaging the environment, as it ultimately violates the very building of human rights itself. Business and Human Rights have always seen as two end of the spectrum, but if one has to dig deeper it will show its traces in each other. No business can thrive by violating the very rights of the human beings, the most important stakeholders of any business. The business of mining if looked at closely, highlights the dire need of balance between development and protection. In the name of development, no company can take away the right of community or their future generations right to breathe freely. Mining in itself is an activity that extracts the resource much more than it can ever replenish.

As Jan Elaisson mentioned that "There can be no peace without development, no development without peace, and no lasting peace or sustainable development without respect for human rights and the rule of law." It is best suited for a mining company which degrades the environment the most but must also strive to help in revival of the resources as much as possible. As no development shall bring in peace if its against the rule of law or disrespects human rights.

The mining companies that contribute to this significant economic development of the states, undertakes few CSR activities, that they pick and choose, sometimes cherry pick couple of the activities from Schedule VII of the Act to spend their two percent of profits on. Now these activities could more or less be anything but it is never proportionate or equal to the environmental harm that the company has done by encouraging degradation, displacement and/or deforestation. The implementation of these CSR activities are not taken seriously and furthermore the impact assessment of these activities aren't done religiously. Gives the mining companies, a safe ground to play the good corporate citizen which would be able to show adherence to CSR provision in letter of law, by just spending the two percent of their average net profit, and not the spirit.

Therefore, in order to have an impactful legislation that yields greater impact one must start from the very beginning, the language of the legislation, supplementary rules should be clear and not left for lacunae and convenience based interpretation that desires mere checking the boxes of compliance and not achieving any actual result.

Savitz and Weber (2006) submit that "a sustainable company is one that generates benefits for its shareholders while protecting the environment and improving the lives of those with whom it interacts". The mining companies have almost explored all avenues of bringing a good name to them, but alas, none of that has materialized in their favor. The only way that these mining companies may give back to society and the environment is by being more aware and active in the changing times and developed technology.

7 Implications and Future Work

Today is the generation of Artificial Intelligence (AI) and if used ethically for the betterment of people and society can do wonders. One of the most crucial aspects of ethical AI is to take the help of AI with regard to navigating implementable solutions for sustainable practices by mining companies via their CSR activities. AI can work for tracing resources and making use of the same responsibly. The utmost need to make use of the present technology without harming the needs of others and eventually saving the resources for generations to come deserves to be highlighted. AI is slowly but steadily being used to track criminals in the criminal justice system. Ethical AI, when paired with corporate funds, can help society develop and make the world sustainable. Especially, Mining companies should make use of AI to enhance their transparency among shareholders and to remove potential bias against them by explaining the reasonableness of their CSR activities. Ethical AI, when integrated to explain CSR activities of the mining companies, can result in upholding the principle of fairness and nondiscrimination among the community and other stakeholders. All onus with regard to accountability shall lie with mining companies, who would then go on to prove how their AI-driven socially responsible work has positive impacts. The concern of data privacy if taken care of, can help mining companies use ethical AI for better branding and visibility as a socially conscious positive sector.

Out of the many ways CSR and AI can be paired, optimizing resources, monitoring environmental management, resource extraction and contributing towards sustainable practices shall serve the best interest of all the concerned stakeholders. Another important role that AI can play is enhance community engagement, thereby leading to active participation and informed decision making, ultimately leading to resolving the issues of the community that are affected at the mining sites, especially. No protection of the environment is going yield results if it doesn't stop violating the rights of human beings around. Ethical AI can be used to highlight potential violations of human rights by mining companies and thereby giving the companies a chance to not risk any human beings life by contributing more to environmental degradation. It is also very helpful in the identification and management of the risk of mining operations being conducted. Risk can also be mapped relating to water quality and water consumption level, resulting into enhanced environment. One of the aspect of AI when used by mining companies is better and wider stakeholder participation and engagement. Many times, these mining companies do not bat an eye towards the under privileged persons, which stays near the mining area, sometimes mining companies actions shows disrespect towards the tribes who have been staying there for centuries, all of these can be worked on if AI is used for fairer and wider stakeholder engagement. One must highlight the multi layers with regard to AI driven solution that is required today to bring out the best CSR practices by mining companies ensuring environment sustainability, first being the AI usage in water management, be it consumption, reduction, improvement or monitoring. Second, to have AI built system to map operations, raise signals for hazards and enhance safety measures and protocols. Third, to enable AI based waste management techniques for processing, reducing waste and promoting the concept of reuse and recycle. And at last, none of these would work if there is no AI based mechanism to ensure active and involved community engagement for efficient, inclusive and informed decision making. It is pertinent to mention that as suggested above, AI if used responsibly and ethically can act as a tool for social good, that navigates through and addresses challenges and create a sustainable world.

Another suggestion would be to ensure that the sector that draws the most benefit from a resource especially natural, returns multifold to the same. No new grant of license should be given to mining companies who haven't been able to assess the impact of their old CSR activities. Environmental upgradation needs to be at the core of the CSR activities of mining companies, they shouldn't restrict themselves to mere compliance of the mining acts across the world but go beyond it, for an actual, sustainable growth, that shall yield better results both for the company and community. The companies must align their activities with Sustainable Development Goals 2030 especially the ones dealing with environment in particular. SDG 13 has been much talked about but less worked on. Climate change is also triggered because of the existence of money-minting mining companies that must now change their approach and work for a sustainable future. For Indian legislation, Schedule VII needs to have the specific requirement of a term in the form of environmental upgradation and must include mandatory activities for companies that contribute to alarming issues, such as Climate change, therefore it is imperative to have a mandatory requirement for mining companies to use their CSR funds in environmental protection and betterment. The mining companies may choose other activities, but the necessity to choose environmentally friendly activities must be prioritized and infact mandated. India became the first to have a legislative provision for CSR, and considering it has the highest population in the world today, it needs to protect the future generation's interests and distribute resources among them. India must also become the first, a torchbearer, in mandating CSR funds to be used first and foremost for environment upgrading activities rather than others, with the help of the advent of technology and in the form of ethical AI, for companies degrading the environment to map activities for restoration, reformation and replenishment of resources rather than for their commercial profits.

References

Bharat Aluminium Company Limited (2023) Annual report 2022–23. https://www.balcoindia.com/annual-report
Bowen HR (1953) Social responsibilities of the businessman. Harper & Row, New York
Carroll AB (1979) A three-dimensional conceptual model of corporate performance. Acad Manag Rev 4(4):497–505
Carroll AB (1999) Corporate social responsibility: evolution of a definitional construct. Bus Soc 38(3):268–295
Chandler D (2016) Strategic corporate social responsibility: sustainable value creation. SAGE Publications.
Davis K (1960) Can business afford to ignore social responsibilities? Calif Manage Rev 2(3):70–76
Down to Earth (n.d.) Odisha's mines may power the country, but what about its people? https://www.downtoearth.org.in/governance/odisha-s-mines-may-power-the-country-but-what-about-its-people--66074
Economic Times (2024, December 24) Lithium mines: India set to open first lithium mines in Chhattisgarh. https://economictimes.indiatimes.com/industry/indl-goods/svs/metals-mining/lithium-mines-india-set-to-open-first-lithium-mines-in-chhattisgarh/articleshow/112514340.cms?from=mdr
Government of Australia (1992–1998) Mining acts of NSW, Queensland, South and Western Australia. State Government Publications
Government of South Africa (2002) Mineral and Petroleum Resources Development Act. https://www.gov.za/
Government of Zambia (2015a) Mines and Minerals Development Act, 2015. https://www.parliament.gov.zm
Government of Zambia (2015b) Section 12, Mines and Minerals Development Act, 2015. https://www.parliament.gov.zm
IISD (n.d.) Sustainable development. International Institute for Sustainable Development. https://www.iisd.org/
Mahanadi Coalfields Limited (n.d.-a) About MCL. https://www.mahanadicoal.in/About/heformation.php
Mahanadi Coalfields Limited (n.d.-b) CSR policy. https://www.mahanadicoal.in/mclcsr/pdf/csrpolicynew.pdf
Mahanadi Coalfields Limited (2023) Annual report 2022–23. https://www.mahanadicoal.in/mclcsr/pdf/annualreport2223.pdf
Ministry of Corporate Affairs (2013a). Companies Act, 2013, Schedule VII: Activities which may be included by companies in their CSR policies. Government of India. https://www.mca.gov.in/
Ministry of Corporate Affairs (2013b) Companies Act, 2013, Section 135: Corporate Social Responsibility. Government of India. https://www.mca.gov.in/
Ministry of Corporate Affairs (2013) Companies Act, 2013, Section 135(1)–(3): CSR Committee requirements. Government of India. https://www.mca.gov.in/
Ministry of Corporate Affairs (2014) Companies (Corporate Social Responsibility Policy) Rules, 2014. Government of India. https://www.mca.gov.in/
Ministry of Environment, Forest and Climate Change (2016) Compensatory Afforestation Fund Act, 2016 and CAMPA Guidelines. Government of India. https://campa.nic.in/
Ministry of Law and Justice (1957) The Mines and Minerals (Development and Regulation) Act, 1957. Government of India. https://www.indiacode.nic.in/

Ministry of Mines (2015) The Mines and Minerals (Development and Regulation) Amendment Rules, 2015. Government of India. https://www.mines.gov.in/

Ministry of Mines (n.d.) National mineral scenario. Government of India. https://mines.gov.in/webportal/nationalmineralscenario

National Aluminium Company Limited (n.d.-a) CSR expenditure report. https://mudira.nalcoindia.co.in:3443/CSR/CSRExpenditureReport

National Aluminium Company Limited (n.d.-b) Home. https://nalcoindia.com/

National Mineral Development Corporation (n.d.-a) Company information. https://www.nmdc.co.in/investors/company-information

National Mineral Development Corporation (n.d.-b) CSR activities. https://www.nmdc.co.in/csr/csr-activities/details-of-csr-activities

Savitz AW, Weber K (2006) The triple bottom line: How today's best-run companies are achieving economic, social, and environmental success-and how you can too. John Wiley & Sons

South Eastern Coalfields Limited (2023) Impact assessment report of CSR activities for FY 2022–23. https://seclcil.in/writereaddata/Impact%20Assessment%20Report%20of%20CSR%20Activities%20of%20FY%202022-23.pdf

Tata Steel (n.d.-a) Company profile. https://www.tatasteel.com/corporate/our-organisation/company-profile/

Tata Steel (n.d.-b) CSR programmes. https://www.tatasteel.com/media/17063/csr-programmes-1.pdf

The New Indian Express (2023, July 19) 15.68% below poverty line in Odisha, says NITI Aayog. https://www.newindianexpress.com/states/odisha/2023/Jul/19/1568-per-centbelow-poverty-line-in-odisha-says-niti-aayog-2596353.html

United Nations (n.d.-a) Goal 13: Climate action. United Nations Sustainable Development Goals. https://sdgs.un.org/goals/goal13

United Nations (n.d.-b) Goal 15: life on land. United Nations Department of Economic and Social Affairs, Sustainable Development.

United Nations (2024) The Sustainable Development Goals Report 2024. United Nations. https://unstats.un.org/sdgs/report/2024/The-Sustainable-Development-Goals-Report-2024.pdf

Developing Thriving Workplaces: An Interplay between Sustainable and Servant Leadership, Organizational Culture and Employee Wellbeing

Roopika Kahera, Khyati Kapil, Anshu Latiyan, and Sachin Ghai

Abstract This chapter investigates the dynamic relationship between sustainable leadership and servant leadership, company culture and employee wellbeing. In a continuously changing, increasingly complicated and volatile business environment, organization face pressure to not only meet performance targets but also maintain long term sustainability, creativity and employee centric values. Sustainable leadership and servant leadership is defined by ethical behaviour and long-term vision and a dedication to stakeholder well-being, is essential for establishing a positive company culture. Together these leadership methods create a supportive workplace culture that emphasizes trust, inclusion, and continual learning. Such cultures promote trust, inclusion, psychological safety, and Purpose—all of which are critical to employee well-being. This chapter guide us to investigates how sustainable leadership and servant leadership affects cultural norms and organizational behaviour, resulting in environments where employees feel appreciated, supported and motivated. The chapter discusses the role of leadership and corporate culture in promoting employee mental health, engagement, work satisfaction, and resilience.by combining theoretical viewpoints, empirical research, and practical experience, it provides a thorough knowledge of how aligning leadership with corporate culture can improve sustainability and human well-being. It finishes with actionable recommendation for leaders looking to integrate Sustainable practices into their daily operation. Furthermore, the chapter emphasize the need of connecting leadership beliefs with managerial action in order to maintain consistency, develop credibility, and strengthen organizational identity. Reflective leadership, which encourages leaders to assess their behaviours and their impact on culture and employee well-being. This strategy encourages the development of a more real and adaptive leadership style. This chapter is fulfilling

R. Kahera (✉) · K. Kapil · A. Latiyan · S. Ghai
Graphic Era (Deemed to be University), Dehradun, India

K. Kapil
e-mail: khyatikapil.mgt@geu.ac.in

S. Ghai
e-mail: profsg.mba@geu.ac.in

A. Singh et al. (eds.), *Human-Centered Sustainability: Bridging Well-Being, Gender Equity, and Technology*, World Sustainability Series,
https://doi.org/10.1007/978-3-032-22020-2_15

the united nation sustainable development goal 3 (Good Health and Wellbeing), as servant and sustainable leadership promotes employee wellbeing.

Keywords Sustainable leadership · Organizational culture · Employee wellbeing · Psychological safety · Ethical leadership · Servant leadership · Mental health

1 Introduction

In today's rapidly changing and unpredictable business world, all businesses are under increasing pressure to maintain high performance while maintaining long-term viability putting employee well-being first. Globalization, digital revolution, and shifting demographics have all added complexity to how firms operate and manage their human capital. As a result, leadership must continuously adapt to meet these changing expectations. This Chapter investigates how sustainable leadership and servant leadership influence organizational culture while also increasing employee mental health, job engagement, satisfaction and resilience. It emphasizes the importance of ethical behaviour, long term vision, stakeholder well-being and the creation of learning focused workplace. As organizational become more diverse and employee expectations develop, traditional leadership approaches that focus solely on profitability are no longer effective. Today's workplace is increasingly seeking purpose-driven culture where personal and company values are in sync. In this setting, leadership style based on ethical ideals, empathy, and inclusivity are not only preferable, but also required for long term organizational success. Sustainable and servant leadership framework offer an integrated approach that not only addresses operational effectiveness but also enhances the human experience at work. These leadership styles promote environment where continuous learning, open communication, and psychological safety are embedded into the cultural fabric of the organization. Through the thoughtful integration of these leadership principles, organizational can build resilient teams capable committed to shared values and collective goals.

Sustainable leadership is a long term, value-driven approach to leadership that emphasize organization endurance, stakeholder equity, and environmental stewardship. It focuses on building system and practices that are ethical, inclusive and forward-thinking. Sustainable leaders aim to preserve the organization legacy while preparing it for future challenges, integrating social responsibility into core strategies (Hargreaves and Fink 2006) Servant Leadership, introduced by Robert K. Greenleaf, invers traditional leadership models by placing the needs of employees and stakeholders first. Servant Leaders empower team members by listening actively, fostering their development, and building a community within the organization (Greenleaf 1997). Servant leadership style is characterized by empathy, Stewardship, and commitment to the growth of people. Both leadership models converge in their ethical underpinning and prioritization stakeholder well-being. When implemented effectively, they can mutually reinforce each other to create resilient and adaptive organization that valve human dignity and long-term impact over short term gains.

Organizational culture is the shared set of beliefs, values and norms that guide behaviour within a company. It influences how employees interact, solve problems, and perceive their roles and responsibilities. Leadership has a profound impact on shaping this culture (Schein 2010). Leaders model desired behaviours and set the tone for communication, collaboration and Innovation. Sustainable and servant leaders contribute to cultivating a culture that is purpose-driven and employee-centric. They ensure that policies and practices reflect fairness, Inclusion, and psychological safety. For instance, sustainable leaders may integrate environmentally conscious initiatives, while servant leaders may establish consistently align their action with shared values, they reinforce a sense of purpose and direction across the organization (Sajjad et al. 2024).

An inclusive Culture shaped by reflective leadership promotes openness and diversity of thought, and mutual respect. These culture traits are critical in fostering high performance, reducing turnover, and encouraging Innovation (Denison 1990). Employee wellbeing is no longer a peripheral concern; it is central to organizational success. It includes physical health, emotional resilience, mental clarity, and a sense of belonging. Organization with high levels of employee wellbeing report improved productivity, reduced absenteeism and greater Innovation (Danna and Griffin 1999).

Sustainable and servant leadership contribute directly to these outcomes by prioritizing the holistic development of employees. Leaders who recognise the interconnectedness of well-being and performance implement policies such as flexible work hours, access to counselling service, wellness programme, and continuous learning opportunities. Servant leadership foster wellbeing by emphasizing empathy, listening and personal development. Employees feel more secure and supported when they are treated as valued contributors rather than mere resource (Linden et al. 2008). Similarly sustainable leadership supports wellbeing by encouraging ethical practices, long-term thinking and accountability. This dual emphasis creates a psychological safe environment where employees can thrive.

Leadership and culture are inextricably linked. The behaviours and values modelled by leadership influence cultural norms, which in turn shape future leadership development within the organisation. This Cyclical relationship highlights the importance of consistency and alignment between stated values and practiced behaviour (Schein 2010). When leadership behaviour contradicts organizational values, credibility and trust suffer. Conversely, when leaders embody the organization's mission and values, they serve as authentic role models and reinforce a unified sense of identity. Sustainable and servant leadership styles emphasize transparency, shared decision-making, and ethical conduct, thereby reinforcing a culture that supports collective goals over individual ambition. A strong aligned culture helps organization navigate periods of change more effectively. It acts as a stabilizing force that keeps employees engaged and focused, even during uncertainty. Leaders who invest in shaping and sustaining such cultures contribute to long-term organizational resilience and adaptability.

In addition to internal transformational, sustainable and servant leadership can extend their influence beyond organizational boundaries through active community engagement. Leaders who engage with external stakeholders, including local communities, non-profits, and policy makes, build trust and enhance the organization's social capital. This commitment to broader social impact not only reinforces ethical leadership but also strengthens the organization's brand and long-term viability (Maak and Pless 2006). Such leadership that looks outward encourages a sense of shared duty among workers, lining up personal and business goals for the greater good of society. When workers see their leaders doing good things for the community, it makes them feel proud and more connected to the organizations goal. Mentorship and coaching programs can help to systematically establish sustainable and servant leadership ideas in business culture. These efforts not only help employees grow, but they also foster future leaders who reflect company values. Mentors and coaches act as role models by sharing expertise, providing, encouraging accountability.

2 Literature Review

A growing body of literature highlights the significant impact of leadership styles on organizational outcomes, particularly in the areas of culture, employee wellbeing, as stated by (Hargreaves and Fink 2006), is rooted in ethical behaviour, long-term orientation, and social responsibility. It aims to build resilient organizations capable of enduring changes while balancing economic, social and environmental imperatives. (Avery and Bergsteiner 2011) further argue that sustainable leadership fosters system thinking, stakeholder involvement, and continuous improvement, which collectively enhance organizational resilience and performance.

In parallel, servant leadership pioneered by (Greenleaf 1997) emphasizes a people-first approach that nurture trust, empathy, and development within teams. This leadership model has gain prominence for its ability to create high-trust, value-driven culture that support employee engagement and motivation (Liden et al. 2008). Servant Leadership act as facilitators rather than commanders, which helps to build a psychologically safe environment that encourages innovation and collaboration. Scholars like (Schein 2010) and (Denison 1990) underscore the critical role leadership plays in shaping and sustaining organizational culture. Tey contend that leaders set the tone for acceptable behaviour and create the conditions under which employees feels either empowered or constrained. When leadership is aligned with ethical and inclusive values as seen in sustainable and servant leadership organizational culture becomes a strategic asset that can drive long-term success.

Employee wellbeing is increasingly featured in organizational studies as a core determinant of productivity and retention. (Danna and griffin 1999) highlight the multifaceted nature of wellbeing, which encompasses emotional, physical, and psychological aspects. Leadership that prioritizes wellbeing through empathy, flexibility and ethical conduct not only enhances morale but also reduces burnout and turnover (Liden et al. 2008) Emotional Intelligence, popularized by (Goleman 1995), is another essential trait that links both leadership styles. Emotional Intelligence equips leaders with the ability to manage interpersonal dynamics effectively, resolve conflicts, and inspire teams with authenticity. (Raelin 2002) introduces the concept of reflective practice, arguing that self-awareness and continuous learning are fundamental for leaders to adapt and improve. These qualities are central to the execution of both sustainable and servant leadership models.

Together, these Theoretical contributions form a cohesive foundation that justifies integrating sustainable and servant leadership principle. The Literature consistently demonstrates that leadership grounded in empathy, ethical behaviour, and long-term vision contributes to building inclusive cultures and improving employee wellbeing core elements necessary for modern organizational success.

2.1 Servant Leadership

Servant Leadership is conceptualized by (Greenleaf 1997), Represents a paradigmatic shift from traditional Hierarchical models toward a leader-as-servant mindset. This approach places the growth, development, and well-being of followers at the forefront of leadership priorities. Servant Leaders focus on empowering there, listening actively, and building a community of trust and collaboration within the organization. Study by (Spears and Lawrence 2001) identified ten core characteristics of servant leadership, including empathy, healing, awareness, persuasion, conceptualization, foresight, stewardship, commitment to the growth of people, and building community. These traits collectively foster an ethical and inclusive environment where employees feel valued and supported. Studied have shown that Servant Leadership correlates positively with job satisfaction, organizational commitment, and performance (Liden et al. 2008; Hunter et al. 2013).

(Liden et al. 2008) further development a multidimensional framework to measure servant leadership, encompassing behaviour such as emotional healing, creating value for the community, and helping subordinates grow and succeed. These behaviours contribute to psychological safety and trust, essential components for fostering Innovation and Engagement. (Van Dierendonck 2011) emphasized the psychological mechanisms through which servant leadership enhances employee well-being, including increased intrinsic motivation and reduced burnout.

Servant Leadership has also been linked to the development of ethical cultures in organizations. (Ehrhart 2004) found that servant leaders serve as moral exemplars, promoting fairness and integrity in decision-making processes. Their commitment to authenticity and inclusivity helps to create environment where diverse voices are heard and respected. Moreover, servant Leadership has been shown to be particularly effective in times of organizational change and uncertainty. By building strong Interpersonal relationship and focusing on the needs of others, servant leaders help employees cope with ambiguity and stress, thereby improving resilience and adaptability (Eva et al. 2019).

2.2 Sustainable Leadership

Sustainable leadership has emerged as a crucial paradigm for long-term organizational success, particularly in environment characterized by rapid change and heightened stakeholder expectations. (Hargreaves and Fink 2006) Pioneered the concept, defining sustainable leadership as leadership that preserves and advances deep learning for all stakeholder's overtime. It emphasizes the development of system that are not only effective in the short term but also ethically resilient, inclusive and future-oriented.

(Avery and Bergsteiner 2011) introduced the "Sustainable Leadership pyramid," which identifies 23 interrelated practices that support both social responsibility and business performance. These include employee development, long term orientation, ethical behaviour, stakeholder engagement, and systemic Innovation. Their research shows that organizational that adopt sustainable leadership practices outperform competitors in resilience, employee satisfaction, and reputation.

Sustainabe Leadership is inherently values based and rooted in ethical conduct, making it especially suitable for addressing complex challenges such as climate change, social inequality, and mental health in the workplace (Fullan 2006). It promotes practices like inclusive decision-making, continuous learning and responsible resource management. These practices are linked to enhance employee engagement, Innovation and retention (Shrivastava 1995). Furthermore, studies by (Kerbs Davies 2009) and (Norcross and Lambert 2011) argue that sustainable leadership builds organizational capacity by fostering distributed leadership, where responsibility and influence are shared across all levels. This not only reduces the risk of leadership burnout but also embeds a sense of ownership and purpose among employees.

The link between sustainable leadership and organizational culture has also been widely acknowledged. Sustainable leaders model transparency, ethical consistency, and long-term thinking, which helps shape a culture of trust and psychological safety (Hargreaves 2006). This culture environment is essential for fostering resilience during organizational transitions and economic uncertainty. In sum, the literature affirms that sustainable leadership is both a moral imperative and a strategic advantage. Its long-term focus, stakeholder-centric mindset, and systemic thinking make it an essential framework for building ethical, adaptive, and high performing organizations.

2.3 Organizational Culture

Organizational culture has been extensively studied as a determinant of organizational effectiveness, employee satisfaction, and adaptability. (Schein 2010) defined organizational culture as a pattern of shared basic assumptions learned by a group as it solves problem of external adaptation and internal integration. These assumptions are taught to new members as the correct way tp perceive, think, and feel in relation to those problem. Culture shapes how employees behave make decision and interact within the workplace.

Aggarwal (2024) emphasized that strong organizational culture, characterized by high levels of involvement, consistency, adaptability, and mission clarity, are linked with enhanced performance and innovation. Organizations with a cohesive and values-driven culture can better align employee behaviours with strategic goal, resulting in improved morale and lower turnover. (Deal and Kennedy 1983) also highlight that culture can act as a powerful control system, guiding day-to-day behaviours more effectively than formal policies. Research has also shown that leadership plays a critical Role in Shaping and sustaining organizational culture. Leaders act as culture carriers, modelling values and behaviours that influence group norms (Schein 2010; Tran 2025). Transformational, servant and sustainable leadership styles are particularly effective in fostering inclusive and ethical cultures that support psychological safety and innovation (Tadesse Bogale and Debela 2024).

Moreover, organizational culture must evolve to meet changing employee expectations and external pressures. (Cameron et al. 2008) introduced the concept of positive organizational scholarship, which focuses on fostering strengths, resilience, and human flourishing with the workplace. Culture that supports learning, collaboration and inclusivity have bee found to contribute positively to and overall wellbeing (Edgar and Geare 2005) In sum, The literature undergoes that organizational culture is a dynamic and multifaceted that profoundly impacts organizational success. Leadership is both a product and a shaper for culture, intentional efforts to align cultural values with leadership behaviour can enhance trust, engagement and long-term sustainability.

2.4 Employee Wellbeing

Employee well-being has emerged as a critical determinant of organisational performance and sustainability. It encompasses multiple dimensions including physical health, mental and emotional Wellness, social connectedness, and sense of purpose and belonging at work (Elufioye et al. 2024). Researchers increasingly emphasis that well-being is not merely an individual concern but a strategic organisational imperative link to engagement, retention and Innovation. According to (Warr 2002), employee wellbeing is influenced by factors such as job characteristics, workplace relationships, and organizational support. Positive work environments that offer autonomy, clarity of role, recognition and developmental opportunities are associated with higher wellbeing outcomes. Moreover, the identifies workplace wellbeing as a key component of public wellness strategies.

Leadership plays a fundamental role in shaping employee wellbeing. Empirical Studies show that supportive leadership styles especially those rooted in empathy, ethical conduct, and inclusivity directly enhance employee's psychological safety and job satisfaction (Kelloway and Barling 2010). Servant leadership fosters wellbeing by prioritizing individual growth, listening and genuine concern for team member (Liden et al. 2008). Similarly, Servant leadership promotes long term well-being through ethical governance, purpose-driven goals and equitable work practices (Avery and Bergsteiner 2011). Moreover, organisational culture that values open communication, trust, and learning tend to nature resilience and reduce burnout (Bakker et al. 2007). Interventions such as mental health programme, flexible working arrangements, and recognition system further support being and create a positive feedback loop between employee satisfaction and organizational effectiveness. In conclusion, Literature undergoes that employee wellbeing is both moral and strategic priority. It is closely linked with leadership behaviour, cultural values, and structural supports all of which can be intentionally shaped to foster healthier, more engaged and more productive workplace.

3 Discussion and Findings

This chapter discussion reveals that integrated sustainable and servant leadership practices significantly enhances organizational culture an employee wellbeing. A cold finding is that leadership approaches emphasise ethics, empathy, and shared purpose foster environment where employees feel secure, valued, and motivated. Search workplaces exhibit reduced turnover, increased productivity, and a higher capacity to innovate. (Danna and Griffin 1999; Linden et al. 2008) One of the key outcomes observed is bad the alignment between leadership behaviour and organisational values, which contributes to cultural coherence and employee trust. Leaders who practice reflective and emotional intelligent leadership bridge the gap between stated values and enacted behaviour (Goleman 1995; Raelin 2002). The authenticity enhances psychological safety and encourages open communication and collaboration. Organizations that successfully adopt these leadership styles often implement structural supports, including mentorship and Wellness programs, flexible work arrangements, and ethical governance policies. These structures institutionalized Care and support, transforming leadership principles into daily practices (Avery and Bersteiner 2011).

Furthermore, findings highlight that employee wellbeing should be viewed as a systematic outcome influenced by leadership, culture and strategic alignment. Wellbeing is not merely a human resource function, but an organizational value embedded in decision making, performance metrics, and leadership evaluations. This integrated approach ensures long term resilience and adaptability in volatile business environment. Community engagement emerges and secondary but significant dimension. Leaders who extend their influence beyond the organization and foster social responsibility cultivate deeper purpose and loyalty among employees. This outward focus also reinforces the organizational ethical stance and public credibility (Maak and Pless 2006). The findings reinforce that emotional intelligence and reflective capacity are critical traits for leaders in modern organization. Training programs that develop these competencies enable sustainable and servant leadership behaviour to flourish across hierarchical levels (Goleman 1995). Search development ensures continuity in leadership values, builds a pipeline of future ready leaders and support consistent culture building. Finally, the interplay of sustainable and servant leadership results in an inclusive, ethical add resilient culture. These leadership frameworks are not mutually exclusive but mutually reinforcing. When adopted in tandem, they create organization where people feel psychological safe culturally align and deeply engaged in collective goals. This alignment enhances performance, innovation and overall organizational health.

3.1 Expanding Leadership Influence through Community Exchange

In addition to internal transformation, sustainable and servant leadership can extend their influence beyond organisational boundaries through active community engagement. Leaders who engage with external stakeholders, including local communities, non-profit organization, and policy makers, build trust and enhance organisation social capital. This commitment to border social impact not only reinforces ethical leadership but also strengthens the organizational brand and long-term viability. (Maak and Plees 2006). Search outward looking leadership foster a sense of collective responsibility among employees, aligning personal and organizational with a societal betterment. When employee observe their leader contributing meaningfully to community welfare at foster pride and deepens their connection to organisational mission.

3.2 The Role of Emotional Intelligence in Leadership

Emotional intelligence plays a crucial role in successful practices of both sustainable and servant leadership leaders with high emotional intelligence are adapted at understanding their own emotion and those of others which allows them to build stronger relationships, manage conflict effectively, and foster positive work environment. Empathy, a core component of emotional intelligence, is essential for servant leaders as they seek to understand and support the need for their team members (Goleman 1995).

Similarly sustainable leaders rely on emotional intelligence to navigate complex ethical dilemmas, communicate vision effectively, and lead with authenticity. Developing emotional intelligence through training and self-reflection enhances a leaders capacity to inspire, motivate and sustain positive culture and organizational change.

3.3 Building Leadership Capacity through Mentorship and Coaching

Sustainable and servant leadership principles can be systematically embedded into organisational culture through mentorship and coaching programs. This initiative not only support employee development but also cultivate future leaders who embody organisational values. Mentors and coaches served as a role model, sharing knowledge, offering guidance and fostering accountability.

3.3.1 Practical Strategies for Leaders

To effectively implement sustainable and servant leadership principle, leaders can adopt the following strategies:

1. Showcase ethical and thoughtful behaviour by maintaining integrity, displaying humility, and accepting Responsibility. Engage in regular self-reflection on your decision and actively seek input from coworkers and teammates.
2. Encourage employee participation by implementing inclusive decision-making processes like cross-functional teams, town hall meeting, and structured suggestion system.
3. Create an inclusive environment by aggressively supporting diversity in recruitment, encouraging open discourse, and maintaining strong anti-discrimination measures.
4. Commit to ongoing learning by providing training and development opportunities that are aligned with both organizational goals and individual career objective.
5. Encourage work life integration by allowing for flexible work arrangements and respecting personal time, which helps to decrease burnout and boost employee morale.
6. Assess and promote staff well-being through frequent surveys, mental health services and department-wide wellness initiatives.
7. Celebrate accomplishment and milestones that are consistent with the organization's fundamental principles and culture goals to recognized and reward positive behaviour.

4 Conclusion

In Today's fast paced and complex world, where uncertainty has become the norm and principles like as empathy and sustainability are finally getting the attention they deserve, this chapter emphasizes the transformational power of combing sustainable leadership with servant leadership. As a duo these two leadership models provide a strong, human centred approach to transforming workplace culture and improving employee well-being. This chapter is also achieving the United nation sustainable development goal 3 (Good Health and Wellbeing) which states that Sustainable and servant leadership are essential, not just complementary. Sustainable leadership prioritizes long term vision and fairness for all stakeholders, whereas servant leadership is founded on empathy, empowerment and a sense of community. Both contradict established power-based leadership paradigms. Their shared principles go beyond intellectual alignment to create a clear roadmap for creating ethical, people-focused firms that cannot just resist but thrive amid adversity.

A significant takeaway from this conversation is the dynamic interplay between leadership and corporate culture, which constantly impacts the other. Leaders shape culture through their daily actions, language and decisions and culture shapes how leadership is perceived and perceived and practiced. When leaders consistently demonstrate the value of sustainable and servant leadership, such as inclusion, accountability, and foresight, they aet a tone that is felt throughout the organization. This consistency builds trust, reinforces organizational identity, and prepare teams to face change with confidence and integrity. One of the most important contributions of these leadership styles is their positive impact on employee morale at work. Many organizations struggle with stress, burnout and a lack of psychological safety. Servant leadership addresses these challenges by putting people first-actively listening, offering support, and empowering them. Meanwhile, sustainable leadership ensure that organizational system and strategies work together to promote long term employee well-being and satisfaction. Together, these initiatives transform well-being from a checklist actively to deeply ingrained culture value.

A Prominent focus in this chapter is the importance of reflection of leadership. Both servant and sustainable leaders are extremely introspective they constantly assess their influence, adhere to their beliefs and seek to understand their own leadership journey. Self-awareness promotes resilience, deepens trust and help people stay focused in their mission, especially during difficult circumstance. However, developing this type of leadership culture extends beyond individual characteristics. Organizations must consciously create mechanisms that encourage and sustain such behaviour. This includes implementing mentoring programs, ethical governance framework, flexible work arrangement, and community engagement initiatives that embody servant and sustainable leadership principles. Furthermore, investing in leadership development programs that focus on emotional intelligence, ethical decision making, and reflective practices is critical for achieving real transformation at all levels.

Organizations that actively engage in social responsibility broaden their reach beyond the workplace, boosting both their culture identity and their employees' sense of belonging. When leaders are seen investing in worthwhile causes and giving back to the community, it strengthens a feeling of shared purpose and promotes greater company pride. Another important takeaway is that employee well-being necessitates a holistic, systemic approach. It's more than just providing wellness benefits; its about assessing leadership behaviours, cultural norms, and workplace dynamics that influence daily experiences. When firms integrate leadership, strategy, and emotional intelligence with their employee true needs, they foster an atmosphere in which individuals may thrive.

Emotional intelligence is recognized as a fundamental characteristic that underpins both servant and sustainable leadership. The ability to perceive, comprehend, and control one's emotion, as well as to be aware of feeling of other, is what distinguishes true, empathic leaders who can make tough but ethical judgement, developing these skills is no longer discretionary; it is a strategic need for firms that want to remain relevant and human-centred. Leaders must continually embody the organization basic principles, particularly in high pressure situations. Their words and actions must be consistent-tools like regular feedback loops, 360-degree evaluations, and cultural audits are critical for holding leaders accountable and ensuring that their behaviour represents the organization commitments. Cultural transition is a gradual, progressive approach to implementing these leadership principles. Small but impactful actions, such as creating peer mentoring programs, organizing servant leadership workshops, or forming sustainability committees, can provide that ground for long-term transformation. To achieve real and long-term development, Leaders must be ready to question obsolete norms, accept vulnerability and set an example.

In essence, sustainable and servant leadership are more than just qualities; they are required for creating ethical, inclusive, and resilient organizations. When leaders combine these leadership styles with a clear cultural vision and a purposeful emphasis on employee well-being, they create cultures in which both people and performance thrive. As the corporate landscape evolves in response to social, technological, and generational trends, firms that prioritize empathy, honesty and long-term thinking will prosper. Companies that embrace sustainable and servant leadership may develop a culture based on trust, compassion, and resilience-one in which success is shared, and employees take pride in their efforts.

References

Aggarwal S (2024) Impact of dimensions of organisational culture on employee satisfaction and performance level in select organisations. IIMB Manage Rev 36(3):230–238

Avery GC, Bergsteiner H (2011) Sustainable leadership practices for enhancing business resilience and performance. Strateg Leadersh 39(3):5–15

Bakker AB, Hakanen JJ, Demerouti E, Xanthopoulou D (2007) Job resources boost work engagement, particularly when job demands are high. J Educ Psychol 99(2):274–284

Cameron KS (2008) Paradox in positive organizational change. J Appl Behav Sci 44(1):7–24

Danna K, Griffin RW (1999) Health and Well-being in the workplace: a review and synthesis of the literature. J Manage 25(3):357–384

Deal TE, Kennedy AA (1983) Culture: a new look through old lenses. J Appl Behav Sci 19(4):498–505

Denison DR (1990) Corporate culture and organisational effectiveness. John Wley & Sons

Edgar F, Geare A (2005) HRM practice and employee attitudes: different measures–different results. Pers Rev 34(5):534–549

Ehrhart MG (2004) Leadership and procedural justice climate as antecedents of unit-level organizational citizenship behavior. Pers Psychol 57(1):61–94

Elufioye OA, Ndubuisi NL, Daraojimba RE, Awonuga KF, Ayanponle LO, Asuzu OF (2024) Reviewing employee Well-being and mental health initiatives in contemporary HR practices. Int J Sci Res Arch 11(1):828–840

Eva N, Robin M, Sendjaya S, Van Dierendonck D, Liden RC (2019) Servant leadership: a systematic review and call for future research. Leadersh Q 30(1):111–132

Fullan M (2006) The future of educational change: system thinkers in action. J Educ Chang 7(3):113–122

Goleman D (1995) Emotional intelligence why it can matter more than IQ. In: Emotional intelligence

Greenleaf RK (1997) Futurist lessons from Thomas Jefferson. Futurist 31(2):68–68

Hargreaves A, Fink D (2006) Redistributed leadership for sustainable professional learning communities. J Sch Leadersh 16(5):550–565

Hunter EM, Neubert MJ, Perry SJ, Witt LA, Penney LM, Weinberger E (2013) Servant leaders inspire servant followers: antecedents and outcomes for employees and the organization. Leadersh Q 24(2):316–331

Kelloway EK, Barling J (2010) Leadership development as an intervention in occupational health psychology. Work Stress 24(3):260–279

Krebs JR, Davies NB (2009) Behavioural ecology: an evolutionary approach. John Wiley & Sons

Liden RC, Wayne SJ, Zhao H, Henderson D (2008) Servant leadership: development of a multidimensional measure and multi-level assessment. Leadersh Q 19(2):161–177

Maak T, Pless NM (2006) Responsible leadership in a stakeholder society–a relational perspective. J Bus Ethics 66:99–115

Norcross JC, Lambert MJ (2011) Psychotherapy relationships that work II. Psychotherapy 48(1):4–8

Raelin JA (2002) I Don't have time to think! versus the art of reflective practice. Reflections: SoL J 4(1):66–79

Sajjad A, Eweje G, Raziq MM (2024) Sustainability leadership: an integrative review and conceptual synthesis. Bus Strateg Environ 33(4):2849–2867

Schein EH (2010) Organizational culture and leadership, vol 2. John Wiley & Sons

Shrivastava P (1995) The role of corporations in achieving ecological sustainability. Acad Manage Rev 20(4):936–960

Spears LC, Lawrence M (2001) Focus on leadership: servant-leadership for the twenty-first century. John Wiley & Sons

Tadesse Bogale A, Debela KL (2024) Organizational culture: a systematic review. Cogent Bus Manage 11(1):2340129

Tran QHN (2025) The role of leadership skills in organisational citizenship behaviour at Vietnamese libraries: organisational culture as a mediator. Glob Knowl Mem Commun 74(1/2):463–479

Van Dierendonck D (2011) Servant leadership: a review and synthesis. J Manage 37(4):1228–1261

Warr P (2002) Psychology at work. Penguin UK

Entrepreneurial Orientation, Sustainable Leadership, and Culture: A Synergistic Path to Innovation and Well-being

Anshu Latiyan, Sachin Ghai, Roopika Kahera, and Khyati Kapil

Abstract This chapter analyses the link effects of entrepreneurial orientation, sustainable leadership, and organizational culture on employee wellbeing. In today's dynamic and competitive climate, firms must innovate while preserving long-term viability and employee-centered Policies. Organizations can achieve great success by adopting an entrepreneurial mindset, which entails being creative, taking measured risks, and being proactive. But when it functions in harmony with long-term, sustainable leadership, its full potential is released. Leaders with a sustainability mindset encourage ethical behaviour, long-term thinking, and the well-being of all stakeholders. A strong foundation for responsible and fruitful innovation is created by this type of leadership. When such leadership is supported by an adaptable and positive corporate culture, it creates an atmosphere at work where innovation and employee well-being can thrive. The manner in which this triadic harmony promotes psychological safety, work satisfaction, engagement, and organisational resilience is examined in this chapter. It provides insights into how businesses might encourage human sustainability and entrepreneurial growth via the use of theoretical models and empirical research. Helpful advice is provided to CEOs and managers who want to create cultures that blend strategic agility with worker engagement. This research revealed how this trio synergy promotes psychological safety, engagement, job satisfaction and organisational durability. Managers and executives who want to create cultures that balance flexibility in strategy with the well-being of employees are given helpful guidance based on empirical research and theoretical models, offering ideas on how organisations may encourage social sustainability and growth in entrepreneurship. This chapter also examines the value of truthful communication, collaboratively

A. Latiyan (✉) · S. Ghai · R. Kahera · K. Kapil
Graphic Era Deemed to be University, Dehradun, India

S. Ghai
e-mail: profsg.mba@geu.ac.in

R. Kahera
e-mail: roopikakahera.phd@geu.ac.in

K. Kapil
e-mail: khyatikapil.mgt@geu.ac.in

A. Singh et al. (eds.), *Human-Centered Sustainability: Bridging Well-Being, Gender Equity, and Technology*, World Sustainability Series,
https://doi.org/10.1007/978-3-032-22020-2_16

making decisions, and real leadership in enhancing the link between well-being and creativity. It focused on the value of flexibility and continuous learning as cultural and organisational traits. Moreover, this chapter provides specific recommendations for embedding sustainability and human development in the core of entrepreneurial efforts, enabling long-term success that is both financially and ethically possible. To improve the relationship between employee well-being and creativity, this part focuses on the value of open communication, involvement in decision-making, and authentic leadership.

Keywords Entrepreneurial orientation · Sustainable leadership · Employee wellbeing · Innovation · Organizational culture · Organizational resilience

1 Introduction

In the fast-paced, uncertain business environment of today, organizations are coming to understand that innovation and employee well-being are both critical for long-term, sustainable success. Businesses that actively encourage an entrepreneurial mindset-characterised by creativity, major risk-taking, and initiative tend to be more competitive and flexible. This approach frequently leads to improved adaptability, ongoing innovation, and strong human capital development when combined with sustainable leadership and a strong, positive organizational culture (Lumpkin and Dess 1996; Covin and Slevin 1993). Deloitte suggested it is now clear that a people-centered approach to organizational development is not an option but rather a strategic imperative due to the mounting demands of globalization, technological disruption, and changing employee expectations. In light of contemporary issues, traditional models that place excessive emphasis on financial performance are failing. Building resilient organizations that can innovate successfully while simultaneously protecting the mental and emotional health of their employees has become more popular as a result (Kelloway and Barling 2010). Three interconnected ideas—enterprise orientation (EO), sustainable leadership, and organizational culture have surfaced as key components of long-term organizational success in this changing corporate environment. Although each of these elements can spur innovation and progress on its own, it is their combination that gives them their true strength. When they worked together, they could produce a synergistic effect that improved employee happiness and organizational performance, establishing the groundwork for inclusive, long-term, and responsible success (Avery and Bergsteiner 2011; Schein 2010). When aligned strategically, they foster a dynamic work environment where innovation thrives alongside employee well-being (Zahra and Covin 1995).

Entrepreneurial orientation (EO) enables firms to identify opportunities in uncertain markets, act decisively, and continuously evolve their value propositions. However, the presence of EO alone does not guarantee success. Its impact is significantly enhanced when guided by leaders who prioritize sustainability and inclusivity. Sustainable leadership goes beyond profit maximization and incorporates ethical behaviour, stakeholder engagement, and long-term thinking. It provides the moral compass and visionary guidance necessary to align entrepreneurial actions with broader organizational goals and societal expectations.

Entrepreneurial orientation—characterized by proactiveness, risk-taking, and innovativeness—acts as the engine that propels an organization toward opportunity and growth. Sustainable leadership, on the other hand, provides the ethical compass and long-term vision necessary to guide entrepreneurial endeavors responsibly (Avery and Bergsteiner 2011). At the same time, organizational culture serves as the underlying framework that shapes employee behaviour, communication, and collaboration. A culture rooted in openness, trust, and continuous learning supports both innovation and well-being. When employees feel psychologically safe and empowered to contribute ideas without fear or failure, they are more likely to engage in innovative behaviours that benefit the organization as a whole. Drawing from theoretical perspectives, real-world examples, and empirical research, the chapter proposed that fostering synergy among EO, sustainable leadership, and organizational culture leads to resilient, innovative, and human-centered organizations.

This chapter investigates the dynamic interplay among organizational culture, sustainable leadership, and entrepreneurial orientation, looking at how these elements work together to produce a potent synergy with businesses. When properly coordinated, these 3 components, 2 more than simply coexist; they strengthen one another to create a unified framework that fosters creativity and promotes worker well-being. For organisations to be strong, adaptable, and passionately dedicated to their goal and people, all three are necessary. In a world where businesses must adapt to continual shifts and an increase in complexity, executives can build organisations that are prepared for the future by recognising and using this cooperation. Such organisations are more capable and able to perform continual creativity, fast reaction to obstacles, and maintain an atmosphere where employees are valued as they can focus better on their work, active, and encouraged to provide significant impacts.

2 Literature Review

Academics and practitioners are increasingly focusing on the relationship between organizational culture, sustainable leadership, and entrepreneurial orientation (EO). In recent years, there has been a growing recognition that these elements are not only important on their own but also intimately interwoven in affecting how firms adapt to change, promote innovation, and sustain long-term performance. Particularly, entrepreneurial orientation is often regarded as a strategic stance that incorporates 3 essential elements: proactivity, risk-taking, and inventiveness. These characteristics help businesses spot fresh opportunities, Outsmart rivals, and come up with innovative solutions in unpredictable situations (Lumpkin and Dess 1996). It is often considered a critical determinant of a firm's capacity for innovation and competitive advantage. Research consistently finds that firms with a strong EO are more agile and better equipped to seize new market opportunities (Covin and

Slevin 1989). However, such orientation can also lead to instability or burnout if not balanced by values that emphasize sustainability and human-centric governance. Sustainable leadership extends beyond traditional leadership models by integrating ethical responsibility, long-term value creation, and stakeholder inclusiveness (Avery and Bergsteiner 2011). Sustainable leaders not only focus on profitability but also on the social and environmental impact of their decisions.

According to D'Amato et al. (2009), Sustainable leadership promotes employee engagement and builds trust across stakeholder networks, providing a counterweight to the potential excesses of an aggressive EO. This kind of leadership is particularly effective when embedded within a culture that values collaboration, continuous learning, and psychological safety. Organizational culture serves as the contextual fabric that binds EO and leadership together. Schein (2010) emphasized that culture shapes behaviour, influences decision-making, and affects how innovation is received and implemented. A supportive culture that values experimentation and employee input can amplify the benefits of EO and sustainable leadership, creating a resilient organization that thrives in dynamic environments. Hofstede (2011) cultural dimensions also show how values like individualism versus collectivism or uncertainty avoidance influence the implementation of entrepreneurial strategies.

Additionally, these 3 constructs confluence greatly enhance worker well-being. According to studies by Ryan and Deci (2001), organizational situations where leadership is moral, culture is inclusive, and innovation is promoted have a beneficial impact on autonomy, competence, and relatedness—all important components of employee well-being. According to more recent studies by Hakanen et al. (2008), Job resources—such as leadership that encourages innovation—are strongly associated with reduced burnout and increased employee engagement. Empirical research also confirms that EO improves organizational outcomes when it is tempered by strong leadership and culture. For example, Gupta et al. (2004) showed that innovative companies with visionary leaders and a strong sense of community had a higher chance of continuing to innovate over time. Similarly, Cameron and Quinn discovered that organizational resilience and flexibility are enhanced when a transformational leadership style is paired with a value-based culture.

In conclusion, the research backs up the claim that EO, sustainable leadership, and culture are intricately linked and not separate success factors. Their alignment fosters an environment where innovation and well-being reinforce one another, providing a strategic pathway to long-term organizational health and performance.

- **Entrepreneurship:**
  Entrepreneurship, at its heart, is a deeply human endeavour. It begins with an idea, often born from lived experiences, frustrations, or aspirations, and transforms into action through passion, determination, and resilience (Shane and Venkataraman 2000). Entrepreneurs are not merely business starters; they are problem solvers, risk-takers, and visionaries who dare to imagine something better and work relentlessly to make it real (Drucker 1985). They navigate uncertainty, face setbacks, and make countless decisions driven by their values, beliefs, and goals (Hisrich et al. 2017). At its best, entrepreneurship creates opportunities not

just for financial gain but for personal fulfilment, social contribution, and community upliftment (Bornstein 2007). It reflects the courage to innovate, the empathy to serve others, and the wisdom to adapt in a constantly changing world.

- **Sustainable Leadership:**

  Sustainable leadership is grounded in a commitment to lead with integrity, foresight, and responsibility. It goes beyond short-term success and profit to include the long-term well-being of people, the plane, and the organization itself (Avery and Bergsteiner 2011). As a stainable leader leads by example—championing ethical practices, nurturing an inclusive environment, and making decisions that honour both current and future generations. Such leadership requires empathy koma resilience, and a strong moral compass (Hargreaves and Fink 2006). It involves listening to diverse voices, fostering trust, and inspiring others to see the big picture. At its core, sustainable leadership is about creating a legacy that prioritises human values alongside innovation and growth, ensuring that progress does not come at the cost of sustainability or social equity (Tideman et al. 2013).

3 Entrepreneurial Orientation: Driving Changes and Opportunities

Entrepreneurial Orientation refers to a firm's strategic posture that is defined by its propensity to innovate, take calculated risks, and act proactively in anticipation of future demands (Lumpkin and Dess 1996). The concept is rooted in the behavioural aspects of strategy and focuses on how organizations behave in the pursuit of new opportunities. A company with a strong EO is not reactive; it actively seeks to shape its environment through innovative decision-making.

Innovativeness, a core dimension of EO, reflects a firm's tendency to support new ideas, experimentation, and creative processes. This includes the development of new products, services, or processes that add value to customers and differentiate the organization in the market. Risk-taking involves the willingness to commit resources to initiatives with uncertain outcomes, a hallmark of entrepreneurial behaviour. Proactiveness indicates a forward-looking perspective, where firms anticipate and act on future needs rather than merely responding to current trends (Covin and Slevin 1989). Three dimensions, while essential for competitiveness, can create tension within an organization if not supported by appropriate leadership and cultural frameworks. A high EO without mechanisms to safeguard employee well-being can lead to burnout, anxiety, and high turnover rates. Therefore, EO must be balanced with leadership that values sustainability and a culture that nurtures and supports employees during periods of rapid change.

4 Sustainable Leadership: Ethical Stewardship for Long-Term Success

Sustainable leadership extends beyond traditional notions of effective leadership by incorporating ethical responsibility, stakeholder inclusion, and a long-term perspective (Hargreaves and Fink 2006). Sustainable leaders focus not only on financial performance but also on social and environmental considerations. They understand that long-term success is built on trust, integrity, and the well-being of all stakeholders, including employees, customers, communities, and the environment.

Key traits of sustainable leaders include transparency, inclusiveness, empathy, and ethical decision-making. They champion practices that ensure the health, safety, and professional development of their teams. Importantly, sustainable leadership encourages participative decision-making and values diversity of thought, which are essential for fostering a culture of innovation. In entrepreneurial organizations, sustainable leadership plays a crucial role in tempering the inherent risks of rapid innovation. While EO pushes boundaries and seeks bold opportunities, sustainable leadership ensures that such efforts are grounded in ethical practice and aligned with the organization's long-term vision. This balance is essential for cultivating trust among employees, managing uncertainty, and embedding resilience within organizational systems.

5 The Cultural Context: Foundation for Organizational Identity

Organizational culture refers to the shared beliefs, values, norms, and behaviours that define how things are done within an organization (Schein 2010). It represents the collective identity of an organization (Schein 2010). It represents the collective identity of an organization and significantly influences employee behaviour, motivation, and performance. Culture acts as a social control system, guiding employees on acceptable behaviour and decision-making.

A culture that supports innovation and sustainability encourages risk-taking, open communication, and continuous learning. Such a culture does not penalize failure but sees it as a stepping stone toward improvement. When employees operate in an environment where they feel psychologically safe, they are more likely to take initiative, share new ideas, and engage fully in their work (Edmondson 1999). Moreover, culture serves as the connective tissue that aligns entrepreneurial orientation with sustainable leadership. Without a supportive culture, even the most visionary leaders and the most entrepreneurial strategies can fail to achieve desired outcomes. A healthy culture promotes collaboration, trust, and mutual respect-all of which are essential for organizational agility and employee well-being.

6 The Synergistic Relationship: Innovation Meets Well-being

The convergence of EO, sustainable leadership, and a positive organizational culture creates a synergistic environment that supports both innovation and employee well-being. This triadic relationship is more than the sum of its parts; it creates a self-reinforcing system where each component strengthens the others.

- **Psychological Safety:**
  When employees feel safe to express themselves without fear of negative consequences, they are more likely to propose innovative ideas and take initiative. Sustainable leaders foster this safety by being approachable, supportive, and transparent (Kahn 1990).
- **Employee Engagement:**
  Engagement levels are higher when employees find meaning in their work, feel valued by their leaders, and operate in a culture that supports their personal and professional growth. This engagement fuels creativity and innovation (Bakker and Demerouti 2008).
- **Job Satisfaction and Retention:**
  A culture that supports autonomy, mastery, and purpose contributes to job satisfaction. When paired with the excitement and challenge of an entrepreneurial environment and the stability of sustainable leadership, retention rates improve (Moosmann 2009).
- **Organizational Resilience:**
  The combined effect of these three elements enables organizations to respond effectively to challenges and recover from setbacks. Resilience is built through shared purpose, collaborative spirit, and adaptive capacity (Lengnick-Hall et al. 2011).

7 Leadership, Authenticity, and Participatory Governance

Authentic leadership, characterized by self-awareness, moral integrity, and relational transparency, is a vital component of sustainable leadership (Avolio and Gardner 2005). Authentic leaders do not pretend to have all the answers; instead, they are honest about challenges and open to learning. This humility builds trust and strengthens team cohesion. Authentic leadership and participatory governance together create a culture of mutual respect and accountability. Employees are more likely to invest emotionally in their work and contribute to the organization's success when they feel seen, heard, and valued. Involving employees in decision-making processes, known as participatory governance, enhances organizational effectiveness. Employees who have a voice in shaping organizational effectiveness. Employees who have a voice in shaping organizational effectiveness. Employees who have a voice in shaping organizational strategy are more committed to its execution. This demographic approach to leadership not only empowers employees but also ensures that diverse perspectives inform strategic decisions, thereby enhancing innovation (Hassan et al. 2013).

8 Organizational Learning and Adaptive Capacity

In today's knowledge-based economy, continuous learning and adaptability are essential organizational competencies. Learning organizations prioritize knowledge acquisition, critical reflection, and experimentation (Senge 1990). They view mistakes not as failures but as opportunities for growth. Adaptive capacity-the ability to change in response to internal and external pressure, closely linked to innovation. Organizations with high adaptive capacity can pivot quickly, reallocate resources efficiently, reward initiative, and tolerate calculated risk-taking (Teece et al. 1997). When learning and adaptability are embedded into the organizational culture, they enhance the impact of EO and sustainable leadership. Leaders can make informed decisions, employees are better equipped to handle complexity, and the organization as a whole becomes more resilient and innovative.

9 Practical Application for Managers and Leaders

To operationalize the synergy between EO, sustainable leadership, and culture, organizations should consider the following actionable strategies:

- **Leadership Development:**
  Invest in programs that build sustainable and authentic leadership capabilities. Focus on ethical decision-making, emotional intelligence, and inclusive leadership.
- **Cultural Audits:**
  Regularly assess organizational culture to ensure alignment with strategic goals. Identify areas where culture may be hindering innovation or employee engagement.
- **Innovation Labs and Cross-functional Teams:**
  Create spaces where employees from different departments can collaborate on new ideas. Encourage experimentation and reward creativity.
- **Employee Wellness Programs:**
  Support mental, emotional, and physical well-being through comprehensive wellness initiatives. Demonstrate a commitment to holistic employee care.
- **Feedback Mechanisms:**
  Implement systems for regular feedback and dialogue. Encourage upward communication and involve employees in shaping organizational priorities.

These Practices not only enhance organizational performance but also cultivate a workplace where people thrive.

10 Challenges and Barriers to Integration

Despite its benefits, integrating EO, sustainable leadership, and culture is not without challenges. Significant barriers may arise from reluctance to change, competing agendas, and a misalignment of leadership and culture. Furthermore, organizations may prioritise short-term performance pressures over long-term innovation and well-being to achieve instant profits. Inconsistent leadership can significantly undermine cultural integrity and erode trust within an organization. When leaders say one thing and do another, it creates confusion and weakens employee confidence. These contradictions often appear in mixed messages during a change initiative, a lack of follow-through on commitments, or shifting priorities without clear communication. Such behaviours leave employees uncertain about expectations and sceptical of leadership intentions. To resolve these problems, the leaders should strive for alignment between their words and actions, communicate transparently, and be accountable for their decisions as building trust requires or needs consistency, clarity and a genuine commitment to organizational values.

Mixed signals discourage innovation:
For instance, a leader may encourage employees to take initiative and innovate, yet penalize them when ideas don't succeed. Because of this paradox, employees are not willing to take risks and they are more creative that helps them in playing safe rather than try new things and come up with different ideas.

- **Credibility gaps undermine engagement:**
  Entrepreneurs frequently convey ideals like cooperation, diversity, and inclusivity. However, when these values are not reflected in actual policies or daily practices—such as unequal opportunities, biased decision making, or lack of employee voice—it creates credibility gaps. Employees notice when birds and actions don't match, and their engagement often declines as a result.
- **Lack of alignment weakens culture:**
  When did she behaviours are inconsistent with the stated cultural values, it sends mixed messages across the organization. Overtime, this can lead to weekend the shared sense of purpose that a strong culture provides.
  To overcome these challenges, organizations must make a conscious effort to maintain consistency between leadership behaviour, strategic direction, and cultural values. This requires:
- **Clear and Honest communication:**
  Leaders must clearly articulate the rationale behind decisions and openly address failures or challenges. Transparency fosters trust and help emphasises understand the bigger picture.

- **Leading by example:**
  Leadership should consistently model the behaviours and values they expect from others. It adds answerability, appreciating efforts and hard work, and respecting employees equally and fairly on the basis of their efforts.
- **Integrating principles into actions:**
  Living out organizational values means going beyond just putting them in a mission statement. These values should be actively reflected in how the organization hires, evaluates performance, develops teams and recognizes achievements.
- **More focus towards Building a learning culture:**
  Companies should emphasis more on learning from both the success and the failure as we learn from our mistakes. When we treat failure as a chance for growth, employees are more willing to take initiative and also play a major role in contributing innovative and creative ideas.

When there's consistency between what an organization says, does and truly values, it helps build a strong, unified culture. This alignment creates a workplace where trust grows, innovation is encouraged, and employees feel supported and motivated.

11 Future Research and Emerging Trends

The way that EO, sustainable leadership, and culture interact will change as the workplace develops further. New opportunities and difficulties are brought about by the workforce's generational transition, digital transformation and the growth of remote work.

On the basis of emerging trends and challenges, future directions can be monitored by some of the following important points:

- **Virtual work environment:**
  Examine how remote and hybrid work settings affect psychological safety, employee engagement, and team cohesion within entrepreneurial organizations.
- **Digital Leadership and Authenticity:**
  To figure out how virtual communication or interaction platforms impact transparency, authenticity and employee trust
- **Cross-Cultural Contexts:**
  Studies could focus on how national culture moderates the effect of sustainable leadership on innovation and well-being.
- **Longitudinal Impact studies:**
  Conduct long-term investigations into the outcomes of integrated EO-leadership-culture frameworks on innovation output, employee satisfaction, and organizational adaptability.
- **Purpose-Driven Business Models:**

Purpose-oriented strategies influence EO and sustainable leadership, especially in sectors driven by social entrepreneurship and environmental missions.

- **Environmental and Social Governance (ESG) Integration:**
  Investigate how organizations incorporate environmental stewardship, ethical practices, and social equity into their EO and leadership strategies.
- **Impact of Generational Diversity:**
  Examine how generational differences in values and work expectations influence the dynamic among EO, leadership, and culture, and how organizations can tailor strategies accordingly.
- **AI and technological influence:**
  Though not the focus of this study, future research could also analyse how emerging technologies like AI and automation reshape organizational culture and leadership dynamics, particularly in entrepreneurial ventures.
- **Psychological resilience in crisis:**
  Assess how organizations with integrated strategies fare during crisis (e.g., economic downturns, pandemics) in terms of maintaining innovation, resilience, and employee well-being.
- **Sustainable HR practices:**
  Investigate the role of human resource management in embedding sustainable leadership and cultural values that shape innovation.
  By addressing these directions, future research can deepen understanding of how businesses can sustain innovation and human well-being in a rapidly changing world. These insights can also inform policy, leadership development, and organizational design for long-term, ethical success.

12 Conclusion

Entrepreneurial orientation, sustainable leadership, and organizational culture are interwoven pillars that together form a robust foundation for innovation, employee well-being, and long-term organizational success. Based on the analysis presented throughout this chapter, several key conclusions can be drawn:

- **Strategic Alignment Enhances Outcomes**
  Innovation and human-centered values coexist in a system that is reinforced when EO, sustainable leadership, and culture are purposefully integrated. Businesses that can effectively strike a balance between these factors can explore innovation concepts while upholding moral standards and worker satisfaction.
- **Employee well-being as a strategic Imperative**
  Without considering the human elements of organizational life, innovation cannot be sustained. Psychological safety, work happiness, and engagement are not only the outcomes of effective leadership and culture, but they also serve as drivers for continued innovation.
- **Leadership as a Cultural Architect:**

Leaders have an important role in shaping and preserving culture. Whether the culture fosters or inhibits entrepreneurial endeavors is determined by honesty, moral integrity, and inclusivity. Thus, emotional intelligence and long-term thinking should be the primary aims of leadership development.

- **Organizational culture as the Anchor:**
  Culture offers change. When firms try out new methods, a strong, values-driven culture acts as a compass. It guarantees that, even as businesses grow, their essential identity stays consistent.
- **Adaptability as a Competitive Advantage:**
  Organizations must be nimble due to the rapid rate of change in today's world. Companies may improve their ability to pivot and expand by cultivating a culture of learning and adaptation, as well as empowering teams via participatory governance.
- **Long-Term Vision Over Short-Term Gains:**
  Organizations that prioritize rapid gains may neglect the long-term worth of employee loyalty, brand reputation, and stakeholder trust. Strategies that promote long-term viability and social responsibility lead to sustainable success.

Finally, by combining entrepreneurial orientation, sustainable leadership, and an adaptive organizational culture, businesses may not only survive in today's complicated environment but also lead with purpose and vision. Businesses that embrace this synergy foster settings in which people are motivated, ideas are nourished, and long-term value is created—for workers, consumers, and society as a whole.

References

Avery GC, Bergsteiner H (2011) Sustainable leadership practices for enhancing business resilience and performance. Strateg Leadersh 39(3):5–15

Avolio BJ, Gardner WL (2005) Authentic leadership development: getting to the root of positive forms of leadership. Leadersh Q 16(3):315–338

Bakker AB, Demerouti E (2008) Towards a model of work engagement. Career Dev Int 13(3):209–223

Bornstein D (2007) How to change the world: social entrepreneurs and the power of new ideas. Oxford University Press

Covin JG, Slevin DP (1989) Strategic management of small firms in hostile and benign environments. Strateg Manage J 10(1):75–87

Covin JG, Slevin DP (1993) A response to Zahra's "critique and extension" of the Covin-Slevin entrepreneurship model. Entrep Theory Pract 17(4):23–28

D'amato A, Henderson S, Florence S (2009) Corporate social responsibility and sustainable business. In: A guide to leadership tasks and functions, vol 102

Dixon S, Irshad H, Pankratz DM, Bornstein J (2019) The 2019 Deloitte City mobility index. Deloitte Insights 18

Drucker PF (1985) Entrepreneurial strategies. Calif Manage Rev 27(2):9–25

Edmondson A (1999) Psychological safety and learning behavior in work teams. Adm Sci Q 44(2):350–383

Gupta V, MacMillan IC, Surie G (2004) Entrepreneurial leadership: developing and measuring a cross-cultural construct. J Bus Ventur 19(2):241–260

Hakanen JJ, Schaufeli WB, Ahola K (2008) The job demands-resources model: a three-year cross-lagged study of burnout, depression, commitment, and work engagement. Work Stress 22(3):224–241

Hargreaves A, Fink D (2006) Redistributed leadership for sustainable professional learning communities. J Sch Leadersh 16(5):550–565

Hassan S, Mahsud R, Yukl G, Prussia GE (2013) Ethical and empowering leadership and leader effectiveness. J Manage Psychol 28(2):133–146

Hisrich RD, Shepherd DA (2017) Entrepreneurship. McGraw-Hill Education

Hofstede G (2011) Dimensionalizing cultures: the Hofstede model in context. Online Read Psychol Cult 2(1):8

Kahn WA (1990) Psychological conditions of personal engagement and disengagement at work. Acad Manage J 33(4):692–724

Kelloway EK, Barling J (2010) Leadership development as an intervention in occupational health psychology. Work Stress 24(3):260–279

Lengnick-Hall CA, Beck TE, Lengnick-Hall ML (2011) Developing a capacity for organizational resilience through strategic human resource management. Hum Resour Manage Rev 21(3):243–255

Lumpkin GT, Dess GG (1996) Clarifying the entrepreneurial orientation construct and linking it to performance. Acad Manage Rev 21(1):135–172

Moosmann F, Pink O, Stiller C (2009) Segmentation of 3D lidar data in non-flat urban environments using a local convexity criterion. In: 2009 IEEE intelligent vehicles symposium. IEEE, pp 215–220

Ryan RM, Deci EL (2001) On happiness and human potentials: a review of research on hedonic and eudaimonic Well-being. Annu Rev Psychol 52(1):141–166

Schein EH (2010) Organizational culture and leadership, vol 2. John Wiley & Sons

Senge MP (1990) Quinta disciplina. Best Seller, São Paulo

Shane S, Venkataraman S (2000) The promise of entrepreneurship as a field of research. Acad Manag Rev 25(1):217–226

Teece DJ, Pisano G, Shuen A (1997) Dynamic capabilities and strategic management. Strateg Manag J 18(7):509–533

Tideman SG, Arts MC, Zandee DP (2013) Sustainable leadership: towards a workable definition. J Corp Citizsh 49:17–33

Wan B, Xiao W, Zhang L, Windley BF, Han C, Quinn CD (2011) Contrasting styles of mineralization in the Chinese Altai and East Junggar, NW China: implications for the accretionary history of the southern Altaids. J Geol Soc 168(6):1311–1321

Zahra SA, Covin JG (1995) Contextual influences on the corporate entrepreneurship-performance relationship: a longitudinal analysis. J Bus Ventur 10(1):43–58

From Challenges to Change: Navigating Sustainable Practices in Municipal Infrastructure and Services

Priya Bijalwan, Ashulekha Gupta, Navneet Rawat, and Rajat Rawal

Abstract Purpose—The chapter aims to explore sustainable municipal service and infrastructure delivery practices in the Global South where institution limitations, climate vulnerability, and urbanization render efficient and equitable governance complicated. It aims to evaluate prominent delivery models, argue governance issues, and discover emerging practices allowing for resilience, inclusivity, and environmental sustainability in fast urbanizing contexts. **Design/methodology/approach**—A qualitative, multi-phase approach is employed integrating bibliometric and thematic analysis using Biblioshiny, comparative case study research in India, Brazil, Malaysia, and South Africa, and policy analysis framework. The triangulated approach allows for the identification of the prominent patterns of governance, measures of sustainability, and challenges of implementation across different urban settings. **Findings**—The research finds three thematic clusters at the core—waste and environmental renewal, circular economy integration, and water and urban infrastructure management. It explores four prevailing service models: public sector-driven, public-private partnerships (PPPs), community-driven, and digital/smart city. The research finds technical innovation is required but success relies significantly on adaptive governance, institutional capacity, and inclusive process. **Research limitations/implications**—The chapter places in perspective the demand for additional empirical studies, i.e., comparative smart city deployment and digital governance initiative assessment across various regional contexts. Secondary data and sparse case

P. Bijalwan
School of Management and Commerce Studies, Shri Guru Ram Rai University, Dehradun, Uttarakhand, India
e-mail: priyabijalwan@sgrru.ac.in

A. Gupta (✉) · N. Rawat · R. Rawal
Department of Management Studies, Graphic Era (Deemed to be University), Dehradun, Uttarakhand, India
e-mail: ashulekha1806@gmail.com

N. Rawat
e-mail: rawat.navneet@geu.ac.in

R. Rawal
e-mail: rajatrawal.phd@geu.ac.in

A. Singh et al. (eds.), *Human-Centered Sustainability: Bridging Well-Being, Gender Equity, and Technology*, World Sustainability Series,
https://doi.org/10.1007/978-3-032-22020-2_17

studies may limit generalizability, requiring more intense, participatory, and longitudinal research. **Practical implications**—Policy recommendations are to increase municipal autonomy, encourage green finance instruments, combine digital platforms with participatory planning, and enhance inter-sector coordination. Techniques such as optimization models and water circularity indices demonstrate the relevance of evidence-based governance to urban sustainability and resilience. **Originality/value**—This chapter presents an interdisciplinary framework linking urban governance, digital innovation, and sustainability. It contributes a unique synthesis of empirical evidence and conceptual insights, offering a roadmap for city planners, policymakers, and researchers aiming to navigate sustainable municipal transitions in diverse and challenging environments.

Keywords Sustainable practices · Municipal services · Public-private partnerships (PPPs) · Governance model · Global south

1 Introduction

Sustainable municipalities play a pivotal role in fostering economic growth, social equity, and environmental protection, as highlighted in various studies Mensah 2019). A comprehensive sustainability model for municipalities emphasizes the integration of sustainability into operations, stakeholder engagement, and compliance with legislative frameworks (Brink and Singh 2024). A sustainable municipality can be articulated as a local governance entity that provides essential services such as water, waste management, energy, and transportation through inclusive, resilient, and resource-efficient frameworks, whilst simultaneously preserving ecological integrity and advancing participatory governance (UN-Habitat 2020; Bulkeley and Castán Broto 2013). Contemporary urban expansion is unparalleled in both magnitude and intricacy. Cities situated in the Global South are particularly burdened as they encounter a multitude of challenges: surging populations, the proliferation of informal settlements, constrained fiscal autonomy, antiquated infrastructure, and an augmented vulnerability to climate-induced disasters (Parnell and Robinson 2012; Satterthwaite 2016). These municipalities frequently operate under conditions of inadequate funding and excessive demands, existing within governance frameworks characterized by fragmentation, political instability, and limited innovative capacity. The imperative to deliver fundamental services is exacerbated by the rising expectations for smarter, greener, and more accountable urban systems (Angelo and Wachsmuth 2020).

The involvement of the private sector in municipal services, including waste management, contributes even more to governance intricacy, requiring a sophisticated level of understanding accountability and corporate social responsibility for promoting urban sustainability (Hughes and Peterson 2018). Cities are also being viewed as increasingly interdependent social-ecological-technological systems, meaning there is a need to have coordinated governance to achieve more comprehensive transformations, along with enhancing resilience to the impacts of climate change (Krueger et al. 2022). Integrated governance necessitates cross-sectoral coordination, harmonizing policies related to housing, energy, transportation, and waste management (OECD 2020). Equitable governance entails the active involvement of vulnerable populations—who are often the most adversely affected by service deficiencies and climate perturbations in the planning and decision-making processes (Mitlin 2008; Watson 2009). Adaptive governance acknowledges the necessity for flexible, learning-oriented systems that can evolve in response to dynamic conditions and citizen input (Gerlak and Heikkila 2024).

This chapter delves into the methodologies employed by municipalities, particularly in the Global South, in confronting the sustainability challenge. It commences with a literature review that contextualizes the concept within current dialogues in public administration, sustainability science, and urban studies. Thereafter, an overview of methodology is given that describes the chapter's dependency on thematic analysis and example-based cases. Most of the chapter critically appraises four well-known service delivery models: traditional public sector practices, public-private partnerships (PPPs), community-based, and innovations featuring digital or smart city paradigms. All the models are discussed in terms of the strengths, weaknesses, and actual usage in the real world. This is preceded by an argument on principal implementation issues and a set of policy and practical recommendations. The chapter concludes with a whole-of-government approach to guiding the municipal service shift to sustainability.

2 Literature Review

Urban sustainability has emerged as a central issue of contention within the global development debate, especially with regard to rapid urbanization and growing environmental pressures (Alizadeh et al. 2024). Theoretical frameworks including resilience, circular economy, and Sustainable Development Goals (SDGs) have significantly impacted academic research and policy formulation for the future of urban spaces (Puntillo 2023). Specifically, Goal 11—"Sustainable Cities and Communities"—emphasizes the need to create inclusive, safe, resilient, and sustainable cities and thus promotes transformative governance, infrastructure, and citizen engagement approaches (UN-Habitat 2020). Resilience theory, which originates in ecological systems thinking, is used in urban contexts to underscore the capacity of urban systems to buffer stressors, respond to various disturbances, and transform in the presence of persistent threats (Meerow et al. 2016). With this comes the paradigm of circular economy, which facilitates closed-loop material loops that reduce waste and promote material recycling in urban-level consumption and production systems (Ellen MacArthur Foundation 2019). In governance, a shift can be seen from traditional bureaucratic forms of governance towards cooperative and multi-actor forms. The conventional models of centralized power, formalized processes, and hierarchical bureaucracies have been faulted for their incapacity and rigidity in facing the complexity of city issues (Pierre and Peters 2000). This contrasts with collaborative models of governance, which emphasize openness, inclusivity, and co-production and thus engage private organizations, neighborhoods, and civil society in decision-making (Ansell and Gash 2008).

It has been noted in the Global South from academic research that challenges in service delivery stem both from historic disparities and structural constraints. Shortages in foundational infrastructure—water supply, sanitation, electricity, and waste management—are also complemented by the growth of slums where delivery of services is irregular or non-existent altogether (Satterthwaite 2016; Mitlin and Walnycki 2020). These urban neighborhoods are often relegated to the periphery in formal governance arrangements, leading to a bifurcated urban governance system that denies large sections of the population equal access to amenities. Researchers have indicated how such structural inequalities increase social exposure and constrain adaptability in climate change (Tschakert et al. 2013). The advent of digital technologies and the discursive framework of smart cities has added new opportunities—and attendant tensions—into the world of urban governance. Smart city technologies such as IoT networks, real-time data platforms, and AI-based infrastructure management systems have been touted to maximize urban systems, enhance service efficiency, and enable data-driven decision-making (Batty et al. 2012). On the other hand, there have been criticisms about the digital divide, danger posed by surveillance, and exclusion of people who are poor from technology-facilitated services (Datta 2015; Söderström et al. 2014). In the Global South, where connectivity is patchy and institutional capacity is generally weak, unreflective adoption of smart city paradigms threatens to amplify existing prevalent urban inequalities.

One of the significant lacunas in the literature is that there is not enough integration of digital governance innovations with participatory planning. Most smart city literature is excessively fixated on technical efficacy and infrastructure modernization, inclined to overlook the social and political aspects of urban transformation. Conversely, participatory urban planning literature focuses on empowerment, deliberation, and fairness but is typically divorced from digital governance innovations (Bittencourt et al. 2024). This chapter contends that sustainable cities must integrate both—utilizing technology not as a silver bullet, but as a tool nested within participatory, place-oriented, and adaptive systems of governance.

3 Methodology

The chapter employs a three-step qualitative methodology in analyzing sustainable municipal infrastructure and practice of service delivery within the framework of cities in the Global South. The methodology is a mix of bibliometric analysis, synthesis based on cases, and policy analysis in exploring the evolving landscape of municipal governance systems. During the first phase, pertinent studies in the literature were searched systematically on the Scopus database, which maintains a complete record of peer-reviewed studies in all disciplines. The search strategy involved keywords and Boolean operators like "sustainable urban infrastructure," "municipal services," "urban governance," and "Global South," with the expectation of gathering a wide and diverse set of publications concerning city-level service delivery sustainability.

Once the relevant literature was accessed, two web-based software applications, Biblioshiny (a web interface to the R package Bibliometrix), were used to conduct bibliometric and thematic analysis (Aria and Cuccurullo 2017). Biblioshiny also permitted an investigation of source influence, author activity, and temporal trends in the literature to enhance our knowledge of research development and direction on this topic. The flow chart diagram in Fig. 1 represents the keyword search parameter technique. Having these results as a foundation, the second stage was towards identification of case study from Global South countries, as presented in the reduced pool of literature. Each article was scrutinized on a careful basis to provide relevant contextual and empirical information. Key variables were coded, such as country and city or region of focus, type of municipal service or infrastructure concerned (e.g., water provision, solid waste, or transport), sustainability-related issues and innovations that arose, and governance arrangements in place. These were arranged in a comparative matrix to enable systematic analysis across cases. This matrix facilitated the creation of patterns, including chronic problems in slums, pressures from donor agencies, and varying roles of community involvement in sustainable urban provision of services (Parnell and Pieterse 2010; Watson 2009). The third and final stage entailed qualitative policy analysis of the municipal reactions and institutional arrangements identified in the case studies. The chosen literature was consulted to obtain policy-relevant data, e.g., verbatim extracts of national or municipal legislation, institutional frameworks, and sustainability planning tools. The study sought to determine how policy design and implementation reflect the tenets of sustainable urban development, i.e., equity, resilience, and efficiency (UN-Habitat 2020). Attention was also given to recording the impact of governance dynamics—decentralization, inter-agency coordination, and public-private partnerships—on the success or failure of these sustainability initiatives (Bulkeley and Betsill 2005; Satterthwaite 2016). Triangulating bibliometric research, case study observation, and policy text, this chapter provides a three-dimensional picture of how Global South cities are managing daunting sustainability imperatives within a context of urbanization and institutional limitations.

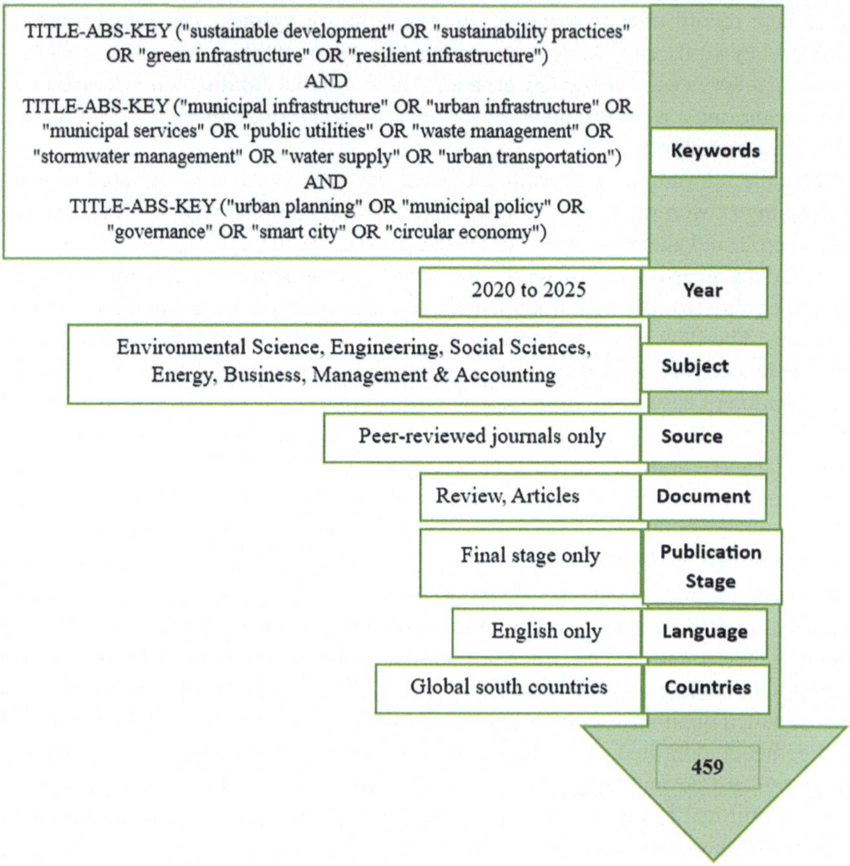

Fig. 1 Flow search parameter for thematic analysis

4 Results and Findings

4.1 Thematic Analysis

With the help of Biblioshiny Thematic network of keywords were identifies as shown in Fig. 2. Based on the keyword network and corresponding document analysis, the themes within the literature on environmental management and urban sustainability in the Global South coalesce around three major clusters (Van Eck NJ et al., 2010) (Table 1):

Theme 1: Waste Treatment, & Environmental Treatment—This thematic category holds research focused on the sustainable management of industrial, municipal, and hazardous waste considering treatment technology, recovery operations, and environmental health. Venes et al. (2025), for example, examined the socio-political

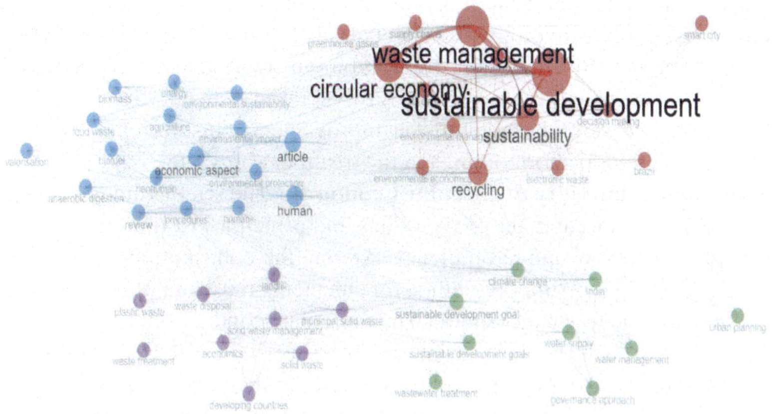

Fig. 2 Thematic map network of keywords

Table 1 Four primary thematic clusters identified based on color

Color	Theme focus	Dominant keywords
red	Waste Management & Environmental Treatment	Waste, biomass, biofuel, nonhuman, bioremediation, wastewater treatment, effluents, industrial waste, nitrogen, ph, chemistry, carbon, greenhouse gases
green	Sustainability & Circular Economy	Sustainable development, recycling, circular economy, economic aspect, life cycle assessment, solid waste, electronic waste, energy utilization, supply chain, environmental economics, governance, policy, city
blue	Water Management & Urban Infrastructure	Water supply, water conservation, urban planning, smart city, urbanization, decision making, India, Brazil, developing countries
purple	Solid Waste Disposal & Landfill Management	Landfill, disposal, waste disposal, solid waste management, municipal solid waste, plastic waste, waste treatment, management goal, developing countries

aspects of municipal solid waste collection in Brazil and unearthed the relationship between the governance of services and the efficiency of waste. Gummadi et al. (2025) proposed a novel method by employing rice-husk derived zeolites for the removal of heavy metals from wastewater and green alternatives in environmental remediation. In the same manner, James et al. (2024) published behavioral dynamics in an article whose gap between environmental consciousness and waste-minimization practices, especially consumer behavior, was being researched. Moving on to green chemistry, Singh et al. (2024) discussed how advances in low-impact chemical process and biodegradable material technology are re-shaping industrial production sustainability avenues. In biotechnology, Baloch et al. (2024) discussed the employment of Arthrospira platensis (Spirulina) to carry out bioremediation and recovery of bioresource from contaminated water systems. Lastly, Prasad et al. (2025) gave a study on urban sanitation interventions targeting the improvement of institutional coordination to handle infrastructural shortcomings in the treatment of waste.

Theme 2: Sustainability & Circular Economy—The second thematic category is the integration of circular economy values and principles to limit environmental footprint and enhance resource efficiency. Among the contributions from Almeida et al. (2025) is a decision-support method for enhancing reuse of tailings run-of-mine waste, with industrial waste incorporated into circular economy systems. Transportation also appears to be the greatest sustainability issue—Jain et al. (2025) talked about how green transport systems can help limit climate change effects in the urban environment (Jain et al., 2025). Also, Gummadi et al. (2025) examined the recycling of compounds obtained from rice husk for heavy metal removal, emphasizing the co-environmental and economic benefits of circular resource loops. James et al. (2024) placed emphasis on the urgent awareness-action gap on the environment, considering that successful circular action relies on consumer behavior change. Venes et al. (2025) analyzed the performance of municipal solid waste collection in Brazilian cities, with a suggestion that circularity can be boosted by more efficient systems of governance. Lastly, Khan et al. (2025) discussed the application of bentonite and fly ash in de-polluting contaminated soil, endorsing sustainable land reclamation practices.

Theme 3: Water Infrastructure & Urban Planning—This topic gives priority to increasing the significance of inter-dependent water networks and resilient urban infrastructure in facing environmental and sustainable development challenges. Prasad et al. (2025) addressed inter-coordination of sanitation planning in urban South Asia, where greater inter-agency coordination was needed to efficiently manage waste and water. With this, Kapoor et al. (2025) utilized hydrological modeling in urban planning and came up with sophisticated measures in stormwater management and urban flooding. Nielsen et al. (2025) incorporated socioeconomic elements into planning water infrastructure resilience through emphasizing the coordination of investments in water infrastructure and community needs. Furthermore, Jain et al. (2025) examined city transport systems in the broader context of infrastructure sustainability, connecting mobility-water nexus in urban planning. Mendoza et al. (2024) also gave bibliometric reviews of circular systems for infrastructure, emphasizing the main challenges faced in designing drainage and water reuse systems for urban areas. Finally, Oliveira et al. (2025) evaluated digital preparedness in circular infrastructure, emphasizing smart technologies used in monitoring water and city utilities.

Theme 4: Solid Waste Disposal & Landfill Management—The third thematic cluster deals with the aspects of policy and operations of solid waste disposal, landfill systems, and waste handling at the municipal level. Papers in this sector revolve around the waste management techniques that take place after waste has already been generated, such as the efficiency in disposal, landfill optimization, and governance of municipal solid waste. To illustrate, Oliveira et al. (2025) demonstrate the increasing relevance of integrated solid waste management frameworks which blend landfill engineering with material recovery strategies. At the same time, Mendoza et al. (2024) point out the environmental effects of disposing of waste improperly, especially in fast developing urban areas where the infrastructure capacity is still limited. Khan et al. (2025) explore environmentally friendly landfill practices, with their research, among other things, dealing with the control of leachate, management of the site, and circular recovery opportunities within waste disposal systems. Besides, Prasad et al. (2025) brings to light the institutional and planning issues related to municipal solid waste handling and recognizes that urban authorities' effective coordination to be the key factor in improving disposal outcomes. Altogether, this third cluster is indicative of a strong research focus on the improvement in efficiency, environmental safety, and governance of solid waste disposal systems to be a necessary complement to upstream waste reduction and recycling strategies.

4.2 Case Study Analysis

For the purpose of thematic findings of bibliometrics and keyword co-occurrence, the subsequent section offers a case study of in-depth on real practice of sustainable urban infrastructure practices in the global south. The chosen case illustrates integrative practice in the form of the themes of Sustainability & Circular Economy, Waste and Environmental Treatment, and Water Management & Urban Infrastructure. Four diverse cases each from both nations have been considered from the period of 2020–2025.

Case 1: India—Arora et al. (2022) present a case study approach to methodologically examine urban water circularity of the city of New Delhi, India. The study is focused on the urban water management industry to learn how cities can develop from linear to circular systems of water use through reuse, recycling, and demand reduction practices. A new water circularity index (WCI) measures current performance and potential for improvement through the supply chain for water—extraction to use, distribution, treatment, and recycling. It points to key sustainability issues, including excessive dependence on centralized supply systems, draining groundwater, and poor wastewater reuse. Governance and institutional complexity are fundamental concerns, with the research identifying that multi-responsibility of municipal authorities and data limitations hinder circular water practices. Major conclusions indicate that more effective policy coordination, investment in reuse facilities, and improved performance metrics can contribute immensely to enhancing water circularity in resource-constrained cities. It is a diagnosis and planning tool used for deployment in all cities of the Global South with fast-growing water demands (Arora et al. 2022).

Case 2: Brazil—The Manaus, Brazil, case study discusses strategies towards promoting a circular economy in regional construction and demolition (C&D) waste management. Located in an urbanizing Amazonian city with logistics and environmental issues at hand, the study enumerates the major challenges to sustainable regulation of waste as informal recycling, inadequate regulatory enforcement, and a fractured value chain. With the aid of a mix of policy analysis and stakeholder interviews, authors suggest a strategic plan with focal points being the necessity of regional planning integration, recycling material incentives, and institution of C&D waste recycling networks. Usually, the research calls for cooperation between urban planning, environmental policy, and the private sector to facilitate circular material flows. These findings apply directly to Global South cities that share comparable problems of infrastructure growth, informal economies, and ecological exposure (Oliveira et al. 2021).

Case 3: Malaysia—A Malaysian case study, Ooi et al. (2021), offers a multi-objective optimisation model to advance municipal solid waste (MSW) management that equilibrates economic and environmental objectives. It analyses various waste treatment routes—landfill, recycling, composting, and energy recovery—over several regions, achieving maximum cost savings and greenhouse gas reduction. Guilt-stricken in Malaysia's recent waste experience, with its extensive use of landfilling and growing urban waste generation, the model indicates the potential of comprehensive solutions that are capable of lowering environmental impact at an elevated level of efficiency. The research establishes how these evidence-based tools are able to enable evidence-based policy decisions and guide the shift towards sustainable waste systems, especially in rapidly emerging economies with strained city infrastructure (Ooi et al. 2021).

Case 4: South Africa—Bloemfontein, South Africa, case study explores the intersection of public health and urban infrastructure in marginalized communities. In a city with spatial inequality and lack of infrastructure, the study looks at how effective networks and community innovation can manage public health vulnerabilities, especially in informal settlements. The authors record interventions that integrate improved sanitation, waste collection, and access to clean water with local hotspots of innovation and inclusive systems of governance. By establishing immediate connections between infrastructure planning and health, the case illustrates the merits of cross-sectoral coordination and adaptive planning frameworks carefully designed to suit poor urban settings. It reaffirms that not just technical upgradation but also resilience-promoted, inclusive types of governance are necessary to attain sustainable development objectives in cities (Rammile et al. 2024).

4.3 Policy Analysis

To advance an understanding of how cities in the Global South address sustainability imperatives, the current section presents a comparative policy examination of four countries—India, Brazil, South Africa, and Malaysia—each of which reflects different governance forms and developmental trends. The examination is based on municipal sustainability policies, institutional frameworks, and implementation tools for urban infrastructure, waste, and climate resilience. Table 2 demonstrates the policy analysis of four south global counties.

Table 2 Policy analysis of case studies in global south cities

Country	Policy	Implementation	Challenge
India	Urban water circularity framework (water circularity index—WCI)—New Delhi	Application of a novel WCI to evaluate water reuse, recycling, and demand reduction. Diagnostic tool for policy design across the water supply chain	Fragmented institutional responsibilities, over-reliance on centralized supply, groundwater depletion, and limited wastewater reuse capacity. Needs improved data access and coordination mechanisms (Arora et al. 2022)
Brazil	Circular economy in Construction & Demolition (C&D) waste—Manaus	Integrated regional planning model emphasizing formalization of recycling networks, regulatory alignment, and public-private collaboration for C&D material reuse	Informal waste handling, weak enforcement, fragmented policy environment, and low private-sector engagement (Oliveira et al. 2021)
Malaysia	Multi-objective optimization for municipal solid waste (MSW) management	Use of optimization models to evaluate landfill, composting, recycling, and energy recovery trade-offs. Supports low-carbon and cost-efficient waste infrastructure strategies	High dependence on landfilling, growing urban waste, limited capacity for integrated MSW solutions, and policy inertia despite data availability (Ooi et al. 2021)
South Africa	Resilient urban infrastructure for public health—Bloemfontein	Community-led governance and infrastructure co-design for sanitation, water, and health services in informal settlements. Prioritizes health-security linkages and participatory resilience planning	Deep spatial inequalities, under-resourced municipalities, coordination gaps, and vulnerable urban health systems (Rammile et al. 2024)

5 Discussion

This chapter shows that a sustainable municipal infrastructure and services in the Global South is a multifaceted undertaking—and one determined by environmental imperatives, governance complexity, and capacities for innovation. Thematic analysis yielded four salient clusters of sustainability: (1) Waste Treatment & Environmental Remediation, (2) Sustainability & Circular Economy, (3) Water Management & Urban Infrastructure, and (4) Solid Waste Disposal & Landfill Management. Each theme represents varying but interrelated priorities among cities facing rapid urbanization and environmental strain. Particularly, circularity and tech integration were common ambitions, but usually defeated by institutional fragmentation and uneven capability.

Drawing on these strands, the case studies offered real-world examples of how cities interpret sustainability principles as context-dependent practices. In New Delhi, using a Water Circularity Index illustrated the effectiveness of diagnostic tools in managing water reuse and governance fragmentation (Arora et al. 2022). Manaus depicted the issue of introducing circularity to construction waste systems, and private sector resistance, policy gaps, and informal recycling are slowing it (Oliveira et al. 2021). Optimization models in Malaysia presented a technical pathway to improving the management of solid waste in a balanced way with environmental and economic objectives (Ooi et al. 2021). Alternatively, Bloemfontein prioritized participatory governance and health, necessitating infrastructure resilience also to be based on equity and inclusion (Rammile et al. 2024). These examples demonstrate that transitions necessitate technology higher; they necessitate contextualized, adaptive, and socially embedded institutional arrangements for governance. Comparative policy analysis also highlighted the divergent pathways of institutional design and delivery modalities in the four nations.

India and Malaysia have most to do with technocratic, centralized solutions, but Brazil and South Africa have more decentralized, inclusive models. But all suffer from chronic policy fragmentation, fiscal reliance, and inter-agency coordination gaps. But they all have the need to more comprehensively integrate the urban policy arenas of water, waste, energy, and public health into a overarching, long-term sustainability strategy. Taken as a whole, the thematic, empirical, and policy analyses provide a full picture of urban sustainability in the Global South. They propose that adaptive and equitable innovations in municipal systems need to combine innovation and inclusiveness, central policy and local adaptability, and technical efficiency and social responsiveness.

6 Implication

The results of this chapter are pertinent to policymakers, urban planners, and development practitioners who are working toward constructing sustainable and resilient municipal systems in the Global South. First, the three theme clusters of circular economy, waste treatment, and water infrastructure shed light not just on the complexity of the priorities for sustainability but also how interdependent they are. They need integrated solutions because isolated approaches will embed inefficiencies and sabotage systemic resilience. Cities need to appreciate the intersectionality of infrastructure networks as well as the cross-cutting advantage of co-managed services. Second, the case studies underscore the importance of context-adaptive and context-sensitive approaches. Mechanisms such as India's Water Circularity Index or Malaysia's optimization models show the value of data-driven planning. These mechanisms, however, are only as effective as the institutions that back their application. This then leads us once more to the significance of local capacity-building via investment in institutional coordination, data systems, and human capital. Third, the policy analysis confirms that while national and provincial models provide significant direction, implementation success is dependent on municipal autonomy, local imagination, and true public participation. Bloemfontein demonstrates the ability of participatory government to address service quality and equity concerns. Too centralized models, however, can limit responsiveness and fail to address appropriately the needs and vulnerabilities of the surrounding environment.

7 Conclusion

his chapter draws attention to the pressing and cumulative nature of sustainability attainment in urban infrastructure and services, especially in Global South cities. From the synthesized assessment of thematic papers, empirical case studies, and policy contexts, it is evident that technological advancement, circular economy principles, and green remediation hold great potential but that these need to be based on adaptive, inclusive, and context-sensitive governance structures. The widespread concerns—ranging from institutional disintegration and restricted fiscal autonomy to infrastructural weaknesses and virtual gaps—demonstrate that one model or policy prescription is not adequate. Rather, sustainable change rests on resolving local ingenuity with national designs, technological savvy with social planning, and short-term resilience with long-range vision. The findings of the study are bound to sustain that healthy municipal sustainability is multi-faceted in scope and demands cross-sectoral linkages and strong local governance. Diagnostic instruments such as Malaysia's optimization models or India's Water Circularity Index show the potential of evidence-based solutions, but these are still contingent on strong institutional capacity and public participation. Likewise, comparative cases show that distributive service delivery cannot entirely be reliant on central planning; instead, it requires decentralization, accountability, and co-production of solutions with poor populations.

Sustainable city government is not nearly so much a technical or financial matter—it is a socio-political necessity. This chapter suggests that municipal sustainability happens through a mixed strategy: one that combines digital innovation with people's participation, builds inter-agency alignment, and makes planning infrastructure about equity. For policymakers, city planners, and practitioners, the future lies in establishing institutions that can learn, change, and transform based on the changing requirements of their cities.

Limitations and Future Direction

Although this chapter gives an overview of sustainable city service models in the Global South, there are several limitations to be noted. To begin with, use of secondary data and published texts—systematically chosen and thematically examined as it is—is constrained by a lack of obtaining precise local observations and possibly by the exclusion of emergent or un-documented bottom-up innovations. Case studies are by definition context-specific; their conclusions can only indirectly be transferred to other socio-political or environmental contexts. Also, the concentration of the chapter on four nations (India, Brazil, Malaysia, and South Africa) is enlightening comparatively but cannot be indicative of the completeness of urban governance contexts or sustainability issues in the whole of the Global South. Moreover, while the research includes bibliometric resources and thematic grouping, it has limitations in terms of temporal horizon and database availability, and hence possibly omits grey literature or non-English studies.

And looking to the future, more participatory and comparative research is called for to bridge the voices of municipal officials', community, and informal sector actors. Longitudinal studies following the development of sustainability interventions over time might cast further light on adaptation and resilience in governance. And there is great room for further examining the interface between digital innovation and inclusive planning, i.e., how data technologies can complement—rather than replace—community participation. Subsequent research would investigate more integrated models that connect municipal infrastructure planning with health, education, and climate resilience objectives within a city-wide strategy. Finally, the development of regional knowledge platforms to enable cross-city learning, particularly between secondary and medium-sized cities, would develop capacity for scaling up sustainable practice across cities.

References

Almeida J, Costa F, Lopes M (2025) Systematic selection of waste from run-of-mine tailings based on economic, environmental, and technical criteria. Sustainability 17(4):1350. https://doi.org/10.3390/su17041350

Angelo H, Wachsmuth D (2020) Why does everyone think cities can save the planet? Urban Stud 57(11):2201–2221. https://doi.org/10.1177/0042098018812729

Ansell C, Gash A (2008) Collaborative governance in theory and practice. J Public Adm Res Theory 18(4):543–571

Aria M, Cuccurullo C (2017) Bibliometrix: an R-tool for comprehensive science mapping analysis. J Informet 11(4):959–975

Alizadeh H, Doostvandi M, Zandsalimi M (2024) Redefining sustainable urban development: a critical analysis in the wake of COVID-19. Sustain Dev 32(6):6163–6181

Arora M, Yeow LW, Cheah L, Derrible S (2022) Assessing water circularity in cities: methodological framework with a case study. Resour Conserv Recycl 178:106042. https://doi.org/10.1016/j.resconrec.2021.106042

Baloch FA, Chen M, Osei K (2024) Biotechnological trends and optimization of *Arthrospira platensis* for environmental treatment. Bioresour Technol 404:131585. https://doi.org/10.1016/j.biortech.2024.131585

Bittencourt JCN, Costa DG, Portugal P, Vasques F (2024) Smart cities for equitable societies: promoting digital equality, trust, and ethical governance. In: 2024 IEEE Smart Cities Futures Summit (SCFC), pp 37–40

Brink D, Singh A (2024) Municipal sustainability strategies and institutional alignment. J Urban Policy Pract 28(1):1–19

Batty M, Axhausen KW, Giannotti F et al (2012) Smart cities of the future. Eur Phys J Special Topics 214:481–518

Bulkeley H, Betsill MM (2005) Rethinking sustainable cities: multilevel governance and the 'urban' politics of climate change. Environ Politics 14(1):42–63

Bulkeley H, Castán Broto V (2013) Governing climate change through cities: infrastructure, innovation and the urban politics of climate change. Routledge

Datta A (2015) New urban utopias of postcolonial India: 'entrepreneurial urbanization' in Dholera smart city, Gujarat. Dialogues Human Geogr 5(1):3–22

Ellen MacArthur Foundation. (2019). Completing the picture: how the circular economy tackles climate change

Gerlak AK, Heikkila T (2024) Learning for environmental governance: insights for a more adaptive future. Cambridge University Press

Gummadi V, Tiwari R, Srivastava N (2025) Application of rice-husk derived zeolites for wastewater treatment. J Environ Manag 343:126258. https://doi.org/10.1016/j.jenvman.2025.126258

Hughes S, Peterson J (2018) Urban climate politics: how private sector actors shape local sustainability. Environ Plann C: Politics Space 36(5):927–944. https://doi.org/10.1177/2399654418758663

Jain A, Patel S, Mehta V (2025) Towards sustainable transport: an analysis of policy implementation. Period Polytech Transp Eng 53(2):36936. https://doi.org/10.3311/PPtr.36936

James A, Kumar S, Lin B (2024) From awareness to action: understanding the resistance to sustainable waste behaviors. Bus Strateg Environ 33(2):3934. https://doi.org/10.1002/bse.3934

Kapoor R, Singh V, Hassan A (2025) Advancing urban development: applications of hydrological modeling. Smart Cities 8(2):0051. https://doi.org/10.3390/smartcities8020051

Khan R, Sharma D, Ali M (2025) Assessing the effectiveness of bentonite-FA and gypsum for heavy metal immobilization. Environ Res Commun 7(1):addd64. https://doi.org/10.1088/2631-8695/addd64

Krueger R, Schulz D, Gibbs D, MacLeod G (2022) The new urban governance of climate change: experimentation and the politics of city-region resilience. Urban Stud 59(5):857–876. https://doi.org/10.1177/00420980211019624

Meerow S, Newell JP, Stults M (2016) Defining urban resilience: a review. Landsc Urban Plan 147:38–49

Mendoza L, Alvarez C, Tan R (2024) A bibliometric review of barriers to circular infrastructure planning. Circ Econ Sustain 4:50. https://doi.org/10.1007/s44274-024-00050-4

Mensah J (2019) Sustainable development: meaning, history, principles, pillars, and implications for human action: literature review. Cogent Soc Sci 5(1):1653531

Mitlin D (2008) With and beyond the state—co-production as a route to political influence, power and transformation for grassroots organizations. Environ Urban 20(2):339–360. https://doi.org/10.1177/0956247808096117

Mitlin D, Walnycki A (2020) Informality as experimentation: water utilities' strategies for cost recovery and institutional change in urban Kenya and Bolivia. Environ Urban 32(2):365–382. https://doi.org/10.1177/0956247820927263

Nielsen H, Costa L, Dube M (2025) Integrating resilience and socioeconomic demands in urban infrastructure design. Environ Sci Pol 146:104048. https://doi.org/10.1016/j.envsci.2025.104048

OECD (2020) The circular economy in cities and regions. OECD Publishing. https://doi.org/10.1787/10ac6ae4-en

Oliveira MDPSL, de Oliveira EA, Fonseca AM (2021) Strategies to promote circular economy in the management of construction and demolition waste at the regional level: a case study in Manaus, Brazil. Clean Techn Environ Policy 23:2713–2725. https://doi.org/10.1007/s10098-021-02074-7

Oliveira M, Bianchi D, Cho Y (2025) Circular economy readiness and digital solutions in water infrastructure. Technol Forecast Soc Chang 200:124242. https://doi.org/10.1016/j.techfore.2025.124242

Ooi JK, Woon KS, Hashim H (2021) A multi-objective model to optimize country-scale municipal solid waste management with economic and environmental objectives: a case study in Malaysia. J Clean Prod 316:128366. https://doi.org/10.1016/j.jclepro.2021.128366
Parnell S, Pieterse E (2010) The 'right to the city': institutional imperatives of a developmental state. Int J Urban Reg Res 34(1):146–162. https://doi.org/10.1111/j.1468-2427.2010.00954.x
Parnell S, Robinson J (2012) (re)theorizing cities from the global south: looking beyond neoliberalism. Urban Geogr 33(4):593–617. https://doi.org/10.2747/0272-3638.33.4.593
Pierre J, Peters BG (2000) Governance, politics and the state. Palgrave Macmillan
Prasad N, Deen A, Yadav T (2025) Improving sanitation strategies through coordination in urban South Asia. Water Policy 27(3):168. https://doi.org/10.2166/wp.2025.168
Puntillo P (2023) Circular economy business models: towards achieving sustainable development goals in the waste management sector—empirical evidence and theoretical implications. Corp Soc Responsib Environ Manag 30(2):941–954
Rammile S, Bhanye JI, Lehobo T, Hugo M (2024) Building resilient urban infrastructure and innovation for urban public health: a case study of Bloemfontein, South Africa. In: Sustainable development goals and urban health: strides, challenges and way forward for poor neighborhoods. Springer Nature, Switzerland, pp 69–96. https://doi.org/10.1007/978-3-031-32514-4_5
Satterthwaite D (2016) Missing the millennium development goal targets for water and sanitation in urban areas. Environ Urban 28(1):99–118. https://doi.org/10.1177/0956247816628435
Singh R, Zhang L, Bhatia R (2024) Green chemistry revolutionizing sustainability in industrial waste management. Sustain Chem Pharm 32:101774. https://doi.org/10.1016/j.scp.2024.101774
Söderström O, Paasche T, Klauser F (2014) Smart cities as corporate storytelling. City 18(3):307–320. https://doi.org/10.1080/13604813.2014.906716
Tschakert P, Van Oort B, St. Clair AL, LaMadrid A (2013) Inequality and transformation analyses: a complementary lens for addressing vulnerability to climate change. Clim Dev 5(4):340–350
UN-Habitat. (2020). World cities report 2020: the value of sustainable urbanization. United Nations Human Settlements Programme. https://unhabitat.org
Van Eck NJ, Waltman L (2010) Software survey: VOSviewer, a computer program for bibliometric mapping. Scientometrics 84(2):523–538. https://doi.org/10.1007/s11192-009-0146-3
Venes M, Silva J, Correia P (2025) The impacts of municipal solid waste collection governance in Brazilian cities. Int J Environ Sustain Dev 24(1):145338. https://doi.org/10.1504/IJESD.2025.145338
Watson V (2009) Seeing from the south: refocusing urban planning on the globe's central urban issues. Urban Stud 46(11):2259–2275. https://doi.org/10.1177/0042098009342595

Integrating Human Resource Management and Corporate Social Responsibility: A Review and Empirical Amalgamation

Kamlesh Kumar Maurya

Abstract The current research examines the intertwined connection between Corporate Social Responsibility (CSR) and Human Resource Management (HRM), highlighting HRM practices and policies' critical contribution to organizational sustainability. In a structured literature review of peer-reviewed articles from various sources like Google Scholar, Web of Science, Scopus, and ResearchGate, the review detects significant themes, such as green HRM, employee engagement, strategic management, ethical leadership, and corporate social responsibility. This chapter provides a comprehensive exploration of the interaction between corporate social responsibility (CSR) and human resource management (HRM, with a focus is to investigate and highlight the joint ability of these in promoting organizational sustainability, which is one of the core pursuits of the United Nations' Sustainable Development Goals (SDGs). Results of the review indicate increased scholarly interest in integrating HRM-CSR. The research illustrates growing academic interest in combining HRM–CSR, with publication rates highest in 2023. Keyword terms such as 'green', 'performance', and 'sustainability' indicate ongoing research interests. The review highlights the strategic importance of integrating CSR into HRM practice and offers policy implications, outlines implementation challenges, and proposes lines for future research. The findings of the present study are arranged in line to help policymakers, HR-OB practitioners, and researchers in creating socially responsible and sustainable organizations.

Keywords Corporate social responsibility · SDGs · SDG 8 (Decent Work and Economic Growth) · Green HRM · Employee engagement · Strategic management · Human resource management

K. K. Maurya (✉)
School of Business, UPES, Dehradun, India
e-mail: maurya.kamlesh31@gmail.com

A. Singh et al. (eds.), *Human-Centered Sustainability: Bridging Well-Being, Gender Equity, and Technology*, World Sustainability Series,
https://doi.org/10.1007/978-3-032-22020-2_18

1 Introduction

In the current business environment, organizations are expected to increase alignment of their operations with the Sustainable Development Goals. An area that is currently attracting a lot of attention is where Human Resource Management (HRM) meets Corporate Social Responsibility (CSR). This emerging area highlights the strategic contribution of HRM to encouraging socially responsible and sustainable business practices (Podgorodnichenko et al. 2021a, b). The current research explores the relationship between CSR and HRM, with emphasis on the ways HR policies and practices affect, and are affected by, CSR efforts, especially in promoting organizational sustainability.

This crossroads—commonly referred to as the CSR-HRM interface—is a result of the growing perception of employees as internal stakeholders and participants in sustainability initiatives (Sarvaiya et al. 2021). Fundamental HR roles, such as recruitment, training, performance management, and employee engagement, are being leveraged to instill CSR values into the organizational culture (Renwick 2018). While the theoretical and empirical bases in this field are expanding, there are still significant gaps in knowledge regarding how HRM can function strategically versus operationally, in facilitating CSR adoption in various contexts and industries.

This chapter provides a comprehensive exploration of the interaction between corporate social responsibility (CSR) and human resource management (HRM, with a focus is to investigate and highlight the joint ability of these in promoting organizational sustainability, which is one of the core pursuits of the United Nations' Sustainable Development Goals (SDGs). There is a direct alignment between SDG 8 (Decent Work and Economic Growth) and the implementation of organizational practices in collaboration with CSR and HRM. By fostering an inclusive, ethical, and sustainable work culture and working environment, organizations may serve the wider perspective of the most vulnerable and valuable stakeholders in the organizational context.

Themes like green HRM, employee engagement, strategic management, and ethical leadership through the structured literature review process provide not only a mirror of the growing academic interest and motivation towards building responsible organizations, but they also aim to integrate commitment towards another very important SDG, i.e., SDG 12 (Responsible Consumption and Production) and further SDG 13 (Climate Action) by designing and implementing the environmentally responsible human resource strategies.

The present chapter draws attention to how human resource policies can play a key role in placing social responsibility at the heart of an organization. It encourages a shift in how management views employees, not just as contributors to productivity, but as essential participants in the journey toward sustainable development. This shift is vital for creating workplaces that are ethical, environmentally aware, and socially responsible. The review explores how HR strategies—when aligned with values like fairness, ethics, and environmental care—can lead to long-term benefits not only for organizations but also for society at large. It reflects a growing awareness that the success of any sustainability effort depends heavily on how people are treated, empowered, and engaged within their workplace.

The purpose behind this research is to respond to the pressing need for different disciplines to work together in support of global development goals. The study shows a noticeable increase in interest among scholars, especially in recent years, in linking business performance with broader social and environmental concerns. This interest is seen in frequent discussions around terms such as "green," "performance," and "sustainability." The research calls for improved policies and more cooperation among experts to connect corporate responsibility with human resource management in a meaningful way. It also points out common barriers, such as weak leadership support and unclear policies, which make such integration difficult. By speaking directly to decision-makers, HR professionals, and academics, the study offers guidance on how to create workplaces that are both responsible and resilient—an essential step toward a more inclusive and sustainable future.

Human Resource Management

Human Resource Management entails a formalized and integrated way of dealing with an organization's human resources, generally its most valuable resource. HRM comprises a wide range of activities, such as recruitment, selection, training and development, performance appraisal, employee relations, and strategic workforce planning (Boxall et al. 2011; Jamali et al. 2015a, b). All these activities combine to assist an organization in reaching its goals through the maximization of employees' performance and participation.

In this research, HRM is defined as the collection of policies and practices used by organizations to recruit, develop, motivate, and retain employees in a manner that aligns with CSR goals. These include:

- Green recruitment and selection based on CSR values
- Training and development around CSR
- Integrating CSR measures into performance management systems
- Employee participation in CSR activities (e.g., volunteering, sustainability initiatives)
- Communication practices aimed at fostering CSR awareness within the organization

Corporate Social Responsibility

CSR is understood to be the intentional incorporation of social, environmental, and economic issues into the values, culture, operations, and stakeholder engagement of an organization (Aguinis 2011a, b). CSR is understood as organizational behavior that reflects on stakeholder needs and upholds the triple bottom line—people, planet, and profit (Carroll and Shabana 2010).

Operationally, CSR within this research comprises internal and external programs such as:

- Environmental sustainability initiatives (e.g., waste management, carbon footprint mitigation)
- Social responsibility initiatives (e.g., diversity, health and safety, community involvement)
- Code of conduct and ethical business practices
- Embedding CSR in strategy, job description, and employee performance appraisal

The extent to which these are integrated into HRM practices is evaluated using content analysis of organizational reports and secondary data from published research (Sarwar et al. 2025; Sarvaiya et al. 2021).

Sustainable Performance

Sustainable performance is the capacity of an organization to sustain financial sustainability while at the same time promoting environmental stewardship and social equity (Elkington 1998; Latan et al. 2018). It suggests a long-term strategic focus in which business objectives converge with sustainable development objectives (SDGs). This variable is captured by indicators that represent the triple bottom line:

- Financial: Return on investment (ROI), customer retention, and profitability
- Environmental: Resource consumption reduction, emissions, and waste reduction
- Social: Employee well-being, diversity, labor rights, and community development

In the context of the studies reviewed herein, sustainable performance is frequently measured via the use of performance data, self-reported results in surveys, and CSR reporting indices (Sarwar et al. 2025; Podgorodnichenko et al. 2021a, b).

Strategic Versus Operational HR Roles in CSR

Following Golden and Ramanujam's (1985) typology, HR roles in CSR can be classified as:

Strategic: HR is engaged in CSR strategy development, policy incorporation, and cultural change.
Operational: HR facilitates CSR implementation through tactical activities such as training, compliance, and communication.

The study assesses HR roles against:

- Document evidence of HR's involvement in CSR decision-making
- The level to which HR policies are fashioned to drive CSR behaviors and performance

Whether HRM functions as a strategic partner or administrative facilitator (Sarvaiya and Wu 2014; Gond et al. 2011a, b).

2 Rationale for the Study

The justification for this research lies in the need to close the theory–practice gap in the CSR-HRM literature. As noted by Podgorodnichenko et al. (2021a, b), despite general agreement on the significance of CSR-HRM integration, disjunctures remain as to how academic insights are translated into practice, particularly mechanisms and contextual factors. In addition, most of the literature that is in place is fragmented, typically concentrating either on external CSR commitments or on individual HR interventions, without an overarching perspective of how they interact in the greater strategic management paradigm.

It has been illustrated through research that CSR strategies are increasingly reliant on the internalization of sustainable practices and values by employees, a function inherently coupled with HRM (Jamali et al. 2015a, b; Sarvaiya et al. 2017). However, empirical data indicate that HR's role is often confined to operational assistance and not strategic collaboration (Sarvaiya et al. 2017). Moreover, increased pressure is being demanded for holistic strategies that consider how CSR integrates within organizational culture through HR interventions that are aligned with the triple bottom line—economic, social, and environmental performance (Aguinis 2011a, b; Renwick et al. 2013).

Theoretically, the Resource-Based View (RBV) highlights the significance of intangible resources—human capital, organizational culture, and ethical leadership—as drivers of long-term competitive advantage (Barney 1991) Grant 1991). HRM, as a key guardian of these intangible resources, is critical in facilitating organizations to develop and execute effective CSR strategies. As illustrated in Sarwar et al. (2024a, b), CSR and HRM together shape sustainable performance, especially in hospitality environments in the UK, Italy, and Pakistan, supporting the RBV proposition.

3 Objectives of the Study

Given the dynamic and multifaceted nature of the HRM-CSR relationship, this study aims to achieve the following objectives:

- To explore the strategic and operational functions of HRM in CSR adoption.
- To explore the effect of HRM practices on CSR performance and organizational sustainability.
- To investigate contextual drivers of the CSR-HRM interface.
- To determine research-practice disconnects in CSR-HRM integration.
- To suggest an integrative model connecting HRM practices with CSR dimensions and sustainable development goals (SDGs).

4 Methodology

This research follows a mixed-methods strategy by blending an exhaustive literature review of available studies and empirical examination in investigating the correlation between Human Resource Management (HRM) and Corporate Social Responsibility (CSR). Based on a sample of 103 peer-reviewed, Scopus-listed journals published from the year 2005 to 2024, the study identifies how academic research and comprehension on this topic evolved over time.

To determine important trends and themes, the research analyzed article titles, abstracts, journal sources, and publication dates. Through keyword frequency analysis and thematic coding, it outlined the key areas of interest in the literature. Qualitative and quantitative methods were employed, which emphasized common topics like green HRM, employee engagement, CSR-HR alignment, and sustainable organizational performance. Visual aids such as pie charts and trend graphs were employed to directly present how these themes have changed in the last two decades (Podgorodnichenko et al. 2021a, b) Sarwar et al. 2024a, b).

The overall objective of this review is to assess how HRM practices have been theoretically and empirically associated with CSR goals in academic studies. By synthesizing observations from the data gathered, the research provides a twofold view: one evaluating how academic literature has acknowledged the HRM-CSR relationship, and another that looks at its application and practicality for contemporary organizations striving for sustainability. This holistic perspective is part of the increasing demand for more robust frameworks linking internal HR practices to wider social and environmental obligations (Sarvaiya et al. 2017; Renwick et al. 2016).

5 Literature Review

This review-based study consolidates knowledge from a formal literature review, comprising peer-reviewed academic sources acquired via gateways such as Google Scholar, Scopus, Web of Science, and ResearchGate. The main topics which were ascertained are green HRM, employee involvement, ethical leadership, strategic HRM, and processes of CSR application. The research is not single-industry- or single-place-specific but rather takes into consideration cross-sectoral as well as global perspectives in order to diversify the study.

Emerging Interface Between HRM and CSR

Throughout the last decades, the crossroads of Human Resource Management (HRM) and Corporate Social Responsibility (CSR) has been a focus of both research and practice. While CSR emphasizes an ethical and sustainable corporate responsibility toward stakeholders and the environment (Aguinis 2011a, b; Carroll and Shabana 2010), HRM focuses on leveraging human capital in order to pursue strategic goals

(Boxall and Purcell 2011). The convergence of these two disciplines has generated new notions such as "sustainable HRM" and "green HRM," which attempt to bring people management in line with CSR principles (Renwick et al. 2016; Podgorodnichenko et al. 2021a, b).

The Resource-Based View (RBV) theory has been applied in numerous literature reviews to confirm that human capital is a significant intangible asset that can generate sustainable competitive advantage if combined with CSR (Barney 1991; Sarwar et al., 2024a, b). HRM is a channel through which CSR initiatives are converted into action by employees to achieve both operational fit and cultural transformation (Sarvaiya et al. 2017). For example, CSR-specific training programs, sustainable recruitment, and CSR-specific performance appraisals have all been seen to increase the commitment of employees to social and environmental goals (Ren and Jackson 2020).

Recent empirical work substantiates this connection. For instance, some 2024 Scopus-indexed dataset papers here cited mention several examples where green HR practices were seen to be critical for enhancing social responsibility outcomes within organizations. Shaaban (2024a, b, c) explains the significance of green HRM practices as significant predictors of socially responsible employee behavior and emphasizes HRM's facilitatory role in spreading CSR principles at the ground level.

6 Research Trends and Thematic Concentrations

A bibliometric study of 103 Scopus-published articles between 2005 and 2024 indicates a consistent rise in publications after 2015, with dense concentration since 2020. Most studies are theoretically themed around constructs like green HRM, ethical leadership, employee engagement, sustainability orientation, and CSR performance. For example, Yang et al. (2024) studied green transformational leadership in manufacturing and discovered that such leadership enhances environmental consciousness and mediates the link between HRM practices and CSR outcomes.

There is a sharp rise in research publications after 2020, indicating increased scholarly attention toward HRM–CSR integration. Additionally, contemporary studies have examined the motivational and behavioral consequences of CSR-integrated HRM. Katfi et al. (2024) contend that corporate ethics perception plays a critical mediating role in employees' involvement in CSR activities, further highlighting the significance of HRM in determining ethical work environments. Likewise, Zihan et al. (2024) emphasize that institutional implementation of green HRM in Malaysian companies enhances greater consistency between corporate strategy and sustainable development objectives (SDGs) (Table 1).

Despite such advancements, thematic gaps are apparent in the analysis. Most noteworthy among these are the underexamined roles of mediators and moderators, evidenced by empty or non-coded columns in the dataset. There is also relatively little emphasis on targeted industries, with the majority of studies taking a general or cross-sectoral view. The variable application of theoretical frameworks—like Stakeholder

Table 1 Publication trend over time

Year	No. of publications
2005	1
2010	1
2015	3
2018	5
2019	6
2020	10
2021	11
2022	14
2023	31
2024	31

Table 2 Key thematic areas in literature (word frequency from titles)

Theme	Frequency
Green HRM	High
Sustainable HRM	Medium
Employee engagement	Medium
Ethical leadership	Medium
CSR strategy	Medum
Organizational performance	Low

Theory or Institutional Theory—also curtails the explanatory power of many studies to depth. Green HRM and employee-focused practices are dominant themes in current literature (Table 2).

HRM as a Strategic Driver of CSR

While there are scholars who place HRM as an operational facilitator of CSR, others recommend its strategic position in policy development and cultural assimilation (Jamali et al. 2015a, b; Gond et al. 2011a, b). Sarvaiya et al. (2017) indicate that HR professionals' involvement in designing CSR strategy is a function of organizational structure and the formal existence of CSR departments. Where HRM is incorporated early in the strategy cycle, results are more solid, since employees view CSR as not just a compliance issue but as an organizational core value (Table 3).

The review also suggests an increasing interest in aligning HRM practices with environmental sustainability, a notion often termed as green HRM. HRM functions directly support CSR integration at both policy and practice levels.

Various studies, such as those of Din et al. (2024a, b), reveal that green training, eco-sensitive job design, and performance appraisals with a focus on sustainability promote organizational citizenship behavior for the environment (OCBE). These behaviors are essential for organizations that seek to internalize sustainability beyond marketing rhetoric.

Table 3 Integration dimensions of HRM and CSR

HRM practice	CSR application
Recruitment & Selection	Hiring with sustainability
Training & Development	CSR training modules
Performance Appraisal	CSR-linked KPIs
Employee Engagement	Volunteering, eco-initiatives
Communication	CSR awareness campaigns

Table 4 Research gaps observed

Underexplored area	Description
Mediators & moderators	Rarely identified in empirical studies
Theoretical frameworks	Limited use of RBV, Stakeholder theory
Industry-specific insights	Most studies are general or cross-sectoral
Longitudinal studies	Short-term focus dominates

Nonetheless, there is variation in how strategic this alignment is in reality. Podgorodnichenko et al. (2021a, b) note that a great deal of academic literature does not easily translate to managerial practice—a phenomenon referred to as the "research-practice gap." Particularly, most studies do not indicate how CSR principles can be institutionalized within HR systems or do not assess the long-term effect of such integration on business results (Table 4).

Future research should aim for deeper theoretical framing and longitudinal, contextual analyses.

7 Methodological Landscape and Theoretical Underpinnings

The methodological analysis of the dataset reveals the dominance of empirical research employing quantitative methods, including survey-based measurements of employee attitudes. Yet, most abstracts provide too little information regarding research design, theoretical framework, or contextual variables. This is problematic for the generalizability and applicability of results. For instance, although many articles report positive correlations between CSR and employee outcomes, few adequately control for mediating variables like organizational support, leadership style, or cultural context.

In addition, theoretical models like Stakeholder Theory (Freeman and Phillips 2002), Institutional Theory (Scott 2005), and Social Exchange Theory (Cook et al. 2013) remain underutilized in most reviewed articles. Such theoretical scarcity erodes the interpretive power of studies that strive to connect HRM mechanisms to CSR outcomes.

Table 5 HRM's role in CSR—strategic versus operational

Role type	Description	Practical approach
Strategic	Co-designing CSR policy with top management	Aligning CSR with HR strategy
Operational	Implementing training and engagement programs	CSR orientation sessions

Conversely, other works provide more combined solutions. As an illustration, Sarwar et al. (2024a, b) utilize the Resource-Based View empirically to ascertain the HRM–CSR–sustainable performance nexus in Pakistani, Italian, and British hotel industries. Its findings verify that strategic HRM practices, complemented by CSR goals, positively impact superior sustainable performance—particularly in higher-rung establishments.

8 Gaps and Future Research Directions

Although the literature verifies the strategic significance of HRM in facilitating CSR outcomes, a number of areas are yet to be explored. Firstly, additional longitudinal and cross-cultural research is required to ensure the long-term effectiveness of combined CSR-HRM approaches. Secondly, industry-level research, especially in high-impact industries such as manufacturing, hospitality, and technology, can reveal contextual variations. Thirdly, the impact of leadership—especially ethical and transformational leadership—as a moderator in the CSR-HRM-performance relationship requires further examination.

Finally, it is important that future studies prioritize the creation of a synthesizing framework that brings together major HRM practices—like green recruitment, CSR training, sustainable performance appraisal, and employee voice—together with quantifiable CSR outcomes. This framework needs to be anchored in strong theoretical frameworks and empirically validated using mixed-method approaches to ensure greater practical significance (Table 5).

9 Findings and Suggestions

The contribution of this study lies in its effort to advance a more nuanced understanding of how HRM can serve not merely as an administrative supporter but as a strategic enabler of CSR initiatives. By addressing both conceptual and practical dimensions, the study aims to inform academic scholarship, guide HR and CSR practitioners, and contribute to the development of sustainable organizations capable of meeting both stakeholder expectations and societal demands.

This study sought to synthesize scholarly research on the interface of Human Resource Management (HRM) and Corporate Social Responsibility (CSR) based on a systematic review of 103 Scopus-listed publications published between 2005 and 2024. The study found the major themes, emerging trends, theoretical gaps, and practical insights pertinent to the changing CSR-HRM interface.

Strategic and Operational Functions of HRM in CSR Adoption

The review shows that HRM has both operational and strategic functions in the adoption and implementation of CSR. Operationally, HRM facilitates CSR through practices like green recruitment, CSR-focused employee training, and incorporating CSR metrics into performance appraisals (Shaaban, 2024a, b, c; Zihan and Makhbul 2024). Strategically, HR professionals are increasingly involved in policy-making, especially when CSR is incorporated into organizational culture and leadership frameworks (Sarvaiya et al. 2017).

Despite such developments, most organizations constrain HR's function to operational assistance, enforcing top-down CSR decisions rather than enabling HRM as a strategic ally. Examples such as Sarwar et al. (2024a, b) indicate that strategic alignment is stronger in larger or globally networked organizations, particularly in sectors such as hospitality and manufacturing. Findings suggest the following insights to fill the gap in the academic and application levels.

- Enable HR leaders to make input in CSR policymaking and strategic planning exercises.
- Institutionalize dual roles for HR to ensure both day-to-day implementation and top-level strategy input, particularly in companies seeking ESG ratings or SDG alignment.
- Embed CSR KPIs within HR scorecards and leadership performance systems to institutionalize HR's strategic contributions.

Impact of HRM Practices on CSR Performance and Organizational Sustainability

HRM practices have a direct impact on CSR results and, by implication, sustainable organizational performance. Sustainability training, green work systems, and inclusive engagement mechanisms have been found to have a direct positive impact on employee involvement in CSR activities (Yang et al. 2024). Moreover, when employees feel that HR policies are ethically sound and socially responsible, they become more committed to the organization's CSR vision (Katfi et al. 2024).

But research tends to be short of longitudinal data correlating HRM practices to tangible sustainability measures such as carbon footprint reduction, employee well-being, or financial returns—pointing to a measurement gap.

Below are the recommendations based on the literature review and analysis

- Design HRM systems with triple bottom line outcomes (social, environmental, economic).
- Use performance dashboards with SDG-linked indicators to measure impact of HRM-driven CSR interventions.

- Enrich research with sustainability audits, reporting HR's impact on environmental compliance, waste minimization, and social outreach measures.

Contextual Drivers of the CSR-HRM Interface

Organizational structure, leadership style, and industry type are strong contextual drivers that impact the intensity and direction of CSR-HRM integration. A separate CSR department, for example, tends to increase HR's strategic participation (Sarvaiya et al. 2017). Manufacturing and hospitality industries—common objects of the examined studies—are affected in their adoption of CSR by international exposure as well as by requirements of compliance (Din et al. 2024a, b).

Cultural context is also significant. CSR strategies based on community development in collectivist cultures are more meaningful and likely to be endorsed through HR-driven engagement initiatives. Nevertheless, industry-specific and culture-specific understandings are scarce in extant literature.

Based on the inputs and insights from analysis of the literature, the following are the few recommendations which practitioners of the field may use in future aspects of the focus.

- Contextualize HRM-CSR strategies in terms of industry requirements and cultural ethos.
- Promote sectoral benchmarking to establish best practices and context-relevant CSR-HRM models.
- Encourage cross-cultural studies to analyze whether CSR agendas and HRM responsibilities differ between different national and industrial boundaries.

Research-Practice Disconnects in CSR-HRM Integration

One of the most prevalent findings is the gap between research based on academics and applications of HRM in practice regarding CSR. Podgorodnichenko et al. (2021a, b) pointed out that while theoretical evidence indicates the contribution of HRM to CSR, practitioners remain unaware or lack training in applying these practices in everyday work. Additionally, topics like mediators, moderators, and longitudinal effects are infrequently addressed in empirical studies, and thus the practical applicability of findings gets restricted.

Further, jargon and access barriers hinder practitioners from taking academic knowledge into practice. For example, terms such as "CSR-HRM synergy" or "employee sustainability orientation" stay within the domain of academics.

Recommendations:

- Close the gap between practice and academia with executive education, practitioner toolkits, and more accessible models.
- Request journals and conferences to release practitioner briefs giving summaries of main findings in non-technical language.
- Have practical validation processes incorporated in research (e.g., pilot application in organizations) to make it applicable.

Table 6 Proposed integrative model

HRM practice	CSR domain	Linked SDGs
Green recruitment	Environmental responsibility	SDG 13: Climate action
Employee training	Ethical & social awareness	SDG 4: Quality education
Performance management	CSR outcome evaluation	SDG 8: Decent work & growth
Employee engagement	Community outreach	SDG 11: Sustainable cities
HR communication	Transparency & compliance	SDG 16: Peace & justice

Suggesting an Integrative Model Connecting HRM, CSR, and SDGs

The review supports the development of an integrative framework that connects HRM functions to CSR domains and SDG targets (Table 6).

10 Conclusions

This research investigated the complex and dynamic interaction between Human Resource Management (HRM) and Corporate Social Responsibility (CSR) using a systematic literature review and analysis of Scopus-indexed articles between 2005 and 2024. The results confirm that HRM has tremendous potential as both a strategic ally and an operational facilitator in promoting CSR objectives and organizational sustainability. Although HR practices like green recruitment, CSR-driven training, and employee engagement are increasingly used to aid CSR initiatives, their impact depends on situational factors such as organizational structure, leadership, industry dynamics, and cultural orientation.

Notwithstanding the increase in studies focused on the interface of CSR and HRM, the gap between research knowledge and practice remains an important issue of concern. A large proportion of empirical work has no explicit details on how implementation should occur or how research is presented for it to influence practitioners' decisions. This further creates the imperative to develop more holistic, multidisciplinary, and context-oriented methods of research for CSR-HRM.

To fill these gaps, the research suggested an integrative model connecting HRM activities to CSR areas and Sustainable Development Goals (SDGs), providing a pragmatic framework for organizations pursuing responsible and resilient development. In the future, more in-depth theoretical foundation, longitudinal examination, and closer collaboration between academia and industry are needed to turn HRM into a strategic force for sustainable and ethical business practices.

References

Aguinis H (2011a) Organizational responsibility: doing good and doing well. In: Washington DC (ed) APA handbook of industrial and organizational psychology, vol 3. American Psychological Association, pp 855–879

Aguinis H (2011) Organizational responsibility: doing good and doing well

Barney J (1991) Firm resources and sustained competitive advantage. J Manag 17(1):99–120

Benuyenah V (2023) Rethinking recruitment ethically through the lens of corporate social responsibility (CSR). In: Evidence-based HRM: a global forum for empirical scholarship, vol 11, no 3. Emerald Publishing Limited, pp 372–376

Bizri R, Wahbi M, Al Jardali H (2021) The impact of CSR best practices on job performance: the mediating roles of affective commitment and work engagement. J Organizational Effectiveness People Performance 8(1):129–148

Boxall P, Purcell J (2011) Strategy and human resource management, 3rd edn. Palgrave Macmillan, Basingstoke

Boxall P, Ang SH, Bartram T (2011) Analysing the 'black box' of HRM: uncovering HR goals, mediators, and outcomes in a standardized service environment. J Manage Stud 48(7):1504–1532

Carroll AB, Shabana KM (2010) The business case for corporate social responsibility: a review of concepts, research and practice. Int J Manag Rev 12(1):85–105

Cook KS, Cheshire C, Rice ER, Nakagawa S (2013) Social exchange theory. Springer, Netherlands, pp 61–88

Crichton R, Shrivastava P, Walker T, Farhidi F, Renwick D, Ellegate N (2024) Going green in the Norwegian fossil fuel sector? The case of sustainability culture at Equinor. German J Human Resource Manage 38(2):140–158

Din AU, Yang Y, Yan R, Wei A, Ali M (2024 Growing success with sustainability: The influence of green HRM, innovation, and competitive advantage on environmental performance in the manufacturing industry. Heliyon 10(10)

Din AU, Yang Y, Ud Din QM, Khan IU (2024) Green leadership in manufacturing industry: unveiling the mediating role of green transformational behavior in CSR initiatives. Heliyon [Online] Available at: https://www.sciencedirect.com/. Accessed 15 Apr 2025

Elkington J (1997) The triple bottom line. Environ Manage Readings Cases 2:49–66

Freeman RE, Phillips RA (2002) Stakeholder theory: a libertarian defense. Bus Ethics Q 12(3):331–349

Gond JP, Igalens J, Swaen V, El Akremi A (2011) The human resources contribution to responsible leadership: an exploration of the CSR–HR interface. In: Responsible leadership. Springer Netherlands, Dordrecht, pp 115–132

Gond JP, Igalens J, Swaen V, El Akremi A (2011b) The human resources contribution to responsible leadership: an exploration of the CSR–HR interface. J Bus Ethics 98(1):115–132

Grant RM (1991) The resource-based theory of competitive advantage: implications for strategy formulation. Calif Manage Rev 33(3):114–135

Gyensare MA, Adomako S, Amankwah-Amoah J (2024) Green HRM practices, employee well-being, and sustainable work behavior: examining the moderating role of resource commitment. Bus Strateg Environ 33(4):3129–3141

Hameed R, Mahmood A, Shoaib M (2022) The role of green human resource practices in fostering green corporate social responsibility. Front Psychol 13:792343

Hronová Š, Špaček M (2021) Sustainable HRM practices in corporate reporting. Economies 9(2):75

Jamali DR, El Dirani AM, Harwood IA (2015a) Exploring human resource management roles in corporate social responsibility: the CSR-HRM co-creation model. Bus Ethics Eur Rev 24(2):125–143

Jamali D, El Dirani AM, Harwood IA (2015b) Exploring human resource management roles in corporate social responsibility: the CSR–HRM co-creation model. Bus Ethics Eur Rev 24(2):125–143

Jang S, Ardichvili A (2020) Examining the link between corporate social responsibility and human resources: implications for HRD research and practice. Hum Resour Dev Rev 19(2):183–211

Karatas-Ozkan M, Özgören Ç, Yamak S, Ibrahim S, Tunalıoğlu MN, Pinnington A, Nicolopoulou K, Baruch Y (2022) Dual nature of the relationship between corporate social responsibility and human resource management: a blessing or a curse? Corp Soc Responsib Environ Manag 29(5):1578–1594

Katfi A, El Mnouer O, Katfi H, Katfi O (2024) Influence of employees' perception of corporate social responsibility practices on organizational commitment: the case of Moroccan companies. Organizational Cultures 24(1):45

Khan U, Liu W (2023) Does environmental responsible effect human resources management practice on firm effectiveness and green technology innovation? Environ Sci Pollut Res 30(13):36160–36175

Latan H, Jabbour CJC, de Sousa Jabbour ABL, Wamba SF, Shahbaz M (2018) Effects of environmental strategy, environmental uncertainty and top management's commitment on corporate environmental performance: the role of environmental management accounting. J Clean Prod 180:297–306

Nakra N, Kashyap V (2023) Investigating the link between socially-responsible HRM and organizational sustainability performance–an HRD perspective. Eur J Training Dev (ahead-of-print)

Nguyen TLP, Nguyen TTH, Ključnikov A (2023) The influence of socially responsible human resource management on green behaviours in the aviation industry. J Competitiveness

Patil BS, Priya MSR (2023) The synergy of human resource management, corporate social responsibility as an organizational soft power and sustainable performance. In: Global perspectives on soft power management in business. IGI Global, pp 389–413

Podgorodnichenko N, Edgar F, Akmal A (2021a) An integrative literature review of the CSR–HRM nexus: learning from research–practice gaps. Hum Resour Manag Rev 31(1):100839. https://doi.org/10.1016/j.hrmr.2021.100839

Podgorodnichenko N, Edgar F, Akmal A, McAndrew I (2021b) Sustainability through sensemaking: human resource professionals' engagement and enactment of corporate social responsibility. J Clean Prod 293:126150

Portocarrero FF, Winkler ALP, Pearce JL (2023) Broadening our understanding of human resource management for improved environmental performance. Bus Soc 62(1):14–53

Rashmi S, Preeti T (2023) The effect of artificial intelligence-enabled work from the home culture in strengthening the green HRM. In: AIP conference proceedings, vol 2721, no 1. AIP Publishing

Rawshdeh ZA, Makhbul ZKM, Rawshdeh M, Sinniah S (2023) Perceived socially responsible-HRM on talent retention: the mediating effect of trust and motivation and the moderating effect of other-regarding value orientation. Front Psychol 13:1087065

Ren S, Jackson SE (2020) HRM and sustainable development: insights from international experiences. Hum Resour Manag Rev 30(3):100688

Renwick DW (2018) Towards an understanding of green human resource management. In: Contemporary developments in green human resource management research. Routledge, pp 1–20

Renwick D, Redman T, Maguire S (2013) Managing HR for environmental sustainability. Empl Relat 35(1):1–18

Renwick DWS, Jabbour CJC, Muller-Camen M, Redman T, Wilkinson A (2016) Contemporary developments in green (environmental) HRM scholarship. Int J Human Resource Manage 27(2):114–128

Sarvaiya H, Arrowsmith J (2023) Exploring the context and interface of corporate social responsibility and HRM. Asia Pacific J Human Resources 61(2):300–323

Sarvaiya H, Wu M (2014) An integrated approach for corporate social responsibility and corporate sustainability. Asian Soc Sci 10(17):57–70

Sarvaiya H, Arrowsmith J, Eweje G (2021) Exploring HRM involvement in CSR: variation of Ulrich's HR roles by organisational context. Int J Human Resource Manage 32(21):4429–4462

Sarvaiya H, Eweje G, Arrowsmith J (2017) The roles of HRM in CSR: strategic partnership or operational support? J Bus Ethics 153(4):875–894. https://doi.org/10.1007/s10551-016-3402-5
Sarwar H, Ishaq MI, Franzoni S (2024a) Influence of HRM on CSR and performance of upscale hotels in developed and developing countries. Environ Dev Sustain 26(1):335–357
Sarwar H, Ishaq MI, Franzoni S (2024b) Influence of HRM on CSR and performance of upscale hotels in developed and developing countries. Environ Dev Sustain 26:335–357. https://doi.org/10.1007/s10668-022-02711-x
Sarwar U, Baig W, Rahi S, Sattar S (2025) The ethical workplace and sustainability: understanding employee green behavior through a moderated mediation framework
Scott WR (2005) Institutional theory: contributing to a theoretical research program. Great Minds Manage Process Theory Dev 37(2):460–484
Shaaban S (2024a) Performance optimization of an integrated solar combined cycle for the cogeneration of electricity and fresh water. Renew Energy 227:120482
Shaaban S (2024) The impact of green HR practices on socially responsible HRM: an empirical study from HR practitioners. Int J Sustain Policy Pract 20(1)
Shaaban S (2024) The impact of green HR practices on socially responsible HRM: an empirical study from HR practitioners. Int J Sustain Policy Pract 20(1)
Sun J, Sarfraz M, Ivascu L, Ozturk I (2023) Unveiling green synergies: sustainable performance through human resource management, CSR, and corporate image under a mediated moderation framework. Environ Sci Pollut Res 30(45):101392–101409
Tanveer MI, Yusliza MY, Ngah AH, Khan MAK (2023) Mapping the link between CSR and sustainability performance through GHRM practices in hotel industry. J Clean Prod 429:139258
Tien Thanh P, Thu Ha N, Thi Hong Ngoc P, Thuy Ha LT (2024) Corporate social responsibility, corporate reputation and intention to apply for a job: evidence from students in an emerging economy. Int J Organizational Anal 32(1):17–34
Tsymbaliuk S, Vasylyk A, Stoliaruk K (2021) Green human resource management: how to implement environmental issues into HR practices. In: E3S web of conferences, vol 255, p 01037. EDP Sciences
Yang Y, Jiang Y (2024) The impact of suppliers' CSR controversies on buyers' market value: the moderating role of social capital. J Purch Supply Manag 30(1):100904
Yang Y, Din AU, Din QMU, Khan IU (2024) Green leadership in manufacturing industry: unveiling the green Revolution's impact on organizational performance. Heliyon 10(6)
Zihan W, Makhbul ZKM (2024) Green human resource management as a catalyst for sustainable performance: unveiling the role of green innovations. Sustainability 16(4):1453
Zihan W, Makhbul ZKM, Alam SS (2024) Green human resource management in practice: assessing the impact of readiness and corporate social responsibility on organizational change. Sustainability 16(3):1153

Governance Guardian: Legal Frameworks for Corporate Accountability and Sustainability

Akanksha Singh Fouzdar

Abstract The business model has been a constant transformation from ages. Initially there was lot of reliance on sole proprietorship model, slowly and gradually business model started to get more complex driven by need for expansion. The level of business started to grow for which larger amounts of resources were needed and hence corporate model of business started to get more popular. The legislative framework provides norms, guidelines code of conduct and the line of action that must be follow by the organization. The present chapter delves is to elucidate the legal frameworks that govern corporate governance that support corporate sustainability and accountability; is examined by the complex terrain of these framework. The first section of the chapter securitizes the intricacies of corporate governance legal frameworks according to time line that highlighting significant advancements and their implications for modern governance practices. The crucial components have been identified and discussed thoroughly that formed the corporate governance and examine the impact of the same in the emerging era. Apart from that the present chapter assesses the potential of strictly implemented regulatory requirements that discourage corporate misconduct and wrong practices. It also looks for the integration of Environmental, Social, and Governance (ESG) into legislature framework that underlines the importance of sustainability in corporate governance. Few case studies have been added to showcase the challenges and failure faced by the corporations. Further, the chapter sum up with conclusion and recommendation for strengthening legal frameworks to enhance corporate governance globally and provide useful insight to the academician, decision makers and industry professional who are concerned with advancing sustainable business practices.

Keywords Corporate governance · Corporate social responsibility · Law and ethics · Corporate accountability · Corporate sustainability · ESG

A. S. Fouzdar (✉)
Institute of Business Management, GLA University, Mathura, India
e-mail: akanksha.singh@gla.ac.in

A. Singh et al. (eds.), *Human-Centered Sustainability: Bridging Well-Being, Gender Equity, and Technology*, World Sustainability Series,
https://doi.org/10.1007/978-3-032-22020-2_19

1 Overview of Corporate Governance

The business model landscape has changed dramatically over the years and is now developed from simple sole proprietorships to increasingly more complex corporate structures. Initially, the sole proprietorship model, purely because of its simplicity, dominated the market (Bui and Krajcsák 2023). This kind of business allowed direct control by an individual who could make quick decisions and keep all the profits. However, as businesses sought to expand and needed larger resources, the corporate model came into existence. It was because of the requirement of having more capital, specialized management, and a proper system of managing risks related to large-scale ventures (Affes and Jarboui 2023).

The corporate model provides significant benefits that help deal with resource constraints and growth facilitation. Capital can be pooled from several investors to allow large investments in infrastructure, technology, and human resources. Complexity generates agency issues in terms of the theory of agency, whereby the directors are regarded as agents, dealing with corporate matters for the benefit of different stakeholders (Safari and Parker 2024). These would include shareholders, creditors, suppliers, customers, as well as the greater public at large (Costa 2017). Under those circumstances, financing needs from shareholders' long-term financing, but then the directors acquire responsibilities not only in the interest of shareholders but also to satisfy the welfare needs of all stakeholders.

The fiduciary duty of directors requires them to act in the best interest of the company and its stakeholders. However, several cases have been documented where directors breach such duties in favor of their personal interests above those of their stakeholders. Such breaches raise serious concerns over corporate governance, and calls for enhanced oversight and accountability have ensued. Such instances have once again highlighted the importance of effective mechanisms in corporate governance for protecting the interest of stakeholders and sound management practices (Solikhah and Maulina 2021).

1.1 Corporate Governance Mechanism

The mechanism of corporate governance is planned in such a way that it is transparent, legal compliance, and accountable within the structures of the corporation. The framework compels directors and management to be in conformity with legal standards and act with appropriate diligence while carrying out their functions. In this manner, clear guidelines and expectations provide corporate governance with the scope to reduce the risks arising through mishandling, malpractices, and unethical practices. Simultaneously, these frameworks assure the stakeholders that their interests are being protected, so trust and confidence evolve in the corporate management itself (Saha and Kabra 2022).

The corporate governance implications are thus of prime importance going beyond mere compliance to address the global challenges facing the world, especially sustainability and corporate social responsibility. This is because, nowadays, with rising awareness among the public about issues affecting the environment and society at large, the stakeholders of this modern world, from consumers and investors to regulatory bodies, are demanding that corporations sustain themselves in their operations. This has brought about incorporating sustainability into corporate governance models to enable corporations to take into consideration and report not just on the bottom line, but on their responsibility to society and the environment.

Now, the study of the legal framework of corporate governance and accountability as well as sustainability becomes very crucial. There are different kinds of legal frameworks issued by other countries to address such an issue. This indicates an awareness that companies should be responsibly operated. For example, the standalone requirement concerning ESG disclosures has emerged into common practice, therefore requiring companies to provide transparent sustainability reports of efforts and results (Pesqueux 2005).

Beyond the generation of profits, the United Nations' Sustainable Development Goals-the international standards-now take on a new meaning in terms of corporate accountability, ensuring that these big business corporations create value and avoid creating destruction. It encourages companies to devise long-term strategies that are in line with global sustainability objectives while fostering moral practices that benefit shareholders while making the bigger society better off (Lazarides 2011).

Against this general backdrop of increased complexity in the landscape of corporate governance, with greater and more intensive interaction between legal frameworks, corporate accountability, and sustainability, the present study attempts to investigate the legal frameworks underpinning corporate governance and their involvement in promoting accountability and sustainable practice. In an effort to meet this, we study the nexus of governance, accountability, and sustainability in trying to find how effective legal structures can better help mitigate risks for stakeholder interest protection and for promoting sustainable practices within a globally interconnected and increasingly complex economy (Nour et al. 2024).

Hence, this research will add onto the larger discourse of corporate governance by showing that right legal frameworks are necessary for promoting ethical conduct and sustainability. From this premise, this research study further aims at reminding corporations, policymakers, and academicians of the renewed need to create a healthy integration of governance and sustainability within their strategies going forward (Arjoon 2005).

1.2 Global Impact: Enhancing Corporate Governance for a Sustainable Future

1.2.1 In Consideration of this, the Significance of Research on the Relevance to a Global Audience Forms Great Findings, Especially Considering the Fact that Matters of Corporate Governance and Sustainability Cross Borders. The Dynamic Nature of Business Operations across Other Borders Calls for Different Legal and Regulatory Frameworks Surrounding Corporate Governance (Guluma 2021). Indeed, this Study Really Assists in Reflecting the Opportunities Engaged by Multinational Companies while Navigating Various Forms of Governance Standards in Support of Ethical Practices. Dwelling on the Increasing Significance of Regulatory Harmonization for Transnational Compliance, this Article Takes into Account the International Frameworks of the European Union's Corporate Sustainability Reporting Directive (CSRD) As Well as the UN Guiding Principles on Business and Human Rights (Almaqtari et al. 2020)

The study also zeroe in on the accountability of corporates in international supply chains and underlines due diligence in avoiding human rights violations and environmental degradation, especially in areas where monitoring is weaker. Additionally, the work demonstrates how corporate governance is now integrated into global issues like climate change and SDGs, which propel the need for sustainability as an imperative in doing business (Singhania et al. 2022).

For stakeholders — investors, consumers, and policymakers among them-the study reveals the increased clamor for transparency and accountability in the activities of firms. The expanded legal frameworks thus empower stakeholder groups as they make companies transparent regarding their key disclosures on sustainability, hence calling for responsible global business practices. In conclusion, the study gives businesses and governments a framework on which to harmonize their governance practices with global sustainability targets (Affes and Jarboui 2023).

2 Literature Review

Corporate governance has evolved significantly since its inception, with all of it directed to business practices with more accountability and sustainability. This chapter surveys the literature related to legal frameworks underpinning corporate accountability, which we submit forms a key determinant in ethical behavior and compliance by corporations. We also explore the emerging principle of sustainability embedded within such frameworks, an expression of stakeholder expectations and regulatory needs. Moreover, the chapter discusses how the standards of governance have evolved historically and the challenge they present to an organization's effective governance practices. Through this all-inclusive review, we intend to establish a basis by which the complexities of corporate governance may be understood in the terms of contemporary business challenges.

2.1 Importance of Legal Frameworks in Corporate Accountability

Over the decades, the relevance of corporate governance has been realized and advocated by various research studies. It serves many purposes to ensure higher standards of corporate accountability. Corporate governance legal framework specifies and underlines the rule which any corporate entity is obliged to follow with due compliance. Any non-adherence is treated as a lapse and is entitled to some kind of penalty (Laine et al. 2021). This way this acts as a mechanism for protection of stakeholders. The corporate accountability ensured this way ensures that agents (directors and management group) are held accountable for their actions and decisions which affect stakeholders at large (Nunhes et al. 2020).

Over the past many years, corporate governance has also acted as a method for preventing the misconducts like fraudulent activities and other similar malpractices with the proper legal methods in place (Nunhes et al. 2020). Besides that, it also ensures the optimum transparency which fosters the higher confidence among investors and stakeholders at large. This transparency is the core of strengthened governance mechanism. In such environment, robust mechanism of corporate governance also entitles to better decision-making and a better environment overall (Hickman and Petrin 2021).

If significance of legal frameworks in corporate accounting is observed, this also facilitates the fair competition as such legal and statutory norms are common to everyone and this levels the playfield for all corporate entities in an economic system (Lazarides 2011). No corporate entity is entitled to take any advantage of any norms and capital market regulators (like SEBI in case of India and SEC in case of USA) are directly involved in the process mechanism and ensuring due adherence and compliance (Mohamad and Sori 2011). Thus overall, this all supports economic stability by creating an environment which is conducive to fair and transparent business practices with no mala fide intentions. Thus, corporate governance mechanism ensures better practices related to internal as well as external environment of the business unit. (Jain et al. 2014; Claessens and Yurtoğlu 2013).

2.2 Sustainability Integration with Corporate Governance

Corporate governance frameworks ensure that companies are reporting the practices duly in their annual reports. However, sustainability of the framework is a key factor in achieving long-term value creation (Shrivastava and Addas 2014). Corporate governance facilitates the set up in a corporate entity where long-term profitability and viability drives the entire decision-making process balancing all three legs of business viz. People, Profit and Planet (Scherer and Voegtlin 2020). Companies now a days are making conscious and deliberate attempts for environmental stewardship by reducing carbon footprints and committing to sustainable practices. At the same time, corporate social responsibility norms are also there which targets better impact on society at large and community engagement (Aras and Crowther 2008).

Sustainability in corporate governance also requires a better internal system which is conducive to prioritizing the interests of all stakeholders including customers, suppliers as well as employees. This fosters an environment of trust and confidence which makes any practice realistic, acceptable and sustainable in long run (Jaimes-Valdez et al. 2017). Now a days, companies are placing due focus on corporate risk management for the range of business risks involved in business functioning and a better communication about potential risks and risk management practices to all stakeholders makes the corporate governance mechanism more sustainable (Michelon and Parbonetti 2012). The major implication lies in sustainable reporting and disclosures also which provides clear and accurate substantial information to all concerned stakeholders (Naciti et al. 2022).

2.3 Evolution of Governance and Accountability Standards

The evolution of corporate governance has followed the contemporary business development. The early governance models were informally formulated and focused on interest of investors and had minimal regulatory interventions (Martynova and Renneboog 2011). The industry revolution and change in business models and structure led to development of formal governance structures and accountability mechanisms to protect investors. Then, in the emergence of financial scandals and market crashes, structured regulations were introduced which kept on transforming post world war as well (Scherer and Voegtlin 2020).

The biggest milestone in corporate governance framework was seen in 1990s when corporate governance codes like the Sarbanes-Oxley Act (US) and the Cadbury Report (UK) were introduced which specifically underlined the responsibilities of board of directors, specifications about the internal control and auditing mechanisms (Jaimes-Valdez et al. 2017). The further rise of globalization in 2000s has led to the requirement of standardization of norms across various jurisdictions. There was evolution of OECD principles as well as International Corporate Governance

Network (ICGN) which have laid the foundation of global best practices for corporate governance. In last decade, the focus has shifted to environmental, social and governance issues (ESG) ensuring that stakeholders are interests are guarded duly (Mohamad and Sori 2011; Bruner and Sjåfjell 2019).

3 Objectives and Methodology

3.1 Objective of the Study

Objectives of this study are enumerated as follows".

1. ***To analyze the Evolution of Corporate Governance Legal Frameworks***
   Explore key historical developments shaping modern corporate governance and accountability regulations globally.
2. **To examine the Integration of Sustainability within Corporate Governance**
   Investigate how legal frameworks incorporate sustainability principles into corporate governance and ethical business practices.
3. **To evaluate the Impact of Legal Frameworks on Corporate Accountability**
   Assess how legal frameworks influence corporate transparency, ethical behaviour, and sustainability across different industries and regions.

3.2 Methodology

This study adopts a literature review and case study method of analyzing legal frameworks for corporate accountability and sustainability. Sources drawn from relevant areas have used a systematic sampling procedure, emphasizing post-2008 material and diversity of opinion, to include academic research, regulatory reports, and international guidelines in an effort to understand not just the historical development but also the current trends in governance.

The case studies were sampled using criteria such as diversified industry, geographic representation, and the significance of the cases to the key governance issues. Cases from various industries such as finance, energy, and technology were analyzed by using suitable cases from different regions: North America, Europe, and Asia. The chosen cases again focus on significant events-based information on corporate scandals and climate-related accountability as reflections of real-world challenges and successes. This method, in this way, allows for a balanced and robust approach towards integration of theoretical insights and practical examples, which obviously resonates with the very objective of looking into corporate governance frameworks.

4 Historical Evolution of Legal Frameworks for Corporate Governance

The political, social, and economic shifts are the major constraints that influenced the development of corporate governance. It is indispensable to grasp this growth to appreciate the current state of corporate legal frameworks. The present chapter covers the major changes in corporate governance from the distant past to the current, accenting noteworthy decisions and occurrences that have shaped contemporary norms (Effah et al. 2023). The above-mentioned phases and crucial laws exemplify the fluid character of corporate governance, mirroring evolving societal norms and the continuous endeavor to harmonize the concerns of diverse stakeholders inside the corporate structure (Pesqueux 2005).

4.1 Early Beginnings: The Roots of Corporate Governance

The evolution of corporate governance can be seen from ancient Rome, when country's servicers supervised public works and were often constituted as legal organizations. However, the ides of a corporation are given or came from medieval Europe which is particular about the development of trade associations and joint stock companies (Igbinenikaro and Adewusi 2024). The Dutch East India Company (VOC) (1602) was the first international company that claim the formal corporate sustainability and accountability structure where it has formal setup of board of directors, timely scheduled meetings of shareholders and has established the milestone for modern corporate governance (Weiss 2023) (Fig. 1).

4.2 The 18th and 19th Centuries: The Rise of the Corporation

After industrial revolution the focus has been shifted to the growth of corporate and given straight message about the importance of formal governance structure. The decisive case of Dartmouth College v. Woodward (1819) established the corporate charters' inviolability. As per the opinion of U.S Supreme court, a corporate charter is a legally binding agreement between the state and the corporation that is safeguarded by the U.S. Constitution's Contract Clause. This choice gave firms a degree of predictability and stability that is essential for long-term investment and planning (Tricker 2020).

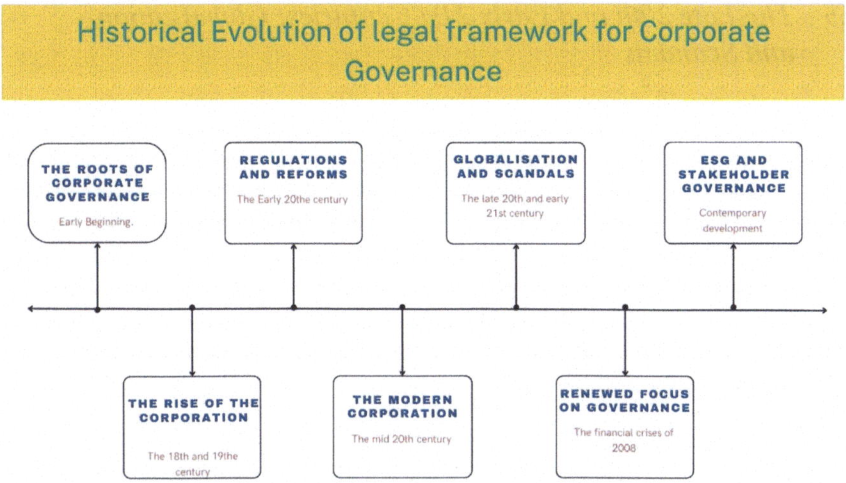

Fig. 1 Historical evolution of legal framework for corporate governance. *Source* Author's Compilation

4.3 *The Early Twentieth Century: Regulation and Reform*

The twentieth century epitome for legislative changes as corporate worries has increased. In 1914, the Federal Trade Commission (FTC) and Clayton Antitrust Act was enacted with the aim of preventing monopolistic activities and encouragement of fair competition. After the crash of stock market (1929) and Great Depression were brought many reforms. The Securities Act 1933 and Securities Exchange Act 1934 was passed to look after the interest of investors by introduction of all mandatory disclosure standards (Jain et al. 2014).

4.4 *The Mid-Twentieth Century: The Modern Corporation*

The Mid-twentieth century was remarked for division of ownership and control which is profoundly known as modern corporation. Adolf Berle and Gardiner Means in this book "The Modern Corporation and Private Property (1932)" contended that "enormous firms were effectively run by managers rather than by shareholders, which could give rise to conflicts of interest." The comprehensive set of principles of corporate governance released by American Law Institute in 1994 which significantly point out about the shareholder rights, Directors' fiduciary duties and responsibility and independent board structure (Schymik 2018).

4.5 *The Late 20th and Early 21st Centuries: Globalization and Scandals*

In the late twentieth century and start of twenty-first century, the rise of multinational corporations (MNCs) and interference of global financial markets drove the globalized corporate governance. The business scandals like Enron, Volkswagen and Tyco left the ice breaking governance flaws. Following this seriousness U.S. Congress passed the Sarbanes-Oxley Act (SOX) in 2002 with the aim to curb the stricter restrictions for improving the corporate responsibility. For this the act incorporated certain clauses like independent audit committee, strict penalties and appointment of Chief executive officers and Chief finance officer in order to certify and authenticate final accounts (Guha et al. 2019).

4.6 *The Financial Crisis of 2008: Renewed Focus on Governance*

Predefined risks and shortcomings of corporate governance were also high lightened by the global financial crises, 2008 majorly in the financial industry (Schymik 2018). The financial crises 2008 impacted many nations and similarly nations were responded in positive notion of enactment required to improve the efficacy of governance procedure. The Dodd-Frank Wall Street Reform and Consumer Protection Act (2010) established regulations in the US to enhance financial institutions' risk management, accountability, and transparency (Martynova and Renneboog 2011).

4.7 *Contemporary Developments: ESG and Stakeholder Governance*

The significance of environmental, social, and governance (ESG) integration has been increased now a days. The sustainable and responsible practices is pronounced in the eyes of authorities and investors as they are taking initiatives like application of United nation principles (UN) principles and promote those firm which have key objective in the favour of social and environmental and profit motive is secondary (Miller and Loman 2014; Bedenik and Barišić 2019). Through the Berman v/s Neo LLC 2020 case is the milestone in the corporate governance through this Environmental, social and governance (ESG) concern was raised and law amended accordingly (Waddock et al. 2002).

Therefore, the ESG and other factors such as political, social, and economic made responsible to have positive amendment on legal framework of corporate governance. The antiquity of the Dutch east India company is complete with noteworthy events and according to the modern era and issues like ESG. Through this implementation of changes and considering the future trends also legal framework has gaining appreciation over existing corporate governance and able to modified historical backdrops (Scherer and Voegtlin 2020).

5 Key Components of Legal Frameworks for Corporate Governance

These prominent components should be followed while designing legal framework for corporate governance (Fig. 2):

5.1 Regulatory Bodies and Legislation

The law and regulatory legislation considered as the key component of legal framework for corporate governance as enforcement of rules, norms and standard is the vital task of it. Stringent rules and norms for financial reporting, independent board and their monitoring, appointment of external auditor for internal control were some of norms of the Dodd-Frank Act (2010) and Sarbanes-Oxley Act; 2002 (Luo 2005).

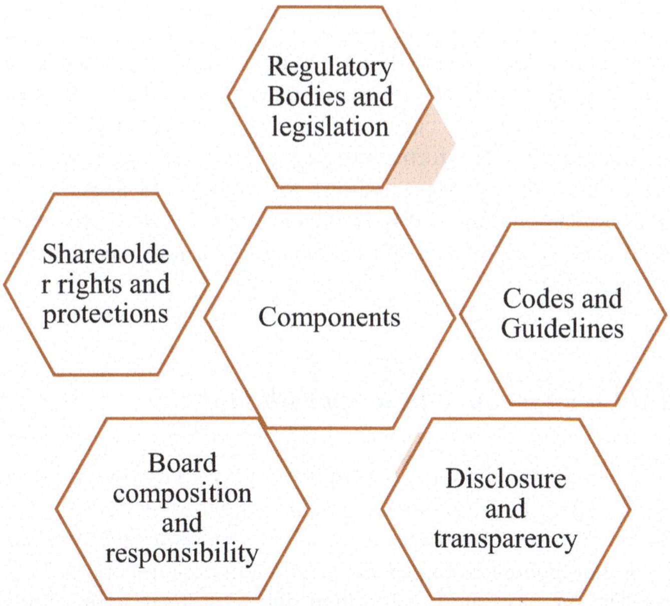

Fig. 2 Key components of legal frameworks for corporate governance. *Source* Author's Compilation

The Enron and WorldCom Scandal gave rise to the Sarbanes-Oxley Act, presented comprehensive rules for improvement of corporate accountability and sustainability such as public appearance of financial reporting and strengthen the corporate accountability and sustainability (Cooper and Owen 2007). They have stronger rules for audit standard so that it help to provide shield to the shareholders from dishonest financial practices. So likely, the Dodd-Frank Act required to lower systemic risks in the financial system by extending regulatory monitoring following the global financial crises in 2008.

5.2 Corporate Governance Codes and Guidelines

In addition to the mandatory legislation, codes and guidelines has also developed for the firms that offer voluntary suggestions to the firm for best practices with in the organisation. The involvement of stakeholders in decision making and risk management with their capacity, and structure independent board. Although adherence to these codes and guidelines are totally voluntary but they function like crucial standard for governance of corporate (Hoffmann et al. 2018).

5.3 Disclosure and Transparency Requirements

This is crucial to have disclosure and transparency component for the sound governance of corporates. Comprehensive financial reporting and disclosure standards are enforced by legal frameworks to ensure that investors and stakeholders are provided with accurate and up-to-date information. These policies protect against corporate fraud and uphold investor trust by promoting transparency in business processes. This creates trust between corporate and stakeholders when they have all required information in their eyes, it allows to take frequent and well-informed decision (Global CSR Disclosure 2023).

5.4 Board Composition and Responsibilities

The board composition and their responsibilities play a key role in governance of corporates for bringing best legal practices and improving supervision and decision making. The board composition dictates the fiduciary duties and responsibility of the directors so that guarantee to provide best practices and work in the interest of shareholders. The directors are loyal to their duties, perform with proper care and monitor them unbiasedly to facilitate lower occurrence of conflicts of interest and they diverse viewpoints and opinion or have expertise views if shareholder takes part in decision making process and formed the diverse and independent board (McCahery and Vermeulen 2006; Jain et al. 2014).

5.5 Shareholder Rights and Protections

The goal of corporate governance is to protect the rights of shareholders and protect them from fraudulent business practices. Shareholders take part in decision making and get timely update about companies' issues and solve their issues with the proper interaction with management. Schymik (2018) defined some of the ways such as proxy voting, resolution team of shareholders and shareholder activism for giving shareholder a power to impact governance procedure and hold management responsible. The shareholder have power to present their opinion and interest if they are unsatisfied with certain policies of the corporate.

6 Impact of Legal Frameworks on Corporate Accountability and Sustainability

6.1 Corporate Accountability

The corporate accountability pointed out clear fines, penalties and lawsuits and figure out the companies' misconduct and noncompliance activities. Under this corporate is accountable for prevention of shareholders rights and outlining clear expectations for corporate behaviour (Tricker 2020). Furthers, corporate accountability serves to improve the corporate integrity by following and fulfilling legal guidelines, law and ethical obligations so that result in reduction in fraudulent behaviour. Consequently, the legal system affects corporate accountability, as is covered here:

6.1.1 Enhancing Transparency and Disclosure

Following the corporate accountability ensure the transparency and disclosure practices for the shareholders. The transparency and disclosure practices expanded mainly by the enactment of Sarbanes-Oxley Act (SOX) in the year 2002. According to Deumes et al. (2012), SOX enforces strict internal controls and financial reporting regulations, which subsequently reduce corporate fraud and increase investor confidence. The act's stringent requirements for independent auditors and accurate financial reporting have made it more challenging for companies to conceal unethical activity (Fig. 3).

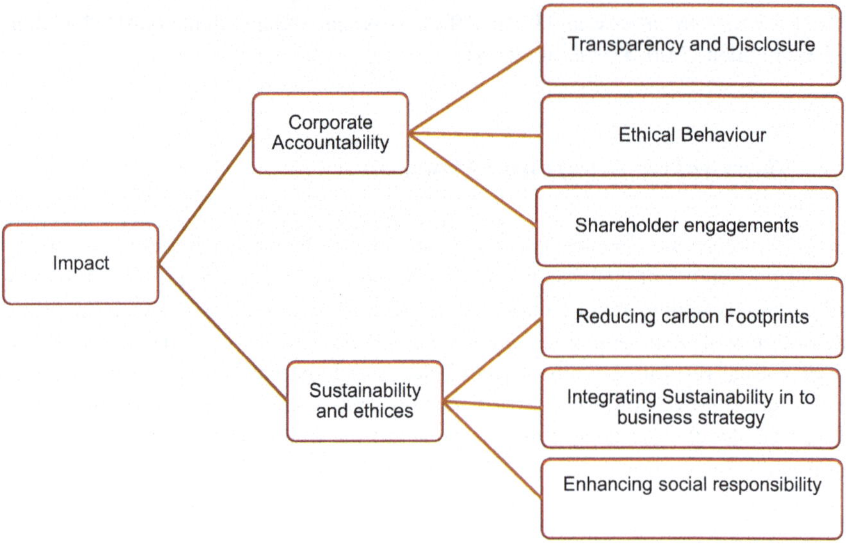

Fig. 3 Impact of legal frameworks on corporate accountability and sustainability. *Source* Author's Compilation

6.1.2 Promoting Ethical Behaviour

The UK Bribery Act 2010 has passed to encourage moral behaviour with the stakeholders. The business has strict anti-corruption practices that provide protection against bribery and corruption. With strict regulation and legislation, UK based corporate follow and upload updated and strict guidelines for moral conduct for best business practices and enhancing global corporate accountability and responsibility (Arjoon 2005).

6.1.3 Encouraging Stakeholder Engagement

Governance codes encourage companies and stakeholders to participate in business practices like UK corporate governance code given that power to communicate with their shareholders. Shareholders engagement includes board composition, discussion about shareholders rights and cultural transparency and accountability. Therefore, it improved the communication between the shareholders and ensure corporate's decision are in favour of everyone interest (Mallin 2016).

6.2 *Sustainability and Ethical Practices*

The integration and introduction of Environmental, Social, and Governance (ESG) principles and their application to legal framework showing the significance of sustainability. The long-lasting corporate adoption and ethical projects is sparked by legal incentives and requirements related to sustainable and ethical business practices. The EU's Non-Financial Reporting Directive, for instance, requires large corporations to report information on social and environmental issues to promote more accountability and transparency in these areas.

There is growing recognition now a days of the significance of sustainability activities in corporate governance. Apart from creation of legal framework it also thinks and promotes values for society and shareholders. Standard of ESG incentivize corporate to integrate sustainability in their main operations and visible impact can see as the corporates following the ESG standards are facing lower risks and promoting sustainable development (Felo 2011). Here are a few notable consequences that have been discussed:

6.2.1 Reducing Environmental Footprint

Companies are reducing the environmental footprints by following the rules pertaining to environment. The Emissions Trading System (ETS) are in role for reducing greenhouse gas emission, it worked for European union (Konadu et al. 2021). On the similar ground Ellerman et al. (2016), explained that ETS has significantly reduced emissions from energy-intensive industries, and appreciated the market-based methods for bringing sustainability (Albitar et al. 2023).

6.2.2 Integrating Sustainability into Business Strategy

The companies are forced to incorporate sustainable business practices into the business strategy. As per the KPMG research approximately 80% businesses presented report on sustainability by following the Global Reporting Initiative (GRI). Due to popularity and wide spread adoption, environmental and social reporting has been disclosed by many corporates in transparent manner and make informed stakeholders about crucial issues of businesses in order to take wise decision (Naciti et al. 2022).

6.2.3 Enhancing Social Responsibility

Labour regulations and human rights strengthened the social responsibility parameter of sustainability. The UK Slavery Act has been passed by the UK government for providing protection against trafficking and slavery. LeBaron and Ruhmkorf's (2017) this has led to improved labour practices and increased supplier chain inspection.

More accountability is being placed on businesses for the working conditions of their suppliers, which promotes social justice and equity (Gold and Taib 2023).

7 Challenges and Failures of Corporate Governance

There have been certain challenges in achieving the desired outcomes from corporate governance system. At times, this system has failed in acting as guardian to protect stakeholders from the misconduct from key managerial personnel. There have been instances of conflicting interest between management and stakeholders where management overlooked the overall interests of stakeholders and upheld their direct or indirect interests. Real independence of directors is a major concern then only directors can ensure no insufficient oversight of management of business entity (Manda and Khaliq 2024; Gopalakrishnan et al. 2008).

The lack of transparency and insufficient disclosure also prompted corporate governance failures at few instances. Non-compliance to legal norms may lead to penalties and Loss of reputation resulting to loss of stakeholders' confidence in overall corporate governance system (Lenz and Chesshire 2023). Another issue that hampers the corporate governance system has been the short term focus by the management as this might yield short term results but does not serve the purpose in long run. Another major reason for governance failures has been varied cultural and Ethical issues, especially in case of MNCs operating in varied geographical setup. (Mujkic and Klingner 2019; Shapira and Zingales 2017).

The major corporate governance failures are listed below along with the major highlight of the failure (Table 1):

All these governance failures indicate that the major worry point has been on intended breach of trust and role of auditors in the process (Bruner and Sjåfjell 2019). These failures also provide for improving compliance norms to ensure that same breach is not repeated in future (Wahyuningrum et al. 2023).

8 Conclusion

Corporate governance ensures accountability towards stakeholder and sustainability towards environmentally friendly operation. This chapter offers a thorough clutch of these governance that influence the corporate to behave with in the legal framework. The present chapter analyzed the historical development, its key components, and its impact of governance. It illustrates the success and failure or challenges through the real cases of corporates and studies and providing insightful advice for bolstering legislative systems around the world.

Through this study it underlines that the corporations ought to underscore that how important to have strong governance framework, corporate culture accountability and maintain transparency. To prevent governance disasters and promote the best

Table 1 Major corporate governance failures

S. No.	Name of company	Year of failure	Major issue
1	Enron corporation	2001	Widespread corporate corruption and poor governance mechanism led to accounting fraud and misleading financial statements, eventually it turned into one of the largest bankruptcies
2	WorldCom	2002	Governance lapses led to a high-volume accounting fraud in which there was overstatement of assets by billions of dollars, finally leading to bankruptcy.
3	Lehman Brothers	2008	Company made risky investment in subprime mortgages and had exposure to inadequate risk management practices which eventually resulted in the collapse of firm
4	Satyam computer services	2009	Serious governance lapses inspite of being awarded for best corporate governance practices, auditors role was in question mark as well, major issue was inflating profits and falsification of accounts
5	Pamalat	2003	Financial fraud including the misrepresentation of accounts and hiding debt
6	Tyco international	2002	CEO and other top officials looted the company for personal financial gains, misuse of corporate funds via unauthorized transactions
7	Volkswagen	2015	A software was used to manipulate emission tests, a serious case of ethical lapses and non-compliance to regulations.
8	Olympus corporation	2011	Firm concealed losses worth $1.7 billion via a series of fraudulent accounting transaction in books of accounts
9	Wirecard	2020	Euro 1.9 billion was found to be missing from company's accounts, which resulted in insolvency of firm
10	Wells Fargo	2016	Employees found guilty of millions of unauthorized bank accounts to meet sales targets

practices in corporate governance, they should also emphasize how important it is to impose strict regulatory frameworks to avert governance calamities and advance optimal practices in corporate governance. In order to accomplish this, the corporate incorporate board independence and strong internal control and effective engagement of stakeholders in decision making process by adhering laws, standards, guidelines and code and conducts which ultimately promote accountability and transparency.

9 Recommendations and Suggestions for Strengthening Legal Frameworks

It is suggested to implement suggested views and ideas to improve the efficacy and effectiveness of the legal framework. Certain rules and standards must be harmonized globally for improvement of corporate governance. Harmonized and systematic governance supports accountability and sustainability which is incentivized by international collaboration and legislative authorities like International Organization of Securities Commissions (IOSCO) and the Organization of Economic collaboration and Development.

For promoting and encouraging ethical behaviour and stakeholder trust, it was suggested some initiatives like disclosure of financial and non-financial information, discussion and talks with stakeholder and regular reporting on governance procedure. So, it is advisable for the businesses or corporates that stakeholder involvement and openness is necessary (Jain et al. 2014). Secondly it is recommended to employ cybersecurity, digital transformation and climate change issues must be included in legal framework and accordingly modified that. With the similar views researchers also advised to employ emerging issues and modified governance standards and provide transparent direction for best practices.

Thirdly it is mandatory for the public corporate that they should annually disclosed the sustainability report by following internationally recognized standards like Sustainability Accounting Standards Board (SASB) and the Global Reporting Initiative (GRI) with the integration of Environmental, social, and governance (ESG) measures (Cole et al. 2021). For the reporting purpose proper and standardized format should be followed. Further, it is also recommended that the corporates should discloses the financial, non-financial and sustainability reporting together so they suggested the Integrated reporting that is contended by the International integrated reporting council (IIRC 2020). Use independent auditor committee, verification of sustainability reports for improving and suggesting correctness (Rossi 2010). By doing so, stakeholders' trust can be increased and greenwashing can be avoided (Deumes et al. (2012)).

Another, the board composition should have equal number male and female. Therefore gender diversity is recommended to have stronger governance practices, including higher levels of monitoring and accountability Adams and Ferreira (2009).

From a regulatory perspective, regulatory organizations should receive increased funding to strengthen their ability to oversee and enforce compliance (Transparency International 2019). For that purpose, number of inspectors should be employed, provide them training and updated and refresher courses for emerging technologies. This is very challenging that global corporate responsibility and sustainability norms are not unified; leads to arbitrage process and inconsistent enforcement. Such as the IRENA, 2018 to increase the use of renewable energy government offers different incentives like tax breaks and subsidies offer by the policy makers. The Fair Labor Association (FLA 2015) suggested to raise minimum wages, worker welfare and raised their living standard in order to bring workers satisfaction which ultimately raise the productivity of business product.

10 Implications and Scope for Future Research

10.1 Implications of Legal Frameworks on Corporate Governance and Sustainability

Legal frameworks for corporate governance and sustainability have wide-ranging implications for business, its stakeholders, and society.

Business Conduct: Such frameworks make corporations more responsible, especially in terms of ESG standards. This transition into responsible practices assures companies that, over time, they focus on creating sustainable value since financial performance will align with sustainability and responsible business practices.

Stakeholder Engagement: Legal frameworks are also redrawing the way shareholders and broader stakeholders interact with corporations. Enhanced protections and transparency measures increase the influence of a stakeholder in corporate decisions. Increased participation means that corporations will always be answerable for their actions, which engenders more impressive ethical behavior and sustainable long-run mechanisms of action.

Global Economic Impact: On the macroeconomic level, these frameworks contribute to a just and sustainable global economy. Stronger ESG regulation nations are attracting responsible investments, stimulating economic growth while also helping to deal with more urgent global problems, such as climate change and inequality.

10.2 Scope for Future Research

Even though corporate governance frameworks have made significant strides, however, many areas remain relatively underexplored, which offers plenty of scope for future research:

Harmonization of Global Frameworks: Corporate governance and sustainability frameworks ought to be harmonized across different jurisdictions. Multi-national firms find themselves operating under numerous different regulatory environments, thus creating inconsistent compliance and enforcement. International standardization, guided by the forces of entities like the UN or OECD, can create unified and comprehensive guidelines for corporate governance. This paper would address the issue of regulatory arbitrage, stemming from companies exploiting weaker governance standards in a particular region.

Effectiveness of ESG Disclosures: Despite the growing rate of ESG reporting, many remain skeptical about how far these disclosures really push the corporate accountability agenda. Future research could look into whether and how the mandatory reporting of ESG impacts the long-term sustainability performance of a company. Comparative studies of reporting frameworks, including GRI, SASB, and TCFD may look at how these ultimately improve transparency and avoid "greenwashing." Researches on innovation in digital models, such as AI and blockchain, may also be very enlightening in terms of how the accuracy and access to sustainability data may be revolutionized in the near future.

Strengthening Enforcement Mechanism: This is one area where compliance to corporate law depends absolutely. Future research should delve into detailing the effectiveness of the present day enforcement mechanisms, be it legal and market-based and how penalties can be strengthened to encourage responsible behavior. The role of nongovernmental actors also needs to be researched in holding companies accountable, mainly through the consumers and investors and other civil society.

Technological Innovation and Governance: As researchers redefine the corporate governance model in new ways with AI, big data, blockchain, and others, they could find scope in those areas for these innovations to transform the formulation of transparency, accountability, and risk management. The ethics of AI use in decision-making and digital channels for stakeholder engagement provide an area to build upon further research. Technology can help reduce greenwashing through a process of monitoring and verification of corporate sustainability claims.

Emerging Markets and Corporate Governance: The main concentration of research on corporate governance is in the developed markets. However, such markets are uniquely challenging. Future research will focus on how the governance frameworks have to be adapted to fit in with the socio-economic contexts of such regions. In addition, research may come along to find out how emerging markets adapt to the global ESG standards and how governance reforms affect sustainable development in those regions.

Filling these gaps will significantly help future research add strength and flexibility to the legal frameworks of corporate accountability and sustainability.

References

Adams RB, Ferreira D (2009) Women in the boardroom and their impact on governance and performance. J Finan Econ 94(2):291–309

Affes W, Jarboui A (2023) The impact of corporate governance on financial performance: a cross-sector study. Int J Disclosure Governance 20(4):374–394

Albitar K, Borgi H, Khan M, Zahra A (2023) Business environmental innovation and CO_2 emissions: the moderating role of environmental governance. Bus Strateg Environ 32(4):1996–2007

Almaqtari FA, Al-Hattami HM, Al-Nuzaili KM, Al-Bukhrani MA (2020) Corporate governance in India: a systematic review and synthesis for future research. Cogent Bus Manag 7(1):1803579

Aras G, Crowther D (2008) Governance and sustainability: an investigation into the relationship between corporate governance and corporate sustainability. Manag Decis 46(3):433–448

Arjoon S (2005) Corporate governance: an ethical perspective. J Bus Ethics 61:343–352

Bedenik N, Barisic P (2019) Nonfinancial reporting: theoretical and empirical evidence. Sustain Manage Pract:1–22

Bruner, C. M., & Sjåfjell, B. (2019). Corporate law, corporate governance and the pursuit of sustainability

Bui H, Krajcsák Z (2023) The impacts of corporate governance on firms' performance: from theories and approaches to empirical findings. J Finan Regulation Compliance 32(1):18–46

Cole R, Johan S, Schweizer D (2021) Corporate failures: declines, collapses, and scandals. J Corp Finan 67:101872

Cooper SM, Owen DL (2007) Corporate social reporting and stakeholder accountability: the missing link. Acc Organ Soc 32(7–8):649–667

da Costa APP (2017) Corporate governance and fraud: evolution and considerations. Corporate Governance Strategic Decis Making 23(1):15–23

Deumes R, Schelleman C, Vander Bauwhede H, Vanstraelen A (2012) Audit firm governance: do transparency reports reveal audit quality? Audit J Pract Theory 31(4):193–214

Effah NAA, Asiedu M, Otchere OAS (2023) Improvements or deteriorations? A bibliometric analysis of corporate governance and disclosure research (1990–2020). J Bus Socio-Econ Dev 3(2):118–133

Felo AJ (2011) Corporate governance and business ethics. In: Corporate governance and business ethics. Springer Netherlands, Dordrecht, pp 281–296

Global CSR Disclosure. (2023, January 1). https://iri.hks.harvard.edu/csr

Global Reporting Initiative (GRI). (2019). GRI Standards. Global Reporting Initiative

Gold NO, Taib FM (2023) Corporate governance and extent of corporate sustainability practice: the role of investor activism. Social Responsibility J 19(1):184–210

Gopalakrishnan S, Mangaliso MP, Butterfield DA (2008) Managing ethically in times of transformation: challenges and opportunities. Group Org Manag 33(6):756–759

Guha SK, Samanta N, Majumdar A, Singh M, Bharadwaj A (2019) Evolution of corporate governance in India and its impact on the growth of the financial market: an empirical analysis (1995–2014). Corporate Governance: Int J Bus Soc 19(5):945–984

Guluma TF (2021) The impact of corporate governance measures on firm performance: the influences of managerial overconfidence. Future Bus J 7(1):50

Hickman E, Petrin M (2021) Trustworthy AI and corporate governance: the EU's ethics guidelines for trustworthy artificial intelligence from a company law perspective. Eur Bus Organ Law Rev 22:593–625

Hoffmann E, Dietsche C, Christine H (2018) Between mandatory and voluntary: non-financial reporting by German companies. In: Sustainability management forum, vol 26, no 1–4). Springer Nature BV, pp 47–63

Igbinenikaro E, Adewusi AO (2024) Navigating the legal complexities of artificial intelligence in global trade agreements. Int J Appl Res Soc Sci 6(4):488–505

Jaimes-Valdez MA, Jacobo-Hernandez CA, Ochoa-Jimenez S (2017) Corporate governance: international context and trends from 2005 to 2015. Int J Bus Manage 12(3):158–169

Jain S, Blackford B, Dabney D, Small J (2014) What is the optimal balance in the relative roles of management, directors, and investors in the governance of public coroporations? Directors, and Investors in the Governance of Public Coroporations
Konadu R, Ahinful GS, Owusu-Agyei S (2021) Corporate governance pillars and business sustainability: does stakeholder engagement matter? Int J Discl Gov 18(3):269–289
Laine M, Tregidga H, Unerman J (2021) Sustainability accounting and accountability. Routledge
Lazarides T (2011) Corporate governance legal and regulatory framework's effectiveness in Greece: a response. J Finan Regulation Compliance 19(3):244–253
Lenz R, Chesshire J (2023) Rethinking internal audit: governance needs gardening. EDPACS 68(3):7–15
Luo Y (2005) Corporate governance and accountability in multinational enterprises: concepts and agenda. J Int Manag 11(1):1–18
Mallin CA (2016) Corporate governance. Oxford University Press
Manda VK, Khaliq LN (2024) Corporate governance failure and the regulatory fight against it: an examination of select instances from India. In: Cases on uncovering corporate governance challenges in Asian markets. IGI Global, pp 24–54
Martynova M, Renneboog L (2011) Evidence on the international evolution and convergence of corporate governance regulations. J Corp Finan 17(5):1531–1557
McCahery JA, Vermeulen EP (2006) Corporate governance and innovation-venture capital, joint ventures, and family businesses. ECGI-Law Working Pap 65
Means G (2017) The modern corporation and private property. Routledge
Michelon G, Parbonetti A (2012) The effect of corporate governance on sustainability disclosure. J Manage Governance 16:477–509
Miller, A. C., & Loman, M. (2014). Measuring and reporting sustainability: the role of the public sector (white paper# 2)
Mohamad S, Muhamad Sori Z (2011) Corporate governance from a global perspective. Available at SSRN 1817082
Mujkic E, Klingner D (2019) Dieselgate: how hubris and bad leadership caused the biggest scandal in automotive history. Public Integrity 21(4):365–377
Naciti V, Cesaroni F, Pulejo L (2022) Corporate governance and sustainability: a review of the existing literature. J Manag Gov:1–20
Nour AI, Najjar M, Al Koni S, Abudiak A, Noor MI, Shahwan R (2024) The impact of corporate governance mechanisms on corporate failure: an empirical evidence from Palestine exchange. J Accounting Emerging Econ 14(4):771–790
Nunhes TV, Bernardo M, Oliveira OJD (2020) Rethinking the way of doing business: a reframe of management structures for developing corporate sustainability. Sustainability 12(3):1177
Pesqueux Y (2005) Corporate governance and accounting systems: a critical perspective. Crit Perspect Account 16(6):797–823
Roe MJ (2003) Political determinants of corporate governance: political context, corporate impact. Oxford University Press, USA
Rossi CL (2010) Compliance: an over-looked business strategy. Int J Soc Econ 37(10):816–831
Safari M, Parker LD (2024) Understanding multiple accountability logics within corporate governance policy discourse: resistance, compromise, or selective coupling? Eur Accounting Rev 33(4):1467–1496
Saha R, Kabra KC (2022) Corporate governance and voluntary disclosure: evidence from India. J Finan Reporting Accounting 20(1):127–160
Sarbanes-Oxley Act of 2002, Pub.L. 107–204, 116 Stat. 745, enacted July 30, 2002
Scherer AG, Voegtlin C (2020) Corporate governance for responsible innovation: approaches to corporate governance and their implications for sustainable development. Acad Manag Perspect 34(2):182–208
Schymik J (2018) Globalization and the evolution of corporate governance. Eur Econ Rev 102:39–61
Shapira R, Zingales L (2017) Is pollution value-maximizing? The DuPont case, vol No. w23866. National Bureau of Economic Research

Shrivastava P, Addas A (2014) The impact of corporate governance on sustainability performance. J Sustain Financ Invest 4(1):21–37

Singhania S, Singh RK, Singh AK, Sardana V (2022) Corporate governance and risk management: a bibliometric mapping for future research agenda. Indian J Corporate Governance 15(2):223–255

Solikhah B, Maulina U (2021) Factors influencing environment disclosure quality and the moderating role of corporate governance. Cogent Bus Manag 8(1):1876543

Tricker B (2020) The evolution of corporate governance. Cambridge University Press

Waddock SA, Bodwell C, Graves SB (2002) Responsibility: the new business imperative. Acad Manag Perspect 16(2):132–148

Wahyuningrum IFS, Chegenizadeh A, Humaira NG, Budihardjo MA, Nikraz H (2023) Corporate governance research in Asian countries: a bibliometric and content analysis (2001–2021). Sustainability 15(8):6381

Weiss EB (2023) The emerging structure of international environmental law. In: The global environment. Routledge, pp 98–115

The Role of Technology in Today's Workplace: Mental Health, Productivity, and Work-Life Balance

Deeksha Dwivedi

Abstract The integration of technology in the modern work environment has changed the way organizations work. This chapter focuses on the role of technology in today's workplace, specifically discussing how it affects mental health, work-life balance, and productivity. This paper has four key objectives. To discuss the impact of technology on employee's mental health, to explore how the workplace uses technology to promote work-life balance, to determine the implications of workplace technology on employee productivity, and to investigate the impact of technology overload on employee's health. The findings showed that technology can be beneficial for employee's mental health if used properly. Remote work technology and flexible work arrangements have encouraged employees to work in ways that better balance their personal and professional lives. Remote work and technical collaboration tools allow employees to work from any location in the world. Findings show that most of the employees love flexibility with their schedule. Stress, burnout, and technology overload are increasing day by day. Technology has so many positive impacts on employee productivity, but it also has some negative impacts, like stress and burnout. Remote culture and constant availability, and the expectation from leaders to complete the task fast, lead to stress and burnout. Organizations are trying to create policies that can help employee's well-being. Organizations are taking initiatives, like taking a break between work and providing training on how to use technology effectively without being burnt out. Technology plays an important role in today's world. Technology should be used responsibly to support mental health, work-life balance, and productivity.

Keywords Technology · Mental health · Work-life balance · Digital burnout · Artificial Intelligence · Stress management

D. Dwivedi (✉)
Leslie's, Phoenix, USA
e-mail: deekshadwivedi17@gmail.com

A. Singh et al. (eds.), *Human-Centered Sustainability: Bridging Well-Being, Gender Equity, and Technology*, World Sustainability Series,
https://doi.org/10.1007/978-3-032-22020-2_20

1 Introduction

Technology has transformed workplaces over recent decades. It enhances efficiency, productivity, monitoring, idea generation, and flexibility. Organizational systems have drastically changed across all business operations, from hiring and onboarding new employees to project management to taking decisions with the help of data analytics and data visualizations. Emerging technologies like augmented reality, advanced AI, and the Internet of Things (IoT) continue to reshape workplace environments. Such technologies offer unprecedented opportunities for efficiency gains and improved employee engagement. However, organizations must recognize that these technologies should function as enablers rather than substitutes for human capabilities when developing their workforce strategies.

Technology has improved employee productivity. It helped in streamlining the business and operation process. Automation is minimizing repetitive work time with better efficiency, and real-time data and collaboration help the team to take better decisions on time. Technology and automation tools help employees with strategic planning and decision-making, which directly helps business operations. Collaboration platforms and cloud technologies have changed data sharing and communication, especially in the remote and hybrid work environment. Such tools facilitate seamless communication across different locations and regions, enabling employees to maintain productivity while benefiting from increased flexibility.

Latest technologies like artificial intelligence and machine learning are transforming job roles and responsibilities by taking care of repetitive jobs. Artificial intelligence-enabled chatbots are now providing 24/7 customer service, while machine learning processes large amounts of data to get insight to take strategic decisions for leadership. Such technological advancements increase productivity and create new employment opportunities for the latest emerging technologies.

Cloud-based technology and mobile solutions provide increased work opportunities, greater flexibility, and ease of remote work opportunities that are improving both work-life balance and job satisfaction. Successful organizations find a strategic balance between technology and human capabilities (i.e., creativity, emotional intelligence and interpersonal communication). Organizations that emphasize continuous learning, ongoing digital literacy improve employee resiliency to technological changes while allowing employees to focus on wellbeing and engaging collaboratively.

Recent studies show that AI integration into leadership processes, raises emotional intelligence and promotes adaptive decision-making to support a learning organization in a technology-intensive organization (Dwivedi 2025).

Digital tools and technologies are impacting employee's mental health. Digital tools and virtual platforms help organizations in stress management, mental health monitoring, and providing mental health support to the employees. Digital tools and applications, along with virtual services, are helping organizations to support anxiety and stress. Research shows that tools and technologies can help organizations to identify mental health issues on time and support employees in their well-being.

Technology has a significant impact on mental well-being. Work overload, digital overload, constant email alerts, project notifications and pressures to complete the task on time can enhance anxiety, stress, and burnout in employees. The Happy Wisdom Framework shows happiness strategies to promote organizational well-being and counteract technology-induced stress (Agarwal and Dwivedi 2025). Digital detox strategies effectively mitigate FOMO and technostress (Mewafarosh 2024).

2 Review of Literature

Organizations have undergone profound transformation in recent decades, and we are now seeing the rise of transformations brought on by the rapid improvements in artificial intelligence (AI). AI, as it relates to machine learning and many other technologies, including robotics, digital platforms, and other emerging technologies that affect business, has transformed and will continue to transform work and how we understand work going forward.

Productivity gains from technology have been significant. Organizations similarly benefit from predictable workflows and improved choices. Davenport noted organizations are increasingly using advanced analytics and tools to improve talent management (Davenport 2010). Organizations are increasingly using wearables to gather information about employee well-being and engagement to help managers create situations to help them enhance productivity in their teams (Moore 2016).

Digital platforms, tools, and communications have radically transformed work. Organizations can currently use real-time communications to support their work across teams dispersed hundreds or thousands of miles away from each other (Jesuthasan 2017).

Technology is transforming workplace learning through personalized and flexible training experiences. Employees can access IT learning platforms that provide on-demand, individualized training resources. Contemporary work conditions impose new pressures that require continuous skill development in instability and uncertainty. Organizations are now introducing game characteristics in learning modules to assist employees in being engaged and remembering the information they learnt (Holland 2016).

Marler emphasized the pivotal role leadership will play in managing teams via technology. Leaders need to develop the digital literacy capabilities to manage technology while building an inclusive, collaborative organizational culture (Marler 2016).

Technology has changed how HR functions. HR's increasing reliance on and use of data, where HR managers are using analytics, which is more data-driven, when making strategic decisions regarding hiring talent, performance management, and employee engagement (Stone 2015). For example, predictive analytics algorithms can identify high-potential talent that HR managers may not identify, even before the interview phase starts.

On the one hand, data analytics tools are valuable for optimizing leadership decisions. And on the other hand, if HR leaders do not pay careful attention to how data is affecting workplace decisions by managers or the work over the long term, this is a challenge. One increasingly pervasive concern in organizational life is information overload, which is symbolic of a more generalized workplace dilemma where managers need to coordinate and balance productivity with cognition (Dwivedi 2025). HR departments must continue to regulate the unethical use of data in decision-making (Jesuthasan 2017). HR departments should not be overly emotional in determining which platform to use because organizations must be transparent and fair to have the trust and confidence of employees operating within the technology-enabled workplace and the quality of organizational decision-making.

Automation can facilitate the replacement of those repetitive tasks prevalent in manufacturing, logistics, and back-office work. Essentially, technology eliminates jobs arbitrarily while creating new jobs, generally within the technical fields. According to Higgins, augmented reality (AR) and virtual reality (VR) may be disruptive technologies of the future and for industries such as healthcare and construction, where they have improved the execution of tasks that require enhanced accuracy and productivity (Higgins 2018).

Emerging technologies, particularly blockchain, have changed how transactions can be made in a secure environment for data exchanges (Wright 2018). Organizations need to apply technology responsibly after development has taken place and develop a plan to manage the negative effects of adopting technology on employees (Bondarouk 2016). This literature review provided the basis for future studies to understand the effects of emerging technology on work settings.

3 Methodology and Objectives

This study employs a qualitative research approach to gain in-depth insights into the role of technology in today's workplace. The research objectives are as follows:

1. Assess the Impact of Technology on Employee's Mental Health in the Organization
2. Examining the Role of Technology in Promoting Work-Life Balance
3. Assess the Impact of Technology on Employee Productivity
4. Evaluating the Impact of Technology Overload on Employee Well-Being

4 Impact of Technology on Employee's Mental Health in the Organization

The integration of technology within workplace environments has had a significant impact on employee mental health. The digital age has further augmented the already present aspects of mental health assistance to employees. Technology has also brought mental wellbeing tools and apps such as mindfulness apps, tele-therapy sessions, and communities of support networking online. These tools can be flexible and can often instantaneously support someone that is experiencing stress, anxiety, or various other mental health issues. Employers are now also recognizing the importance of these tools and integrating them into the employer wellness ecosystem to improve employee psychological health literacy and resilience. The application of digital technology in a workspace can either positively or negatively affect mental health. If appropriately implemented, the best digital suite can affect mental health by supporting psychological well-being but can also affect mental health adversely if not used properly (Wang 2022).

However, technology also adds new stressors that can negatively impact mental health. Technology allows for increased connectivity and communication via smartphones, emails, and instant messages. This high-pressure environment of being constantly available to respond to work-related issues, worries, and requests can lead to digital overload or "technostress", stress, and burnout.

The collective impact of technology on mental health mostly depends on the strategies organizations choose to confront the issues that arise from technology. Organizations that craft digital communication policies, support structured breaks, and promote a culture that respects personal time mitigate the stressor consequences of technology use see positive impacts of technology. Teaching managers to identify digital fatigue and mental distress and offering mental health resources should build a supportive strategy for mental health. From this, mental health may benefit from technology instead of being harmed by it, stimulating and encouraging workplace wellness and productivity.

4.1 Improved Access to Mental Health Resources

Digital mental health solutions have enabled employees to receive psychological support within work settings. These digital solutions deliver on-demand resources such as cognitive behavioural therapy (CBT) programs, mindfulness programs, and interventions targeting stress management. As technology-driven workplaces become enforced with higher workloads as well as continuous distractions and interruptions, stress management systems have become highly relevant (Dwivedi 2018).

The availability of digital health technology has advanced accessibility for employees to receive mental health resources. This access translates to timely interventions and appropriate support that can have lasting influence and positive effects on workplace well-being (Smith and Jones 2021). Research demonstrates that employees are more willing to engage with interventions that promote performance enhancement and improvements in well-being rather than solutions that target mental illness. Digital tools are flexible and allow employees to engage in mental health interventions at their own pace and time. It is also broadly accepted that, while

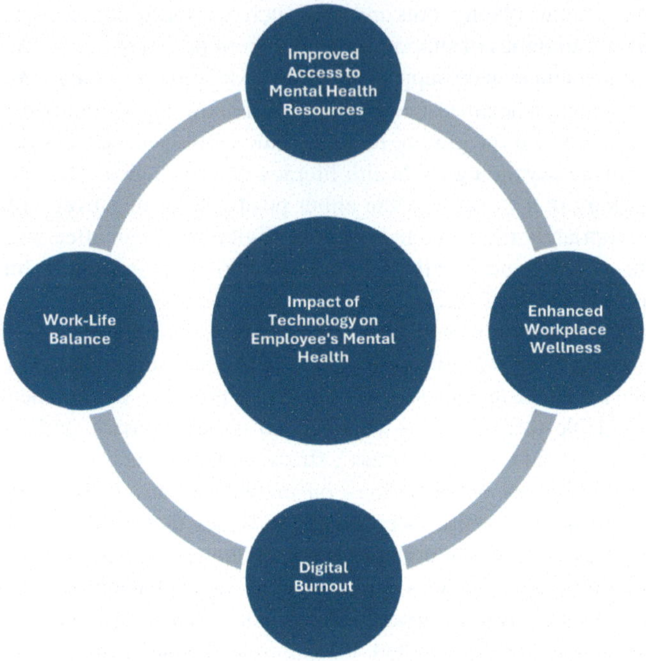

Fig. 1 Impact of technology on employee's mental health in the organization

information technology can increase productivity, it may also impose stress-inducing time pressures on the employee (Dwivedi 2015) (Fig. 1).

4.2 Enhanced Workplace Wellness

Digital wellness platforms have become effective tools for organizations that want to enhance employee well-being through technology. Digital wellness platforms are an intersection between technology and health promotion, allowing for all-encompassing wellness initiatives that meet the physical and psychological needs of employees. Research from Brown and Green (2020) finds that the implementation of digital wellness solutions in workplace settings leads to mindfulness practices, workplace stress reduction, and higher levels of employee happiness & resilience. To implement these platforms properly is an innovative wrinkle to an organization's wellness culture.

Health care organizations have set a great example of successful use of IT infrastructure in support of wellness initiatives. By providing digital platforms that include mental health apps, real-time health monitoring, and preloaded mindfulness applications, these organizations have successfully irrevocably changed their wellness approaches on a larger scale.

The accessibility of digital wellness resources means employees really can use and access these always, which could bring a level of flexibility that many traditional wellness programs can't. The teams and employees who use a variety of digital wellness solutions expose themselves to more adaptive coping strategies that can be utilized more broadly in their workplace and personal lives. The efficiency of using digital solutions promotes increased adherence and creates a more positive and predictable workplace across the board. Organizations that embrace a wide array of digital wellness programming report benefits such as better employee morale, improved productivity, and reduced absenteeism. These long-lasting effects show that we should define workplace wellness with a wider lens in respect to both physical and mental health considerations.

4.3 Digital Burnout

Smartphones, laptops, and collaborative tools create valuable channels of communication and collaboration; they also create shared experiences of physical and emotional exhaustion, which is digital burnout. Employees receive an overwhelming number of emails, notifications, and data streams that create constant distractions, which often create a lack of focus and contribute to stress and have an impact on mental and physical health. Research by Maslach (2016) indicates that demanding unlimited digital connections is a significant contributor to employee stress and fatigue and can accelerate conditions leading to burnout.

Digital burnout is progressive and passes through distinct stages. Initially, longing for new technologies and digital tools, employees may be enthusiastic about the change and resources. This enthusiasm then dissolves into chronic symptoms of burnout, including signs of exhaustion and decreasing performance.

The implicit demands of constant availability and responsiveness, often augmented by electronic tracking and surveillance systems, add to the pressure and stress of work. The "always on" mentality collapses boundaries between work and home, and it diminishes individuals' ability to effectively unplug from work responsibilities and recharge appropriately.

As organizations continue to leverage digital technology for productivity and collaboration purposes, they also should become aware of the potentially negative effects of constant exposure to technology and, in turn, be able to mitigate those effects to sustain employee well-being and organizational effectiveness.

4.4 Work-Life Balance

The role of technology has fundamentally changed. Employees are finding it harder to disconnect from work when they are out of the office or their official working hours, and the expectation is that they should always be reachable or responsive to work communications. The only difference is that there is now a perception of a more pressing need for real-time responses even when it is outside of "normal" working hours. This kind of intrusion into personal time generates a problematic habit for employees because they are engaging with work for much of the time, and over time they are likely seeking to respond at the same time, which inevitably leads to being overworked, increased stress, and ultimately burnout. Many of these technologies have promoted work-life balance or can help support work-life, but Allen (2014) points out that most of those technologies now often create the pressure for organizations to infringe on a person's life outside of the boundaries of work or create new demands that directly undermine employee well-being.

While remote work arrangements may have little side effects on increased job satisfaction, they have the unintended consequence of minimizing work visibility and prompting pressures for workers to showcase productivity through more work effort. Remote workers often find themselves working during non-traditional hours and unable to set up boundaries between their work time and personal responsibilities.

The undermining of boundary settings increases the probability of less satisfaction, burnout experiences, and identified work-life balance among employees. As organizations continue to adopt technology-enabled work arrangements, recognizing and responding to these subsequent and unintended circumstances is crucial to sustainable workforce well-being and productivity.

5 Role of Technology in Promoting Work-Life Balance

Technology plays a major role in advancing work-life balance in general. For example, technology enables flexible work options like remote work, and it has provided tools like video conferencing, project management, or instant messaging that allow employees to work together or collaborate near and far with less transit time and greater productivity. Mobile apps and shared calendars offer organization of a busy work and personal life schedule while clearly differentiating work and personal tasks. Therefore, while technology provides many opportunities for work-life balance, it is essential we are aware of how and when we use different technologies so we do not experience excessive fatigue or burnout, as well as optimize our overall sense of well-being.

5.1 Enabling Remote Work and Flexible Schedules

Technology has given employees greater power to work remotely, providing a unique level of flexibility to work out schedules around personal commitments. Technology also allows employees to access flexible options for scheduling. Employees can also work when it works best for them, optimizing their productivity to align with their natural rhythms, while still meeting personal needs. For example, a worker may find they are more productive working in the evening rather than in the morning and can arrange schedules to accommodate those needs, which ultimately improves job satisfaction and job performance. Performance management systems need to adapt to the influence of club's technology (Dwivedi 2014).

Remote employees are more engaged, as they have the flexibility to create their own routines to maximize productivity and work-life balance. Flexible schedules also improve attendance; employees are much more likely to attend work for a short time if they are not feeling well or have a personal appointment, rather than take the day off.

5.2 Enhanced Task Management and Collaboration

Task management tools have changed how collaborative teams work together. These tools allow for workflows to be more efficient, ensuring that all individuals are productive and can coordinate effectively from any location. These platforms offer numerous features, such as real-time notifications, shared calendars, and task priorities, managing employee's complex obligations with professionalism and thoroughness. Task management with integrated collaboration tools allowing staff to comment and mention colleagues in their progress or even automate their next workflow steps. Employees can recommend task management styles or work streams that would complement their roles in the completion of any large or complex project. Consideration of the digital environment indicates that digitized tools do provide an efficient task management tool and help work collaboratively to ensure workers are able to focus on getting things done (Grant 2013).

Such tools allow employees to recognize efficiencies in completing tasks without sacrificing personal time. For example, users receive real-time notifications that provide information on status changes or that a deadline is approaching too closely and simply stay on top of employees and their accountable responsibilities. Shared calendars and task categorization features (immediate actions required, waiting for action, and long-term project planning) assist users in prioritizing their work continuously. Many tools also allow users to change their view by seeing all the actions taken related to one project or for all activity depending on the different projects. This feature allows the user to manage their workload and decision-making more effectively.

6 Assessing the Impact of Technology on Employee Productivity

Technology has greatly changed the modern workplace. It has changed how employees work, communicate, and interact. Technology has made employees more productive through automation, artificial intelligence (AI), and digital communication technology. For example, automation driven by AI can help to reduce manual work—less repetitive tasks, allowing employees to focus on the strategic and creative elements of their jobs.

6.1 Automation of Repetitive Tasks

Automation has become an essential tool in workplace productivity. Automation allows organizations to ease the operational processes, transforming operational processes such as data entry, invoice processing, scheduling, etc., and helping employees shift effort towards high-value strategic practices. Automation improves productivity and ultimately job satisfaction by alleviating repetitive work.

6.2 AI-Powered Decision-Making

AI has transformed decision-making by analyzing large datasets. By utilizing AI, organizations can reallocate resources to increase efficiency. For example, AI can predict sales trends, find new market opportunities, or recognize potential risks, allowing leaders to make final decisions that are driven by data. AI-influenced tools can help with successful informed decision-making, supporting quicker decisions and higher-quality decisions in the organizational processes (Shrestha 2019).

AI-driven analytics can improve labor productivity. By automating data processing and pattern recognition for the most complex data problems, AI significantly reduced human effort for a task that required many hours. Automating workflow speeds up the decision cycle time, reduces errors and creates a more standardized process. AI requires little or no human interaction to process information quickly and objectively enough to minimize human bias. AI makes more accurate results as well as improves reliability in a decision-making process.

6.3 Enhanced Collaboration Through Digital Tools

Collaboration tools have changed the way teams can work remotely by removing geographical boundaries that allow teams to communicate with one another while working separately. These platforms allow teams to share files in real time, have instant messaging, and conduct video calls so individuals can effectively collaborate on tasks regardless of their geographic location. This ability to collaborate is especially important in a remote working world where keeping everyone on the same page is critical to productivity and morale as teams continue to navigate through limited interactions.

The ability for teams to have the same calendars, share resources without waiting, and work together in remote teams while they are separated by distances are some of the most strategic advantages of a collaborative platform. For example, real-time collaborative documents allow team members to make changes and comments on files when they're working on them, thereby removing the waiting time of having to wait on the return of search and retrieval files we now refer to as back-and-forth email. Similarly, shared calendars and project and task management features provide visibility on the priority timelines for each other on deadlines.

6.4 Flexibility and Remote Work

Remote work policies allow employees to create their workdays according to their personal needs and preferences, resulting in personalized work. Studies have shown that flexible work environments create higher job satisfaction, reduced stress, and improved performance.

Employees can now stay productive from anywhere and at any time. Employees are now using cloud-based project management tools to work on what they need to collaborate on or access documents remotely and complete tasks. The flexibility allows individuals to juggle obligations from work to home (family care, appointments, personal development, etc.) and allows them to feel like they can hold themselves accountable to meaningful work rather than losing out on productivity. To clarify, as digital environments offer flexibility of work commitments, they also offer opportunities for remote working that enhance job satisfaction and productivity within organizations (Bloom 2015).

6.5 Data-Driven AI Insights for Efficiency

Artificial intelligence (AI)-enabled data analytics platforms are indispensable to organization's efforts to increase capacities and reduce inefficiencies. These platforms process streaming volumes of structured and unstructured data and convert that data into real-time information-for-action. Through constant monitoring, AI-based analytics can explicate trends, patterns, or an abnormality that could affect productivity or provide an indication of an emerging problem. Analytics technologies provide data-driven insights that organizations can operate and act on to optimize workflows and improve efficiency (Brynjolfsson and McAfee 2014; Shrestha et al. 2019).

AI-based data analytics technologies can continually promote best practices by prescribing where processes can be streamlined or automated. In the case of workflow analysis, a similar analysis may reveal patterns of repetitive manual tasks that could be automated, allowing an organization or employee to promote higher value utilization. In a similar view, AI-powered analytics platforms could help track organizational key performance indicators (KPIs) in real-time and allow organizational managers or leaders to make better-informed, timely decisions to quickly pivot towards other efforts if warranted.

7 Impact of Technology Overload on Employee Well-Being

Technology overload encompasses the burdens placed on employees by digital tools and platforms, which negatively impact on their health and productivity. Information overload occurs when employees are flooded with emails, notifications, and data trails that hinder their ability to effectively prioritize their tasks or think straight. When they are inundated by information, it can leave employees feeling confused, without the ability to focus, and with no understanding of their priorities. It can also cause employees to miss deadlines and feel overwhelmed.

Constant connectivity fosters a culture in which employees are expected to be "always on". Because employees are always connected, they cannot disconnect properly, resulting in increased stress, burnout, ambiguity, and work-life misalignment to imbalance with their job responsibilities and personal life (Fig. 2).

7.1 Burnout and Fatigue

Digital overload is causing employee burnout in today's technologically demanding work environments. The ongoing barrage of emails, messages, notifications, and work-related digital tasks creates a context where employees find it difficult to unplug, disconnect, and separate from their work. The expectation for employees to remain connected, regardless of time, extends their working hours and obstructs boundaries between their work and daily life, creating ongoing stress with emotional fatigue.

Fig. 2 Impact of technology overload on employee well-being

7.2 *Reduced Productivity*

One of the main concerns is the constant interruptions by notifications from emails, messaging applications, and social media that distract from the employee's ability to do deep work doing complex tasks in a state of concentration. These notifications force the employee's attention on different things, which leads to attention being pulled constantly because their focus has changed, not only shortening attention spans but also lowering cognitive performance as well. Increased reliance on digital tools may produce a noticeable loss of production due to the reduction in focus and mental fatigue (Tarafdar 2015). Constantly checking our phones, finding ourselves multi-tasking, and being connected to various conversations are all examples of the fragmented workflow problem that prevents employees from maintaining their momentum on difficult projects.

7.3 *Organizational Impacts*

The effects of technology overload also impact on an organization, not just individual employees. One significant effect is lower job satisfaction, as employees dealing with continuous digital demands are likely to feel frustrations, disengagement, and lack of motivation. This drop in morale will lead to lower outcomes and lower teamwork sense of purpose. The presence of technology overload and overwhelm in an organization has negative effects by introducing higher rates of absenteeism and disengaged employees (Tarafdar 2015).

If employees are experiencing exhaustion and chronic stress from unforgiving digital demands, they will be more likely to change jobs. With fluctuating turnover, organizations, along with continual recruitment and onboarding costs, will affect not just continuity of workflow, but stability and ultimately growth.

8 Strategies to Enhance Mental Health, Productivity, and Work-Life Balance in the Organization

Organizations seek to establish a workplace environment that promotes employee well-being, productivity, and ownership of work-life balance. By using evidence-based actions, organizations can effectively support employee well-being. Supporting employee well-being means recognizing and supporting employee physical, mental, and emotional health. Employees in good health will be more productive and engaged in their work. Employee well-being can be enhanced by offering regular mental health assessments, stress management workshops, and access to counselling, reducing burnout and ultimately resulting in greater employee well-being.

Maximizing productivity means organizations will need to adopt workflow practices that limit unnecessary burdens. Examples of reducing time on routine tasks with technology so that employees can spend their time doing work that has higher value or free up time altogether, therefore improving efficiencies. Flexibility in work expectations, such as opportunities to work remotely or with flexibility in hours, helps employees balance their work and other responsibilities. This balances the workload and improves performance and satisfaction at work. Maintaining an appropriate work-life balance is about establishing basic boundaries in an employee's work and personal lives. Some of these methods are promoting regular breaks, allowing people to take vacation time and establishing norms around after-hours communication.

8.1 Flexible Work Arrangements

Providing flexibility in the workplace in the form of remote work, hybrid work, and flexible work schedules is an effective way to promote mental health by supporting employees. Flexible work arrangements give employees more control over their workload and support employees in managing their personal responsibilities, thereby reducing the stress caused by traditional work structures. Flexible work options also enable employees to maintain their productivity in the workplace while still attending to important personal matters, such as family obligations, medical appointments, and self-care, all of which are essential for maintaining a job. Flexible work options offer employee's control over their work life while alleviating the pressure of competing demands. Research evidence demonstrates that employees report lower stress, increased engagement, and overall well-being when they have a flexible work option. Flexible and remote work options may provide employees with the opportunity to enhance their mental health and work-life balance while reducing levels of stress and being more productive (Bloom 2015).

8.2 Promoting Digital Wellness

One effective mechanism to help them do this is with policies like "no-email weekends" or "quiet hours". These policies encourage employees to disengage from workplace communications and to fully recover from the demands of the workplace. Establishing boundaries reduces the pressure of having to be constantly available and has a significant impact, allowing employees to recharge mentally and physically. It also reflects positively on the organization as they honor the personal time of those who work there by reducing after-hours electronic communications that interfere with normal recovery time. By offering employees effectively planned-out boundaries around technology, they will benefit from less stress, a higher level of focus and productivity, and more engagement in their work. In addition to boundaries for employees around technology, mindfulness training and regular stress management workshops enhance employee's mental health. Mindfulness is the non-judgmental awareness in the present moment, and it provides people with the opportunity to develop greater awareness and ultimately control over their responses to challenges and stress, thereby increasing their emotional resilience. The early stages of stress management systems can inform the current state of competing workplace challenges, such as the increased stress associated with ICT workload (Dwivedi 2011). Stress management workshops give employees practical and useful tools to effectively manage their pressures of work in the workplace where performed. Encouraging healthy tech habits, being in the habit of mindfulness, and supporting employee's mental health are a few strategies organizations can employ to support their employees with the challenge of burnout, job satisfaction, and the experience of having a better workplace. Organizational commitment to digital wellness can be realized by promoting healthy technology and breaks and downtime, which can help employees from suffering from burnout (Maslach 2016).

8.3 Employee Assistance Programs (EAPs)

Employee assistance programs (EAPs) are helpful programs offering confidential counselling and support for employees dealing with mental health issues. EAPs help employees access professional counsellors to support their mental health needs, including stress, anxiety, depression, and work-life balance challenges faced by employees. EAPs in the workplace can receive the support they need in a confidential and safe setting. Encouraging employees to use EAPs is a good indication of an organization supporting the well-being of employees and promoting a supportive workplace culture. Programs like EAP often provide counselling and workshops, self-help materials, and referral services where employees can work with specialty mental health providers for employees who require ongoing or specialized care. "EAPs provide confidential psychological support for mental health problems, paying special attention to help employees overcome work-related stress as well as to improve employee wellbeing or wellness" (Cooper 2017). Employees who feel supported will be far more likely to feel satisfied in their jobs and remain loyal to the organization they work for. Companies that embrace EAPs in their wellness strategy will be creating a supportive environment where employees will always feel enabled to ask for help when unexpected events occur in their personal or professional lives.

8.4 Leadership Training on Mental Health

It is important for managers to learn to identify the signs of mental health concerns at work to create a culture of support. Training programs should be designed to teach managers how to recognize the early signals of stress, burnout, anxiety or depression. These indicators come in many forms, including observed changes in behavior, observable decreases in productivity or engagement, and pulling away from interactions with their team. By understanding early signs of an employee's mental health struggles, managers can seek to intervene at an earlier stage and offer appropriate help. Training leaders in general mental health awareness and providing information about employee wellbeing can help leaders create supportive workplace environments, reduce or eliminate the stigma about mental health, engage employees positively, and enhance sporting environments or workplace wellbeing (Brown 2020).

9 Strategies to Boost Productivity

There are many ways to improve productivity in a technology-based workplace today while helping mental health and work-life balance. Increasing productivity in today's work environment means using a combination of factors, including technology, well-being, and intentional collaboration. Perhaps the most effective means of boosting productivity is with artificial intelligence (AI) and automation. There have been numerous AI tools developed that now help to automate routine tasks, analyze a large amount of data in a short time, and provide valuable insights for employees to do more creative and strategic work. AI tools can also improve the overall efficiency of projects by predicting timelines, identifying risks, and pairing tasks with the more suitable team members while also improving collaboration.

Flexible work options can also significantly contribute to productivity. Providing flexibility for employees to choose their own work hours and locations places them in control and allows them to operate when their productivity is highest, which means lower burnout and greater enjoyment. Facilitating events that support team-building or cross-functional collaboration helps to create a connection and purpose in belonging to a team and organization. Building personal relationships, even in a technology-orientated environment, is a substantial factor in a productive and engaged workforce.

9.1 Implementing Productivity Management Software

Productivity software provides visibility into team activity, workloads, and goal completion as it relates to time spent on that goal, all in real-time. This allows managers to quickly identify sources of delay or improvement. These software tools all include features like productivity reports, dashboards, and time tracking, making it easy for organizations to measure how employees spend working hours in terms of productive, unproductive, and idle time. Productivity management software is useful for tracking tasks and deadlines and can improve overall efficiency within the organization (Grant 2013).

Productivity tools can also be used for workload management and capacity planning, so leaders can understand how resources are being allocated and how to optimize their people. When transparency is established, productivity software helps leaders make analytics personal with privacy-first analytics, allowing employees to improve their own productivity and work-life balance while the organization builds trust. Finally, using productivity software is an effective and strategic means for optimizing workflows, establishing benchmarks that are meaningful, and enabling long-term success for the organization.

9.2 Automating Repetitive Tasks

Tools such as artificial intelligence (AI) are creating a profound change in the modern workplace by allowing organizations to automate repetitive tasks, streamlining processes and enabling unprecedented levels of efficiency. Essentially, organizations can utilize AI in automation to take over many of the mundane tasks we encounter daily. Ranging from document processing, data entry, and email management to customer support and report generation, AI can automate any lower-value functions, regardless of the scale or complexity. When organizations can replace lower-value tasks with an intelligent system, they reduce human errors, produce more accurate results, and free employees from previously time-consuming processes that involved manual effort and tasks that could take hours.

Perhaps the most significant impact of automation is the level of employee inspiration that is now possible when they can spend most of their time focused on higher-value creative and strategic work. Another benefit of automating repetitive tasks is the ability to save money and outdoor control costs. More organizations can utilize automated systems to scale their operations; to think about it, the less a company must rely on manual labor to perform a task, the more it can scale down labor costs. Automation can eliminate many or all the repetitive processes for workers and reduce the need to do the 'easy or boring work', allowing employees to spend time on more difficult work (Shrestha et al. 2019).

9.3 Encouraging Regular Breaks

Encouraging regular breaks is vital to ensuring productivity and well-being in the workplace. Short breaks will help manage fatigue and can ensure employees maintain focus by providing necessary space to recharge, both physically and mentally. Scheduled breaks during the workday are vital for helping employees be focused and creative. Having scheduled breaks allows time for an employee's mind to turn off, recharge, and then when working again, they can re-engage the task with more energy and an improved focus. This could be as simple as standing up or changing their work position. Intervals of focused work followed by a brief pause can enhance cognition and productivity. Offering employees the opportunity to step away from their screen during breaks will also mitigate strain on their eyes and help relieve stress. A supportive culture surrounding breaks will foster a more engaged, productive, and healthy employee. Breaks or downtime can break fatigue and allow employees to refocus their attention, allowing for greater sustained productivity for longer periods of work time (Maslach 2016).

9.4 Setting Clear Goals and Priorities

Regular reviews of goals are an important means of ensuring that an organization stays aligned with a mission and that employees remain focused on priority tasks. More frequent touchpoints allow employees to see how their actions are contributing to progress, address barriers, and potentially amend goals to reflect priorities and insights that might have changed since the goals were set. This reviewing environment also helps employees to manage competing demands along with their workload more effectively, thus addressing the feeling of being overwhelmed by competing demands. Clearly defined and outlined goals provide direction for employee time and energy related to completing work, which can foster a sense of motivation and accomplishment, enabling a greater productive effort (Grant 2013).

The combination of regularly reviewing goals and prioritization can also help employees understand what area they should focus their energy on to enhance productivity and satisfaction. Additionally, a culture of regular goal reviews also encourages a culture of accountability and continuous improvement, as individuals can see how their work fits within the organizational frame.

10 Strategies to Improve Work-Life Balance

It is crucial for the well-being of employees and organizational productivity for employees to achieve a healthy work-life balance. Individuals and organizations can support achieving work-life balance in several ways. Flexible work arrangements such as remote work, hybrid work, or having flexible hours help employees balance personal and professional demands.

Another approach is to encourage employees to take time off, use their vacation days, and take breaks to help employees de-stress and recharge while reducing burnout and supporting mental health.

An important way to maintain a healthy work-life balance is to set boundaries between work and home life. This might mean putting your out-of-office message on and shutting down work notifications after your workday is done. Employers can help their employees establish a work-life balance by respecting their employees time away from work by refraining from contacting employees outside of work hours. Employees, supported by employers, should be taught time management and prioritizing skills, which can promote improved efficiency and reduce stress for employees.

Employers support an inclusive and supportive culture, where employees can comfortably discuss and acknowledge their workload and ask for help if required. Providing employees with easy access to employee assistance programs, mental health and wellness resources, or programs can promote employee well-being.

Regular reviews of employee workloads, and analysis of shifting one employee's workload to other employees when necessary, can help ensure employees are not overworked. Finally, leadership in organizations should model behaviors that support work-life balance, so employees do not feel guilty or shamed for attempting to achieve a healthy work-life balance. Through the introduction of these initiatives and programs, individuals and organizations alike can create an environment in which work-life balance flourishing, and work-life can coexist.

10.1 Encouraging Use of Paid Leave

A significant number of employees are reluctant to take vacations due to fear of not being current or possibly letting their team down and additional stress, burnout, and the inevitable lack of productivity. It is important to spur your employees to use their paid time off and not to be afraid to do it. Encouraging paid leave helps recovery and avoids burnout, both of which enhance work-life balance (Cooper 2017).

One way to do this is to simply remind employees when they have vacation time that they have accrued but have not used. The management team can create policies that maintain an employee's vacation experience and do not allow interruptions related to work. That would include asking colleagues and managers not to contact employees about work while they are away.

Creating a culture that values and allows time off not only protects the mental health of the employees but may also increase productivity for the entire organization. When employees return to their job well-rested, they will be focused and motivated, and their creativity will be unimpeded. Supporting your employees and their use of vacation is a signal to them that the company supports and prioritizes their well-being; this can lead to higher job satisfaction and retention.

10.2 Flexible Work Schedules

Flexible work schedules provide the option for employees to adjust their start and end times or work locations, an alternative to the typical 9-to-5 workday. Flexible work schedules can take many forms, including remote work schedules, hybrid work schedules, compressed workweeks, and flextime (when employees allocate their hours to their own personal needs).

For employees, flexible work schedules allow them to achieve great advantages, including improving their work-life balance, lowering their stress level, and improving their health. Employees can better manage family responsibilities and attend appointments without taking paid time off. Increasing employee's flexible work schedules allows them to have more autonomy at work, leading to greater job satisfaction, greater motivation, and increased employee engagement. Flexibility in work hours meets personal needs and reduces competing roles stress (Bloom et al. 2015; Allen et al. 2014).

Employers have advantages with flexible work schedules as well. Flexible work schedules can help employers attract and retain top talent, providing a wider pool of talent and diversity to work. Organizations also typically see an increase in productivity, reduced absenteeism, and reduced turnover with employees who are on flexible work schedules. Other benefits for employers might be cost savings in office space, utilities, and office supplies with flexible work schedules. Flexible work schedules create a workforce that is healthier, more satisfied, and more productive while providing organizations to be competitive and adaptable in a rapidly changing work environment.

10.3 Time Management Training

Time management training provides individuals with fundamental strategies to prepare, prioritize, and accomplish work in a timely manner. Well-managed time increases the likelihood that your workplace will be a healthier environment because employees will be less overwhelmed and better able to manage work-life harmony. When employees can properly organize and delegate their work, they are less likely to burn out and more likely to find job satisfaction. Ultimately, organizations will gain more with their productivity, morale, and a lower likelihood of turnover. Time management training improves decision-making skills, as employees will divert less time to less important issues as they are not working under pressure.

11 Findings

Research findings show that technology in the workplace is having both a positive and negative impact on employees. Perhaps one of the most important positive findings is related to technology's role in fostering work-life balance. Remote work technologies have become popular. These include video conferencing platforms, cloud-based collaboration tools, and project management software that allow employees to work from virtually anywhere. The success of working remotely would offer workers some geographical flexibility, and consequently, people can now arrange their workdays around their personal commitments. Technology, when used appropriately, can improve employee's mental health. Technology fosters flexibility for employees, which can alleviate work commute stress and improve the management of their personal obligations. The research also demonstrates that technology has a significant impact on employee productivity. The introduction of automation, artificial intelligence, and advanced analytics has streamlined many routine and repetitive tasks, allowing employees to turn their focus to higher-value activities that involve creativity, critical thinking or problem-solving. For example, an automated scheduling system.

Collaboration technologies have obviously changed teamwork. Employees can work together on projects in real time, regardless of where they are physically located, and are able to share ideas, documents, and comments at a fast pace, all at little cost. This has led to greater efficiency, improved turnaround times and the ability to connect with talent anywhere on the globe.

The findings show that technology can also have some negative impacts on work-life balance if not properly managed. The tools needed for flexibility can also create an "always on" culture where employees feel they must check emails or respond to messages outside of work. Being expected to work in their personal time can increase the line between work and home, creating stress. Organizations that do not set clear expectations regarding after-hours communication run the risk of neglecting the benefits of the technology "flexibility" offers.

Technostress is the result of an increased emphasis on the workload, digital exchange of communication and information, and the pressure and expectation to be always connected and multitask. Negative consequences of technostress can lead to mental and physical health issues.

combination of these factors leads to higher rates of burnout, absenteeism, and dissatisfaction from work. Employees who have a high amount of technostress are more likely to report physical health issues (such as headaches and sleep disruptions) and mental health issues (such as anxiety and depression). Organizations that take a holistic approach to technology management (simultaneously balancing the benefits of digital technologies with proactive efforts to reduce risks) will tend to see high levels of employee engagement, satisfaction, and well-being.

12 Implications and Future Directions

Future research should focus on the consequences of rapidly changing technologies, particularly artificial intelligence (AI) and machine learning, on employee mental health and organizational culture. AI-driven tools and technology are becoming important parts of organization practices and cultures, and we must understand their impact in terms of psychological or social consequences over time. Research must investigate whether increased automation and algorithmic decision-making enhance employee empowerment or increase anxiety.

13 Conclusion

Technology in the workplace presents both positive and negative influences on employees. On the upside, digital and remote work technology has increased flexibility, enabled a better work-life balance, and enhanced productivity by automating various routine tasks, while allowing employees to collaborate with people anywhere in the world. Employees can plan or keep management of personal commitments and reduce time, frustration, and stress from commuting, helping to maintain mental health. On the other hand, these technologies may also destroy the boundaries of work and home so that employees may feel they have to be "on" all the time, and if the organization does not have strict expectations for time from its employees, work-life balance can lead to stress, strain, and technostress-related physical and mental health issues. Organizations should take a holistic approach to making the strategy for appropriate technology use, which will strengthen the culture of well-being and mental health in the workplace.

References

Agarwal S, Dwivedi D (2025) Happy wisdom framework: a step-by-step guide. In: Harnessing happiness and wisdom for organizational well-being. IGI Global Scientific Publishing, pp 263–294

Allen TD, Golden TD, Shockley KM (2014) How effective is telecommuting? Assessing the status of our scientific findings. Psychol Sci Public Interest 16(2):40–68 (Sage Publications)

American Psychological Association (2023) Electronically monitoring your employees? It's impacting their mental health. APA Monitor Psychol 54(7):12–13 (American Psychological Association)

Bloom N, Liang J, Roberts J, Ying ZJ (2015) Does working from home work? Evidence from a Chinese experiment. Quarterly J Econ 130(1):165–218 (Oxford University Press)

Bondarouk T, Brewster C (2016) HRM in technology-driven organizations. J Bus Res (Elsevier Publishing)

Brown KW, Green TD (2020) The essentials of workplace mindfulness. Routledge

Brynjolfsson E, McAfee A (2014) The second machine age: work, progress, and prosperity in a time of brilliant technologies. W. W. Norton & Company

Cooper CL, Quick JC (2017) The handbook of stress and health: a guide to research and practice. Wiley-Blackwell

Davenport TH, Ronanki R (2018) Artificial intelligence for the real world. Harvard Bus Rev 96(1):108–116 (Harvard Business Publishing)

Davenport T et al (2010) Analytics at work: smarter decisions. Harvard Business Review Press

Dwivedi D (2015) Impact of information technology on the productivity of the organization. HR J Manage 7:77

Dwivedi D (2025) Emotional intelligence and artificial intelligence integration strategies for leadership excellence. Adv Res 26(1):84–94

Dwivedi D, Agarwal S (2024) Green technology and its effect on the modern world. In: Green innovations for industrial development and business sustainability. CRC Press, pp 120–126

Dwivedi D, Agarwal S (2025) Evaluating the effectiveness of leadership decision making by data analytics and reporting technologies. In: Advancing the marketing technology (MarTech) revolution. IGI Global Scientific Publishing, pp 207–232

Dwivedi D (2011) Stress management: how to reduce and manage workplace stress. Juris Ray 5
Dwivedi D (2014) New dimensions of HRM: performance management
Dwivedi D (2018) Stress management in software industry
Grant AM, Fried Y, Juillerat T (2013) Work matters: job design in classic and contemporary perspectives. In: Schmitt NW, Highhouse S (eds) Handbook of psychology: industrial and organizational psychology, 2nd edn. Wiley, pp 417–453
Higgins J (2017) Augmented reality applications. Constr Manage J (Emerald Insight Publishing)
Holland P, Bardoel E (2016) Gamification in corporate training. Int J Training Res (Taylor & Francis Publishing Group)
Jesuthasan R (2017) Digital engagement strategies. HR Magazine. Chartered Institute of Personnel Development Press
Marler JH, Parry E (2016) Strategic HRM practices. In: Human resource management review. Elsevier Publishing Group
Maslach C, Leiter MP (2016) Burnout: a brief history and how to prevent it. Harvard University Press
Mewafarosh R, Agarwal S, Dwivedi D (2024) Balancing FOMO and digital detoxification. In: Business drivers in promoting digital detoxification, pp 1–15. IGI Global
Moore P, Piwek L (2016) Employee monitoring via wearables. J Bus Ethics (Springer Publishing Group)
Shrestha YR, Ben-Menahem SM, von Krogh G (2019) Organizational decision-making structures in the age of artificial intelligence. California Manage Rev 61(4):66–83 (University of California Press)
Smith JA, Jones LM (2021) Digital wellness in the modern workplace. Springer
Stone DL et al (2015) Technology-enabled HR functions. Personnel Psychol J (Wiley Publishing Group Inc)
Tarafdar M, Cooper CL, Stich JF (2015) The technostress trifecta-techno eustress, distress and the dark side of technology. Int J Inf Manage 35(6):733–741 (Elsevier)
Ungar R. Technology's effects on mental health. MadeOfMillions.com. 19 Feb, pp 10–18
Wang B, Liu Y, Qian J, Parker SK (2022) Achieving effective remote working during the COVID-19 pandemic: a work design perspective. Appl Psychol 70(1):16–59 (Wiley)
Wright A (2018) Blockchain applications in workplaces. J Financial Innov (Springer Nature Publishing Group Inc)

Digital Transformation, AI, FinTech Innovation and Future of Work

Synergizing Digital HR Practices and Fintech Growth: A Pathway towards SDG 8 in the Digital Banking Industry

Bhanu Pratap Singh Meena and Vinay Kumar Jain

Abstract Digital human resource (HR) practices are growing rapidly with financial technology, or Fintech. This is transforming operational landscapes in digital banking (DB) business sectors. This research looks at the innovative human resources technologies (AI-based recruitment, digital performance management and remote workforce enablement) are synergistically integrated with Fintech to contribute to Sustainable Development Goal 8 (SDG 8) for decent work and economic growth. Through qualitative research, the HR digitization strategy creates empowered and economically inclusive financial services that promote employee engagement, productivity, and retention. Moreover, organizational digital culture, leadership, and technological readiness were incorporated to assess their influence further to this synergy. The study shows that aligning digital HR activities with fintech innovation is a strategic lever for achieving sustainable growth in emerging digital bank ecosystems. The intersection at which this point is located is a transformative pathway through which institutions can harness opportunities to enhance inclusive employment, innovation, and long-term economic growth in alignment with SDG 8.

Keywords Digital synergy · Digital human resources · Digital banking · Fintech · SDG 8

1 Introduction

The life-changing ability digitalization has greatly impacted the international finance world. Thanks to digitalization we are creating a positive impact on the world. One of the major shifts has been the fast growth of financial technology (Fintech), which

B. P. S. Meena (✉) · V. K. Jain
Department of Commerce, Graphic Era (Deemed to be) University, Dehradun, Uttarakhand, India
e-mail: bhanupratappsingghmeena@gmail.com

V. K. Jain
e-mail: vinaykumarjain@geu.ac.in

A. Singh et al. (eds.), *Human-Centered Sustainability: Bridging Well-Being, Gender Equity, and Technology*, World Sustainability Series,
https://doi.org/10.1007/978-3-032-22020-2_21

is fast changing the way banks do business, their customer experience and service delivery. At the same time, digital Human Resource (HR) practices in the financial services sector, especially digital banking, are enabling the organizations to align people processes with technology. When used together, digital HR and Fintech can create sustainable growth and decent work and contribute to the achievement of Sustainable Development Goal 8 (SDG 8), 'Promote sustained, inclusive and sustainable economic growth, full and productive employment and decent work for all.' In the digital banking industry, the overlap of human resources and technology is not something to look forward to—it is something we are already facing. Digital Human Resource practices involve the use of technology-assisted systems related to hiring, onboarding, performance management, employee engagement, learning and development, workforce analytics and more. The employee training and development practices enhance employee experience and build agile innovative culture in organizations by equipping employees with digital skills. On the other hand, the sound waves of fintech by means of AI, blockchain, big data, and cloud computing is causing disruption and evolution in the financial services space. This has helped banks reach more customers, work better, and provide customized and easy-to-use services at a lower cost.

As the banks and other organizations are gearing to become agile and more responsive in a competitive digital economy, the marriage between digital HR and Fintech is acquiring increasing prominence. Digital HR isn't just about automating its traditional functions. It is also capable of recognizing the capacity of HR to partner with Fintech and support innovations and business models. The potential of Fintech can be fully utilized only with a digitally empowered workforce. Here, the role of HR as a catalyst for digital transformation becomes essential. When organizations improve the digital aptitude of employees, support flexible work arrangements, motivate innovation and creativity, and inculcate diversity and inclusivity in the workplace through the adoption of digital HR—it helps create decent work and economic growth (SDG 8).

Further, the contribution of digital HR to building a future-ready workforce is important for financial institutions to be resilient and competitive in the face of technological disruption. The industry of digital banking is primarily dependent on workers who can adapt easily to new technologies and continually learn and innovate processes. HR tech is getting all essential tools to make the HR functions productive, cost-effective, and efficient. The enablement of a digital HR ecosystem is possible with these essential tools which typically include the AI-enabled recruitment tools, e-learning modules, performance dashboards, real-time people analytics, and more. It helps to attract and retain the best talent, foster high-performance culture which goes a long way in ensuring organizational sustainability. While Fintech advancement is mostly related to modern technology, it also has some socio-economic benefits. By giving lower-cost and easy-to-access financial facilities to the underbanked and unbanked population—it enables Financial Inclusion. In doing so, it opens the doors to entrepreneurship, small and medium enterprise (SME) development, and job creation, all of which fall under SDG 8.

In addition, the digital banking revolution and rise in Fintech start-ups are creating new job roles and career paths that are making way for more jobs across various sections of society. Nonetheless, to achieve the right mix of digital HR and Fintech the first step requires strategic planning and investment and subsequent shift—organizational mindset. It is the challenge of data privacy, digital divide, skill gap and resistance to change disrupts the effective implementation of the digital strategy. To ensure the benefits will be distributed fairly and not exacerbate existing inequalities in the labour market, it is also vital to have robust policies, ethical frameworks, and inclusive digital infrastructure in place. This study aims to investigate the ways digital HR practices and Fintech growth converge strategically in the digital banking industry and their contribution to the realization of SDG 8. The study will examine the role of digital HR in creating an agile workforce that can harness and drive Fintech innovation, analyze the implications of Fintech adoption for employment and economic growth and evaluation of their synergy in promoting decent work and inclusive growth. The study believes that the digital transformation in banking is sustainable only when it is technical and human at the same time. Overall, in the digital age, the achievement of SDG 8 can only be done through the integrated development of technology and human capital. As Fintech shapes the future of digital banking in institutions, the HR function has to re-position itself as a strategic enabler of digital innovation and sustainability. Experts predict that digital HR and Fintech can not only improve an organization's performance but also contribute to larger socio-economic goals. By promoting decent work, and productive employment, and fostering inclusive economic growth, this synergy offers an excellent opportunity for a more equitable and sustainable future in the ecosystem of digital banking.

2 Review of Literature

The banking sector has benefited from rapid improvements in financial technologies (Fintech) and the digitization of human resource (HR) practices. When they come together, they have what it takes to make a significant contribution towards achieving Sustainable Development Goals 8 (SDG 8) decent work and economic growth. Digital HR and fintech innovation are becoming the latest things, especially in digital banking. This area is of strategic interest. This has been discussed in the literature both independently as well as together. Hence, providing insight into the challenges, synergies, and impact to economic development. Bhutto et al. (2023) examine the need for FinTech adoption, HR competency, and service innovation for driving the firm growth in the banking sector. According to their research, the strategic combination of technology and skilled employees for service innovation will help create a competitive advantage, operational efficiency, and sustainable development for financial institutions.

According to John et al. (2025), a bibliometric analysis links digital innovation with Sustainable Development Goal (SDG) and its role in green growth. The study shows how digital tech are playing catalyst role in transform policies, innovate practices, advancing global public goods and project sustainable development. It helps runs how digital technologies become a catalyst for sustainable development. Purnamawati et al. (2023) explore the important role of Corporate Social Responsibility (CSR) in the sustainable economic development of villages. The researchers provide evidence that CSR initiatives aimed at rural communities have the potential to drive inclusive growth, sustainable resource management and socio-economic resilience. According to the study, CSR should become part and parcel of local development planning to achieve longevity and public participation. Ul-Haq et al. (2024) examines the difference between economic growth and high-quality economic development on energy insecurity across Chinese provinces. Researchers found that energy insecurity can be aggravated an increase in the pace of growth, but high quality and sustainable development can mitigate this risk. Balanced regional strategies are critical for long term energy security and environmental sustainability, the research points out. Khan et al. (2025) studied how fintech adoption, HR competence, service innovation impacts growth of firm in banking sector using Entropy and TOPSIS method.

Human resource capability and innovation can be significant mediators impacting FinTech growth, the study shows. The importance of strategic alignment for sustainable digital transformation must be stressed. Kuchciak and Warwas (2021) present a strategic plan that aims to introduce HRM for banking 4.0 to utilize digital technologies, data analytics as well as agile HR practices to enhance client and employee experience. As that study explained, as the digital banking ecosystem and customer models become more resilient, human capital will have to adapt through reskilling, adaptive leadership, and an innovation-driven culture. Qudah et al. (2025) study the impact of fintech solutions on HRM efficiency in Jordanian banks. Researchers noted that introducing fintech creates further flexibility in measuring employee performance and taking managerial decisions. The study mentioned that the banking sector may use digital tools to transform existing HR functions and create innovation. Muduli and Choudhury (2024) explore the use of digital technology, workforce agility and its outcomes in India's banking industry. The businesses also had a raft of literature offered that spoke to employees using digital tools can adjust to changing conditions. It also said they become more efficient over all tasks.

According to the study, it needs to configure technology with the capacities of human capital. Hidayat-ur-Rehman and Hossain (2024) study about the bank's sustainable performance depends on fintech adoption, green financing, and competitiveness quality, and if digital transformation is a moderator. The study shows that technology can aid sustainability. Fintech and green finance will enhance the banking performance and one must rely on digital transformation. Campanella et al. (2023) examine the fintech plays an essential role in the financial ecosystem, which is observed and signifies the shift towards the demand and capital imports of such human capital. The research indicates that institutions will require more sophisticated skills to merge technology into strategies. It also indicates that some changes in financial institutions will be necessary due to fintech. Senyo et al. (2024) investigate the role of government policies in shaping the FinTech ecosystem. FinTech innovation is further driven through public sector involvement that includes infrastructure development, regulatory support, and active participation. According to the study, the collaborative governance process fails to enhance the digital financial inclusion speed. These ecosystems can sustain themselves through these same processes. The paper analyze leadership, organizational culture and technostress inhibitors are very important factors for digital innovation in retail banking. A study's effective leadership and culture raises innovation while the technostress manage is essential for employee performance and digital fitness in the changing payment systems.

Sustainable Development Goal 8 (SDG 8)
"Decent Work & Economic Growth" that encourages the sustained economic growth, along with high levels of employment and decent work culture for all. The key targets of SDG 8 include in the given 12 pointer states are:

1. Sustain economic growth (Target 8.1): It achieves at least 7% GDP growth per year in underdeveloped countries.
2. Broadening productivity possibilities through diversification, technological change, and innovation is essential for achieving economic goals (Target 8.2): Strengthen economic productivity with diversification, technology, and innovation.
3. Promote development-oriented policies (Target 8.3): Facilitate work, job-making, launching, inventiveness, and effectiveness.
4. Make consumption and production resource efficient (Target 8.4): Decouple economic growth from environmental degradation.
5. Ensure equal pay for all workers at all levels (Target 8.5): By achieving full and productive employment and equal pay for work of equal value by 2030.
6. Focus on providing opportunities for youth employment (Target 8.6): By a specific date, significantly reduce the youth not in employment, education, or training in all its forms.
7. End modern slavery, trafficking, and child labour (Target 8.7): Abolish forced labor, end modern slavery, and eliminate child labor in all its forms.
8. Combating against abuse and harassment, protected for equal reduction (Target 8.8): This is especially true for migrant workers and those considered to have precarious jobs.
9. Promote sustainable tourism (Target 8.9): Create jobs and promote local culture and products.
10. Strengthen financial institutions (Target 8.10): Make banking and financial services accessible to the poor.
11. Enhance support for Aid for Trade (Target 8.a): Send more money to developing nations, especially the least developed.
12. Create a youth employment strategy across the world (Target 8.b): To implement the International Labour Organization Global Jobs Pact, global spirit must be put in action.

3 Research Objectives

1. The purpose of the research study is to explore the incorporation of digital HR practices in the digital banking sector and how these practices enhance efficiency, engagement, and innovation.
2. To examine how the fintech growth is changing the working of digital banking. Specifically, how it is impacting jobs, services, and skill requirements of the workforce.
3. The purpose of this study is to assess the synergistic relationship existing between digital HR transformation and fintech development. The paper also aims to study how the confluence of the two delivers inclusive and sustainable economic growth, in tune with Sustainable Development Goal 8 (Decent Work and Economic Growth).
4. The goal of this paper is to identify challenges and opportunities digital banking institutions face with regards to digital HR strategies and the use of fintech tools for decent work and economic productivity.
5. To be better prepared for the digital age, readiness must be aligned with digital transformation and SDG (sustainable development goals) to contribute growth to the economy.
6. To offer policy and strategic recommendations to banking and regulatory institutions to strengthen the coherence of digital HR and fintech with the sustainable development and decent work principles.

4 Challenges and Ethical Considerations

The combining of digital HR practices with fintech creativity to development between SDG-8, Decent Work and Economic Growth in the digital banking sector presents a new opportunity. But with convergence comes a lot of challenges and ethical issues that should also be dealt with on an urgent basis.

Challenges: The most important next challenge is the technology gap and digital divide of the manpower. As fintech technology evolves it would require a digitized workspace. Many of the employees particularly in the developing region lack sufficient digital skills and knowledge as well as limited access to opportunities to improve their digital skills or upskill. The absence of digital HR technologies may increase job losses, over-staffing, and inequalities, which may undermine the very essence of SDG 8. Next, the resistance to change in the organizational culture is a roadblock to the smooth adaptation of digital HR technologies. Traditional HR setups may not lend themselves easily to data-driven models and AI-enabled platforms. That would mean structural transformation and proactive leadership. Integration problems between HR analytics and fintech platforms can prevent synergy in a situation involving legacy systems. In addition, digital labor and finance laws are also changing, increasing operational risk. Organizations must continuously adapt to changing compliance directives, data governance norms, and cross-border digital worker policies.

Ethical Considerations: The ethical implications are equally critical. Our primary concern is employee data privacy and surveillance. Digital HR tools collect, store, and analyze a lot of personal and behavioral data. If appropriate safeguards are not in place, it may lead to misuse, surveillance or biased decision-making in hiring, evaluation, or promotion. The algorithmic decision-making raises yet another ethical issue in the case of the HR and financial services. If not trained or audited appropriately to reduce bias, it could additionally reinforce system bias. We need to look for fairness, transparency, and accountably. Furthermore, the economy and fintech platforms are affecting the employer-employee dynamic. Even though, it is flexible, it may compromise job security, social protection, and benefits, which is contrary to the decent work principles in SDG 8. Finally, there is a need for ethical leadership and governance to balance innovation with human dignity and social responsibility. Inclusive strategies focused on employee well-being, equity, and long-term value ought to be more important than technology.

5 Future Implications

Fintech has been growing rapidly. Its digitization of HR is well enough to achieve SDG 8 in the digital banking industry. HR functions are set to emerge as the key enablers of workforce agility, innovation, and inclusive growth as banks are set to employ more automation, AI-based recruitment, and performance analytics. Together banks and fintech will give rise to an adaptive workforce which is digitally savvy to meet the demands of the evolving financial ecosystem. In the coming years, we can expect to see predictive analytics being used more often to identify talent, ongoing learning platforms to train people for fintech jobs, and data-based initiatives to ensure equal opportunities for growth. Also, it can generate jobs primarily for digitally literate youth while improving efficiencies and customer-centric innovations in the banking space. The policy wise, integrating digital HR and fintech potential will create an industry-wide framework that supports decent work, well-being of workers, and promotes gender equality which will link to the SDG 8 agenda directly. In the future, this partnership will transform the digital banking culture to become more sustainable, more resilient, and more human resources during technology disruption.

6 Future Scope

The combined effects of digital human resource (HR) policies and Fintech on sustainability is arguably vast, more so for Sustainable Development Goal 8 (SDG 8) on decent work and economic growth. As digital banking is undergoing immense change, the judicious use of HR technologies such as AI-based recruitment and remote working management, data analytics, e-learning can help institutions inculcate a more agile, optimized, inclusive and future-ready workforce. The advancement of fintech can give rise to new job roles while enhancing entrepreneurship and financial inclusion in the underbanked regions. Future studies can examine these digital HR tools can upskill staff and help to retain staff in fintech banking. Also, research may closely examine the political framework, ethical dimensions, and the influence of org extensions to make digital transformation a human-centric, socially just movement. We can compare two countries for a better understanding of the different effects of HR-Fintech cooperation on the resilience of the economy and sustainable employment. The future directions of digital banking should focus on innovations which are not only technically sound but also aligned with inclusive growth to reinforce the objectives of the SDG 8.

7 Findings

According to the study, the digital human resource (HR) practices and the growing use of Fintech innovation are increasingly modifying the operational processes of digital banks. Participants from HR, technology, and banking background repeatedly stated that digital HR tools like AI-enabled recruitment, performance analytics, digital learning and remote work enablement have enhanced workforce efficiency, transparency, and inclusiveness. It also enables automation powered by fintech and customer-centric innovations, allowing banks to redirect resources toward strategic HR initiatives that create jobs, effectively build skills and decent work environments in line with SDG 8. Digital HR practices mediate the adaptation of talent to rapid changes in the fintech environment, lower workload of manual employees, and develop a more resilient and agile workforce. Stakeholders noted that banks which have developed an integrated digital ecosystem to support their HR function are better equipped to scale operations and retain talent to meet sustainability goals. The combination of HR digitization along with the growth of fintech encourages continuous learning and development which enhances job satisfaction and career growth opportunities for employees in the industry. The results showed that institutions are getting more familiar with the power of digital HR policies and fintech growth trajectories, especially to close employment gaps, empower women in technology, and ensure equality in skills development programs. This type of alignment will ensure greater financial inclusion as well as enhancing the contribution of the sector towards inclusive and sustainable economic growth (SDG 8).

8 Conclusions

The incorporation of digital HR with the growing fintech is a strategic convergence in the digital banking space. By working together like this, we will see better operations and innovations, which will contribute positively towards achieving of Sustainable Development Goal 8 (SDG 8) decent work and economic growth. When the digital banking industry uses digital tools for hiring, employee engagement, upskilling, and performance management, they are better placed to create a resilient future-ready workforce that can meet the demands of a technology-driven economy.

Fintech growth has entailed financial inclusion, increased reach of services in a cost-effective way and convenience of experience to users through agile and digitally enabled frameworks of human resources. Digital HR strategies aligned with fintech innovations promote a culture of continuous learning, inclusivity, and adaptability that ultimately results in job creation, productivity enhancement and sustainable economic development. Basically, the pathway to SDG 8 in digital banking can be done deliberately and collaboratively by utilizing the digital HR and fintech capabilities.

References

Bhutto SA, Jamal Y, Ullah S (2023) FinTech adoption, HR competency potential, service innovation and firm growth in banking sector. Heliyon 9(3)

Campanella F, Serino L, Battisti E, Giakoumelou A, Karasamani I (2023) FinTech in the financial system: towards a capital-intensive and high competence human capital reality? J Bus Res 155:113376

Hidayat-ur-Rehman I, Hossain MN (2024) The impacts of Fintech adoption, green finance, and competitiveness on banks' sustainable performance: digital transformation as moderator. Asia-Pac J Bus Adm

John HB, Agrawal M, Nema P, Hasan A, Rakhra IK, Singh A (2025) Green growth: a bibliometric analysis of digital innovation and sustainable development goals (SDGs). J Innov Entrep 14(1):15

Khan HU, Abbas M, Nazir S, Khan F, Moon YK (2025) The intersection of FinTech adoption, HR competency potential, service innovation, and firm growth in the banking sectors using entropy and TOPSIS. PLoS One 20(1):e0313210

Kuchciak I, Warwas I (2021) Designing a roadmap for human resource management in the banking 4.0. J Risk Financ Manage 14(12):615

Muduli A, Choudhury A (2024) Digital technology adoption, workforce agility and digital technology outcomes in the context of the banking industry of India. J Sci Technol Policy Manage

Purnamawati IGA, Yuniarta GA, Jie F (2023) Strengthening the role of corporate social responsibility in the dimensions of sustainable village economic development. Heliyon 9(4)

Qudah H, Almustafa E, Almnadheh Y (2025) Exploring the relationship between Fintech solutions and HRM efficiency in Jordanian banking. Hum Syst Manag 44(3):408–423

Senyo PK, Karanasios S, Agbloyor EK, Choudrie J (2024) Government-led digital transformation in FinTech ecosystems. J Strateg Inf Syst 33(3):101849

Ul-Haq J, Visas H, Hye QMA, Rehan R, Khanum S (2024) Investigating the unparalleled effects of economic growth and high-quality economic development on energy insecurity in China: a provincial perspective. Environ Sci Pollut Res 31(15):22870–22884

A Systematic Literature Review on: A Study on the Application of Artificial Intelligence (AI) in Banking

Ruby and Smriti Tandon Gupta

Abstract With a focus on technological advancements in the fields of artificial intelligence (AI) and machine learning (ML), this paper seeks to analyze research and provide a systematic literature review on the initiatives taken by the players in the banking sector. Through increased operational effectiveness, better client experiences, and the ability to make decisions based on data, artificial intelligence (AI) is completely changing traditional financial services in the banking industry. The implementation of machine learning algorithms and advanced analytics has improved banks' ability to identify fraud, evaluate risk, and customize services. AI also makes streamlined procedures possible in sectors like loan underwriting and chatbot customer service. The banking sector is constantly changing, and the strategic application of AI technologies offers chances for increased competitiveness, innovation, and regulatory compliance. These advancements have helped the sector grow in the direction of consumer trust and ease of doing business in the digitalized banking era. The large and intricate online rendition of virtual services accompanied by soaring security risks in cyberspace. To examine the application of AI and machine learning in different aspects of the banking industry and how they have changed to meet the various needs of the banking industry, a total of 734 publications were analyzed. The results demonstrate the applications and areas for optimal usage of such contemporary technologies. Finally, it discusses and assesses utilization of these technologies based upon factors including expandability, moving farther, considering these incredibly dynamic times, and supporting banks' core functions as well as how they are handling the opportunities and problems that these new technologies present. Additionally, this SLR offers a summary among the study it may be conducted within the future by the scholars throughout this field.

Keywords Fintech · Financial industry · Learning by machine and artificial intelligence · Forecasting

Ruby (✉) · S. T. Gupta
Department of Management Studies, Graphic Era Deemed to Be University, Dehradun, India
e-mail: ruby09121999@gmail.com

S. T. Gupta
e-mail: Smriti.tandon.mba@geu.ac.in

A. Singh et al. (eds.), *Human-Centered Sustainability: Bridging Well-Being, Gender Equity, and Technology*, World Sustainability Series,
https://doi.org/10.1007/978-3-032-22020-2_22

1 Introduction

Synthetic intelligence (AI) is ordinally understood on as the capacity of machine can carry out mental tasks, we generally connect with human thoughts. This includes observing, making sense of, picking up knowledge, and interacting with the surroundings problems as well as being creative. Still AI is a collection of sophisticated computing tools varying in maturity. Some of among these technologies are decades old but relatively new at the same time. It is not only the evolutionary progress of technology, sometimes based on some older technology, but is now possible, because of the ability to access large volumes of data and the capacity to process the volume of available data.

Figure 1: The shared diagram is taken directly from Kantar IAMAI's ICUBE 2023 report. It correctly depicts the increase in India's active internet users, which went from 574 million (41%) in 2019 to 821 million (55%), in 2023, and emphasizes how adoption in rural areas has been driving this expansion.

The wording of "artificial intelligence" utilizes terms that can sometimes come across as generic as they seem overused or holistically cover a multitude of types of technologies. In most cases, algorithm use only replicates, automates the processing, and mimics a scenario a person may complete. AI is generally taken to mean cognitive technologies that utilize the processing of large volumes, structured or unstructured (big data). Within this case, when it comes to cognitive intelligence, any processing or modelling of unstructured data that replicates or helps to strengthen and improve the mental capacity of people. As these technologies are expanding and developing, here are some technical fields of research with non-definitive hierarchy:

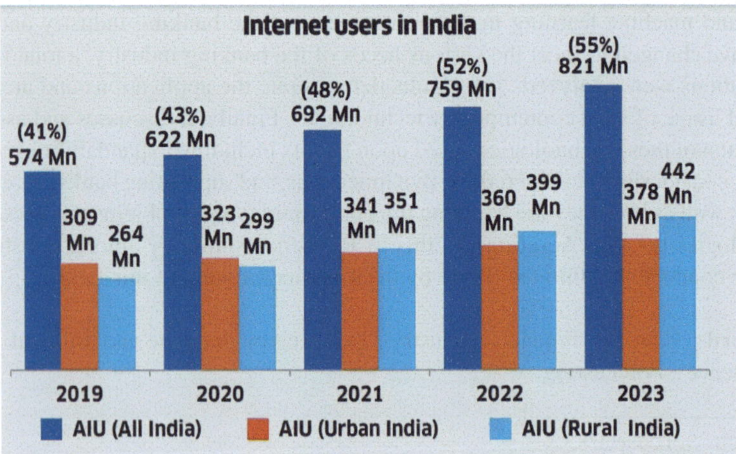

Fig. 1 *Source* Kantar IAMAI's ICUBE 2023 report

1. Language is understood by natural language processing providing significance and intent to language—frequently includes text-to-speech technology and automatic voice recognition.
2. Mental computing—assists within extending a human's capacity to perform mental exercises for Interactive, iterative, evidence-based methods for decision making.
3. Intelligent processing/analytics: simulations and predictive analysis. These facilitate rule-based automated actions, for example, engines that suggest things.
4. Reinforcement learning and deep learning: structural change which shifts the emphasis pattern recognition in machine learning gave way to a sequential, experience-based decision-making process.

Therefore, AI refers to a collection of guidelines, definitions of issues, formulas, and procedures that extract not immediately apparent, significant trends and useful information from massive data sets. The use of artificial intelligence is the machine emulation of human reasoning processes, in the particular computer networks. Second, artificial intelligence that is natural which simply suggests a bottom-up methodology where the computer enhances its performance and capabilities for every task it completes by using self-improvement algorithms. Multiple digital disruption is transforming entire sectors and altering how in which businesses operate (Pio et al. 2020).

"Artificial Intelligence is often defined as the ability of a machine to perform cognitive functions we associate with human minds, such as perceiving, reasoning, learning, interacting with the environment, problem solving, and even exercising creativity". But in reality, artificial intelligence (AI) is a conglomeration of sophisticated computer technologies in differing stages of development (Mehdiabadi et al. 2020).

1.1 Relevance of the Study

- Identify the majority of prevalent fields in which the AI and ML have made a contribution.
- This study helps in understanding bank crisis prediction where the use of AL and ML has been envisioned.
- Management can take the necessary actions to stop asset loss by using ML and AI to accurately predict adverse events in non-performing assets (NPAs) at the sector and individual bank levels.

1.2 AI in Financial Services

The financial sector as a whole has seen a number of advancements in staffing, asset management, customer service, and communications. These days, for example, stock investing and finance all has come to be all about technical skills and divine luck. But in the future, with sentiment analysis, crowdsourced data, and algorithms, we will handle money in a much different way (Kaur et al. 2020).

The following are the main applications of AI in banking:

(1) Tailored Financial Services
(2) Intelligent wallets
(3) Voice-activated banking
(4) Customer service
(5) Digitalization in place of physical branch offices
(6) Cut expenses
(7) Reduce risk
(8) Boost income.

Various AI applications in the financial and banking industries:

(A) **Marketing and Customer Service**: Chatbots are self-learning computers that can have intelligent conversations with people via audio or chat. They are available around-the-clock and are relatively easy to use, but they take a long time to train.
(B) **Financial Product Robotic Advisors**: Algorithm-based online systems made financial recommendations, reinvested dividends, and automatically created and balanced portfolios, among other things. Human intervention would be negligible to non-existent.
(C) **Personalized Financial Services:** Robo-advisors will keep track of clients' objectives and recommend suitable stocks or bonds for purchase or sale; they don't base their recommendations on the clients' risk tolerance.
(D) **Smart Wallets:** Mobile wallets have been enhanced with intelligence to offer smart services like bill payment, events, movie tickets, bus and taxi ticket booking, chat, and more.
(E) **AI Sensation**: facial and voice-based recognition technology for human emotions-advanced facial and voice-based recognition technology of human emotions, the particular type of AI that would enable machines to automatically recognize human emotions recognition technologies.
(F) **Security & Compliance:**

    (1) Fraud detection & Prevention: Reduce the need to inject continuous man hours to detect and prevent security attacks. It is Machine Learning-aided to automate.
    (2) Compliance monitoring: Apply AI to scan lengthy documents and mark possible issues in a matter of seconds, which would have taken hours otherwise.
    (3) Intelligent QRC: A new breed of Artificial Intelligence firms that specialize in maintaining the companies compliant for example no document is missed when filing for something, do risk mitigation by monitoring customer behavior based on empirical data.

(G) **Back-end BPM:**

(1) Robotic Process Automation: Using software robots to automate high-volume, back- office processes. and repeated tasks with the purpose of saving time, boosting effectiveness, & raising precision.
(2) Algorithmic Trading: AI for frequent trading with inputs derived from several financial sources. These are the markets where milliseconds are used to make financial decisions. It has been reported that over 70% of trading takes place globally. Today, algorithms are in charge of all of this.
(3) Investment Research: Let AI Lead on Stock Selection Choices. Cover additional businesses involved in all exchange practices around the globe, researches, and portfolio management.
(4) Human Resources: AI to save hiring manager's time with various recruitment processes e.g. interact with new joiner's shortlist resumes from social media sites, pre-screen candidates over chat, determine candidate dropout chances, etc. (Thach et al. 2021).

Artificial Intelligence refers to the machine abilities that are often used for human thinking, including perception, learning, decision-making, and problem-solving. AI can be developed by several methods such as expert systems, deep learning, machine learning, and rule-based systems. The expert systems, rule-based systems, were some of the early methods of achieving AI, relying on manually entered rules and professional judgment for choices or problem-solving efforts. Conversely, though machine learning employs statistical algorithms for automated learning of patterns in data and makes prediction or choices made in light of that pattern (González-Carrasco et al. 2019). A subset of machine learning is called deep learning, which often involves multi-layer artificial neural networks designed to take out of the data high-level features.

Machine learning and artificial intelligence can be utilized to a considerable extent in the banking sector. Some of the application domains that require high-end techniques include performance of bank branches, credit assessment, client segmentation, and online banking. As Blockchain technologies are gradually being used in the sector of banking, their usage of ML and AI is expected to go up additionally. In addition, with mobile banking, there has also resulted in the use of AI for connecting with a larger customer base. The authors suggest that the adoption of these contemporary technologies is in the direction to improve the experience for the consumers, ease of use in banking, improved risk control, and upcoming back-end operations smart, data analytics, RPA (Robotic Process Automation), chatbot-based new business models, and to make payment operations increased safety (Thomsett et al. 2020).

According to a report by, it has been estimated the fact that by 2030, the banking sector will use AI to save more than $1 trillion, with $31 billion coming from risk management using AI. Institutions in the financial sector have already begun adopting sophisticated methods that employ machine learning in a variety of contexts, such as risk assessment, fraud and churn forecasts, chatbots and virtual assistants, and consumer segmentation modelling as well as data analysis, portfolio management, underwriting, and many more (Mehrotra 2019).

1.3 Research Questions

- Which key domains today have benefited most from the contributions of machine learning (ML) and artificial intelligence (AI)?
- How does this study's vision of the application of AI and machine learning advance our knowledge of banking crisis prediction?
- What steps can management take to avert asset loss by using machine learning and artificial intelligence (ML) to properly forecast adverse events in non-performing assets (NPAs) at the sector and bank levels?

1.4 History and Objective for the Research

While the SLRs inside the wider context of money have contributed to the corpus of information, but are very restricted, of course, mostly since they mainly oriented toward summarizing or characterizing the applications and developments in AI and ML, because there isn't particular compilation of the banking search results (The Ginimachine 2018). It would provide researchers with a deep discussion of the effects of putting it into practice and particular barriers and chances for development of the topic, considering the restricted scope contains an assortment of past studies on ML & AI, particularly inside the banking industry. Review-based research can be useful by including previous details and providing a sketch of major areas that require extensive academic effort (Bilal 2016). We respond to this need by doing an SLR on ML and AI applications in banking. SLRs are a very helpful resource for compiling information in a field of study, pointing out areas of ignorance, and paving the way for fresh lines of inquiry. This work expands on previous SLRs in two distinct ways, hence expanding the corpus of research on AI & ML in banking. Initially it will supply the reader with a thematically organized, state-of-the-art classification of previous studies in respect of their possible applications, limitations, and recommendations. Secondly, in light of our SLR findings, we provide an approach for data synthesis to discover potential gaps which deserve further academic research to progress the physique of knowledge. The following research goals were met in order to make this contribution: goals of the research examined in order to provide this contribution:

1. To outline the greatest related topics several scholarly studies on the application of ML and AI in banking.
2. Investigate and report on the main domains in financial applications of AI and ML.
3. The individuals employed in the banking industry have cited the following as some of the reasons to be wary of AI/ML adoption.
4. Future scope and prospects application of AI/ML to banking.

1.5 How AI in Banking Sector is Aligned with the SDGs?

- **SDG 1: No Poverty**

Financial Inclusion: AI-powered platforms can increase economic empowerment and decrease poverty by providing underserved groups with microloans and other financial services.

Data-driven Insights: Artificial intelligence (AI) can examine enormous datasets to pinpoint populations at risk and customize financial solutions to meet their unique requirements.

- **SDG 8: Economic Development and Decent Work**

Enhanced Productivity: Automation driven by AI can optimize banking processes, freeing up human resources for more difficult jobs and increasing overall effectiveness.

AI may automate certain duties, but it also opens up new career paths in data science, AI development, and customer support.

- **SDG 10: Less Inequality**

Targeted Financial Services: AI can assist banks in resolving disparities in access to financing by enabling them to customize financial services and products to particular groups.

Fair Lending Practices: By using AI algorithms to detect and lessen bias in lending choices, we can encourage more equitable access to credit.

- **SDG 12: Responsible Production and Consumption**

Sustainable Finance: By using AI to assess how loans and investments will affect the environment, banks are able to give priority to sustainable businesses and projects.

ESG Integration: By supporting responsible production and consumption, artificial intelligence (AI) can assist banks in evaluating and incorporating Environmental, Social, and Governance (ESG) considerations into their investment plans.

- **SDG 16: Justice, peace, and strong institutions**

Increased Transparency: AI can increase financial transaction transparency, which lowers the likelihood of fraud and corruption.

Effective Regulatory Compliance: AI-driven technologies may automate risk assessments and compliance checks, guaranteeing banks follow rules and support robust institutions.

Overall Impact

- Better Decision-Making: Artificial intelligence (AI) algorithms are able to examine vast datasets and spot trends and insights that humans might overlook, which improves decision-making across the board in banking.

- Operational Efficiency: By automating repetitive processes, AI can increase overall operational efficiency and free up human resources for more strategic endeavors.
- Sustainable Finance Innovation: Artificial Intelligence (AI) facilitates the creation of new financial services and products that support sustainability, like impact investing and green bonds.

2 Literature Review

In order to clarify the complex effects of artificial intelligence (AI) on the banking industry, this review of the literature summarizes recent research findings with an emphasis on consumer acceptance, operational effectiveness, and the difficulties faced by financial institutions.

1. According to Liu, artificial intelligence (AI) plays a crucial role in helping the Chinese financial industry embrace Central Bank Digital Currencies (CBDCs). By enhancing digital technology understanding, eliminating privacy issues, and guaranteeing simplicity of use, AI increases customers' propensity to use CBDCs, according to the study. Strategic government activities can strengthen the integration of AI technologies in promoting CBDCs, since government backing is regarded as a significant moderating factor that influences the relationship between AI and consumer acceptance.
2. Sharma and Choubey examine the transformative effects of AI and IT in the Indian banking sector. Their qualitative research shows that although the banking industry has made great progress in adopting IT, issues like consumer trust and the need for technical training still exist. The study highlights the significance of resolving these issues in order to create a safe and effective banking ecosystem. Notably, the research suggests that AI-driven innovations, such as digital banking services and blockchain technology, are essential for future developments in the sector.
3. In 2024, Abildtrup looks into how the banking sector is affected by robotic process automation (RPA). According to the report, RPA has the ability to automate repetitive processes, which can improve operational efficiency and lower expenses. Software robots can reduce back-office operations and boost customer satisfaction by simulating human–computer interactions. The results indicate that using RPA is essential for the total transformation of banking procedures in addition to being advantageous for productivity.
4. Naveen Kumar Kokkalakonda investigates how cloud banking infrastructures might incorporate AI technologies like machine learning and natural language processing. AI automation can speed up transaction processing, save operating expenses, and improve security measures, according to the report. Banks can greatly enhance customer service using intelligent virtual assistants by utilizing AI. The paper does, however, also discuss issues with scalability and data

protection, emphasizing the necessity of strong approaches to deal with these problems.

5. In particular, Wang et al. (2021) examines the application of AI-powered financial investment robo-advisors in Pakistan's m-banking industry from the standpoint of the user. The empirical analysis supports the idea that AI can improve user engagement and happiness in financial decision-making processes by showing that customers have a high intention to embrace these AI-driven services. This result is consistent with a larger pattern of rising customer acceptance of AI banking technologies.
6. Uzbekistan's banking industry, where artificial intelligence is significantly advancing, especially in customer care and fraud detection, is examined by Abdurashidova and Balbaa. While adopting AI has tremendous potential to enhance financial services, their mixed-method approach shows that issues like data privacy concerns and a lack of qualified personnel are still common. The study highlights that in order to fully realize the promise of AI technologies, the Uzbek banking industry must keep creating and using them.

3 Research Technique

The principles established in the earlier model research studies are the basis for the SLR conducted in the current study (Fethi et al. 2010). The first methodology that gets in the way is bibliometric analysis, which includes analyzing articles to determine the direction of research in the field as well as the higher-cited research and the publications that pair with it. We proceeded with the procedure into the subsequent stages:

(I) Thoroughly organizing the content that has to be evaluated;
(II) Carrying out the review; and
(III) producing and presenting the review's conclusion.

Figure 2: This self-conceptualized three-phase graphic matches the typical five-phase research approach (conceptual, design, empirical, analytical, and dissemination) that is recommended by major methodological frameworks. It offers a straightforward, rigorous, and logical structure for carrying out a systematic review.

Every step is broken down into a few phases, and the procedure is shown in Fig. 2 so that it is clear what must be done at each step.

We conducted searches using Google Scholar, Science Web (WoS), and Scopus keywords because the studies listed here are of a high caliber and are easily accessible to scholars of study. The writers chose the database because of its wider coverage, even if the ultimate quotations, which show the level of quality of each author's job, are the basis for selection.

A corpus of 997 records was obtained through searches with the following keywords: AI and Banks, Machine Learning, Banking and AI in Banking and Banks, ML in Banking and Banks and AIML in conjunction with Big Data + Banking/Banks, NLP + Banking/Banks, Banking/Banks (2).

Detailed planning	Execution	Compilation and presentation
• Planning about what needs to be reviewed. • What needs to be explored. • Development of the inclusion and exclusion criteria.	• Using the keywords to be searched, filter the research work. • Data cleaning and review of the researched work.	• Compiling of the information on the filtered data. • Presenting the review on various aspects of the research data.

Fig. 2 Self conceptualized by the authors. *Source* Complied by Author

Figure 3: VOSviewer was used to generate this visualization, which displays a co-authorship network with colored gradients denoting publishing years and circles for authors. Lines connecting nodes show co-authorship relationships. By displaying natural clusters or research groups, the spatial arrangement—which is established by VOSviewer's VOS algorithm—captures the collaborative structure and temporal dynamics of an area of study.

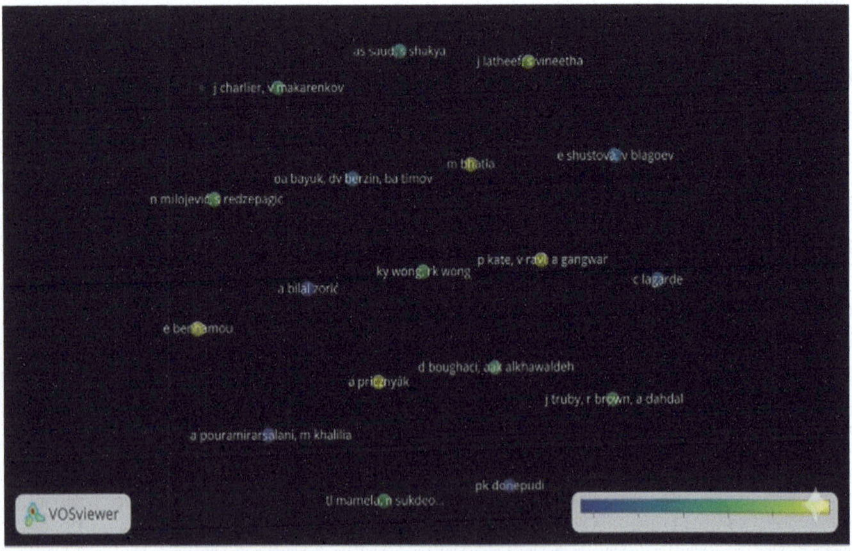

Fig. 3 Authors' overlay visualization. *Source* VOSviewer Software, bibliometric visualization tool

Only 18 writers with two or more published works during the initial phase of the research article shortlisting process utilizing automated tools and a threshold of at least two papers were able to contribute relevant research (Fig. 3). It was carried out with an eye on the caliber of the papers that made the short list for our SLR. We amended the threshold so that it was dependent on the number of citations the research effort had received rather than the number of articles authored because there were very few articles that met this condition to allow for the SLR to be conducted.

Figure 4: The diagram illustrates a clear, three-step review procedure that follows the PRISMA 2020 framework: inclusion, screening, and identification. 299 records are eliminated for lacking citations, non-English languages, missing publication years, and missing source information. The remaining 498 records are deemed irrelevant and are not included. Studies on trends by year, trends by source, and content analysis are all included in the review.

Following thorough scrutiny of the specifics, it emerged that 99 titles did not mention years of publication, 143 lacked source information for publication, and 21 differed in language from the English language. Consequently, a total of 734 articles

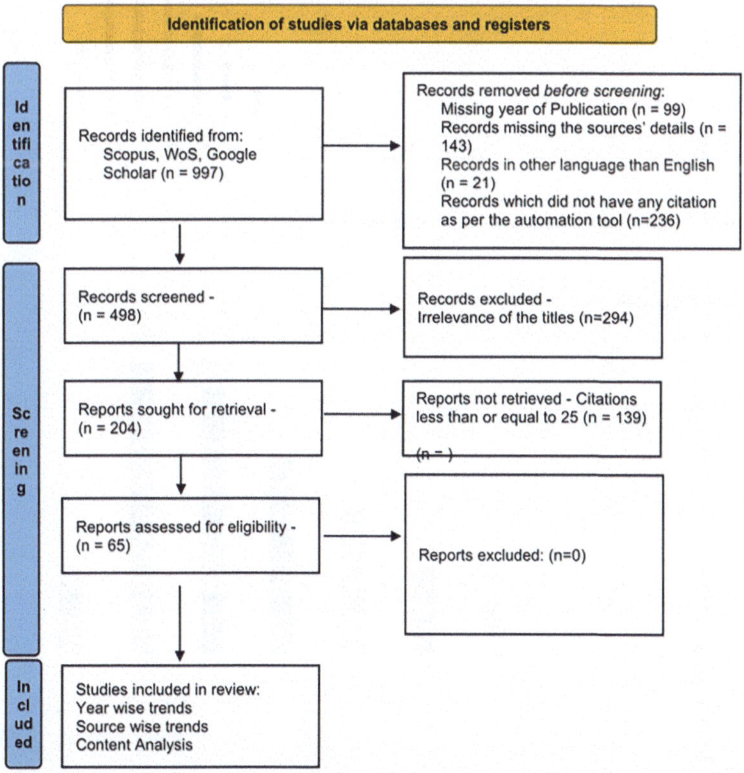

Fig. 4 Data on the selection process of articles (format source—PRISMA). *Source* Complied by Author

were found relevant following the initial filtration stage. Then, on an basis of the articles selection following the selection of the writers and articles having greater citation counts, and thus publication sources with comparatively high citation counts (refer to Fig. 4).

Figure 5: The bar graph, which is probably from databases like Scopus, Web of Science, or Google Scholar, displays a consistent rise in scholarly production over time. This pattern holds true in a number of domains, including university output analyses and literacy studies on climate change.

In the subsequent phase, we examined the following to address Among our research topics are (a) the quantity of papers published annually (Figs. 5 and 6).

Figure 6: According to the chart, the number of original authors contributing to literature increased significantly from 6 in 2013–2014 to 185 in 2022 and 41 in 2023.

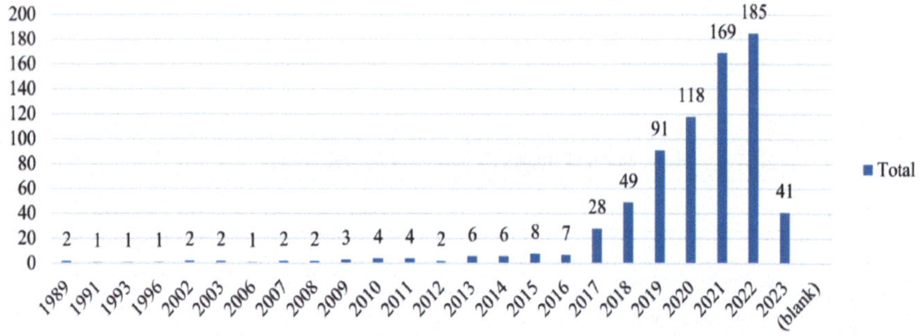

Fig. 5 Total publications by year. *Source* Complied by Author

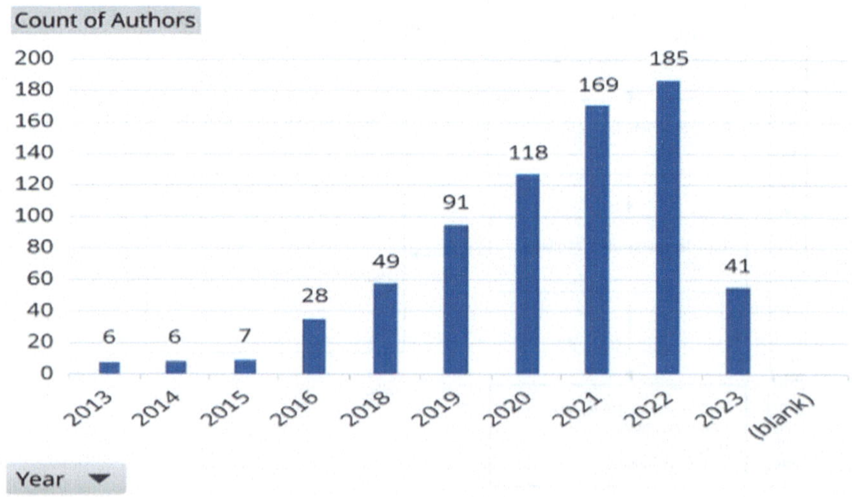

Fig. 6 Publications by year (since 2013). *Source* Compiled by Author

This growth indicates a consistent expansion of authorship and publication over time, which is consistent with bibliometric studies as well as Lotka's and Price's laws.

Table 1: The top contributing authors in banking and business intelligence research are included in the table, with Smoro, P.-Cortez, and Prita at the top with 379 citations. Other noteworthy individuals are J. Wwei; Lcao; Jchen, Mleo; Ssharma; Kmadduley; PCarmona; Fcliment; Amomparler; Hince; Baktari; GGSundarumar; Vvavi; and JWGoodell, Skumar, WMLin, DPattiaik. The study focuses on big data, machine learning, credit risk, and fraud detection.

Figure 7: According to the bar chart, which reflects general academic trends, the number of unique contributing authors in the dataset increased annually, from 6 in 2013 and 2014 to 28 in 2017, 49 in 2018, and 185 in 2022.

Table 1 Lists the top authors and titles with the highest amount of citations, arranged from large to small

Writers	References	Year	The work's business title
Smoro, Pcortez, Prita	379	2015	A review of the literature covering from 2002 until 2013, business intelligence was used in banking
Wwei, Jli, Lcao, You, Jchen	358	2013	Efficient identification of complex online banking fraud using highly unbalanced data
Mleo, Ssharma, Kmaddulety	223	2019	A survey of the literature on machine learning in banking risk management
PCarmona, Fcliment, Amomparler	175	2019	Forecasting collapse in the US financial industry: a radical gradient-boosting strategy
Hince, Baktan	134	2009	management viewpoint on an analysis comparing data mining techniques for credit rating in banking
GGSundarkumar, Vravi	131	2015	Unbalanced data in banking and insurance due to a new hybrid under sampling technique
JWGoodell, Skumar, WMLim, Dpattnaik	126	2021	Fintech and AI: bibliometric analysis reveals themes, research clusters, and underlying principles
Hhassani, Xhuang, Esilva	125	2018	Big Bata mining and digitalization in banking
Kchitra, Bsubashini	117	2013	Data mining methods and their use in the banking industry
AECelik, Ykaratepe	110	2007	Neural network models for assessing and predicting financial crises: application to the Turkish banking industry
Mtavana, ARAbtahi, DDiCaprio, Mpoortarigh	109	2018	A Bayesian network and artificial neural network approach for evaluating liquidity risk in banking
Mromao, Jcosta, CJCosta	104	2019	A case study of robotic process automation in the banking sector

Source Complied by Author

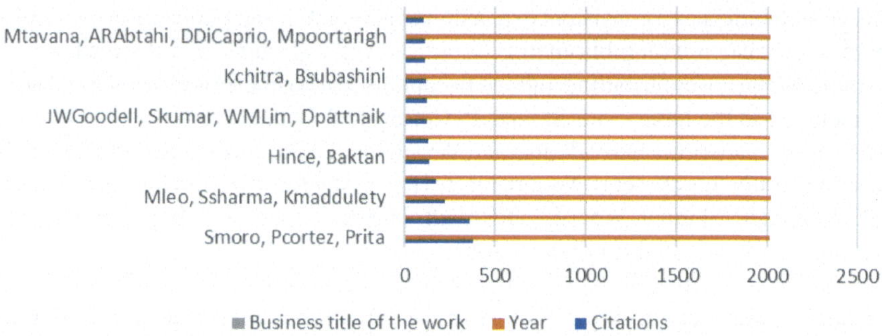

Fig. 7 Authors wise citations. *Source* Complied by Author

Table 2 The wide range of disciplines represented by these citation metrics—including information processing, economics, finance, and artificial intelligence—reinforces "Expert Systems with Applications" as the premier venue for high-impact research.

(b) how many citations the reviewed articles obtained (Table 1); and (c) the scholarly contributions the source/journal made (Fig. 7 and Table 2).

Figure 8: The "Citations" figure displays the amount of citations for AI and machine learning in banking journals; "Expert Systems" has the most, with almost 560 citations.

As we looked at the journal article's citations and decided to raise the bar to more than 25 as any amount over that is thought to be a reliable indicator of a high-quality research paper. This was based on research that showed 769,676 citations were made to a 49,417 papers in all about business research were released between 1998 and 2023. Remarkably, during this time, 32.88% of papers got no citations.

Of the residual 67.12% of the referenced articles, the largest proportion was those that received one citation, which made up 10.60%. Next came 39.75%—a total of 15,085 articles—that had between 1 and 10 citations. The final group whose papers

Table 2 Top journals on the basis of citations

References	Citations
Applications for Expert Systems	560
Hazards	247
Economy of Europe	237
Global Economics and Finance Review	175
Business Economics and Management Journal	134
Artificial Intelligence Uses in Engineering	131
Large-scale information and mental processing	125
The Development and Economic Cooperation Journal	123

Source Complied by Author

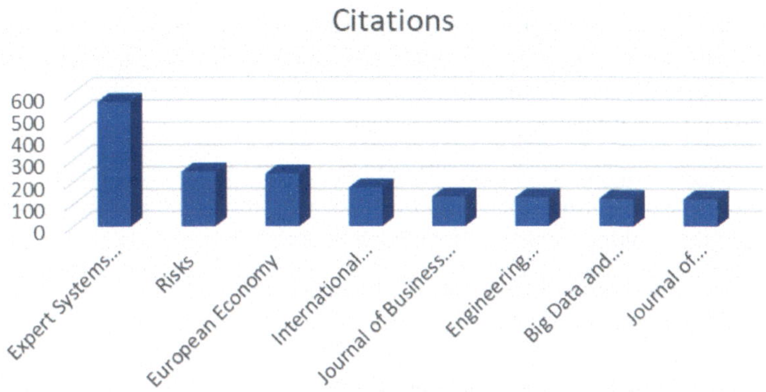

Fig. 8 Source wise citations. *Source* Complied by Author

were cited more than ten times totalled 27.37%. Of these, 2.78%—that is, 1,372 articles—received more than 100 citations per article. In general, it seemed that the bulk of the papers (67.12%) affected worldwide business community by means of citations additionally implementation. The mean over the course of the study period, there were 15.58 citations per manuscript (Ramesh and Misra 2020).

4 Content Analysis

From our research on banking operations, and based on the information gleaned from a few literary works, we list here the domains in which banks have previously implemented or may implement in the future: artificial intelligence and machine learning (Suprabha et al. 2021) and these are:

(A) CRM stands for customer acquisition and retention

- (1) Churning of customers
- (2) Transaction security
- (3) Marketing goods made by third parties
- (4) Investing and retail banking

(B) Evaluation of credit
(C) Inclusion in digital finance
(D) Microfinance
(E) Stopping fraud and online crimes
(F) Risk evaluation and crisis forecasting for banks, including apex bank rates, regulations, etc.
(G) Handling Non-Performing Assets

(H) The bank's treasury operation on interbank services between front-end and back-end operations of the clearinghouse, which solves the line issue that arises within actual banks (Pathak and Singh 2000)

(I) Financial consultants in the form of banks.

In order to respond to the inquiry, "What can AI and ML do in Banking?", writers enumerate the justifications that would explain why these technologies should be used within the banking industry, of which include:

- There are several difficulties facing the financial sector.
- Process-driven features that don't sacrifice efficiency;
- Self-service;
- Efficient and timely decision-making;
- Massive data volumes managed while preserving process integrity and speed.

During the 2007–2018 fiscal year, the banking industry's total advances and total deposits grew at a CAGR of 10.94 and 11.66%, respectively. India is the fourth-largest emerging country in terms of its proportion of the global retail credit market. The sole system identified by the Faster Payments Innovation Index as a level 5 system is India's Immediate Payment Service (IMPS). The customer-focused operations of Indian companies, such as cross-selling, developments, and procedures, have been incorporated with AI (Gawrysiak et al. 2019).

4.1 Client-Focused Services

This could include serving the customers to a point at which they could forecast potential consumer needs and market those products appropriately. The researchers in this study have solved this problem by creating a method that projects consumer preferences for personal loans using time series data (Ładyżyński et al. 2019). The system employed random forest and deep neural network classification methods. It is advised that consumers use a mechanism to recognize credit products that uses both random forests and deep neural networks categorization. Their results show that the likelihood of a product being purchased is determined by extracting important patterns from past customer behavior. Based on information taken from the banking database, the authors of a related study believed which "intelligent recommender systems" may become accustomed to suggest a goods to a customer. One such application is "CEBRA (CasE-Based Reasoning Application)." The method makes use of information from user profiles, ratings, and contextual data. With increased end-user satisfaction, it might assist the team in promoting bank goods (Hernández-nieves et al. 2021).

The paper provides examples of how technology is helping financial service companies advance. To investigate the factors that contributed to increasing consumer trust in Ukrainian banks, machine learning algorithms were employed.

An effective predictive segmentation model for credit cards clients based on their value to the financial institution is proposed by one study among the many applications of machine learning. This means that traditional methods like NNs, DTs, NNs, and random forest algorithms produce poor classification, whereas the DT approach using support vector machines can create express classification laws and regulations provide customers using appropriate marketing tactics from a variety of market areas (Suhel et al. 2020).

4.2 Risk Management and Credit Evaluation

One common method of near analytical credit risk assessment is credit scoring. Artificial intelligence may be applied to have better models that are used for credit scores. For credit award decisions that use hybrid models for credit scoring using ensemble learning, 10 classifier agents for their ensemble model are suggested. The classification accuracy was assessed using SVM, NN, and DTs. The goal was to create a model that could be compared to the current models and determine whether the new model could minimize the loss resulting from incorrect credit scoring (Alkhawaldeh and Boughaci, 2020).

The advancement of machine learning and artificial intelligence has reached an extent wherein one could say that their output is akin to the human mind. It has accomplished some pretty good work in image categorization, speech recognition, and even automated translations, but areas like models for credit risk coupled with experts in human were constantly favored in the financial sectors (Munkhdalai et al. 2019). In their research, the authors created an example that will provide an calculate the PD of the borrower, thus saving several losses because it will identify correctly who to lend to. Models in their study, like results from XGBoost and deep neural networks may be as promising as those from other machine learning techniques (Guns et al. 2020).

4.3 Microfinance

Online crowdlending has only now begun to pick up steam as the costs of transactions seem to be too high compared to loan sizes. Mobile payments are on the rise due to easy execution also. In microfinance, credit scoring addresses information asymmetry issues, as small borrowers may be unaware of numerous criteria when making loan decisions. The issues brought on by information asymmetry are resolved by credit scoring models. Institutions in the microfinance sector use credit scoring to increase financial inclusion and boost front-end team productivity. However, the recently created credit scoring tools would need to properly assess the risk before lending to new clients. Some are even assessing borrowers' creditworthiness using information from social media and internet sources. For new borrowers, it still entails a risk assessment. In order to assess a potential borrower's riskiness, a number of new credit scoring platforms are currently gathering information from social networks (Herrmann and Ashta 2021). When lending decisions need to be made and the credit banking risk for equity loans needs to be assessed, big data techniques are also useful. They may break the financial data set into the numerous kinds of risk groupings (Vaca et al. 2018).

4.4 *Financial Inclusion Through Digital Means*

We discovered research in the literature that looks into how AI affects digital financial inclusion. People from economies focused at the base of the pyramid and moved into the segment that is financially active. In addition to FinTech enterprises that strive for this kind of inclusivity. According to the research, as it is concluded the fields of risk reduction, cybersecurity, information asymmetry resolution, and voice help are all significantly impacted by AI through automated dialogue systems. Authors have recommended using AI technologies to accelerate the adoption of digital finance. For participants in the financial sector, confirming an individual's creditworthiness potential customer during selling a piece of credit has constantly been a major issue. The researchers intend to use a simultaneous social spider algorithm-based representation to verify economic and social background in this regard. The authors employed widely utilized algorithms for credit rating and contrasted nine internal and external metrics. The lack of adequate data, according to the authors, also prevented the model from being developed to the point where it could be effective. But they would like to carry out more research using big data sets (Qu and Cheng 2019). Additionally, FinTech companies must be certain that AI and its technologies are being used for credit rating (Mhlanga 2020).

4.5 *Preventing Scams and Online Crimes*

In the credit card industry, frauds can be classified according to many factors, such as whether they happen at the application level or whether the cards are printed manually or electronically. It is possible to identify a fraudulent credit card transaction using deep learning techniques (Kadam et al. 2020). Along with the ML algorithms, NN (Neural Network), including the Support vector machine and k-Nearest Neighbour techniques, is used for this purpose. ANN-based models have been developed and are nearly foolproof for credit card fraud detection (Błaszczyński et al. 2021). The same risk is involved with regard to loans issued to the customer. Different hybrid machine-learning models used for proper credit ratings have been proven to work well in such cases. These authors have put forward four types of hybrid models that is suitable for many clustering and classification applications when combined with support vector machines (SVM) and self-organizing maps (SOM) (Kuiper et al. 2021).

According to a study, financial institutions should employ AI for fraud detection since of its accuracy, and the same is used for verifying the dependability of reports. Knowledge graph models were adapted with traditional tools. A small number of additional features were implemented that could impact classifier models' ability to analyze FSFD data (Lutfi and Salameh 2021).

According to the previous studies, it is claimed that there is an analysis procedure that is dependent on the most effective machine learning model. They can therefore be used by enterprises to detect fraudulent activities that could have the potential to jeopardize their cloud-based analytics platform. The researchers have created an algorithm for one of the works that can identify fraudulent transactions instantly, enabling the business to take timely help detect the potential frauds (Shen et al. 2021).

4.6 Evaluation of Bank Risk and Forecasting of Crises

This applies in the management of risk of the bank. Since the crisis of the global financial, the working of this capital adequacy bank has been on with stringent regulations and nowadays, with technology, it drives how banks operate. AI has the ability to completely transform the banking industry. Its methods for voice recognition and machine learning, natural language processing, whereas deep learning can imitate people thought processes and evaluate large amounts of data much more quickly (Swankie and Broby 2019). The study's authors examine the risks associated with operations, credit, liquidity, and reputation and come to the conclusion that AI may be able to reduce them. Blockchain and AI together have the potential to completely change how banks handle risk fronts. Big Data processing, cybersecurity, preventing fraud, AML procedures, anticipating and modeling risks, and conformity with regulations are all important, but they also require infrastructure, an organization, and the integration of modern technological into the center of their regulations and plans (Momparler et al. 2018).

The clustering technique was used to categorize the performance of the same institutions. Their train and test accuracy was comparatively high once the ANN was trained, which encouraged them to keep using the same model to forecast that institution wouldn't perform well in the future (Maddulety et al. 2019). It would be encouraging if other experiment results of ML methods' decision trees, after modifications, are used to forecast those possibly shaky credit entities. Supervised Machine Learning techniques, which are based on Neural networks, boosting, and bagging are applied as well for modelling credit risk in financial institutions (Dzhaparov 2020).

This group of scholars provides for a ML-based toolbox that could identify both those infrequent occurrences and those that could trigger emergencies (Kulaev and Domashova 2020). A list regarding macroeconomic variables, including credit-to-GDP disparity and stock values, inflation both short- and long-term rates interest rates, leads & lags, data filtering techniques, and the study of time series, were compiled (Hu et al. 2021). Among the rare models that were created, the top model can be useful determine the one likelihood of a banking crisis up until six a quarter ahead of duration and corresponding to numerous economies (Coffinet and Kien 2019).

Correspondingly, primary prediction instrument used by econometricians to anticipate and shield the impending the financial crisis is the System for Early Warning (EWS). The data is taken to forecast the emergency based on the shift in financial metrics and the ensuing effects. In their study, the authors found that the experts voting EWS framework is more effective at inferring crisis data in a variety of situations (Wang et al. 2021). However, machine learning-based EWS models still require some finetuning. These are of significant value in terms of anticipating the systematic crisis of the banking domain (Messai and Gallali 2015). The reference logit models, created in the earlier work, rely on the information efficiently and prove that it might have actually been able to predict the crisis of 2008 for the majority of nations (Schweinitz et al. 2019).

According to one study of this type, a statistical model founded on RBF (Radial Basis Function) and PUNN (Product Unit Neural Networks) networks can identify financial crises, which are now a common occurrence and can be classified as banking crises. Since the model was trained using 18 years of pertinent data, the suggested methods outperform other statistical and artificial intelligence approaches already in use (Gutiérrez et al. 2010). In subsequent studies, the researchers used the AI and ML models they had created to forecast bank failure (Paule-Vianez et al. 2020). By assessing the indicators of financial hardship, they could determine which institutions were at high danger (Karatepe and Celik 2007).

4.7 Taking Care of Nonperforming Assets, or NPAs

Given the fact that NPAs possess started posing one significant issue facing the banking sector, banks would require an effective system for determining clients' creditworthiness not only at the lending period, as well as through the evaluation of the clients' anticipated future financials. If banks could use AIML to evaluate the companies' financial stress, then the losses considering the bank suffers can be highly reduced (Liu et al. 2012). Though AI is integrated into every aspect of banking operations, their application in resolving NPAs has not been studied so far. The current study compares the performance of algorithms for machine learning and regression methods in estimating the percentage of recovery of nonperforming mortgages. Techniques including cubism, random forests, and enhanced trees were applied in order to examine datasets (Bellotti et al. 2021). Variables from loan contracts and some time series were employed to increase the effectiveness of the algorithms used, and the results produced accurate estimates.

4.8 The Function of Banks as Advisors in Finance

Acceptance in Every Field of Artificial Intelligence: All fields, particularly the financial sector, are beginning to embrace artificial intelligence. AI's safety and security features are the primary reason for its increasing acceptance in banking, stock trading, and insurance. As we've shown, artificial intelligence (AI) can offer instruments for preventing fraud while staying within the compliance and legal parameters established by various agencies. Every step of the adoption of new technology in financial services and products is expected to bring with it more risks and challenges, which must be addressed while taking into account the intentions of providers of services and the consumers' acknowledgment of the same, given their varied topography and educational backgrounds (Ghosh and Chanda 2020).

Numerous projects undertaken by State Bank of India, including Shikhar—the credit card issuing model, fast digital loans on Yono, branch foot traffic reduction, app-based lending through "Krishi," and ATM win back, have effectively included AI, ML, and DA. The Punjab National Bank, or PNB, also employs technology-based methods to anticipate credit card failure, analyze the "National Automated Clearing House (NACH)" mandate, forecast slippages in the MUDRA category for loans, and market third-party products. ICICI Bank specializes in "Zero Credit Touch" (ZCT) methods that leverage AI and ML to provide credit facilities to consumers without requiring further information from them. These technologies have been used by HDFC to unify their business, competency and skill mapping, and the end-to-end flow of data. Citi Bank separates people who are very likely to purchase its specialty products using LR-based AI and ML models. In order to reduce idle cash, moreover, machine learning models are widely utilized to optimize the amount of money at the level of the ATM. Credit scores are calculated using TransUnion CIBIL, which has AI built in. Information is gathered from social media, customer history, SMSs, and address stability. The more effectively Indian banks use AI and ML, the more they increase client engagement, offer pertinent goods and services, and streamline the banking process while reducing the risk of payment fraud. The AI-powered chatbots may operate around the clock and are also available in vernacular languages. It will support the goal of financial inclusion by encouraging consumers to contact banks for even the smallest needs. AI that prevents deception as well boosts assurance and faith between those who are not just yet enrolled in the system of banking. These Indian platforms for which widely use AI to encourage digital payments age and counsel customers on better financial decision-making, include "Unified Payments Interface (UPI)" and "Bharat Interface for Money (BHIM)," which also support financial inclusion" (Ravikumar et al. 2021).

For all financial organizations, extracting information from a large number of textual and visual documents has always been a difficult undertaking. A study provides a deep learning-based algorithm to extract valuable information by combining textual and visual data (Oral 2020). In their trial, they used machine learning (ML) to optimize the information extraction process, and when compared to manual labor, the results were fairly accurate and quick (Cho et al. 2020).

(1) **The study's restrictions**: Limiting to only two or three databases becomes challenging because the current databases are not integrated, leaving researchers constantly torn between them. Furthermore, it was found to be quite challenging to fit even the summaries of significant and highly referenced works into a research paper, which is limited to a certain number of pages. It was believed that the same studies could be carried out by

Figure 9: The graphic illustrates relationships and frequency in a network of keywords that occur frequently in AI and machine learning literature. By graphically grouping related keywords and illustrating their links, it assists in identifying themes and new trends in the banking sector.

Table 3: The table identifies 21 terms that are commonly used in the literature on AI and machine learning in banking, with the most common terms being financial institutions and synthetic wisdom. Other important concepts are financial investment, computer-based learning, and gadgets. This collection sheds light on how AI and machine learning applications are developing.

Figure 10: The graph shows how frequently important terms appear in the literature on artificial intelligence and machine learning in banking. "Artificial Neural" and "Bank" are the most often used terms, with "Machine Learning" coming in second with about 350 references. "Artificial Intelligence," "Banking Industry," and "Computer-Based Studying" are among the other keywords that show up

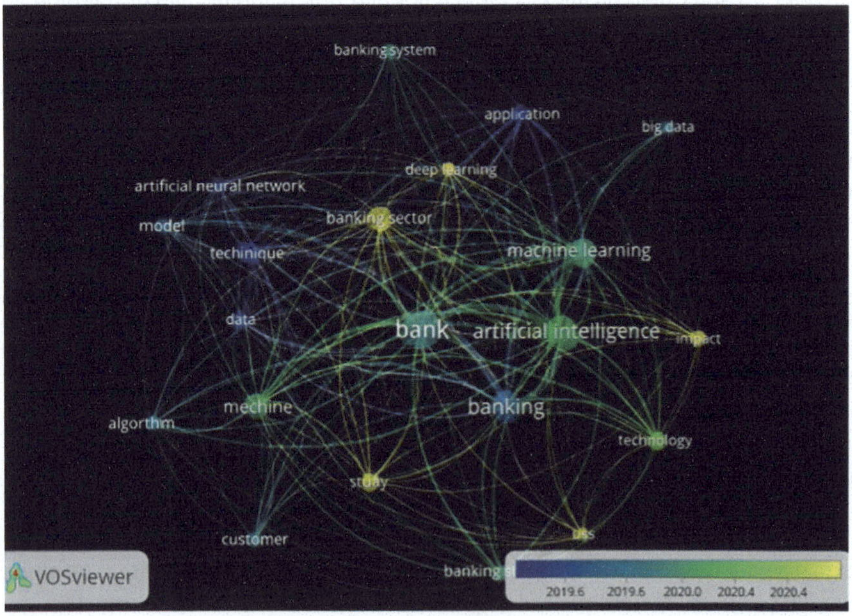

Fig. 9 The network representation illustrates the keywords chosen and their breadth in previous studies. *Source* VOSviewer bibliometric visualisation tool

Table 3 Key words that appear in the articles that made the short list

Id for keyword	The term	Situations
1	An algorithms	75
2	Utilization	100
3	Synthetic wisdom	475
4	A synthetic neural architecture	69
5	The financial institutions	543
6	Investing in monetary	369
7	The monetary sector	90
8	Industry of monetary	140
9	Monetary framework	51
10	Massive information	32
11	A client	65
12	Statistics	73
13	In-depth education	56
14	Influence	49
15	Device	205
16	Computer-based studying	350
17	An example	101
18	Examine	99
19	Methodology	99
20	Technologies	113
21	Utilize	52

Source Complied by Author

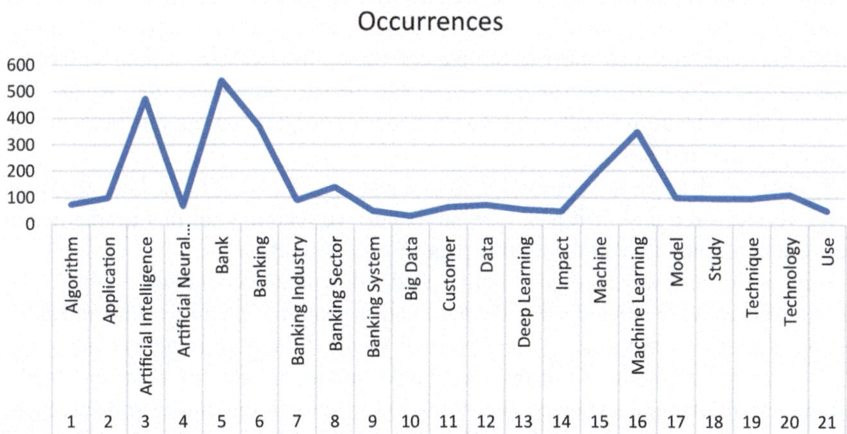

Fig. 10 Key words that appear in the articles that made the short list. *Source* Complied by Author

5 Discussions, Conclusion

Within the banking sector, adoption of AI and ML has been significant for well-known reasons. Following a thorough review of chosen study reports, we were capable of identify the majority of prevalent fields in which these two technologies have made a contribution. Activities pertaining to customers, such as acquisition, retention, churning, improved convenience and experience, and loyalty, have accounted for the majority.

Automated credit scoring methods have been the subject of the most research, with many academics finding that AI and ML are applied at a very high level.

The second topic we looked at was bank crisis prediction, where a lot of research papers, projects, and articles had been published.

Preventing and detecting fraud is another area where the use of these technologies has been envisioned. Additionally, research has been conducted on the likelihood of mortgages becoming delinquent early.

According to numerous research, there is a noticeable application of AI and ML in the field of bank risk management (Swankie and Broby, 2019).

According to the surveys, banks have not used AI or ML very much in their day-to-day daily operations (Balaramachandran and Kumar, 2018). Only one publication on workforce management and queuing problem solving was found in the scant literature on the use of these technologies in operations.

In areas like managing non-performing assets (NPAs) both at the sector and individual bank levels, we think ML and AI, may become accustomed to anticipate unfavorable occurrences effectively in advance & notify management to take the appropriate steps to prevent losing assets (Doumpos M et al. 2023).

Additionally, it engages with the Apex Banks for a variety of purposes, including borrowing and depositing funds for simple periods (Kaur et al. 2020). Frequently shifting interest rate regimes and policies govern the same (Serengil and Ozpinar 2017). Likewise, for such short-term financial needs, the banks negotiate with other commercial banks. When it comes to automated treasury administration, AI and ML may be quite helpful in handling all likely needs or surplus, taking into account the current policies and prices.

In order to determine reasons for implementing AI and ML in these domains hasn't yet occurred and, if it has, to what extent, we recommend that academics in the aforementioned fields take up this paper for investigation.

Some terms that were included in this SLR's downloaded database. with at least 30 occurrences were created using the automation program, factor loading and associated research. Twenty-one key words in all satisfied the requirements (Table 3 in Fig. 7, graphical depiction). For instance, banks and learning by machine and banks and intelligent machines showed very thick links between the dots. Similarly, there are links between banks and AI + ML and banks and banking, but there are also links between the dots with banks and ANN, Deep Learning, or even customers, to mention few of these key words, for which the connections are not as strong.

Nine clusters were discovered to include the keywords. These keywords been in use more frequently within study work completed in this banking industry's use of AI and ML sector since they have an eigenvalue greater than 1. Kaiser's edict states that the quantity of parts in a analysis of principal components (PCA) should be determined if the eigenvalue of those components is greater than 1. Words that contain id numbers 1 through 9 in our study have eigenvalues larger than 1, while those with id numbers 10 through 21 have eigenvalues less than one. These are the terms that, according to earlier study, have not been used very often indicates that there is a research deficit in the area of these keywords.

The best fit between a keyword and a cluster is displayed in this rotated variance components table. For example, the keyword "algorithm" is ideal matched using cluster number five, and so forth. Additionally, a research region can be chosen using the rotational variance components table if the keywords fit into a single cluster. This field of study will probably have more suitable investigation because of the space found there, according to findings of testing of statistics. This is possible to see that groups together with fewer keywords falling into them present a chance to choose a research topic by combining those keywords.

6 Implications of the Study

- According to the report, AI and ML have the ability to boost customer satisfaction, improve risk management, and make banking payment procedures safer.
- According to the report, by 2030, the banking industry will have saved over $1 trillion with AI, of which $31 billion will come from AI-assisted risk management.
- Among the study's practical implications include the use of AI and ML in banking to better customer service, lower risk, and promote financial inclusion. These applications might also boost operational efficiency and cut expenses.
- According to the study, the banking industry can benefit from AI and ML in a number of ways, such as credit rating, fraud detection, risk management, and customer acquisition.
- AI and ML may also be used to detect non-performing assets, enhance client engagement, and optimize treasury management, according to the report.
- The study also reveals that AI and ML may be utilized to expedite banking procedures and provide customized goods and services.
- By grouping phrases that belong to the same cluster, the study's practical applications include guiding researchers in selecting a research topic and offering a way for identifying possible research ideas using keyword analysis utilizing PCA and Kaiser's edict.

7 Scope for Future

- According to the report, future research should concentrate on resolving the gaps and difficulties in the banking industry's adoption of AI and ML, including the requirement for more thorough academic studies to advance the body of knowledge.
- Future research will focus on creating more efficient models with massive data sets and investigating new domains, such voice assistance and cybersecurity, where AI and ML can be used in banking.
- The study recommends that future research look into the research topics found by combining keywords that belong to the same cluster.

References

Alkhawaldeh AAK, Boughaci D (2020) A comparative analysis of suitable machine learning methods for credit rating and bankruptcy prediction in banking and finance. Anal Risk Decis 8:15–24. https://doi.org/10.3233/RDA-18051

Balaramachandran PR, Kumar KN (2018) An investigation on how robotic process automation affects the retail banking sector's client experience. J Internet Bank Commer 23(3):1–27

Bellotti A et al (2021) Using machine learning to predict non-performing loan recovery rates. J Forecast 37(1):428–444. https://doi.org/10.1016/j.ijforecast.2020.06.009

Bilal ZA (2016) Using neural networks to forecast client attrition in the banking sector. Interdiscip Descrip Complex Syst 14(2):116–124. https://doi.org/10.7906/indecs.14.2.1

Błaszczyński J et al (2021) Used a rough set approach based on dominance to detect auto loan fraud. Appl Expert Syst. https://doi.org/10.1016/j.eswa.2020.113740

Cho S, Chai J, Huang J (2020) A study and categorization of the literature on deep learning in banking and finance. Front Bus Res China 14(1). https://doi.org/10.1186/s11782-020-00082-6

Coffinet J, Kien JN (2019) Rare event detection: a machine learning toolkit with a financial crisis application. J Financ Data Sci 5(4):183–207. https://doi.org/10.1016/j.jfds.2020.04.001

Doumpos M et al (2023) Artificial intelligence techniques and operational research in banking. Eur J Oper Res. https://doi.org/10.1016/j.ejor.2022.04.027

Dzhaparov P (2020) Artificial intelligence and blockchain applications for bank risk management. Econ Manag XVII(1):43–57

Eduardo M et al. Use artificial intelligence to forecast service quality in the banking industry. Antverpiensia Linguistica [Preprint] (3)

Gawrysiak P, Żbikowski K, Ładyżyński P (2019) Direct marketing initiatives using random forests and deep learning in retail banking. Appl Expert Syst 134:28–35. https://doi.org/10.1016/j.eswa.2019.05.020

Ghosh S, Chanda D (2020) Banking services and artificial intelligence: a path ahead. Productivity 61(1):11–18

González-Carrasco I et al (2019) Machine learning-based automatic identification of connections among banking transactions. Inf Sci 485:319–346. https://doi.org/10.1016/j.ins.2019.02.030

Guns T, Verbeke W, Coenen L (2020) Machine learning techniques for short-term default probability: a comparison of ranking, regression, and classification techniques. J Oper Res Soc. https://doi.org/10.1080/01605682.1865847

Gutiérrez PA et al (2010) For precise banking crisis identification and prediction, logistic regression is hybridized with product unit and RBF networks. Omega 38(5):333–344. https://doi.org/10.1016/j.omega.2009.11.001

Hernández-nieves et al (2021) CEBRA: a CasE-based reasoning application to recommend banking products. Eng Appl Artif Intell 104(June):104327. https://doi.org/10.1016/j.engappai.2021.104327

Herrmann H, Ashta A (2021) An assessment of the risks and potential for banking, investing, and microfinance associated with artificial intelligence and fintech. Strategy Change 30(3):211–222. https://doi.org/10.1002/jsc.2404

Hu L et al (2021) An overview of supervised machine learning techniques with banking applications. Int Stat Rev 89(3):573–604. https://doi.org/10.1111/insr.12448

Kadam AA, Mundhe KS, Dubey SC (2020) Artificial neural networks and backpropagation are used to detect credit card fraud. In: Proceedings of the international conference on intelligent computing and control systems, ICICCS 2020, pp 268–273. https://doi.org/10.1109/ICICCS48265.2020.9120957

Karatepe Y, Celik AE (2007) Neural network models for assessing and predicting banking crises: a Turkish banking industry application. Appl Expert Syst 33(4):809–815. https://doi.org/10.1016/j.eswa.2006.07.005

Kaur N et al (2020) Banking 4.0: artificial intelligence's impact on the banking sector and how it's transforming contemporary banks. Int J Manag 11(6):577–585. https://doi.org/10.34218/ijm.11.6.2020.049

Kuiper O et al (2021) Investigating explainable AI in the financial industry: supervisory authorities' and banks' viewpoints. arXiv preprint 2111.02244

Kulaev M, Domashova J (2020) Technology of forecasting possibly problematic credit organizations based on machine learning methodologies. Proc Comput Sci. https://doi.org/10.1016/j.procs.2020.02.167

Liu Y-S, Chen M-Y, Shie FS (2012) An application of the American banking sector to the prediction of corporate financial crisis. Appl Neur Comput 21(7):1687–1696. https://doi.org/10.1007/s00521-011-0765-5

Lutfi KM, Salameh RS (2021) Artificial intelligence's contribution to reducing cybercrimes at Jordanian commercial banks. Accounting 7(5):1147–1156. https://doi.org/10.5267/j.ac.2021.2.024

Maddulety K, Leo M, Sharma S (2019) A review of the literature on machine learning in banking risk management. Risks. https://doi.org/10.3390/risks7010029

Mehdiabadi A, Tabatabeinasab M, Spulbar C, Karbassi Yazdi A, Birau R (2020) Are we ready for the challenge of Banks 4.0? Designing a roadmap for banking systems in Industry 4.0. Int J Financ Stud 8(2):32. https://doi.org/10.3390/ijfs8020032

Mehrotra A (2019) Financial services AI must combine automation and human interaction. In: The international conference on automation, computational, and technology management (ICACTM 2019), pp 342–347. https://doi.org/10.1109/ICACTM.2019.87760.

Messai AS, Gallali MI (2015) A microprudential approach to financial leading indicators of banking distress: European evidence. Asian Soc Sci 11(21):78–90. https://doi.org/10.5539/ass.v11n21p78

Mhlanga D (2020) Financial Industry 4.0: artificial intelligence's (AI) effect on digital financial inclusion. Int J Financ Stud 8(3):1–14. https://doi.org/10.3390/ijfs8030045

Momparler A, Carmona P, Climent F (2019) An extreme gradient boosting method for predicting failure in the U.S. banking industry. Int Rev Econ Financ 61(February 2018):304–23. https://doi.org/10.1016/j.iref.2018.03.008

Munkhdalai L et al (2019) An empirical evaluation of machine learning techniques for evaluating the creditworthiness of bank customers. Sustainability (Switzerland) 11(3):1–23. https://doi.org/10.3390/su11030699

Oral B et al (2020) Extraction of information from visually rich and text-intensive banking documents. Inf Process Management. https://doi.org/10.1016/j.ipm.2020.102361

Pasiouras F, Fethi MD (2010) A survey on the use of artificial intelligence and operational research methods to evaluate the performance and efficiency of banks. Europ J Operat Res. https://doi.org/10.1016/j.ejor.2009.08.003

Pathak N, Singh T (2000) Artificial intelligence's growing influence in the Indian banking industry. J Crit Rev. https://doi.org/10.55399/W.GQO2090

Paule-Vianez J, Coca-Pérez JL, Gutiérrez-Fernández M (2020) Financial distress prediction in the Spanish banking system: an artificial neural network application. Appl Econ Anal. 28(82):69–87. https://doi.org/10.1108/AEA-10-2019-0039

Pio L, Cavaliere L, Bhatia S (2020) The use of digital technology in the banking and financial system while socio-distancing limitations are in place—looking ahead. Int J Adv Res Eng Technol (IJARET) 11(6):307–315. https://doi.org/10.34218/IJARET.11.6.2020.027

Qu Y, Cheng M (2020) Does FinTech for banks lower credit risk? Chinese evidence. Pac Basin Financ J 63(December 2019):1–24. https://doi.org/10.1016/j.pacfin.2020.101398

Ramesh DB, Misra R (2020) An examination of highly cited research projects in accounting, management, and business. Philos Pract Libr 2020:1–12

Ravikumar T et al (2021) To bank the unbanked, artificial intelligence is being used. Ann RSCB 25(5):129–132

Schweinitz GV, List S, and Beutel J (2019) Can we anticipate banking crises with the aid of machine learning? J Financ Stabil 45:100693. https://doi.org/10.1016/j.jfs.2019.100693

Serengil SI, Ozpinar A (2017) Machine learning-based workforce optimization for bank operation centers. Int J Interact Multimed Artif Intell 4(6):81. https://doi.org/10.9781/ijimai.2017.07.002

Shen Y et al (2021) Financial feature embedding using knowledge representation learning for the detection of fraud in financial statements. 2021. Procedia Comput Sci. https://doi.org/10.1016/j.procs.04.110

Singh K (2020) AI-powered banking by banks. Int J Adv Res Manag Soc Sci 9(9):1–11

Suhel SF et al (2020) Chatbots that use artificial intelligence language to automate banking conversations. In: IEEE's 8th international conference on dependability, infocom technologies, and optimization (trends and future directions) (ICRITO). https://doi.org/10.1109/ICRITO48877.2020.9197825

Suprabha KR, Kuanr A, Arjun R (2021) A comprehensive review and agenda for the development of banking intelligence in emerging markets. Int J Inf Manag Data Insights 1(2):100026. https://doi.org/10.1016/j.jjimei.2021.100026

Swankie G, Broby D (2019) Investigating how artificial intelligence affects banking risk assessment. Artificial Intelligence Economics, p 18. https://www.researchgate.net/publication/337908452_Analyzing%20the%20Effect%20of%20Artificial%20Intelligence%20on%20the%20Assessment%20of%20Banking%20Risk?C0de8a4b0346-XXX&enrichSource=rgreq-9e9f31b0f708a4b84a50enrichId=NTY4OTM2OThAMTU3NY292ZXJQYWdlOzMzNzkwOD Q1MjtBUzo4MzUyNzgz

Thach NN et al (2021) Industry 4.0 technology quality management and cybersecurity risk management on contemporary banking operations in emerging markets—the Vietnamese case. Int J Qual Res 15(3):845–56. https://doi.org/10.24874/IJQR15.03-10.

The Ginimachine (2018) AI-assisted risk management in banking today. https://ginimachine.com/blog/banking-today-risk-management-with-ai/

Thomsett MC, Wewege L, Lee J (2020) Trends in digital banking and disruptions. J Appl Finance Bank 10(6):1792–6599

Vaca M, Pérez-Martín A, and Pérez-Torregrosa A (2018) Credit banking risk in home equity loans is measured using big data approaches. J Bus Res. 89(August):448–454. https://doi.org/10.1016/j.jbusres.2018.02.008

Wang T et al (2021) An early warning system for systemic banking crises based on machine learning. Appl Econ 53(26):2974–2992. https://doi.org/10.1080/00036846.2020.1870657

Blockchain-Enabled Health Records: A Secure and Patient-Centric Model

M. Thamizhselvi, Veto Dey, and K. A. Preethi

Abstract The Sustainable Development Goals (SDG 3: Good Health and Well-Being SDG 9: Industry Innovation and Infrastructure and SDG 10: Reduced Inequalities) cannot be achieved in the fast-paced digital economy of today without fair access to healthcare. The digital health system needs to promote trust, patient friendliness and inclusion. The traditional electronic health record (EHR) systems are often centralised and impervious, which creates barriers for patients in recovering and controlling their data. To overcome these barriers, this chapter introduces the Blockchain-Enabled Health Records (BEHR) model, an innovative framework to create a transparent, protected, and patient-centric health data system. The model joins blockchain technology, Retrieval-Augmented Generation (RAG), and Large Language Models (LLMs). Blockchain ensures immutability and distributed record-keeping. While the RAG simplifies real-time background-aware reclamation, LLM empowers users to understand complex medical histories through natural language summaries. The main objective of this framework is to enhance inclusiveness and digital health literacy. The model not only promotes digital empowerment, data equity, and inclusive decision-making but also creates an impactful workplace and communities where health access and simplicity are critical for efficiency and well-being. The model exemplifies how digital transformation and AI technologies can drive socially sustainable solutions in healthcare, contributing to a future of transparent, human-centered, and tech-enabled medical systems.

M. Thamizhselvi (✉)
NSB Academy, Bangalore, India
e-mail: thamizhselvim@nsb.edu.in

V. Dey
The Oxford College of Science, Arts, Commerce and Management, Bangalore, India
e-mail: drvetodeyoxford@gmail.com

K. A. Preethi
CMS, Jain Deemed to be University, Bangalore, India
e-mail: preethi_ka@cms.ac.in

A. Singh et al. (eds.), *Human-Centered Sustainability: Bridging Well-Being, Gender Equity, and Technology*, World Sustainability Series,
https://doi.org/10.1007/978-3-032-22020-2_23

Keywords Blockchain-enabled health records (BEHR) model · Retrieval-augmented generation (RAG) · And large language models (LLM) · Vector DB

1 Introduction

In the pursuit of universal health coverage and data equity, healthcare systems worldwide face the dual challenge of improving efficiency and ensuring inclusive access to personal medical information. The existing system of EHRs (Electronic Health Records) is not efficient enough to cater for the needs of the individuals and acts as a significant barrier, especially patients impacting SDG 3 (Good Health and Well-Being), SDG 9 (Industry, Innovation, and Infrastructure), and SDG 10 (Reduced Inequalities). In the modern healthcare system, patients often seek medical attention from different healthcare facilities, general practitioners, specialists in that domain, diagnostic centres and healthcare centres. These kind of visits many times requires routine check-ups, sudden illnesses, chronic condition management, or long-term treatments. Some medical problems are resolved in one appointment, whereas others may require multiple visits, different follow-up checks, tests, and continuous monitoring. Due to the need for healthcare records at various places, a patient's medical records tend to become fragmented and scattered across different healthcare providers (Jensen et al. 2012). Every health care centre may maintain its system for storing medical information, which is often not linked with other systems (Murdoch and Detsky 2013). Maintaining a comprehensive and consistent patient medical record is made extremely difficult by this fragmented data. Ineffective data exchange has negative business effects, including increased expenses, patient dropouts, repeat diagnostics, and non-compliance with data protection laws like GDPR and HIPAA. This poses problems for healthcare organisations looking to expand digitally, as well as for the continuity of care.

Nonetheless, when a patient transfers between medical facilities or sees a new doctor for a second opinion, they usually need to carry previous reports, prescriptions, or test results with them to the hospital. They are frequently forced to gather these records by hand, keep hard copies, or depend on one hospital to forward records to another. This method is laborious, sometimes inconvenient, and prone to mistakes or data errors (Adler-Milstein and Jha 2017). Poor continuity of care, postponed treatment decisions, and repeated diagnostic testing can result from this gap in a unified health record system (Roehrs et al. 2017). Patients also frequently lack control over the storage, sharing, and accessibility of their private health information. When taken together, these challenges highlight the critical need for a secure, patient-centred, and centralised method of effectively managing medical records among various service providers.

A blockchain-enabled health records system that manages a more secure transparent and patient-focused solution is the main focus of this research in order to

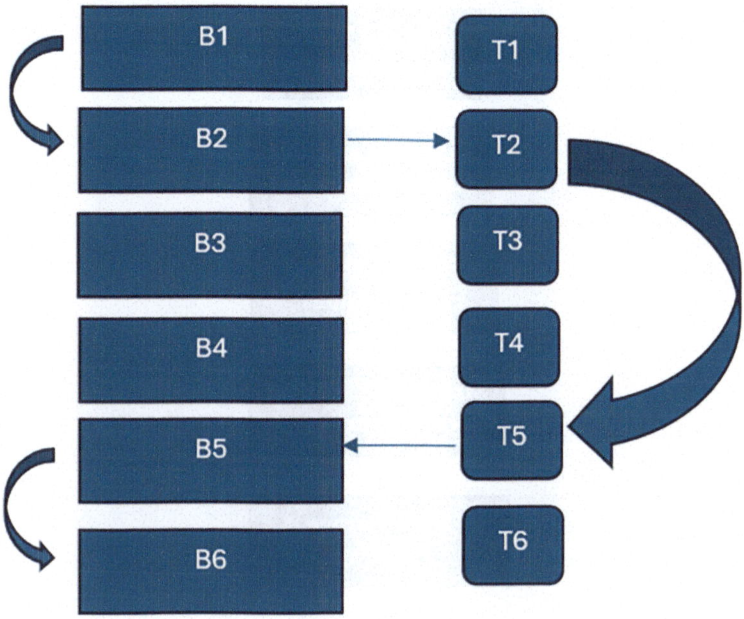

Fig. 1 Multidimensional—Blockchain and time stamp

address these issues. Every hospital visit and other health-related activity is monitored by the system, which subsequently stores the data in a blockchain structure as a distinct block. Both chronological and contextual data arrangement are made possible by the fact that these blocks are grouped according to the type of medical condition or treatment, in addition to being sequenced in the precise pattern of occurrence using timestamps (Azaria et al. 2016). The design prioritises keeping a comprehensive, well-structured and impenetrable medical history. The proposed system's key components are data confidentiality, data protection and data retrieval.

Authorised healthcare providers can access the data whenever needed, and data retrieval improves accuracy and maintains data updates, facilitating faster and more informed medical decisions (Kuo et al. 2017). Data protection uses blockchain's secure and unchangeable nature to prevent unauthorised changes, loss, or misuse of the data (Zhang et al. 2018). Data confidentiality gives patients the power to decide who can see or access their medical records, allowing them to maintain control over their personal health information (Roehrs et al. 2019). This research suggests a novel combination of Blockchain Retrieval-Augmented Generation (RAG) and Large Language Models (LLM) to provide a tamper-proof, context-aware and patient-controlled data management solution in contrast to current digital record systems that are either centralised or inaccessible. This is among the first frameworks to combine all three to increase transparency and accessibility in healthcare.

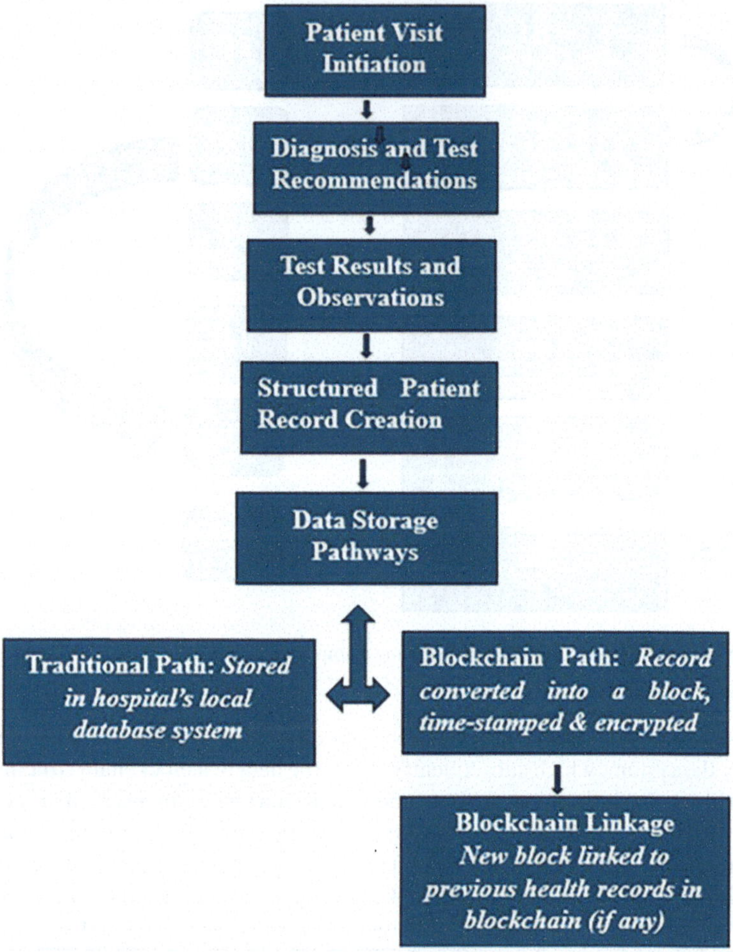

Fig. 2 Model 1: Patient record generation and blockchain integration

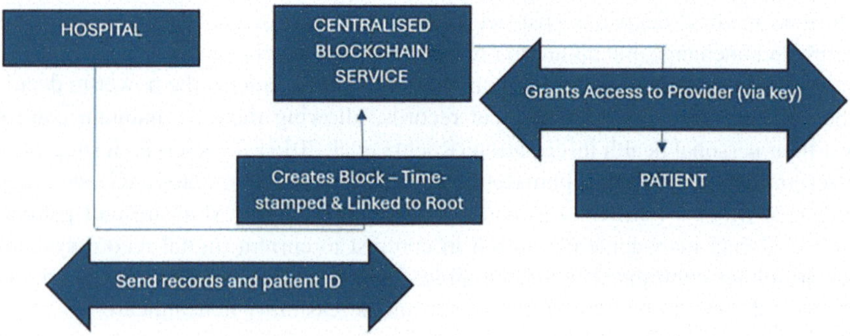

Fig. 3 Model 2: Blockchain integration via root ID

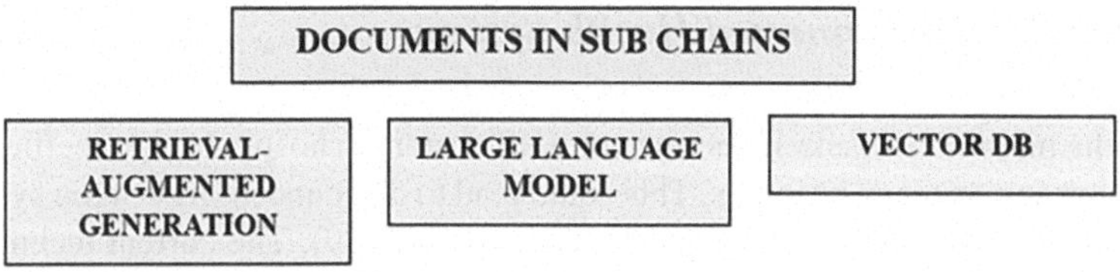

Fig. 4 Integration of AI into blockchain-enabled health records

With the support of advanced technologies such as Large Language Models (LLMs) and Retrieval-Augmented Generation (RAG), the system also offers intelligent access to information using simple, conversational queries (Lewis et al. 2020). This helps in improving the efficiency of healthcare services and allows patients to play an active role in managing their health data. This proposed system envisions a future-ready healthcare ecosystem where patients are empowered with full control over their data, care decisions are better informed, and digital infrastructure supports inclusive, equitable health outcomes across borders.

2 Literature of Review

2.1 Blockchain and Healthcare: Moving toward Decentralized Systems

To simplify man's life there are many new technologies that are emerging. Blockchain is one such recent important technology. Nowadays, for better safety and security, blockchain technology is implemented in many fields. Healthcare industry is one of the prime industry to used these technology. The blockchain technology in the medical sector is making transformative changes in particular to protection. One of the major differences between existing electronic health record systems (EHR) and blockchain technology is that it is fully centralised. It supports the most secure way for data viewing, sharing, and updating (Agbo et al. 2019). This provides better control over patients' records and helps in maintaining data privacy, improves data transparency, and avoids the chance of data being misused (Al Omar et al. 2017).

Maintaining the security of data is a persistent problem in the medical industry. As per the studies, in 2021 over 50 million people in the United States had impacted from data breaches in the medical industry (HIPAA Journal 2022). With the help of cryptographic hashing and consensus-based validation of the data, the blockchain method is very well helping to bring down the risk. This method supports a single centralised point of control and strengthens the data security and monitoring (Engelhardt 2017).

2.2 Solving Disconnected Health Systems

One of the major drawbacks in the current EHR system in hospitals is its inefficiency to function across many hospitals. This could lead to disconnection between systems and result in many testing and gaps (McGhin et al. 2019). The current technology based on blockchain solutions supports data to be exchanged across many hospitals without compromising its safety and security. For example, MedRec and FHIR chain back hospitals to give the best practice of data protection and sharing (Azaria et al. 2016; Zhang and Liu 2010).

2.3 Smart Contracts for Managing Access

With the integrating of AI and blockchain technology, the digital agreements are stored as smart contracts, as they have self-executing capacity. It can be viewed ad patient-friendly and it helps the patients to manage the consent by letting them decide who can view or use the record. These contracts will strengthen the process of following privacy laws and also support in adjusting permissions when required without any delays (Koumaditis et al. 2020). It enabled patients to acquire more power to access their data. This authority helps them to have control over their health records. The researchers have found that this technology also helps patients to track the data when used, to identify who did it and the reason for doing so (Yue et al. 2016).

2.4 Improving Blockchain Networks for Health Data Sharing

The majority of healthcare blockchains now operate alone, restricting their value. As per Belchior et al. (2021), the disruption in connectivity between these systems is a significant challenge. New initiatives such as Polkadot and Cosmos focus on building networks that allow data to float between blockchains. These approaches are critical for ensuring that patients' data follows them wherever they go and regardless of which provider they use.

2.5 Using Blockchain with AI and Federated Learning

The prime focus of this research is to protect patient privacy, federated learning combining blockchain with artificial intelligence (AI). If we need to access the data without causing any alteration in the data at any place, federated learning using AI and block chain would be the best. The block chain technology helps to secure the data, helps to keep it private and also safe. The hospitals and doctors also can retrieve accurate predictions and suggestions for treatment without exposing any sensitive information about the patient (Nguyen et al. 2021).

2.6 Patient-Centered Access and Control

The integration of blockchain methodology with the interplanetary file system (IPFS) will provide patients full influence over their medical records. Some researches have discussed about how patients get to decide who can view their electronic health records (EHRS). By using security methods like SPAKE and smart contracts patients can maintain the privacy of holding their medical records (Sonkamble et al. 2023). Recently the hospitals started storing the patients records in cloud and encrypted, and given smart contract control to access it. The studies had identified that some system includes extra protection to protect hackers and still could perform the best (Yuan et al. 2022). Even blockchain technology also is proved to work better which ensures both safety and privacy for the patients (Abu Taleb 2023).

2.7 Security and Privacy in Health Records

The security and privacy of health records are equally important as a bank record of any individual, considering the importance of patients' health safety and protection. If it is not protected securely and is prone to tampering, hacking, or missing any information, it endangers the patient's life. The blockchain technology is one of the highly accepted technologies, especially to protect the patient's data from tampering. The health records in the hospitals are safeguarded using encryption methods like SHA-256, AES, RSA and DSA. The concept of smart contracts is created to prohibit the persons who do not have access to such records (Sonkamble et al. 2023). It is to ensure to avoidance of cyber-attacks, either in the form of data stealing or any unauthorised data transfer.

2.8 Data Sharing and Interoperability

Another major problem encountered while analysing the maintainability and protection of patients health records, data sharing and interoperability issue was also identified. The hospitals although nowadays use many advanced technologies like HER systems that makes the life easy for individuals, still the data shareability among the hospitals remains challenging. The blockchain technology can used to overcome this issue as it is a shared platform. Hence eventhough hospitals are not able to share the data, this blockchain technology enables secured exchange of data among the hospitals. It is a system that gives authority to the authorised person to access the records when needed (Mamun et al. n.d.). This would be helpful even for insurance formalities favouring both hospitals and patients and reduces lot of time consumption. A real time update of information is possible using blockchain technology when it is connected to devices smart electronic gadgets.

3 Problem Statement and Objectives

The objective of this research is to develop a Blockchain-Enabled Health Records (BEHR) model that addresses these drawbacks through:

- Secure and tamper-proof information storage using decentralized blockchain architecture.
- Condition-linked organization of patient medical records, improving continuity of care.
- AI-driven record retrieval via Retrieval-Augmented Generation (RAG) and Large Language Models (LLMs) to enhance data accessibility and usability for both patients and healthcare providers.
- Patient-centered control over the exchange and accessibility of health data, guaranteeing adherence to privacy laws.

The further section of the research presents the detailed design and functionalities of the three proposed models to achieve these objectives.

4 Decentralized Health Record Management through Blockchain, Temporal Sequencing, and AI-Driven Intelligence

In the current era of medical care, patients often present with diverse medical conditions to different hospitals and outpatient clinics. A visit is sometimes all it takes to heal a health problem, but some ailments need constant attention, several visits and thorough testing. Therefore, when patients change healthcare providers, it is common that their medical information will not be able to be gathered and organized together (Raghupathi and Raghupathi 2014). In most cases, patients have to compile their own medical histories or hospitals must transfer records when people seek a second opinion or when they go to a different health care provider. In addition to being time-consuming and inconvenient, this procedure is also prone to errors, such as doctors sharing inaccurate or incomplete data (Adler-Milstein and Jha 2017). The standard of care that patients receive can be directly impacted by missing information.

According to the suggested model, each time a patient visits a hospital, their visit is documented as a distinct block (B) on a blockchain, which can be arranged according to medical significance and timestamp (T). Additionally, visits that take place at different hospitals or on different days are grouped together. Through the condition-centric blockchain we are developing, this is leading to a more detailed and knowledgeable image of the patient's medical history (Azaria et al. 2016). Blocks of the various health issues are kept apart to make the patient track more readable and organized.

Medical professionals can examine a patient's medical history based on the order of visits and the condition being given, recognizing to this time-stamped, condition-linked structure. This helps to prevent duplication of tests and medications by giving clinicians the ability to see if patients have had specific prescriptions or procedures in previous visits (even at different hospitals).

Additionally, the blockchain interface's usability is improved by the integration of large language models (LLMs) and retrieval augmented generation (RAG). In order to help HCPs in obtaining comprehensive medical records, LLMs modify simple, conversational questions by understanding advanced patient data, which includes words spoken (Singhal et al. 2022). By identifying the most pertinent portions of the patient's medical records from the blockchain, RAG combines into the system to produce summaries or responses that support clinical decision making (Lewis et al. 2020).

This removes the previous waiting time for doctors after inter-hospital data exchange and manual search in different systems for essential medical information. And when the records are poorly structured and span over many clinical institutions, the use of LLM with RAG is found very effective in querying Medline data (Shen et al. 2023).

Empowering patients with blockchain-powered (AI) solutions to integrate into their healthcare workflow. This also leads to better continuity of care by ensuring that healthcare providers have access to accurate and complete medical history. Because of blockchain, patients can own their health data and selectively and securely share it with any provider they wish and access their medical records as needed, ensuring that the health information they receive is current, extensive, and transportable.

The proposed blockchain-enabled health record management system addresses patient data fragmentation by creating a secure, organized, and readily available medical data bank. The use of LLMs combined with RAG simplifies the implementation of information search and improves clinical decisions, and the use of T makes sure that the chronological order and the data are correct. Ultimately, it's a way to create a more efficient health care system, end duplication of care, and give patients more control of their private health.

In a health information system based on blockchain, every patient encounter with a health professional is captured as an inalterable block, time-stamped to preserve order (Azaria et al. 2016; Yue et al. 2016). For example, patient Sarah may travel to various hospitals, each of which is recorded in a different block (block 1 to block 6) with a respective time stamp (T1, T2, T3, T4, T5, T6). These visits can be for new health issues or existing conditions. The same patient may visit a hospital for different problems at different times. The above diagram depicts the same. For example, B1 at a time of T1 may represent a patient visiting for diabetes, while B2 and B5 can indicate a problem of the gastrointestinal system for which the same patient might have visited. The above diagram graphically illustrates the connection between the blocks and how a patient may visit a doctor for different reasons.

The proposed blockchain model provides a patient-centric, tamper-resistant, time-stamped data of the patient to the medical professionals, thereby ensuring continuous care and support for decision-making. The main purpose of inculcating the blockchain model is to maintain the confidentiality of the medical data of the visitors to hospitals and also promptness in retrieving it. The time stamp method is implemented to record the date and time of the day when the visit or treatment occurred (Liang et al. 2017). The time stamps will also help to figure out how the treatment is working, the progression of getting well, the tests, the procedures, complications if any etc. Sometimes for a situation like chronic illness the patients have visit different providers in different settings. The time stamps will help to connect everything. It helps the doctors to plan the treatment, for long-term management and care of conditions. Moreover, using blockchain technology will help to detect any kind of attempt to tamper as well because of its sequential flow. This is the most interesting feature of blockchain technology which supports data reliability, safety, authenticity, dependability of patients health information (Zhuang et al. 2020).

To explain how the blockchain technology can be used to store the patient's data which is collected during his hospital visits is indicated in the above flow chart. The design is to confirm that the patient's data are reliable, consistent, and patient focused (Azaria et al. 2016; Zhuang et al. 2020). It also explains how a hospital can move from traditional system of maintain records of the patient in a decentralised system, which help patients also to view whenever they want. The system also keeps intact to maintain security, accessibility and integrity of the medical information. The integration of technology is to constantly maintain the data reliability, untampered storage of data, patient ownership and easy accessibility in maintaining the health record.

To begin with, the process commences when a patient visits a hospital for any healthcare advice to hospitals. After the consultation, the physician would suggest and carry out diagnostic tests to get an advanced understanding of the patient's illness. Using the test, a complete patient record is created using the test results, clinical observations, and recommended treatments. This document becomes a vital component for the patient's long-term medical history as it includes both diagnostic information and treatment recommendations (Liang et al. 2017).

Hospitals typically keep these documents in electronic health records (EHRs) or local information systems. As it is important to increase security and patients' dependency on hospitals because of the centralised database and provide good control for patients over their own records, the same records are uploaded in the blockchain network. Every hospital will add a new block that is cryptographically encrypted, timestamped, and connected to the previous records if needed. As per Yue et al. (2016), this technology results in a trustworthy and constant sequence of medical instances.

The blockchain technology generally allows for the customary and chronological retrieval of patient records, especially it helps during emergency care transmissions, multi-centre medications, and second opinions. This system is framed to ensure data privacy, ease of accessibility, and unbreakable by using blockchain concepts. The method reduces unnecessary diagnostic testing, enhancing continuity of care, and accelerating administrative procedures related to patient record access or transfer may also result in a more effective and patient-centred healthcare ecosystem (Zhang et al. 2018).

Root ID Created on First Visit

By offering a more structured and centralized blockchain-based service, Model 2 broadens on the fundamental idea of creating patient records that was outlined in Model 1. This service serves as a safe conduit between patients and hospitals, guaranteeing that all substantial medical incidents are securely and inconceivably documented.

The way it works is that the hospital collects the pertinent health-related data every time a patient comes in for a check-up, diagnostic test, or medical care of any kind. The hospital provides that data to a central blockchain service instead of keeping it in its own internal system. For that particular event, this service creates a new digital "block" and associates it with the patient's distinct identity within the system.

When a patient initially logs into this blockchain-based system, a unique identity referred to as the Root ID is generated. All subsequent block (or medical events) is firmly linked to that Root ID. The data held in a block cannot be amended or removed once it has been inserted into the blockchain. As the blockchain ensures that all patient data remains permanent, secure, and trustworthy, it makes it more powerful. It is a technology that wouldn't allow any kind of voluntary or involuntary modification or deletion of the information. It protects the patient data, from any alteration, even if someone would like to do it with harmful intent.

The fact that Model 2 preserves the patient's autonomy and confidentiality is among its most significant features. A secure digital key is provided to the patient upon their entry into the system. They may regulate who has access to their medical records with this key. The Patients can use the key to grant access to their blockchain data to an individual doctor, for instance, if they wish to see a different physician or seek a second opinion from a different hospital. As a result, patients hold control over their health information, deciding when, how, and with whom to share it.

The pace and convenience of medical records anytime, especially during urgent situations, is a massive benefit of this model. Healthcare experts can indeed access important patient data when necessary because the records are stored on a centralised blockchain rather than in a single hospital. It amplifies the doctor's ability to respond quickly and efficiently to any emergencies, as they have swift access to the medical history of the patients, such as allergies, previous surgeries or any current medications. In addition to that, this model reduces the workload on hospitals. The hospitals need not worry about physically sending or demanding patient records from other institutions. The patient's records are already available in the blockchain system, which helps to streamline their operations, saves plenty of time, and also reduces delay in administrative work.

The Model 2 provides a more advanced and patient-friendly way of managing health care records. It also confirms that the medical information of patients is securely saved, accurately recorded and easily accessible when required. Most prominently, this model enables patients to have control over their health data, confirming privacy and supporting better communication between health care workers. Henceforth, this serves as a platform for building trust between patients and hospitals. The patients can be more connected with the hospitals, and hospitals also can improve the efficiency of care and the responsive healthcare ecosystem.

5 AI-Augmented Record Retrieval (LLM, RAG, Vector DB)

Integrating advanced Artificial Intelligence (AI) technologies has become inevitable to enhance the efficacy, availability, and intelligence of blockchain-based healthcare data systems. The research is proposing a model incorporating Retrieval-Augmented Generation (RAG), Large Language Models (LLMs) and Vector databases. The integration of RAG and LLM would assist both the individual and hospitals more efficiently to store, accessing, retrieve, and fully understand the medical data without seeking help.

In The designed blockchain framework, each medical occurance—such as a consultation by doctors, diagnostic test process, or treatment—is properly stored as a document within a particular segment of the blockchain called a "sub-chain." These sub-chains further assist in organizing patient data based on the timeline of events and specific medical requirement. However not all medical record adhere a clear or directly connected sequential steps. A patient may need multiple doctors for different issues or return for treatment of the same illness across over the time periods. While the blockchain usually make sure that all records are saved safely and timestamped, browsing through these scattered entries manually can be exhausting, especially for patients who do not have medical proficiency.

Large Language Models (LLMs) are initiated as a conversational interface to make these records more easy and user centric. The AI systems help the user to ask complicated questions in simple manner by understanding and processing human language. Example could be, a patient asked, "What treatment did I receive for my heart condition during last few years?" or may be "there is any improvement in blood pressure since last year?" Scanning will be done by LLM through the relevant documents which is stored in the blockchain, it analyze the context, further generates a briefed answer. This process will help user to overpass the need for user to go manually check each record and this will make entire process easier and quicker.

The addition of feature like RAG Retrieval-Augmented Generation (RAG) make it very powerful. The normal AI models are relying only on their pre-filled knowledge, RAG retrieves real-time and relevant information from the sub chains of blockchain and, if allowed, from external trusted medical database example as clinical guidelines or research information's.

The combination of the retrieved data with the potential of the LLM to produce responses that are both factually accurate and tailored to the patient's specific need. For example, if a patient is availing the diabetes treatment, RAG can generate a outline that includes recent advancements in diabetes care and the patient's personal health details. The major factor enabling this proficient retrieval is the Vector Database. Unlike traditional databases rely search based on exact keywords, vector databases make use of mathematical framework to confirm the meaning of the words.

AI help in doing semantic search, which is numerical representation of the document, these are stored in "embeddings". The idea behind the system is not to just find the accurate matches but to give relevant and concrete information swiftly. LLMs, RAG and Vector data database all together combines and changes the static secure storage system into modern, intelligent, accessible and more secure program.

It give authority to patient to have the access of their own medial history data in clear and meaningful manner. Moreover it helps the doctor in making quick and better informed decisions by giving instant details into patient healthy history especially when in emergencies or while patients follow up visits with different health practitioner. This combined method drastically minimizes the time required to retrieve and interpret the stored health information, moreover the data is more protected, and usable. It perfectly provide the solution for patient centric, transparent and intelligent healthcare data storage system. By combining the immutability and security of blockchain with the adaptability and intelligence of AI, this model provides a future-ready solution that addresses many of the existing challenges in health data management.

6 Challenges in Using Blockchain for Health Records in India

Blockchain technology can make health records more secure, transparent, and controlled by patients. However, there are still many challenges to using it effectively in India's healthcare system. These challenges include technical issues, costs, system compatibility, legal rules, and public understanding.

6.1 Scalability

India has a large and diverse population, which means there will be a huge amount of health data to manage. Blockchain systems can become slow and overloaded when handling large volumes of data. To solve this, sensitive or large data can be stored outside the blockchain (off-chain), while only important summaries or access details are stored on the blockchain. Newer technologies like Proof-of-Stake and sharding can also help handle more transactions quickly and efficiently.

6.2 Adoption by Healthcare Providers

Hospitals and clinics—especially in rural and small-town India—may find it difficult to adopt blockchain. Reasons include high costs, lack of trained staff, and fear of changing current systems. Many small healthcare centres still use paper records or basic digital systems. To encourage adoption, pilot projects, support from the government, and training programs for doctors and staff are essential. Showing the long-term benefits, such as better security and easier access to records, can also help.

6.3 Integration with Existing Systems

Most healthcare institutions in India already use some form of electronic or digital record system. These may not work well with blockchain, as blockchain follows a completely different structure. To fix this, software tools (APIs or middleware) can help connect the old systems with the new blockchain platform. Following national standards like the Ayushman Bharat Digital Mission and global data standards like HL7 FHIR will help improve compatibility and make integration smoother.

6.4 Legal and Regulatory Rules

Health data is sensitive and protected under laws like India's Digital Personal Data Protection Act, 2023. These laws allow people to change or delete their data if needed, but blockchain doesn't allow easy changes or deletion because it is designed to be permanent. To follow the law, health systems can store patient data outside the blockchain in a secure way, while using the blockchain to record who accessed the data and when. Working with regulators and updating policies to match Indian laws will be very important.

6.5 Patient Awareness and Trust

People living in rural area in India might not be aware of blockchain and how it works. They might have concerned about their privacy of data and the approachability of that data. The trust of the patients can be achieved by educating them in basic language or explaining in more simpler way, making them understand that data access is based on their approval. It will not be accessed without their authority.

7 Discussion

7.1 Blockchain's Role in Enhancing Trust and Accountability

Blockchain revolutionized the way healthcare data was managed by focusing on trust through decentralization. In past years healthcare institutions had the authority to access the data.

This monopoly of data could delay treatment, obscure transparency, and keep patients to potential data misuse (Dubovitskaya et al. 2018). With blockchain, the immutability of data logs and visibility into access histories introduce accountability mechanisms that were lacking earlierThe introduction of new framework aligns with evolving ethical expectations in healthcare data management which focus on patient autonomy and well informed consent.By merging the data sharing policies directly in to smart contracts, individual no longer need to depend on intermediaries for executing their medial information records. (Rifi et al. 2017).

The amount of potential block chain technology has in healthcare has its own challenges to overcome, one of the barrier in implementation is its scalability. Layer-2 solutions, like sidechains and state channels, are being explored to offload transactions from the main chain while preserving security (Xu et al. 2019).

Hybrid blockchain models is combining on-chain metadata with off-chain data storage are proving effective in maintaining performance with safety. The financial cost for implementing blockchain systems is another point to be considered, in low-resource healthcare structure. When analysing the long-term costs of data breaches and redundant testing, Blockchain could provide more sustainable framework (Agbo et al. 2019).

A contraction in blockchain healthcare applications is the root cause problem between data immutability and data erasure rights under regulations like GDPR (General Data Protection Regulation) and the Indian DPDP (Digital Personal Data Protection) Act. Blockchain with its design does not allow for data deletion, compliance can be attained by storing encrypted references on-chain while the actual data remains be in off-chain, mutable databases (Zhang et al. 2021).

These models must include robust key revocation mechanisms to render the data inaccessible upon patient request. While blockchain give patients full authority, emergency situations pose ethical dilemmas. In cases where patients are unconscious, rigid access protocols may delay life-saving interventions. To resolve this, emergency override mechanisms governed by community consensus or predefined smart contract conditions are being proposed (Tanwar et al. 2020). However, ethical guidelines must be clearly defined to prevent misuse of such backdoors.

One of the most advanced developments is the use of AI, including LLMs and conversational agents, to facilitate patient interaction with blockchain-stored health data. It helps the patient to receive contextual summaries or trend visualizations via natural language queries instead of navigating raw data logs (Xia et al. 2017). When AI is blended with blockchain technology it not only improves the usability of information but also gives the liberty to the patients to access their health-related records and, most importantly, with less knowledge on medical information.

7.2 Global Comparison: BEHR Implementation Across Countries

While blockchain's application in healthcare have advantage across the globe, the feasibility, speed of adoption, and regulatory readiness significantly differ by country. In high-income nations such as Estonia, blockchain-based health records have already been successfully implemented at the national level. Estonia's eHealth system, built on a permissioned blockchain, allows safe access and sharing of patient health data across all public and private healthcare providers. Their centralized digital identity infrastructure and strong data protection regulations have facilitated this transformation, exhibiting the benefits of robust government support and digital maturity.

In contrast, in countries like the United States, the adoption of blockchain in healthcare remains scattered. Although there are pilot projects and private sector initiatives—such as Hashed Health and Healthereum—interoperability issues still exist due to a lack of universal data standards and a decentralized healthcare system. Additionally, the U.S. regulatory environment, governed by HIPAA and the HITECH Act, imposes strict regulations that make full-scale implementation very difficult. However, these same regulations drive innovation in patient-centric solutions that emphasize privacy and security, matching well with the principles of blockchain.

Developing countries, such as India and several African nations, face both opportunities and constraints. On one hand, blockchain presents a way to "leapfrog" over legacy health IT systems; on the other, barriers such as low digital literacy, infrastructure gaps, and funding constraints hinder its scalability.

Government schemes like Kenya's blockchain-powered health ID systems and India's Ayushman Bharat digital mission extended the openness for these development in health management ecosystem. Public and private partnership can open the new avenue of success for these simplifies user interface, moreover the stringent government regulations can take these initiatives to successful implementation. The potential of blockchain-enabled health record systems is widely known but their execution is depend on country's national infrastructure, legal frameworks, and digital advancement. Country like Estonia and other developing nations where these initiatives are at pilot-stage, country's like the U.S. can guide implementation policies to the emerging economies, ensuring that the BEHR model is both inclusive and flexible.

8 Conclusion

Entry of blockchain technology into healthcare data management system open avenues for more secured, transparent, and patient-centered medical record frameworks. Traditional Electronic Health Record (EHR) systems are often seen as centralized, scattered into sub chains, and more prone to data breaches or unwanted alterations, whereas blockchain proposes a decentralized, immutable, and trust-enhancing models. (Angraal et al. 2017). These framework give surety that patient information will be secure and tamper-proof which will be also easily accessible in real time by authorized users, enhancing advanced coordination and trust among healthcare stakeholders (Azaria et al. 2016).

Developing nations like India, where healthcare facilities is at developing stage and digital familiarity is uneven, blockchain show both a significant opportunity and challenges in implementation (Ministry of Health and Family Welfare 2020) like low awareness among patients, the huge cost involved in technological upgrades, hindrance from healthcare workers, and barriers in cordinating blockchain with existing EHR systems remain the prime issues. (Kuo et al. 2017). The laws and regulatory compliance, specialty focusing on data privacy laws like India's Digital Personal Data Protection (DPDP) Act or international frameworks like the GDPR, must be taken care meticulously (Radanović and Likić 2018). To handle these challenges requires coordinated efforts from strategy makers, technology providers, researchers, and healthcare providers.

Successful implementation should focus on future strategies on scalable blockchain solutions, customer-friendly interfaces for healthcare providers, middleware to link existing systems, and awareness initiatives to create digital literacy among patients (Bhardwaj and Pradhan 2020). The process of adoption must be gradual, supported by pilot programs and sharp administrativ frameworks which will enable experimentation with keeping patient data safety and maintaining data integrity (Vanschoren et al. 2014). Blockchain technology offer a very safe, decentralized, and patient-oriented data storing platform for managing their health record. The merging of blockchain with IoT devices, advanced encryption techniques, and smart contracts advance the level of security and interoperability of health records, moreover it ensures maintaining the regulatory requirements. Though the potential challenges like the scalability, interoperability, and patient adoption should be taken care and to understand the importance of blockchain-based health record systems. Future research directions should focus on advancing encryption techniques, optimizing system performance, developing interoperability standards, and designing patient-centric systems to ensure the widespread adoption of blockchain technology in healthcare.

References

Abu Taleb H (2023) Probabilistic analysis of batter-scale slopes: bridging the gap between factor of safety and probability of failure. In: SSIM 2023: third international slope stability in mining conference

Adler-Milstein J, Jha AK (2017) HITECH act drove large gains in hospital electronic health record adoption. Health Aff 36(8):1416–1422. https://doi.org/10.1377/hlthaff.2016.1651

Agbo CC, Mahmoud QH, Eklund JM (2019) Blockchain technology in healthcare: a systematic review. Healthcare 7(2):56. https://doi.org/10.3390/healthcare7020056

Al Omar A, Rahman MS, Basu A, Kiyomoto S (2017) Medibchain: a blockchain based privacy preserving platform for healthcare data. In: Security, privacy, and anonymity in computation, communication, and storage. Springer, pp 534–543. https://doi.org/10.1007/978-3-319-72395-2_46

Angraal S, Krumholz HM, Schulz WL (2017) Blockchain technology: applications in health care. Circ Cardiovasc Qual Outcomes 10(9):e003800. https://doi.org/10.1161/CIRCOUTCOMES.117.003800

Azaria A, Ekblaw A, Vieira T, Lippman A (2016) MedRec: using blockchain for medical data access and permission management. In: 2016 2nd international conference on open and big data (OBD). IEEE, pp 25–30. https://doi.org/10.1109/OBD.2016.11

Bhardwaj A, Pradhan S (2020) Blockchain in healthcare: applications, challenges, and future perspectives. Int J Med Inform 141:104244. https://doi.org/10.1016/j.ijmedinf.2020.104244

Belchior R, Vasconcelos A, Guerreiro S, Correia M (2021) A survey on blockchain interoperability: past, present, and future trends. ACM Comput Surv (CSUR) 54(8):1–41. https://doi.org/10.1145/3468628

Dubovitskaya A, Xu Z, Ryu S, Schumacher M, Wang F (2018) Secure and trustable electronic medical records sharing using blockchain. In: AMIA annual symposium proceedings, vol 2017, p 650

Engelhardt MA (2017) Hitching healthcare to the chain: an introduction to blockchain technology in the healthcare sector. Technol Innov Manag Rev 7(10):22–34. https://doi.org/10.22215/timreview/1111

HIPAA Journal (2022) 2021 healthcare data breach report. https://www.hipaajournal.com/2021-healthcare-data-breach-report/

Jensen PB, Jensen LJ, Brunak S (2012) Mining electronic health records: towards better research applications and clinical care. Nat Rev Genet 13(6):395–405. https://doi.org/10.1038/nrg3208

Koumaditis K, Themistocleous M, Mantzana V (2020) Digital transformation and capabilities for healthcare organizations. Health Policy Technol 9(1):100389. https://doi.org/10.1016/j.hlpt.2019.100389

Kuo T-T, Kim H-E, Ohno-Machado L (2017) Blockchain distributed ledger technologies for biomedical and health care applications. J Am Med Inform Assoc 24(6):1211–1220. https://doi.org/10.1093/jamia/ocx068

Lewis P, Perez E, Piktus A, Petroni F, Karpukhin V, Goyal N et al (2020) Retrieval-augmented generation for knowledge-intensive NLP tasks. Adv Neural Inf Proces Syst 33:9459–9474. https://proceedings.neurips.cc/paper/2020/file/6b493230205f780e1bc26945df7481e5-Paper.pdf

Liang X, Zhao J, Shetty S, Li D (2017) Integrating blockchain for data sharing and collaboration in mobile healthcare applications. In: 2017 IEEE 28th annual international symposium on personal, indoor, and Mobile radio communications (PIMRC), pp 1–5. https://doi.org/10.1109/PIMRC.2017.8292361

McGhin T, Choo KKR, Liu CZ, He D (2019) Blockchain in healthcare applications: research challenges and opportunities. J Netw Comput Appl 135:62–75. https://doi.org/10.1016/j.jnca.2019.02.027

Ministry of Health and Family Welfare (2020) National Digital Health Mission: strategy overview. Government of India. https://ndhm.gov.in

Murdoch TB, Detsky AS (2013) The inevitable application of big data to health care. JAMA 309(13):1351–1352. https://doi.org/10.1001/jama.2013.393

Nguyen DC, Ding M, Pathirana PN, Seneviratne A, Zomaya AY (2021) Federated learning for COVID-19 detection with generative adversarial networks in edge cloud computing. IEEE Internet Things J 9(12):10257–10271. https://doi.org/10.1109/JIOT.2021.3111452

Radanović I, Likić R (2018) Opportunities for use of blockchain technology in medicine. Appl Health Econ Health Policy 16:583–590. https://doi.org/10.1007/s40258-018-0412-8

Raghupathi W, Raghupathi V (2014) Big data analytics in healthcare: promise and potential. Health Information Science and Systems 2(1):3. https://doi.org/10.1186/2047-2501-2-3

Roehrs A, da Costa CA, Righi RDR, de Oliveira KSF (2017) Personal health records: a systematic literature review. J Biomed Inform 70:103–132. https://doi.org/10.1016/j.jbi.2017.04.006

Roehrs A, da Costa CA, Righi RDR, da Silva VF (2019) A survey on health data-centric privacy regulations, technologies, and challenges. Futur Gener Comput Syst 92:652–675. https://doi.org/10.1016/j.future.2018.01.014

Rifi N, Rachkidi E, Agoulmine N, Taher NC (2017) Towards using blockchain technology for eHealth data access management. In: 2017 fourth international conference on advances in biomedical engineering (ICABME). IEEE, pp 1–4. https://doi.org/10.1109/ICABME.2017.8167555

Shen Y, Heacock L, Elias J, Hentel KD, Reig B, Shih G, Wang J (2023) ChatGPT and other large language models are double-edged swords. Radiol Artif Intell 5(2):e220020. https://doi.org/10.1148/ryai.220020

Singhal K, Azizi S, Tu T, Mahdavi SS, Lee J, Ou J et al (2022) Large language models encode clinical knowledge. arXiv preprint arXiv:2212.13138. https://arxiv.org/abs/2212.13138

Sonkamble SB, Jadhav P, Hake A, Sanjay V (2023) A review on soil classification using machine learning and crop suggestions. Int J Sci Res Comput Sci Eng Inf Technol 9(1):113–116. https://doi.org/10.32628/CSEIT2390120

Tanwar S, Parekh K, Evans R (2020) Blockchain-based electronic healthcare record system for healthcare 4.0 applications. J Inf Secur Appl 50:102407. https://doi.org/10.1016/j.jisa.2019.102407

Vanschoren J, van Rijn JN, Bischl B, Torgo L (2014) OpenML: networked science in machine learning. ACM SIGKDD Explor Newsletter 15(2):49–60. https://doi.org/10.1145/2641190.2641198

Xia QI, Sifah EB, Asamoah KO, Gao J, Du X, Guizani M (2017) MeDShare: trust-less medical data sharing among cloud service providers via blockchain. IEEE Access 5:14757–14767. https://doi.org/10.1109/ACCESS.2017.2731387

Xu X, Weber I, Staples M (2019) Architecture for blockchain applications. Spring. https://doi.org/10.1007/978-3-030-03035-4

Yuan Z-Y, Wu Y-K, Yan F, Murphy RW, Papenfuss TJ, Wake DB, Zhang Y-P, Che J (2022) Comparative multi-locus assessment of modern Asian newts in southern China suggests a shared biogeographic history. Zool Res 43(5):706–718. https://doi.org/10.24272/j.issn.2095-8137.2022.154

Yue X, Wang H, Jin D, Li M, Jiang W (2016) Healthcare data gateways: found healthcare intelligence on blockchain with novel privacy risk control. J Med Syst 40(10):218. https://doi.org/10.1007/s10916-016-0574-6

Zhang R, Liu L (2010) Security models and requirements for healthcare application clouds. In: 2010 IEEE 3rd international conference on cloud computing. IEEE, pp 268–275. https://doi.org/10.1109/CLOUD.2010.41

Zhang X, Wang L, Liu J, Meng F (2021) Blockchain-based EHR sharing scheme for emergency medical services. Comput Methods Prog Biomed 200:105906. https://doi.org/10.1016/j.cmpb.2020.105906

Zhuang Y, Sheets LR, Chen YW (2020) Applying blockchain technology in health information exchange: architecture and challenges. IEEE Access 8:135632–135647. https://doi.org/10.1109/ACCESS.2020.3010467

Zhang P, Schmidt DC, White J, Lenz G (2018) Blockchain technology use cases in healthcare. In: Advances in computers, vol 111. Elsevier, pp 1–41. https://doi.org/10.1016/bs.adcom.2018.03.006

Supply Chain Dynamics and Sustainability Challenges in Rural Economies: A Study of NWFP Collection Among Odisha's Tribal Communities

Jyoti Prakash Mohanty and Ratnakar Mishra

Abstract Harvesting, processing and marketing of NWFP has been a major livelihood option for people living in and around the forests. In the forests of Odisha, Sal and Siali leaves are a major source of income for the tribal people. Although there are no authentic figures available on the quantity of leaves collected and marketed annually, a rough estimate shows that total trade from leaf plates and cups is about Rs. 1000 crores in the state. Despite a huge turnover from the trade, tribal women of some tribal populated districts in Odisha are earning a megure monthly income. The traders and middlemen are earning good profit, and the primary collectors were gaining less. DRDA under SGSY, in co-ordination with DSMS, is constantly giving training to SHGs and women cooperatives on making leaf plates and cups. SHPI also facilitated the promotion of fine machine-stitched and pressing plates. Due to these trainings in production and constant market support, the income of tribal leaf harvesters is showing an upward trend in the current days in the State of Odisha. As per the recent data released by Government of Odisha around 10,000 rural poor women beneficiaries in the last 10 years period have got better self-employment with better returns in this domain of sustainability. This survey paper presents success stories of two tribal women leaf plucker SHGs from the underdeveloped cluster of Kandhmal district of Odisha. These leaf plucker SHGs have received training in plate stitching and are now controlling their entire supply chain or in other words, from the stage of plucking to processing, then to selling their products in organized market melas these two women are fully in command bypassing the middlemen hence improving their and others' livelihood in a substantial proportion.

Keywords Rural supply chain · Sustainability · Tribal in Odisha · NWFP

J. P. Mohanty
NRLM, Bhubaneswar, Odisha, India

R. Mishra (✉)
Graphic Era Deemed to be University, Dehradun, Uttarakhand, India
e-mail: ratnakar05@gmail.com

A. Singh et al. (eds.), *Human-Centered Sustainability: Bridging Well-Being, Gender Equity, and Technology*, World Sustainability Series,
https://doi.org/10.1007/978-3-032-22020-2_24

1 Introduction

Non-Wood Forest Products (NWFPs) present a significant opportunity to generate large-scale employment, alleviate poverty, and empower tribal and economically disadvantaged populations in India's less developed regions. Beyond their role in ensuring food security, NWFPs offer a pathway for establishing environmentally sustainable micro, small, and medium-sized enterprises at the local level. For generations, the collection, processing, and marketing of NWFPs have been a primary means of sustaining livelihoods for tribal communities residing in and around forest areas. It is estimated that approximately 80% of these communities' daily income is derived from activities related to NWFPs. The NWFP sector in India is a substantial source of employment, generating an estimated ten million workdays annually.

In the eastern Indian state of Odisha, Sal and Siali leaves are particularly important sources of income for local communities. Sal trees, which yield valuable products, are predominantly found in the districts of Mayurbhanj, Keojhar, Kandhmal, Balasore, Deogarh, and Nayagarh. While the production of these leaves is concentrated in these specific districts, the processing infrastructure is largely situated in the Balasore and Mayurbhanj districts. Notably, the largest market for Sal leaf cups and plates is located in Betanati, within the Mayurbhanj district. A striking paradox exists in the absence of precise data regarding the annual quantities of leaves collected and traded. Nevertheless, rough estimations indicate that the total trade volume of leaf plates and cups is around Rs. 1000 crores, with the Mayurbhanj district alone accounting for 50% of the market share. It is important to emphasize the cultural significance of Sal and Siali plants, which are often venerated by tribal communities as part of their traditional practices. This reverence underscores the communities' reliance on these plants for a variety of products that sustain their livelihoods. The production of leaf plates and cups, a practice also known as "Donna Pattal" making, serves as a supplementary livelihood activity for the tribal people in this region, complementing their engagement in agriculture or farming. The consistent demand for these items provides additional income for a large number of forest dwellers and generates profits for numerous traders in Odisha, both large and small (New Scheme 2011).

The processing of leaves into cups and plates is predominantly carried out in urban centers such as Betanati, Basta, Karanjia, and Keonjhor. This is largely due to the lack of access to electricity and technological support in rural villages. Small and medium-sized traders purchase the leaves from village communities and process them at these centralized locations. The finished products are then sold to traders in other states for domestic consumption, and some are even exported to other countries. This pattern highlights a distinct rural-urban dynamic in the NWFP supply chain.

In the Kandhamal district of Odisha, tribal communities constitute 52% of the total population and reside in 71% of the district's forest-covered land. The local economy of this district heavily relies on the sale of Non-Wood Forest Products (NWFPs) by the tribal population in local markets. Siali leaf is a particularly important contributor to the local economy. According to the 2011 census, Kandhamal district has a population of 731,952, with Scheduled Tribes, particularly the Kandha tribes, comprising more than 50% of the population. Overall, the district is classified as one of the less developed districts in the state of Odisha, and a significant portion of the tribal population depends on the NWFP resources that are abundant in the forests (Ranjan et al. 2002).

2 Challenges in NWFP Collection and the Need for Empowerment

In India, women make substantial contributions to the economy through their labor and skills. However, their contributions often remain underrecognized and undervalued in official economic data. According to United Nations statistics, women contribute 67% of the world's total working hours across all genders, yet they earn only 10% of the world's income. In rural areas, women's economic participation is particularly evident in their involvement in various entrepreneurial activities, including poultry farming, agriculture, horticulture, sheep rearing, dairy work, and woodworking. Despite their considerable potential, women often lag behind their male counterparts in effectively commercializing their entrepreneurial skills. This disparity is often attributed to limited access to financial resources and a lack of marketing expertise (Fox and Dwyer 1999).

A confluence of challenges has contributed to a decline in the interest among primary collectors, many of whom are women, in NWFP collection. Two significant factors are uncertainty in market demand and declining production levels. Compounding these issues, government initiatives such as the Mahatma Gandhi National Rural Employment Guarantee Act (MGNREGA) have, in some areas, reduced the incentives for primary collectors to engage in NWFP collection. MGNREGA provides guaranteed employment opportunities that may be perceived as requiring less physical labor compared to NWFP collection. This situation mirrors the labor challenges currently faced in India's agricultural sector, where government-sponsored employment programs can influence traditional labor patterns (Girija 1995).

Both governmental and non-governmental agencies have attempted to address these challenges and improve the conditions of leaf collectors through the formation of Self-Help Groups (SHGs), Self-Help Cooperatives, and producer companies. These collective initiatives have provided NWFP collectors, a significant proportion of whom are women, with a renewed sense of identity and enhanced self-confidence. However, despite ongoing support from the government, these collectors still require more comprehensive support mechanisms and an enabling environment to achieve greater financial stability and growth (Cinamon and Rich 2002).

NWFPs represent a trade sector that is largely driven by women and does not typically require specialized skills. A substantial number of women's SHGs are involved in the collection of forest products. However, a significant percentage of these groups have become inactive over time due to the unsustainable income generated from these activities. To address this issue and enhance the economic viability of NWFP collection, training programs have been introduced to equip women with skills in using sewing machines to improve the production quality of leaf plates. Machine-thread stitching not only increases production speed but also enhances the aesthetic appeal of the plates, making them more marketable. Efforts are also underway to establish linkages with banks and financial institutions to ensure a smooth flow of working capital to both SHGs and individual members, enabling them to purchase stitching machines and other necessary equipment. Furthermore, market linkages are being facilitated for women's groups, empowering them to exert greater control over the entire value chain, from the initial collection stage to the effective marketing of the final product (Duxbury and Higgins 1991).

3 The Role of NGOs and SHGs in Promoting Rural Development in Odisha

The promotion of micro-enterprises is fundamental to initiating a process of sustainable development at the village level. However, for such initiatives to be effective, they must be based on a realistic and thorough assessment of available resources, existing skill sets, infrastructure capabilities, and prevailing market conditions. This comprehensive assessment enables the development of tailored strategies that effectively utilize resources and empower economically disadvantaged populations in a sustainable manner. Non-governmental organizations (NGOs) have played a crucial role in facilitating the growth and development of rural micro-enterprises. NGOs have made substantial contributions to poverty reduction, the creation of gainful employment opportunities, and the reduction of economic and social inequality (Manimekalai and Rajeswari 2001).

Poverty alleviation initiatives undertaken by NGOs have introduced innovative ways of thinking and novel practices that emphasize the active participation of poor and marginalized communities. The concept of microfinance is based on the principle of self-organization among impoverished communities, driven by their inherent desire and capacity to improve their living conditions through collective action (Bhose n.d.). The provision of accessible and low-interest credit to poor individuals, enabling them to pursue economic activities of their choice and improve their overall well-being, has been recognized as an important strategy for poverty alleviation by both the Indian government and the Reserve Bank of India. The Indian government has been proactive in formulating policies that address women's empowerment, with active involvement from voluntary organizations and the corporate sector (Soundari and Sudhir 2003).

Microfinance has emerged as a significant tool for addressing the issue of limited access to credit facilities for the poor, as well as mitigating the pervasive problems of poverty and unemployment. Professor Mohammed Yunus, in Bangladesh, pioneered the microcredit movement by introducing the concept of providing small loans to poor women in the late 1970s through the establishment of the Grameen Bank. His initiatives achieved remarkable success in alleviating poverty among socio-economically disadvantaged populations. This microcredit movement has had a profound impact, contributing to the development of numerous micro-enterprise initiatives across India, including in Kandhamal and Odisha as a whole (Nabavi 2009).

Non-governmental organizations (NGOs) and voluntary organizations (VOs) are actively involved in efforts aimed at the upliftment of the poor, the marginalized, the underprivileged, and the downtrodden segments of society. These organizations typically operate with a high degree of flexibility and autonomy, enabling swift decision-making processes and efficient implementation of programs. The development sector, in which these organizations operate, has made substantial contributions in various areas, including poverty reduction, addressing social deprivation and caste discrimination, creating sustainable livelihood options, promoting health education and awareness, mobilizing communities, delivering essential services, advocating for the rights of marginalized groups, and providing skills training. NGOs often serve as an effective and non-political bridge between the government and the people, facilitating communication and collaboration (Minimol and Makesh 2012).

Specifically, in the context of women's empowerment, NGOs play a critical role in facilitating the formation of Self-Help Groups (SHGs) and motivating women to engage in income-generating programs that align with their existing skills, available resources, and potential for economic empowerment and growth. It is encouraging to note that NGOs in India are actively working to transform the SHG movement into a widespread and impactful social force. Many government agencies view NGOs as a viable and cost-effective means of delivering essential services to communities. The promotion, financing, and strengthening of women's Self-Help Groups is widely regarded as a highly effective approach to enhancing women's productive capacity and breaking the cycle of poverty. In this context, micro-enterprise and microfinance development have emerged as key strategies for addressing the interconnected issues of poverty and unemployment.

However, while microfinance has progressed successfully in empowering women in the state, the effective utilization of credit for generating adequate income and sustainable employment has become a cause of concern for many stakeholders (Shylendra 2007). Moreover, there is a growing concern that micro-enterprise development programs are not achieving their desired outcomes to the fullest extent. This is often attributed to the lack of adequate and timely access to credit facilities. A critical missing link is that micro-enterprise and microfinance initiatives are frequently addressed in isolation, without sufficient synergy between the two. Although major institutions such as NABARD, SIDBI, CARE, RGVN, RMK, and NBCFDC are implementing significant programs aimed at promoting SHGs, these interventions often lack specific and strategic components dedicated to supporting business growth and the promotion of micro-enterprises (Mahajan 2005).

Over the past decade, NGOs have undertaken numerous initiatives to promote livelihoods through SHGs in the non-farm sector, including activities related to NWFPs. These activities have undoubtedly enhanced the economic base of the members of these Self-Help Groups. However, in addition to the benefits that have accrued to the families involved in these income-generating activities, certain challenges persist (Moyle et al. 2006).

For example, organizations like Samanwita are involved in the production and marketing of a diverse range of products, including turmeric powder, chili powder, powdered spices, leaf plates, as well as activities like poultry rearing, tasar yarn production, vermi-composting, mulberry sericulture, and the cultivation of oyster mushrooms through Self-Help Groups. However, these groups often face difficulties in meeting market demand in terms of production volume, even though the quality of their products is comparable to other similar products in the market. Furthermore, there are instances where the quality of the products does not consistently meet the expectations of the market (Valliammal and Mohan 2015).

Odisha is making significant strides towards achieving its goals of empowerment. Increased access to financial institutions has enabled economically disadvantaged women to improve their economic standing, leading to the initiation of various economic activities by them. However, many of these activities do not generate the desired levels of income (Crowell and Burgess 1996). In reality, poor individuals require not only financial support but also a range of non-financial support services to sustain their economic activities and ensure their long-term viability. It has been observed that women entrepreneurs, in particular, often encounter challenges such as limited decision-making power in financial matters, inadequate business management skills, and difficulties in maintaining consistent work schedules (Karuppannan 2014).

The continuous decline in agricultural outputs and the shrinking size of farm holdings has made it increasingly urgent for rural populations to diversify their sources of income and explore alternative livelihood options (Malhotra and Bhattacharya 2010). While Self-Help Groups face certain constraints in effectively managing micro-enterprises, they have significantly contributed to enhancing the income of poor and vulnerable households by undertaking various micro-enterprise ventures across the Kandhamal district. To ensure the sustainability of micro-enterprise promotion efforts, it is essential to address several key factors: capitalizing on the rising demand for new goods and services that offer livelihood opportunities outside of traditional farming practices, enhancing the capacity of SHG members through targeted training and exposure programs to improve their effectiveness in managing micro-enterprise operations, and implementing systematic marketing strategies for products made by Self-Help Groups, particularly NWFPs, to ensure that members receive fair prices for their products, thereby contributing to increased income for both the groups and the individual members. Organizations such as PRADAN and KASAM are actively involved in promoting the effective marketing of goods produced by these groups. It is crucial to ensure that these goods and products align with market demands by incorporating innovative features that mitigate the risk of declining demand and enhance their competitiveness. In this context, NGOs are playing an important role in identifying environmentally friendly and sustainable livelihood opportunities and developing the value chain for NWFPs, from the initial production stages to the final sales and distribution channels (Basu and Srivastava 2005).

4 Identification of Eco-friendly Sustainable Livelihood Opportunities through NWFP

Leaf cup plates have been a familiar and integral part of Indian society for centuries. The specific type of leaf used to make these plates often varies depending on local geographic conditions and the availability of plant resources in a particular region. In Odisha, Siali and Sal leaves are the primary materials used to prepare leaf plates. These plates have been widely used in various social and cultural settings, including marriage ceremonies, birth and death rituals, and religious practices in temples.

Siali leaf (Bauhinia Vahlii) is an important Non-Wood Forest Product (NWFP) in central and southern Odisha. Siali leaves are used to make plates and cups for serving food. These leaf-based products are commonly used by local grocery shops, small hotels, and other food establishments, and they also serve as packing material for certain goods. In local markets, Siali leaf is often considered more valuable and fetches a higher price than Sal leaf, reflecting its desirability and utility.

The process of making leaf plates involves several distinct stages, beginning with the initial collection of leaves from the forests, followed by stitching and drying, meticulous sorting and cutting, careful packaging, and finally, distribution to various markets. Traditionally, the leaves are dried in the sun for a period of one or two days. They are then stitched together manually using thin sticks to create a larger plate by joining together three or four individual leaves. These plates are then tied in bundles, often containing hundreds of plates, and sold in nearby weekly markets to local traders and middlemen.

In the past, the market for leaf-based products primarily offered only two basic options: loose leaves and roughly stitched leaf plates. However, the product range has expanded significantly in recent years to include more value-added products that cater to diverse consumer preferences. These value-added products include medium-stitched plates, fine-stitched plates, machine-stitched plates, and molded buffet plates. These innovations have not only enhanced the quality and appeal of leaf plates but have also increased the income-generating potential for the local communities involved in their production. For example, by selling loose leaves and rough plates, local community members could typically earn around Rs. 1000–1200 per month. However, through the value addition of leaves, families now have the potential to earn significantly more, with monthly incomes ranging from Rs. 2500 to Rs. 3500.

5 Developing the Value Chain for NWFP from Production to Sales

The districts of Kandhamal and Nayagarh collectively have a substantial number of registered NGOs and Community-Based Organizations (CBOs), with estimates suggesting at least 4000 such entities operating in the region. Some of these organizations have been actively involved for a considerable period in attempting to implement interventions aimed at enhancing the income of poor people engaged in leaf stitching activities. However, despite these efforts, the implementation of value addition steps in the NWFP supply chain remains a challenge for many of these communities.

While tribal women, who are often the primary collectors and producers of leaf-based products, earn relatively low incomes, middlemen and traders often capture a significant portion of the profit by exporting these plates to other districts and regions. To address this imbalance and empower local producers, government initiatives such as the District-Based Development Agency (DBDA), under the Swarna Jayanti Gram Swarozgar Yojana (SGSY), have collaborated with District Supply and Marketing Societies (DSMS) to provide training to Self-Help Groups (SHGs) and women's cooperatives on improved techniques for making leaf plates and cups. These training programs have focused on promoting the production of finely stitched plates, machine-stitched plates, and pressed plates, which command higher market prices. As a result of this training, access to upgraded technology, and the provision of assured market support, tribal collectives have been able to achieve higher incomes and improve their economic well-being.

The trade of leaf plates in the region typically operates through two primary channels. First, the simply stitched plates, which require less processing, fetch a relatively low price of around 24 paise per plate. Second, some NGOs and CBOs have established pressing units, which apply pressure to the crudely stitched plates to give them a more uniform and refined shape. Furthermore, layers of paper or hardboard are sometimes added to these pressed plates to create a sturdier "buffet plate" that can be conveniently held in one hand while standing and eating. This diversification of product offerings has allowed producers to cater to a wider range of customer preferences and needs.

The price variations among the different types of leaf plates reflect the value added at each stage of processing:

- Old version plates (roughly stitched): 6 paise per plate
- Thread-stitched plates: 24 paise per plate
- Plate-size cut plates: 50 paise per plate
- Buffet plates: Re. 1 per plate.

Each of these types of plates has found its own niche in the market, appealing to the specific requirements and individual preferences of customers. The interventions by NGOs in promoting value-added leaf plate production have resulted in significant income gains for the beneficiaries, with each beneficiary earning up to Rs. 1500 per month for approximately nine months of the year. These initiatives have enabled around 1000 rural women, many of whom previously faced economic hardship, to achieve better self-employment opportunities with improved financial returns. The capacity-building efforts undertaken by NGOs have been instrumental in driving this positive change and empowering these women.

6 How Capacity Building Initiatives by NGOs Support the Supply Chain

Entrepreneurship is widely recognized as a crucial driver of economic development. It has been well established that the level of economic growth in a region is, to a large extent, dependent on the level of entrepreneurial activity within that region. An entrepreneur is an individual who initiates and develops a business entity with the primary goal of achieving prosperity and growth. The business world is populated by such individuals who identify opportunities and take calculated risks.

Every individual, at some point in their lives, exhibits entrepreneurial qualities. The ability to recognize and capitalize on opportunities, even in everyday situations, is a hallmark of an entrepreneurial mindset. The continuous flow of events and experiences in our daily lives often presents valuable ideas for establishing successful businesses across various sectors. The key is to identify these opportunities and act upon them promptly.

However, it has been observed that there is a relative scarcity of individuals who actively pursue entrepreneurial ventures by starting their own businesses or consistently demonstrating entrepreneurial behaviours. A number of reasons have been proposed to explain this phenomenon, ranging from a fear of incurring debt and taking risks to a lack of access to adequate financial resources, limited business acumen, and insufficient skills. This perception of a lack of entrepreneurs and entrepreneurial spirit has led to the development of various entrepreneurship development programs and initiatives designed to raise awareness about the potential of entrepreneurship and foster entrepreneurial activity.

In India, traditional cultural norms often define women's roles primarily within the domestic sphere, limiting their participation in activities outside the home. Although the Constitution of India guarantees equal rights to women in all respects, including the right to vote, societal practices often treat women as second-class citizens. During their childhood, women are typically dependent on their parents, after marriage, they become dependent on their husbands, and in old age, they may rely on their children for support. This pattern of dependency has been a common phenomenon for women in India. A major factor contributing to this situation is the lack of economic independence among women.

Given this historical and social context, it is perhaps not surprising that entrepreneurship in India has been a predominantly male-dominated phenomenon. However, as a result of sustained advocacy and persistent efforts to raise awareness, women have made significant strides in education, particularly in recent decades. The present era has brought some positive changes for women, with the government implementing specific measures aimed at promoting their economic empowerment. These initiatives have paved the way for a gradual shift in women's roles, enabling them to transition from traditional domestic responsibilities to engaging in a wider range of professional activities and contributing to the formal economy.

A Self-Help Group (SHG) is a village-based financial intermediary, typically composed of 10–20 local women. While most self-help groups are located in India,

SHGs can also be found in other countries, particularly in South Asia and Southeast Asia, where they play a crucial role in community development and financial inclusion.

The Self-Help Group movement in India was launched by the National Bank for Agriculture and Rural Development (NABARD) in 1992. The primary objective of this initiative was to provide financial assistance to the members of these groups through their collective savings and by leveraging external funds from financial institutions. In essence, Self-Help Groups act as intermediaries, connecting banks and other financial institutions with individual beneficiaries to facilitate access to credit. The SHG model is an extension of integrated development programs previously implemented by NGOs with support from various donor agencies. These SHGs are essentially the primary drivers of women-centric economic and social functions at the grassroots level.

In addition to NABARD, several other financial institutions, including the Small Industries Development Bank of India (SIDBI), the Housing and Urban Development Corporation (HUDCO), Friends of Women's World Banking (FWWB), and the National Backward Classes Finance & Development Corporation (NBCFDC), have extended support to microfinance institutions in the form of donations, grants, and loans. These funds are then on-lent to SHGs and federations of SHGs, further expanding their reach and impact.

To promote women's entrepreneurship, the Government of India launched a scheme called Trade-Related Entrepreneurship Assistance and Development of Women (TREAD) in 1998. The primary objective of this scheme was to empower women by developing their entrepreneurial skills and addressing the various constraints they face in trade and business activities. In May 2004, a revised version of the TREAD scheme was implemented by the Small Industries Development Organization (SIDO). This revised scheme provided support for market development and facilitated access to financial loans through NGOs. The assistance provided under the TREAD scheme is specifically targeted towards self-employment ventures undertaken by women in any kind of non-farm activity, promoting diversification of income sources and economic opportunities.

This study aims to understand the effectiveness of Self-Help Groups in promoting women's entrepreneurship in Odisha and to assess their overall contribution to enhancing the economic independence of women. There is a growing recognition of the need to provide targeted skill development and entrepreneurship development training, particularly for first-generation entrepreneurs, to foster economic growth at both the regional and national levels. A key element of these training programs is a focus on motivating individuals to start their own businesses and embrace an entrepreneurial mindset. A wide range of organizations and institutions are involved in the delivery of such programs, including government agencies, local enterprise agencies, Chambers of Commerce and professional associations, and business schools. A central premise underlying these programs is the belief that entrepreneurship is a learned phenomenon. In other words, entrepreneurs are not born with innate abilities but rather develop their entrepreneurial skills and mindset through their experiences, as they grow and learn, and are influenced by teachers, parents, mentors, and role models. It is argued that individuals who express an interest in entrepreneurship, as well as current entrepreneurs, can be encouraged and supported, rather than discouraged, as is sometimes the case in traditional educational institutions and training organizations. However, a general consensus on the core content that should be included in entrepreneurship education and training has not kept pace with the compelling and growing need for such education, particularly within the education delivery community. In particular, many schools and curricula have inadvertently aligned themselves with the more established and structured field of management education, while attempting to address the more complex and less understood goals of entrepreneurship education. It is clear that entrepreneurship is distinct from, or at least goes beyond, management; there is a unique set of qualities and skills that extends beyond the effective coordination of resources.

Traditionally, entrepreneurial activities have been associated with the creation of small businesses. However, there is an increasing recognition that these activities and skills can be applied in various contexts, such as large corporate organizations and the non-profit sector, to drive innovation and achieve strategic objectives. Opportunity recognition and evaluation are the cornerstones of entrepreneurship. This process is both an art and a science, blending keen observation, insightful market analysis, and adaptability. From this blend emerges both the identification of unmet needs in the marketplace and the formulation of an idea for a service or product that effectively addresses that need at an acceptable price point.

The first set of skills that entrepreneurs should possess is the ability to perceive the unusual within the usual, to see the extraordinary in the ordinary. This enables them to identify opportunities that transform commonplace elements into something unique and unexpected, whether they are establishing their own business venture or leading a new venture team within a large organization. This combination of intuitive skills and applied knowledge can be effectively taught and refined through targeted training and educational programs. Once an entrepreneur has identified a promising opportunity, they must possess the willingness and capability to marshal the necessary resources to pursue that opportunity, even in the absence of any guarantees regarding the outcome or potential rewards. This involves a willingness to embrace risk and navigate uncertainty.

Therefore, the second critical area of skill and knowledge for entrepreneurs is the ability to transform a promising idea into a viable and well-structured plan. They must also be able to articulate and communicate that plan with sufficient conviction and passion to persuade potential investors and other stakeholders to provide the resources needed to create the new enterprise. This range of abilities can also be significantly enhanced through effective education and training interventions.

A third very important set of skills for entrepreneurs includes those needed to build a cohesive team and establish an entire organization capable of delivering the product or service to the target market that initially inspired the recognition of the opportunity. This development stage is typically associated with the transition from a start-up venture to a fully articulated and complex enterprise structure. However, obtaining the necessary resources and securing the support of both internal and external stakeholders to create a new venture can be more challenging for those establishing a social enterprise or working within a new product division within a larger organization. In such cases, there is a subtle but important distinction between entrepreneurship and being entrepreneurial. Therefore, training interventions designed to foster entrepreneurship through a new venture creation framework may not be entirely appropriate in these contexts. This is because entrepreneurship and management are highly interrelated in contexts such as large organizations, and entrepreneurship education programs that do not adequately address both skill sets may be less effective in fostering entrepreneurial activity.

Unfortunately, these essential abilities are often lacking in the rural districts of Odisha, where efforts are underway to empower women and promote economic development.

It appears that entrepreneurship education should focus on practical skills built upon a foundation of theoretical understanding. A key challenge is determining the best ways to impart these skills and the entrepreneurial mindset. Research from the 1980s and 1990s indicates that traditional teaching methods often fall short in developing entrepreneurial abilities and thinking. This difficulty arises partly from the wide variety of situations entrepreneurs face, each with unique opportunity costs and varying factors like technology, innovation, risk, and financial implications. The motivations of a social entrepreneur also differ significantly from those of a university graduate starting a biotech company. Furthermore, the entrepreneurial landscape is constantly changing. Consequently, we only see a fleeting glimpse of it at any given time.

Historically, entrepreneurship education has been compartmentalized, with universities providing theoretical frameworks and training organizations offering practical application skills, often informally. Students have typically moved between these two settings sequentially. To create a continuous learning experience, closer collaboration and integration are needed among business schools, technology parks, and government agencies. This also requires the coordinated involvement of lecturers, experienced entrepreneurs, consultants, and investors. While various teaching tools like lectures, readings, case studies, and simulations can be used, evidence suggests that hands-on experience leads to the best learning retention. Students who develop business plans, participate in decision-making games, or launch actual small businesses learn significantly more than those who only attend traditional lectures.

Generating innovative ideas requires investment in various forms of capital: natural, human, intellectual, and financial. Once developed, intellectual property must be protected. The entrepreneur plays a crucial role in converting innovative ideas into tangible value, working in conjunction with leadership and management. This entrepreneurial role encompasses entrepreneurial, leadership, and management competencies. For impoverished, often tribal, families in remote areas, business plans can instill this understanding. A business plan should detail all essential aspects of a new company, including its description, organizational structure, goals, marketing and financial strategies, and more. It serves as a roadmap for achieving objectives and is often required by lenders for business loans, reflecting the owner's competence.

Thorough research into the chosen business is essential. For instance, simply knowing the difference between annual and perennial flowers is insufficient for running a flower shop; one must understand various species, their growing conditions, and native varieties. Attending relevant seminars and courses can enhance business knowledge. While experience is invaluable, one shouldn't wait years to learn fundamental aspects. Consulting experienced business owners can provide valuable insights and practical tips.

Selecting an appropriate business name is crucial for memorability, ease of spelling and pronunciation, relevance to the business, and the desired image. It should strike a balance between creativity and professionalism and facilitate the development of a catchy tagline.

To ensure a competent workforce, a rigorous hiring process is necessary, including background checks, thorough interviews, and employment screening.

Entrepreneurship education and training should primarily emphasize process and action-oriented approaches. Effective learning occurs when students and aspiring entrepreneurs actively participate in the process. Accredited programs should foster dialogue and involve all students in knowledge creation, moving beyond mere lectures. To provide practical knowledge of entrepreneurship, experiential learning and action-based methods are essential. A multidisciplinary approach is also crucial, as identifying, evaluating, and capitalizing on business opportunities requires diverse knowledge and skills in technical, financial, legal, and market aspects. Therefore, effective education will equip prospective entrepreneurs to handle the complexities and uncertainties of entrepreneurship, ideally drawing upon various disciplines and teaching methods to understand and manage the process wisely. While there's general agreement on how to cultivate entrepreneurial behaviors, there's a lack of consensus on measuring the outcomes of such activities, particularly in defining entrepreneurship education and training and developing a consistent framework for training organizations and educational institutions to promote more entrepreneurial activity.

The success stories of two tribal women's collectives from an underdeveloped region of Odisha's Kandhamal district illustrate the impact of training. These groups, trained in plate stitching, now manage the entire value chain, from collecting materials to producing and selling their products.

The Gatungia Mahila Khali Cooperative Ltd., formed in 2011 with 30 members, aims to improve the income of rural tribal families, especially women, through value addition, aggregation, and marketing of locally available "Siali Leaves." Initially focused on basic products, skill training and technology enabled them to produce higher-value items like machine-stitched and molded buffet plates, significantly increasing the members' monthly income. The cooperative has shown substantial growth in production and turnover over the past two years.

The Maa Baraladevi Leaf Plate Cooperative, started in 2009, now involves around 84 members engaged in similar value-added Siali leaf plate production and sales.

Their objective is also to enhance the income of rural tribal women. The cooperative has demonstrated consistent growth in production, turnover, and income for its members over the last three years.

7 Conclusion

In conclusion, micro-enterprises play a vital role in poverty reduction and improving the economic status of rural women. Non-timber forest products (NWFPs) are significant in creating sustainable livelihoods for poor tribal communities living near forests. India's rich biodiversity offers numerous valuable NWFPs like Sal and Siali, actively traded by micro-enterprises. Harvesting and utilizing NWFPs such as Siali leaves can support poverty alleviation efforts. Women are particularly reliant on NWFPs due to the nature of their production and processing. However, this sector often lacks organization and basic infrastructure.

As demonstrated by the study, training and empowering women have been crucial in developing sustainable NWFP-based livelihoods in tribal areas of Khandmal. These efforts have not only improved incomes but also fostered social and cultural change, with women recognizing the importance of the relationship between forests and forest dwellers. Empowerment and training have strengthened their food security and income-generating activities.

With adequate support, these initiatives can serve as models for other regions, promoting both economic well-being and environmentally friendly practices, leading to environmental awareness. Government policies and stakeholder support can facilitate broader change.

There is a need to involve financial institutions in promoting community-based micro-enterprises with benefit-sharing mechanisms. Institutions like NABARD and other public sector banks can play a role in NWFP enterprise development. Establishing forward and backward linkages and organizing existing trade practices are also necessary.

References

Basu P, Srivastava P (2005) Exploring possibilities: microfinance and rural credit access for the poor in India. Econ Polit Wkly, 1747–1756

Bhose SGR (n.d.) Joel (2003), NGOs and rural development-theory and practice. Concept Publishing Company, New Delhi

Cinamon RG, Rich Y (2002) Profiles of attribution of importance to life roles and their implications for the work–family conflict. J Couns Psychol 49(2):212

Crowell NA, Burgess AW (1996) Understanding violence against women. National Academy Press Washington, DC

Duxbury LE, Higgins CA (1991) Gender differences in work-family conflict. J Appl Psychol 76(1):60

Fox ML, Dwyer DJ (1999) An investigation of the effects of time and involvement in the relationship between stressors and work–family conflict. J Occup Health Psychol 4(2):164
Girija S (1995) Group approach to empowerment of rural women-IFAD Experience in Tamil Nadu state. BIRD, Lucknow, GOI (1991), Rep. Working Group on Organizing Self-Help Groups
Karuppannan R (2014) Role of non-governmental organisations in micro finance through SHGs—a study in Vellore District of Tamil Nadu. Available at SSRN 2513715
Mahajan V (2005) From microcredit to livelihood finance. Econ Polit Wkly:4416–4419
Malhotra KC, Bhattacharya P (2010) Forest and livelihood. Published by CESS, Hyderabad, p 246
Manimekalai M, Rajeswari G (2001) Nature and performance of informal self-help groups—a case from Tamil Nadu. Indian J Agric Econ 56(3):453
Minimol MC, Makesh KG (2012) Empowering rural women in Kerala: a study on the role of Self Help Groups (SHGs). Int J Sociol Anthropol 4(9):270
Moyle TL, Dollard M, Biswas SN (2006) Personal and economic empowerment in rural Indian women: a self-help group approach. Int J Rural Manag 2(2):245–266
Nabavi SAH (2009) Poverty and micro enterprise development. Eur J Soc Sci 9(1):120–128
Ranjan J, Sanghi NK, Jawali BG, Rao AP, Reddy K (2002) Organization of men SHGs in rural areas—a felt need of the hour. Training 1(3)
New Scheme (2011) Report of the sub-group-II on NTFP and their sustainable management in the 12th 5-year plan
Shylendra HS (2007) Microfinance bill: missing the forest for the trees. Econ Polit Wkly:2910–2914
Soundari MH, Sudhir MA (2003) Gender disparity in Tamil Nadu. Social Welfare-Delhi 49(12):24–27
Valliammal M, Mohan AC (2015) Role of non-governmental organizations in microfinance through SHGs—a study in Kanchipuram District of Tamil Nadu. Int J Manage Dev Stud 4(2):126–137

Sustainable Banking and NPA Management: Aligning Credit Risk Management in Indian Banks with India's SDG Commitments

Vineet Prabha, Vinay Kandpal, and Abdullah Malik

Abstract India's banking industry has a critical role to play in social and economic development, especially in those related to the SDGs, which aim to bring about inclusive economic growth, climate action, and infrastructure development. Yet, macro challenges such as high NPAs, unequal ESG uptake, and disparate credit risk governance have been roadblocks to this potential. This review provides a critical perspective on how Indian banks are integrating their operations with the SDG framework, with particular emphasis on sustainable, credit risk and regulation. A systematic review method was used, based on 30 more recent high-quality sources on the period between 2015 and 2025. Literature was collected from scholarly databases like Scopus, Web of Science, JSTOR, SAGE, Google Scholar and institutional sources and from RBI, SEBI, NABARD, etc. Five clusters were developed as themes—SDG-aligned banking, green finance and ESG, NPA management, regulatory framework, and blended finance models were the areas in which the insights were extracted. Findings indicate fragmented but emerging momentum in sustainability integration, and also highlight significant voids in ESG standardization, rural banking inclusion and sector-specific SDG impact assessment. The research also highlights the importance of coordinated policy intervention, enhanced disclosure frameworks and the formulation of sustainable risk metrics. For India's banks, the time to shift from being passive intermediaries to active agents of change by putting sustainability at the heart of their financial DNA to deliver against the 2030 Agenda is now.

Keywords Climate finance · Credit risk · ESG integration · Financial inclusion · Green banking · Sustainable development goals

V. Prabha (✉) · V. Kandpal
Department of Management Studies, Graphic Era (Deemed to be University), Dehradun, Uttarakhand, India

V. Prabha
Faculty, Graphic Era Hill University, Dehradun, India

A. Malik
JAIN Online, JAIN (Deemed-to-be University), Bengaluru, Karnataka, India

A. Singh et al. (eds.), *Human-Centered Sustainability: Bridging Well-Being, Gender Equity, and Technology*, World Sustainability Series,
https://doi.org/10.1007/978-3-032-22020-2_25

1 Introduction

The 17 Sustainable Development Goals (SDGs) are the world's best plan to build a better world for people and our planet by 2030. This set of 17 interconnected goals provides a blueprint for global sustainable development and requires mobilisation across the economic sectors and the finance sector. Among the various institutions, which are to implement this transformative vision, the banking sector is epitomized to have key role particularly in the case of developing countries like India, where the banks not only stand for financial intermediation but also work as developmental agencies mobilizing the resources for the national needs (Saxena et al. 2021; Yadav et al. 2024). Since India's SDG obligations are heavily connected to financial budget and institutional restructuring, it is important to analyse to what extent Indian banks are converging with these goals, more so with respect to dealing with NPAs and promoting sustainable credit risk practices (Mittal et al. 2023).

Banks are trustees of the public deposits, conductors of capital formation and intermediaries for financial intermediation. They affect the orientation of credit, as well as the character of economic growth. In countries like India, where banks have traditionally only been seen as tools for national planning by state-led development models, banks' role in the financing of infrastructure, agriculture, micro-enterprises and social services has a strong connection with the advancement on SDGs like SDG 1 (No Poverty), SDG 8 (Decent work and Economic Growth), and SDG 9 (Industry, Innovation, and Infrastructure) (Patni 2023; Kaur 2020). But this potential is dampened by systemic inefficiencies, of which the chronic problem of high NPAs in the Indian banking sector is the most prominent (Akhtar et al. 2024; Yoganandham 2025).

The issue of NPAs is not just technical, but profoundly developmental. Non-performing loans are indicative of poor credit risk assessment, poor institutional governance, and borrowers' non-accountability (Andreoli 2021). High levels of NPAs force governments to provide public funds for bank recapitalisation, at the expense of public sector investments in social sectors such as education, health, and climate resilience (Misra 2021; Damodaran and van den Heuvel 2023). In other words, NPAs represent a systemic drain in the development financing ecosystem, thereby eroding India's financial capability to achieve SDG's targets (Poddar et al. 2023).

The developmental thrust of banks is substantially undermined by NPAs. Banks that build up toxic assets are less capable of lending because risk aversion increases among them, with a subsequent reduction in credit flows to the weaker and underserved segments of the economy. For instance, inadequate funding for women entrepreneurs and climate-resilient agricultural practices results in the underperformance on SDG 5 (gender equality) and SDG 13 (climate action) (Verma et al. 2024; Baglari et al. 2023). In that sense, credit risk management is not only essential for financial reasons but also vital for sustainability (Freytag 2020; SV and Rajitha 2023).

Around the world, financial institutions are increasingly including Environmental, Social, and Governance (ESG) parameters into the way they lend their money and invest their funds. Nevertheless, with regard to structured sustainability-linked finance Indias banking firms have lagged behind (Vaze et al. 2022; Bhatnagar and Sharma 2021). This provides a double gap: exposure to asset quality shocks, and foregoing opportunities to place banks at the forefront to act as a catalyst for sustainable development. Narrowing this gap entails integrating SDGs logic into credit appraisal, portfolio management, and risk avoidance (Manaktala 2020). For instance, integrating climate risk in the credit scoring models, and giving priority to green sectors in this manner, can contribute to the long term bank stability and development impact (Bhatnagar and Sharma 2021).

In India, the Reserve Bank of India (RBI) and the Securities and Exchange Board of India (SEBI) have also undertaken significant initiatives for sustainability, such as the Sustainable Finance Roadmap and the BRSR (Business Responsibility and Sustainability Reporting) guidelines (Datta 2024; Jain 2020). However, the execution of ESG is still dimmed amidst regulatory incongruence, ESG untaught, and inadequacy of disclosure frameworks (Khan and Akhtar 2025). At the same time, the likes of PMJDY, MUDRA, green bonds, etc. signal a movement towards more inclusive and environment-cognizant finance (Yadav et al. 2024; Mittal et al. 2023). Banks such as NABARD and SIDBI lent cash to promote rural credit, microfinancing in renewable energy and sustainable agriculture (Damodaran and van den Heuvel 2023).

Yet despite these efforts, a few implementation hurdles remain. There is no standardized ESG assessment protocol in Indian banks making benchmarks and performance measurements challenging (Kaur 2020; Datta 2024). Greenwashing and the absence of standardized metrics, coupled with institutional resistance to ESG reforms, have also contributed to the weakening of the developmental role of the banking sector.

In this background, the study attempts to review and critically evaluate the dynamic relationship between sustainable banking and NPA management in India. It assesses how Indian banks are integrating sustainability dimensions—including green finance, inclusive lending, and ESG factors in their credit risk governance and management of asset quality. It also analyses the regulatory framework, institutional innovations and sectoral financing flows to see how coherent the architecture is with India's SDG roadmap (Verma et al. 2024; Akhtar et al. 2024).

The review answers the following questions:

- What would be the impact of NPAs on the potential of Indian banking to achieve SDGs?
- What are the sustainable finance tool used by Indian banks, and how relevant are they?
- What has the role of RBIs, SEBIs been in promoting the integration of sustainability?
- What data, policy, and/or institutional gaps impede alignment with the SDGs?

This article deals with commercial banks (both public and private sectors), cooperative banks, regional rural banks, and development finance institutions. It synthesizes findings from scholarly journals, regulatory reports, and case stories, as well as official reports, through a qualitative thematic analysis (Khandbahale et al. 2025; Poddar et al. 2023).

This review is all the more pertinent due to the present development scenario in India. First, in a post-COVID-19 world where the economy is looking for a green recovery, financial stability needs to be linked with sustainability (Yoganandham 2025; Manaktala 2020). Second, India's G20 presidency and Paris Climate Accord commitments necessitate refocusing of domestic banking to back global and national climate goals (Bhatnagar and Sharma 2021). Third, RBI's creation of a sustainable finance unit and climate-risk disclosure standards reflects an evolving policy commitment (Vaze et al. 2022; Khan and Akhtar 2025).

Theoretically, the present review makes a significant contribution to the literature as it integrates research streams related to NPAs, sustainable banking and ESG governance which are often studied in isolation (Saxena et al. 2021; Kaur 2020). From a policy standpoint, it offers practical policy advice on how to mainstream sustainability in Indian banking practices right from ESG scoring to green bond benchmarking in India (Freytag 2020).

The significance of this review lies in the transformation of the NPAs and credit risk from a purely financial problem to a systemic developmental challenge. High NPAs slow the flow of capital into green sectors, delay inclusive financing and thus impede India's attainment of SDGs (Patni 2023; Misra 2021). By locating Indian banks as development actors, and not only profit-maximising institutions, this review envisions banking in the SDG period more strategically. India's banking industry is at an inflection point, with the winds of financial modernization and sustainability blowing against it. How it responds to NPAs and ESG integration will decide not just financial stability, but will also determine India's path towards inclusive and sustainable development. Endorsed with over 100 references, this scoping review will support academic scholars, policymakers, and banking practitioners who are determined to integrate the banking system with both the national and global context of sustainable development.

2 Methods

This paper uses a purposive qualitative systematic review to combine and assess academic literature, regulatory documents, institutional reports and empirical studies on the confluence of sustainable banking, credit risk governance and the Sustainable Development Goals (SDGs) in India. In light of the multiple and interdisciplinary aspects of the subject -including development finance, environmental policy, risk management, and institutional governance- the review is organized in order to manage to minimize the limitations of coverage with respect to that of the thematic focus.

2.1 Databases Used

A wide selection of databases and repositories was systematically applied to guarantee extensive, solid and interdisciplinary coverage of academic and policy literature. The significant about data bases were Scopus (it offered peer-reviewed and indexed analysis in banking, sustainability, and credit danger), Web of Science. Relevant literature published in Springer and Emerald was searched for full-text academic literature on green finance, Environmental, Social and Governance (ESG) related activities and SDG- aligned banking operations. SAGE Journals had been particularly useful to source Indian context-specific studies on financial inclusion, sustainable banking and institutional governance.

In addition to the core academic foundation, several secondary and grey literature sources were incorporated. To support searches of more recent working papers, doctoral theses and institutional reports, the selection process was complemented by the use of Google Scholar, which can offer access to documents not collected in more standard databases. Preprints, technical reports and author-deposited articles in Research Gate, and working papers and policy briefs on economic and financial governance in SSRN (Social Science Research Network). Historical and theoretical perspectives were supplemented with material from JSTOR, particularly with respect to early works on development finance and institutional economics. References from important institutional sources such as volumes from the Reserve Bank of India (RBI), Securities and Exchange Board of India (SEBI) reports, the National Bank for Agriculture and Rural Development (NABARD) bulletin, and development performance indicators from the NITI Aayog SDG Index, were also studied. This 'umbrella' and multi-sourced search design was to include a range of policy-intensive and emergent literature around sustainable banking in India—a topic that has often spanned academia, practice and governance, and which often goes beyond even peer-reviewed works (Table 1).

2.2 Inclusion and Exclusion Criteria

To ensure methodological rigor and relevance, the following criteria were applied to filter the retrieved sources (Table 2):

This stepped inclusion/exclusion matrix also helped to ensure that the selected literature was not only academically robust, but also contextual and temporal in line with the phase of SDG implementation in India.

Table 1 Databases and repositories used for literature review on sustainable banking in India

Source/Database	Type	Focus area
Scopus	Scholarly database	Peer-reviewed articles on banking, sustainability, and credit risk
Web of science	Scholarly database	High-impact, interdisciplinary studies
Springer link	Academic Publisher	Full-text literature on green finance, ESG practices, and SDG-aligned banking
Emerald insight	Academic Publisher	Research on financial services, sustainability, and governance
SAGE journals	Academic Publisher	Indian context-specific studies on financial inclusion, sustainable banking, institutional governance
Google scholar	Broad search engine	Working papers, theses, institutional documents not indexed in traditional databases
Research gate	Scholarly network	Preprints, technical studies, and author-deposited manuscripts
SSRN (social science research network)	Working paper repository	Policy briefs and working papers in finance, economics, and governance
JSTOR	Scholarly archive	Foundational literature in development finance and institutional economics
Reserve Bank of India (RBI)	Institutional source	Reports, circulars, and publications on monetary and banking policy
Securities and exchange Board of India (SEBI)	Institutional source	Regulatory updates, ESG disclosures, and market governance reports
NABARD (National Bank for agriculture and rural development)	Institutional source	Bulletins on rural finance, sustainable development, and microfinance
NITI Aayog SDG index	Government report	Data and performance metrics on India's SDG progress, including sustainable finance indicators

Source The Authors'

2.3 Search Terms

A comprehensive keyword strategy was carefully developed and carried out to find the most pertinent and high-quality literature for this review. The methodology extensively employed the Boolean operators (AND, OR), truncation and wildcard techniques to refine or broaden the search as appropriate. This was useful to adjust

Table 2 Inclusion and exclusion criteria for literature selection

Criteria	Included	Excluded
Time frame	Studies published between 2015 and 2025, coinciding with SDG adoption	Publications prior to 2015 unless they provided foundational theory
Geographical scope	India-specific studies or global studies with direct relevance to India	Studies from unrelated geographies without application to Indian context
Topic relevance	Studies with explicit focus on SDGs, ESG finance, NPAs, or credit risk	General banking literature not related to sustainability or governance
Document type	Peer-reviewed journal articles, government reports, institutional white papers	Blog posts, non-reviewed content, promotional material
Language	English only	Non-English sources
Access type	Open access or accessible through institutional databases	Paywalled content without abstract or summary

Source The Authors'

search depth for precision in topic relevance. The primary search terms developed for this study were designed to capture the main thematic crossover points—sustainable banking, credit risk management, and India's Sustainable Development Goals (SDGs).

The primary keyword combinations employed included:

(a) "Sustainable banking AND India,"
(b) "Non-performing assets AND credit risk AND SDGs,"
(c) "Green finance AND Indian banking sector,"
(d) "ESG integration AND financial institutions India,"
(e) "Financial inclusion AND Sustainable Development Goals,"
(f) "RBI AND green bonds AND policy framework,"
(g) "Public-private partnerships AND sustainable finance India," and
(h) "Cooperative banks AND rural development AND SDGs."

These structured groups were further tailored to reflect the search functionalities of individual databases. This included using subject filters and discipline-driven key words on Scopus and Web of Science, and approximating this on a generic level based on MeSH (Medical Subject Headings) terms for platforms focusing more closely on public policy and health. Secondary search terms besides these primary ones included synonyms and related concepts, such as "development banking", "inclusive finance" and "climate finance" to widen the scope of the results in order to reflect the diversity of words used to describe these concepts in different disciplines or databases. A publication time-window was bordered from January 2015 to April 2025 to assure that the literature was current as well as contextually in line with the timeline set for SDG implementation. This time-period encompasses post-SDG adoption literature and also leaves room for future estimates and prospective policy analysis. Collected references, whether or not they were in full article format or were merely abstracts and bibliographic citations, were systematically compiled, organized, and assessed communally through the Zotero citation management system. This allowed for efficient organization and checking of duplicates, and labelling for thematic analysis in later phases of the review.

2.4 Screening Process

A robust 3-stage screening protocol was followed to ensure that only theoretically relevant, high quality, contextually appropriate literature was presented in this systematic review. This technique supported the reduction of a large list of initial papers to yield a focused set of sources that specifically touched on the essential themes of the study, which were sustainable banking, credit risk management, and SDG alignment within India.

The first stage was to conduct title and abstract screening of all articles retrieved initially (about 128 in number) for their relevance to the theme, the presence of the keywords and geographical focus. Articles that did not show sufficient evidence for a relationship with banking, sustainability, credit risk or the SDGs, or which did not relate to Indian context, were removed. This initial screening resulted in 74 papers being identified as possibly relevant. The next step was full-text review. Every other study was secondary scrutinised to have a clear idea whether it is methodologically sound, empirical based study and related to Indian banking sector or not. Speculative, non-framework driven, and non-evidence-based work were excluded. The next step reduced the pool further to 42 articles that satisfied all core inclusion criteria and contributed to the analysis. A third step in forward and backward citation tracking was used to increase the inclusiveness of the review. Using backward tracking, the reference lists of the selected studies were reviewed for seminal works or missed foundational references. Forward tracking, in contrast, identified newer research that cited pivotal articles already under review. This snowballing approach was especially valuable in finding high-impact grey literature such as policy briefs, regulatory reports, or case studies— especially from agencies such as the Reserve Bank of India and NABARD. This criterion led to a further eight studies being added.

In summary, the 30 final studies included were filtered from the original set of 128 sources based on their empirical weight, policy significance, and capability to contribute to an emerging picture of the increasingly entangled relationship between sustainability and risk management in Indian banking. The results of these studies underpin all the thematic analysis and findings that are discussed further in this review.

2.5 Quality Appraisal Tools

Since a wide range of sources were reviewed ranging from peer-reviewed academic literature to institutional reports and policy documents, a differentiated strategy for quality assessment was applied to guarantee methodological soundness and content validity. For journal articles (peer reviewed), the Critical Appraisal Skills Program (CASP) checklist was used. This well-established tool helped assess key elements such as methodological soundness, clarity of research objectives, validity of findings, and potential bias. Only articles demonstrating high levels of analytical rigor and transparency in research design were included.

For policies and institutional white papers, a specialty rubric was created to assess the quality of information and recommendations according to three major criteria:

(a) Credibility of Source, preference being given to the institutions such as RBI, NABARD, SEBI and other nationally recognized institutions;
(b) The use of empirical evidence and/or proper referencing—this demonstrated data rather than anecdote; Objectivity (neutral and clear) in policy—recommendations and assessments were independent, clearly put and evidence-based.

For grey literature and working papers, which may lack formal peer review, the evaluation was on:

(a) Authority of author, including institutional position or previously published work;
(b) Thematic relevance to core topical areas, including discussion on non-performing Assets (NPAs), integration of ESG, SDG-linked banking reforms; and
(c) Consistency and transparency, ensuring that assumptions and findings were built up in a coherent way and sufficiently supported.

Among all types of documents, only those where scoring was above 70% on the evaluation rubric (on the review form) were included at the end. Disagreements in quality assessment referring to the review's internal (study) validity were resolved by consensus and pooling among the reviewers in order to prevent reviewer's bias. In these instances, a second scholarly source was consulted for cross-reference. To obtain literature relevant to sustainable banking, credit risk management and SDG realization in India a robust and transparent methodology was followed to gather, screen and assess literature. Through the use of a structured keyword strategy with multi-phase screening filtering procedures and various types of quality appraisal tools, the approach endeavoured that the aspects of the RER were representative, of high quality and analytically pertinent. The choice of papers provides a complete base to understand the problems and prospects of India's banking sector in its shift towards a sustainable and inclusive development framework.

3 Thematic Discussion Findings

The outcomes of high-quality 30 selected studies are presented in the following summary account covering five main themes apparent through the review process. The themes capture the interplay of India's banking sector with the Sustainable Development Goals (SDGs), Non-Performing Assets (NPAs) and changing sustainability mandates. The study achieves sensitive comparisons across methods, policy orientations, and organizational responses, and suggests research and policy agendas.

3.1 Sustainable Banking and India's SDG Commitments

One of the common themes in the literature is the relationship between the concept of sustainable banking and the attainment of India's SDGs. A number of studies (e.g., Saxena et al. 2021; Kaur 2020; Patni 2023) contend that the banking industry—particularly the public financial institutions—has a strategic developmental responsibility to finance inclusive growth, green infrastructure, and poverty reduction.

Certainly, schemes such as PSL and PMJDY have contributed to financial inclusion (Yadav et al. 2024), yet there is variation in the degree to which banks overlay the overall SDG framework when it comes to operational strategy. Public sector banks (PSBs) are more usually forced to do SDG-compliant finance but are usually not institutionally innovative/efficient compared to their private sector peers (Khan and Akhtar 2025; Mittal et al. 2023). On the other hand, private banks innovate more in ESG-based products but concentrate on high-margin urban markets, leaving countryside markets underserved.

Despite a powerful policy will there is no such nation-wide strategy, which aligns banking products, services or lending targets along certain specific SDGs. Some of these measurement frameworks are even siloed and seldom integrated into performance scorecards.

3.2 Green Finance and ESG Integration

The theme of green finance is dominant in more recent studies, reflecting India's growing urgency in financing its clean energy transition. Research by Manaktala (2020) and Datta (2024) highlights the rise of green bonds, especially those issued by SBI, EXIM Bank, and private players like Yes Bank. Yet the literature indicates that India's green finance ecosystem is nascent and underregulated. ESG integration remains voluntary and non-standardized, with SEBI's BRSR (Business Responsibility and Sustainability Reporting) still in early adoption. There are methodological inconsistencies in how ESG factors are scored or disclosed, limiting investor confidence and public accountability (Freytag 2020; Vaze et al. 2022).

Absence of a homogenous green finance taxonomy for the Indian industry. ESG disclosure is voluntary for most financial institutions and greenwashing risks are inadequately addressed by policy.

3.3 NPAs and Sustainable Credit Risk Governance

A second robust thematic cluster has to do with NPAs and credit risk governance, and includes authors like Misra (2021), Akhtar et al. (2024) and Yoganandham (2025) argues that NPA ratios limit banks capacity to support developmental objectives. NPAs not only limit the liquidity of banks and add to provisioning costs but also reduce the appetite for long-term infrastructure or inclusive credit expansion. Many PSBs suffer higher NPAs in sectors such as MSMEs and the agriculture sectors critical for SDG attainment (Verma et al. 2024). As such, the sustainability of these institutions depends on the extent to which NPAs are effectively managed through the use of early-warning systems, borrower profiling and underwriting, and ESG assignment. There's increasing demand to incorporate climate risk and social vulnerability in credit risk models. However, there is no established approach to doing it in India, and most banks are using antiquated risk scoring models that do not measure up to emerging environmental and social risks.

There is also little innovation in credit risk tools with respect to the consideration of environmental or climate-related financial risks. In terms of NPA control, regulatory interventions are reactive rather than preventive.

3.4 Institutional and Regulatory Frameworks

Some papers (Jain 2020; Damodaran and van den Heuvel, 2023) explore how sustainable finance policies are being constructed by Indian regulators including the RBI and SEBI. Key steps involve the RBI's Sustainable Finance Department and SEBI's mandatory ESG reporting from the top 1000 listed companies.

Institutional coordination is still limited, however. For instance, there is no specific road map on sustainable finance at the juncture of fiscal policy, financial regulation and development planning. Also, the smaller banks and cooperative societies in such rural regions do not have the expertise and resources to adhere to the sustainability norms, leading to further financial exclusion (SV and Rajitha, 2023; Bhatnagar and Sharma 2021).

Geographical Imbalance: It is also noteworthy that, even though urban and metro-based banks have started to embrace sustainability guidelines, Regional Rural Banks (RRBs) and Co-operative banks that serve the major portion of the underbanked population in India, are far behind the urban and metro-based banks (Poddar et al. 2023).

Centralized regulatory environment and non-uniform institutionalization have created geographical and transactional schism in achieving sustainability targets in the context of the Indian banking industry.

3.5 *Public–Private Partnerships and Blended Finance*

The literature also touches on public–private partnerships (PPPs) and blended finance mechanisms as platforms for the scaling-up of SDG-related investments. Vaze et al. (2022) illustrate instances where both commercial banks, development financial institutions, and foreign agencies have come together to co-finance renewable energy, affordable housing, and digital infrastructure projects. Although promising, PPP is still project-oriented and divided. We do not have enough institutions that would formalize a partnership among the public banks, private investment and multilateral institutions. Further, the risk-sharing arrangements of blended finance can be opaque or politically sensitive, and end up not being used.

Gap Identified: Institutional risk-aversion and regulatory uncertainty constrain the potential scalability of PPP and blended finance solutions to finance high-impact, sustainability-tied infrastructure.

The review notes that India's banking sector has recorded tangible but uneven strides in nurturance of sustainable development imperatives. There has been progress in the key areas of green finance and financial inclusion, in particular among bigger and more urban focused banks. But systemic challenges such as inadequate NPA management, ESG standards gaps, lack of regulatory coordination, geographic imbalances still hamper the potential for the sector to be an SDG delivery vehicle. What is required is a concerted policy effort aiming at a policy framework, capacity building and technological upgradation, assuring sustainable banking development in India at large, in a manner that halts banks to react and instead become proactive enablers of the Indian sustainable development goals.

4 Limitations

Although this review is a culmination of the literature on the nexus between sustainable banking, credit risk governance and SDGs in India, there are some limitations that must be kept in mind. They derive from limitations in sourcing, screening, and analysing an extensive and constantly evolving body of knowledge that crosses academic, regulatory, and institutional boundaries.

A major constraint is the issue of prevailing biases in the published literature. Much of the research on sustainable banking in India is single domain-centric like environmental finance, financial inclusion, or digital banking. As a result, discussions about green bonds, ESG disclosures, fintech and the like tend to take center-stage, with more urgent topics like the governance of systemic credit risk and the role of non-performing assets (NPAs) in undermining financing for development or the contribution of rural cooperative banks to sustainability all too often falling by the wayside. With such a skewed focus, it has been difficult to gain a comprehensive insight into how sustainability connectedness varies across the entire range of banking activities and type of institutions operating in India.

Further, many of the publications in the field are still anecdotal accounts or descriptive case studies with few using rigorous empirical techniques, such as longitudinal analyses. A number of 'expert' policy reports and institutional papers, though often valuable, are not peer-reviewed and may be influenced by political or organizational interests. Government reports, for example, regularly highlight the success of programmes such as Pradhan Mantri Jan Dhan Yojana (PMJDY) and MUDRA yet give little space to the shortcomings in implementation, spatial disparities, and sustainability of these initiatives in the long run.

Regarding search strategy, despite a substantial and multi-platform search strategy, some results could have been left out because of limitations in the scope. First, the study was restricted to English language studies only and may have missed regionally relevant studies that were published in Hindi or other Indian languages, especially those from state level research institutes or cooperative banking networks. Second, some eligible articles were not available in full text or free access, and thus not included unless they were summarized or cited by open-access articles. Third, even if a variety of databases were used (Scopus, Web of Science, Google Scholar, JSTOR, SSRN, ResearchGate), differences between them in indexing policies might have biased disciplinary perspective, with an over-representation of the economic and financial literature (at the expenses of a more comprehensive input on social policy, rural development, or public administration). Finally, this review was restricted to articles written between 2015 and early 2025, which corresponds to SDG timeline.

Therefore, earlier formative studies or policy drafts released after 2025 might have been missed. The review also identified some important shortcomings in the existing literature. For example, few articles examine how the sectoral impact of banking activities relates to the SDGs (e.g., SDG 3, Good Health and Well-Being or SDG 11, Sustainable Cities and Communities). Livelihood contribution of regional rural banks, cooperatives as well as microfinance organisations suffer from limitations, despite their large presence in India's financial landscape. A lack of standardised models that integrate ESG into the models used by banks to assess credit risk is another frequently raised concern, leaving banks without many tools to adequately gauge environmental or social risks. Additionally, most of the prior analyses are cross-sectional, providing static pictures rather than tracing how policies or practices change over time. Third, the literature reviewed above does not adequately capture the perspectives of important actors on the ground (borrowers, loan officers, regulators, and local communities) that would enable us to better ground our theoretical model in the reality of the field. Acknowledging these limitations, the review maintains methodological transparency and indicating key domains for further research. These involve establishing empirically-driven ESG risk models, focused investigations for rural financial institutions, longitudinal evaluations of green finance achievements, and incorporating stakeholder perceptions. Via the lens of qualitative and mixed methods. Filling such gaps is extremely important for taking forward academic knowledge and policy innovation addressing sustainable banking in India.

5 Conclusion and Future Directions

This review has assessed the emerging interface of sustainable banking, credit risk governance and India's journey towards achieving the SDGs. Using a curated collection of academic research, institution-sponsored and regulatory reports, the review reveals advances as well as ongoing shortfalls in how Indian banking institutions bridge national development objectives and sustainability-synchronised financial practices. One of the major learnings is that although Indian banks, in particular public sector banks, have progressed significantly in the areas of financial inclusion, green bond issuance, and interventions in the microfinance space, the systematic factors for incorporating ESG and a long-term sustainability framework into credit risk assessment and portfolio governance continue to be a big gap. While efforts such as SEBI's BRSR framework, RBI's Sustainable Finance Roadmap, and a diverse number of governments supported inclusion programs represent promising intentions, their implementation is uneven across institutions and locations. Especially, rural financial institutions and co-operative banks are under-represented in SDG-linked finance, though they have an important role in reaching out to underserved areas.

One of the other major implications arising from decisions on NPA is this continued detriment to the development potential of Indian banking, and thus that of the Indian economy as a whole. Further, as far as the NPA constrains the credit extension and flow to important SDG-related sectors like agriculture, MSMEs, renewable energy, affordable housing, and so forth. This problem is even worse by the lack of risk models that include environmental and social vulnerabilities. Further, the prevalence of descriptive studies and the absence of sector-specific indicators or empirical models constrain access for both researchers and policymakers to evidence-based information for the design of sustainability transitions for banking.

5.1 Policy and Practice Implications

For policymakers and regulators, the review highlights an immediate imperative to:

(a) To formulate Standard ESG Assessment models for India, particularly for credit risk assessments and NPA-prone portfolios.
(b) Enforce SDG-related lending disclosures for all licensed banks as opposed to just top-tier public/private banks.
(c) Foster capacities at regional rural banks and cooperative banks to use sustainability tools, digital infrastructure and blended finance instruments.

(d) Provide tax incentives, credit enhancement and/or regulatory capital relief to banks that use green financial instruments such as sustainability-linked loans through tax breaks, credit enhancement, or regulatory capital relief.
(e) Promote cross-sectoral cooperation between financial service providers, development agencies and technology providers that invent new types of products and new forms of delivery.

5.2 *Recommendations for Future Research*

To enhance the body of evidence in this context and inform practice, the review suggests various avenues for future research:

To enhance the evidence base and inform practice, the review identified potential areas for future research:

(a) Impact monitoring over time: Subsequent research should follow the long-term impact of ESG integration, green financing and financial inclusion schemes on both SDG outcomes and financial stability.
(b) SDG mapping by sector: Research should capture the impacts along the SDG dimensions (i.e., target) of banking interventions (e.g., SDG 4—Quality Education, SDG 11—Sustainable Cities) using fine data.
(c) Decentralized finance analysis: It is required to analyse how cooperative banks, RRBs and fintech-driven NBFCs contribute to or obstruct sustainability.
(d) Borrower and community perspectives: Incorporating stakeholder voices most notably that of borrowers, loan officers, and rural women entrepreneurs, can infuse practical depth to current sustainability and risk models.
(e) Comparative international benchmarking: Benchmarking India's sustainable banking framework with other emerging economies could highlight best practices, policy transfers, and innovation gaps.

The banking sector in India is uniquely placed to be a driver of positive transformation by building a fairer, climate-resilient and inclusive economy. But implementing this vision involves a tactical shift from retroactive compliance to proactive integration of sustainability into the credit system, governance processes and institutional culture. Now, as India moves into the final stretch of 2030, the year of the SDG deadline, it is time to reimagine its banks as drivers of growth as opposed to enablers, and put them at the helm of moulding a sustainable growth and development trajectory for the country.

References

Akhtar S, Azmi SN, Khan PA, Jan AA, Ansari Z (2024) Unveiling the financial landscape: analyzing profitability, productivity, and efficiency of banks in an emerging economy using the CAMELS

framework and panel analysis. Cogent Bus Manag 11(1):2399747. https://doi.org/10.1080/23311975.2024.2399747
Andreoli, E. (2021) Bridging the gap: ethical finance and microcredit through Bangladesh and India from 1928 to today. University of Padua. Retrieved from https://thesis.unipd.it/handle/20.500.12608/12345
Baglari, N., Roy, T. S., Rubert, E. J., Priya, M. (2023) SHODHCHOLISTAN: a multidisciplinary framework for sustainable development modeling, 1st Edn, ISBN: 978-81-981681-1-5. Retrieved from https://www.andhraloyolacollege.ac.in/mba_books/SHODHCHOLISTAN%20A%20MULTIDISCIPLINARY%20FRAMEWORK%20FOR%20SUSTAINABLE%20DEVELOPMENT%20MODELING.pdf
Bhatnagar S, Sharma D (2021) Green financing in India: identifying future scope for innovation in financial system. Int J Green Econ 15(3):185–212. https://doi.org/10.1504/IJGE.2021.120875
Damodaran A, van den Heuvel O (2023) India's low carbon value chain, green debt, and global climate finance architecture. IIMB Manag Rev 35(2):97–107. https://doi.org/10.1016/j.iimb.2023.03.005
Datta S (2024). An empirical assessment of India's position in global sustainable bond market. 119925
Freytag J (2020) Challenges for green finance in India: an analysis of deficiencies in India's green financial market. DiVA Portal. Retrieved from https://www.diva-portal.org/smash/record.jsf?pid=diva2%3A1441234
Jain, S. (2020). Financing India's green transition. Observer Research Foundation Issue Brief, 338. Retrieved from https://www.orfonline.org/research/financing-indias-green-transition-60753
Kaur H (2020) Achieving sustainable developments goals in India. In: Sjåfjell B, Bruner CM (eds) Cambridge handbook of corporate law, corporate governance and sustainability. Cambridge University Press 2019, University of Oslo Faculty of Law Research Paper No. 2019–37
Khan I, Akhtar MA (2025) An impact analysis of global governance and economic indicators in shaping bank lending in India. BIJ 32(10):3877–3900. https://doi.org/10.1108/BIJ-04-2024-0317
Khandbahale S, Yelamanchili R, Pachorkar S (2025) Boardroom crossroads: CEO prosecution vs bank reputation. Emerald Emerg Mark Case Stud. https://doi.org/10.1108/EEMCS-05-2024-0212
Manaktala S (2020) Green bonds in sustainable finance: exploring the case of India. SSRN. https://ssrn.com/abstract=3644116
Misra P (2021) Inclusion of environmental & social risk factors in infrastructure financing: creating a better development finance institution for India, Centre for Financial Accountability. Retrieved from https://www.cenfa.org/wp-content/uploads/2021/10/Creating-a-better-Development-Finance-Institution-for-India-Report-by-Praachi-Misra.pdf
Mittal S, Chaudhry S, Bhadauria SS (2023). Green banking–the path leading to sustainable economic growth. In: Contemporary studies in economic and financial analysis Retrieved from https://doi.org/10.1108/S1569-37592023000110B013
Patni R (2023) Poverty alleviation with sustainable development in India. In: Dutta V, Ghosh P (eds) Sustainability: science, policy, and practice in India. Sustainable development goals series. Springer, Cham. https://doi.org/10.1007/978-3-031-50132-6_2)
Poddar S, Bhowmik D, Bhaumik A, Tisdell C (2023). Sustainable development and informatics: emerging issues. Lincoln Research Press. ISBN: 978-0-6488798-1-4
Saxena D, Dhall N, Malik R (2021) Sustainable banking: a roadmap to sustainable development. Corp Gov Sustain Rev 5(3):42–56. https://doi.org/10.22495/cgsrv5i3p4
SV PK, Rajitha M (2023) A study of green banking practices in domestic and international banking sectors. Adv Res Money Bank 5(1):78–89
Vaze P, Kumar N, Colenbrander S, Burge L, Sharma N (2022). Identifying, managing and disclosing climate-related financial risks: options for the Reserve Bank of India. Retrieved from https://www.climatebonds.net/files/documents/publications/Identifying-managing-and-disclosing-climate-related-financial-risks-options-for-the-Reserve-Bank-of-India.pdf

Verma M, Verma A, Patil P, Pillai R, Mandala GN, Bansod S (2024) The rise and fall of Chanda Kochhar: Icici's governance breakdown and its linkage to SDG 5, 8 and 16. J Lifestyle SDGs Rev 4(4):e02351. https://doi.org/10.47172/2965-730X.SDGsReview.v4.n04.pe02351
Yadav RA, Premalatha KP, Patil S (2024) Advancing sustainable banking and financial inclusion in India through ESG integration in technological disruptions. Community Pract 20(11):260–272
Yoganandham G (2025) India's economic resilience: managing financial stability amid rising debt, falling savings, and liquidity crunch for sustainable growth—an empirical analysis. XIV(III):99–112

Women Empowerment Through Financial Inclusion, Literacy, and Gender-Responsive Policies

Rakesh Kumar, Pranshu Gupta, and K. M. Poonam

Abstract The chapter explores the major factors that could impact women empowerment. These factors include-the Education and skills training, social norms and cultural barriers, access to financial services and literacy, gender responsive policies etc. Through this chapter, Sustainable Development Goals 4, 5, 8, and 10 are also addressed, as the concept discussed directly relates to these goals. The aim of this chapter is to explore each factor in detail for women empowerment, i.e., financial inclusion, financial literacy, Gender responsive policies and discuss the triple effect of these factors and their relationship in order to empower women and create opportunities for them. The study analyses the effects of financial literacy, access to banking and credit, and supportive legislative frameworks on women's economic, social, and personal empowerment. It concludes that women's empowerment is multifaceted and must be approached in tandem across all three areas. The study suggests future research directions focusing on digital financial instruments, targeted interventions for rural women, and the strengthening of gender-responsive policy frameworks for sustainable empowerment.

Keywords Self help groups (SHGs) · Gender equality · Financial inclusion · Financial literacy · Women empowerment · GDP · Financial management

1 Introduction

The effort to extend banking services to the disadvantaged and low-income population in India, particularly the women who comprise a significant segment of the population, is currently a priority. It promotes social as well as economic development by engaging the excluded groups in the process of development. Women are a majority of the beneficiaries of financial inclusion, according to studies, and Self-Help Groups (SHGs) play a critical role in this initiative. These organizations

R. Kumar · P. Gupta (✉) · K. M. Poonam
Department of Management Studies, Graphic Era (Deemed to be) University, Dehradun, India
e-mail: Pranshugupta2020@gmail.com

A. Singh et al. (eds.), *Human-Centered Sustainability: Bridging Well-Being, Gender Equity, and Technology*, World Sustainability Series,
https://doi.org/10.1007/978-3-032-22020-2_26

enhance the ability of women and empower them bettering their communities. To deal with the existing gaps in empowerment exercises of women, this paper examines the impacts of SHGs on empowerment of women, the social economic forces involved and what further can be done to further enhance these programs. A comprehensive access to a broad range of high-quality financial services and products has attracted significant interest in the scholarly world, the governments, and among the masses. This is because it is understood that the world has massive financial inclusion inequalities that are impeding the achievement of economic growth and stability (Cabeza-García et al. 2019).

Much stress has been created on the financial empowerment of women and it has shown vast disparities between genders. A complex evaluation method is suggested and significant aspects are located through a systematic review of 101 articles provided by Scopus (2001–2023). Among the issues, microfinance, self-help groups (SHGs), and gender inequality can be distinguished including individual, home, institutional/community, and other factors. Key antecedents are socioeconomic and demographic factors, unpaid care work, SHG membership, microcredit and family care. Kabeer based his idea on the division of financial empowerment into three categories and views the evaluation as consisting of: resources, agency, and accomplishments. The outlined gaps emphasize the necessity of gender-responsible policies to address the lack of technology and financial literacy to foster equitable

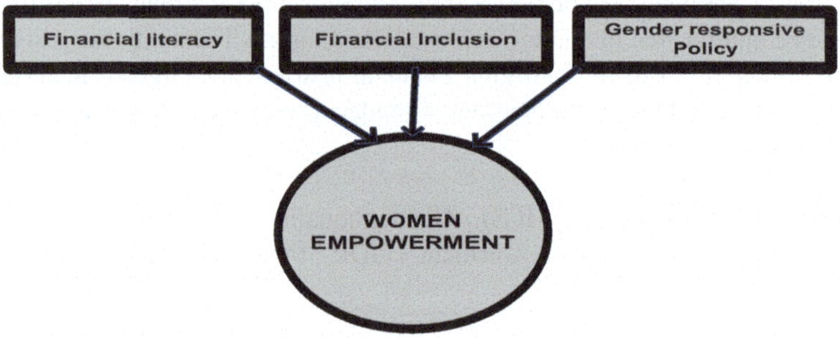

Fig. 1 Indicators of women empowerment

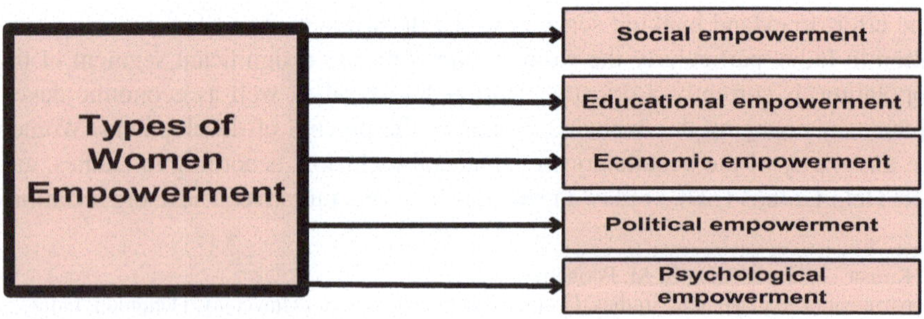

Fig. 2 Types of women empowerments

Table 1 Various government schemes for women empowerment

S. No.	Ministry By	Scheme	Objective	Focus
1	Ministry of Agriculture and farmers welfare	Mahila Kisan Sashaktikaran Pariyojana (MKSP)	Aimed to increase women farmers' income and productivity through the promotion of agricultural-centered livelihood options	Women gain access to financial aid, technical training, and an enhanced understanding of sustainable agriculture
2	Ministry of Agriculture and farmers welfare	Rashtriya Mahila Kisan Yojana	Encourages female farmers to assume leadership positions in the agricultural industry	Concentrating on enhancing agricultural production and skill development
3	Ministry of Health and Family Welfare	Janani Suraksha Yojana (JSY)	Under the National Rural Health Mission (NRHM), JSY is a safe motherhood intervention that offers pregnant women financial support for institutional delivery	JSY provides financial aid to women from families below the poverty line with the goal of lowering maternal and newborn mortality
4	Ministry of Health and Family Welfare	Pradhan Mantri Matru Vandana Yojana (PMMVY)	Gives expectant and nursing mothers a financial incentive to enhance their nutritional and overall health	For qualified women in their first pregnancy, the program offers ₹5000 to help them enhance both their own and their unborn child's health
5	Ministry of Home Affairs	Nari Shakti Puraskar	The Ministry of Home Affairs presents this award every year to recognize women who have made significant contributions to society	Acknowledges the contributions made by women in the social, political, and economic spheres
6	Ministry of Home Affairs	Mahila police volunteer (MPV) scheme	The goal of this program, which is run in conjunction with the MHA, is to develop and empower Mahila police volunteers who act as a conduit between the community and the police	Fighting crime against women and supporting women in need

(continued)

Table 1 (continued)

S. No.	Ministry By	Scheme	Objective	Focus
7	Ministry of Micro, Small and Medium Enterprises (MSME)	Prime minister's employment generation programme (PMEGP)	PMEGP is a credit-linked subsidy program designed to encourage young people, particularly women, in rural and urban regions to work for themselves and start their own businesses	Women entrepreneurs receive increased support through a program that provides a 25% subsidy for starting new microbusinesses, compared to the usual 15%, facilitating easier business launches for women
8	Ministry of Micro, Small and Medium Enterprises (MSME)	Mahila coir Yojana	The program supports women in the coir sector in Kerala by providing financial assistance for training and working capital to female business owners	Women are empowered to work independently and boost their incomes through training, raw materials, and financial support for coir-based enterprises
9	Ministry of Micro, Small and Medium Enterprises (MSME	Udyogini scheme	This program empowers women by providing financial support to rural and urban women entrepreneurs for starting microbusinesses in various industries, including manufacturing, services, and trade	Specifically provides loans and financial assistance to female entrepreneurs to promote their participation in the MSME sector, especially those from disadvantaged areas
10	Ministry of Micro, Small and Medium Enterprises (MSME	SIDBI—Women entrepreneur fund	A fund from the small industries development Bank of India (SIDBI) supports female entrepreneurs in starting or expanding micro, small, and medium-sized businesses (MSMEs)	

(continued)

Table 1 (continued)

S. No.	Ministry By	Scheme	Objective	Focus
11	Ministry of Micro, Small and Medium Enterprises (MSME	National Small Industries Corporation (NSIC) schemes	NSIC provides a variety of programs to support MSMEs by giving them access to resources like capital, technology, and market connections	NSIC provides programs to empower female entrepreneurs by offering business opportunities, financial support, and training, aiming to enhance women's participation in the MSME sector
12	Ministry of Micro, Small and Medium Enterprises (MSME)	Atal innovation Mission (AIM)	The Indian government is taking this action to encourage entrepreneurship and innovation in the nation. To promote innovation in the MSME sector, the ministry of MSME collaborates with AIM	Women innovators and entrepreneurs are encouraged to participate in special programs under AIM, such as funding, mentorship, and capacity-building initiatives aimed at women-led enterprises
13	Ministry of Steel	Women in steel program	Intends to boost the number of women working in the historically male-dominated steel sector	Encourages women to work in steel manufacturing and allied industries by offering leadership and skill development programs
14	Ministry of Social Justice and Empowerment	New Swarnima scheme for women	To encourage self-employment and financial independence, give term loans to women entrepreneurs from underprivileged backgrounds	• loan amount: Up to Rs 2,00,000 with an annual interest rate of 5%. • execution:Overseen by state channelizing agencies and the National Backward Classes Finance and development corporation (NBCFDC)

(continued)

Table 1 (continued)

S. No.	Ministry By	Scheme	Objective	Focus
15	Ministry of Social Justice and Empowerment	Mahila Samridhi Yojana (MSY)	Is a program aimed at assisting female entrepreneurs from underprivileged backgrounds, introduced by the Ministry of Social Justice and Empowerment as part of the National Backward Classes Finance and development corporation (NBCFDC). It provides microfinance to these entrepreneurs directly or through self-help groups (SHGs), implemented by various channel partners nationwide	1. Self-help groups (SHGs): It seeks to support and boost SHGs, especially those that comprise women. The autonomous, economically disadvantaged category consists of these individuals 2. Women from economically and socially disadvantaged groups (targeted groups): Women from disadvantaged groups are taken into account, especially those who belong to the SC or ST categories

Reference: https://www.ncw.gov.in/publications/women-centric-schemes-by-different-ministries-of-government-of-india-goi/

growth, as well as potential opportunities in the future of research. This paper in a special manner, offers a complete strategy and methodology of evaluating the financial empowerment of women (Gupta et al. 2025). Women empowerment is the process that provides women with the tools and opportunities to exercise their rights and be actively involved in decision making and be in charge of their life. The significance of this world problem in sustainable development and gender equality is recognized by many organizations and governments. The concept, relevance, historical development, and barriers to women's empowerment are all examined in this review of the literature. In addition, the analysis highlights effective tactics for strengthening women's empowerment and ends with suggestions for researchers, civil society organizations, and legislators who want to further this cause. In order to achieve sustainable development goals, it highlights how important it is to increase women's economic, social, and political influence (Ahmad and Sudha n.d.).

In order to comprehend women's empowerment, this term highlights the necessity of examining the structural circumstances and ideological/cultural constructions affecting women's lives. Instead of using a Western viewpoint, it emphasizes the centrality of women's self-perceptions in relationships, especially with males, and the necessity of contextualizing women's empowerment via their actual experiences. Women's perspectives are therefore crucial to this investigation (Erman et al. 2002) (Fig. 1).

1.1 Components

- **Women Empowerment**

Giving women the rights, tools, and chances to make wise life decisions is the main goal of the complex idea of women's empowerment. Economic independence, access to education, political engagement, and the capacity to question social norms are all included. Empowerment of women enhances individual agency and society through diversity and equality, leading to significant contribution to resilience and economic growth of the community. Ultimately, women empowerment results in more powerful and inclusive societies (Chakraborty et al. 2025). Monetary inclusions, employment verification and compensation transparency still remain thorns in the flesh of women as far as their economic empowerment is concerned. Because of the old hiring and banking processes, women face an unequal barrier to access credit and fair payment (Parthiban et al. 2025).

- **Financial Inclusion**

Finance is also a major aspect of economic growth. Inclusive growth of such countries as India is supported by the concept of financial inclusion as it is the means of involving marginalized groups. By making these populations have easier access to the banking system, the country will be able to achieve higher economic growth (Goel and Sharma 2017). Financial inclusion that brings both the poor and the rich closer to accessing financial services is increasingly being recognized as a significant driver in economic development and the reduction of poverty. It is a big government agenda of bettering society and an aim of bridging the gap between the rich and the poor. This research employs multiple regression analysis and secondary data to research the effect of financial inclusiveness on economic growth in a period of seven years. While the expansion of ATMs has no discernible effect on India's GDP, the results suggest a positive and significant association between the number of bank branches and the credit deposit ratio with GDP (Iqbal and Sami 2017).

- **Financial Literacy**

The Reserve Bank of India (2013) defines financial literacy as the comprehension of financial market goods with an emphasis on risks and rewards for making well-informed decisions. In a similar vein, the OECD (2012) characterizes it as a combination of financial awareness, knowledge, skills, attitude, and behavior required for making wise financial decisions and improving personal financial well-being. Through the introduction of financial literacy initiatives, workshops, and courses, the Indian government and other regulatory agencies are continuously striving for expansion. India has a large number of users of online financial services, from mobile banking to online payments and insurance, which has improved the nation's knowledge of finances as recognition and convenience of banking and insurance have grown. Three factors are used to gauge young people's financial literacy: financial understanding, mindset, and behaviors. In every Indian city, the working youth exhibit a comparable degree of financial literacy. Financial conduct and financial knowledge are positively correlated, however there appears to be a negative link between financial conduct and a financial viewpoint (Agarwal et al. 2015).

- **Gender-Responsive Policies**

Individual and household portfolio allocation decisions are influenced by gender, according to research, with women showing greater risk aversion and preferring low-risk assets than men. Important research demonstrates that risk aversion and education levels are correlated, frequently demonstrating that lower risk aversion is linked to higher education. Furthermore, there is proof that investing in riskier assets is positively correlated with financial literacy. The purpose of this paper is to investigate whether differences in financial education and knowledge have an impact on the observed gender differences in risk aversion (Hibbert et al. 2013). The gender disparities are related to individual characteristics rather than gender itself by examining how explanatory variables affect financial risk tolerance differently for men and women. Net worth and income uncertainty are important moderating factors, with the former influencing the association between risk tolerance and gender. In order to better understand the tolerance for risk across genders, financial fiduciaries are encouraged to take these disparities into account (Fisher and Yao 2017).

1.2 Objectives of the Study

(a) Education and skills training and the empowerment of women economically.
(b) The role of social norms and cultural barriers in determining whether women have access to financial services and literacy.
(c) To investigate the contribution of gender-responsive policies towards women empowerment.

1.3 Significance

SDG 4 (Quality Education) To define strengths and weaknesses, assess possible dangers and opportunities of the financial industry in terms of the SDGs, and create innovative financial products that can fulfil these purposes, financial companies need to standardize SDG budgeting and reporting. In addition, it is essential to establish financial frameworks so as to minimize risks associated with them and to harmonize financial regulations with SDGs (Taghizadeh-Hesary et al. 2022).

SDG 5 (Gender Equality)

Gender gap is a significant issue to financial inclusion; however, based on the 2017 Global Findex data, only 65% of women and 72% of men have bank accounts. Such a difference attracts the concern of gender inequality problems and its impact on the economy. Such a gap is one of the major barriers to the empowerment and economic participation of women, as financial inclusion is critical as it helps to make savings and enhance the growth of the company (Perrin and Weill 2022).

SDG 8 (Decent Work and Economic Growth):

The probable effect of financial inclusion and equitable employment procedures on economic growth, and lastly, relating our results to Sustainable Development Goal 8 (SDG 8) of the UN. Women's empowerment will boost economic growth through financial inclusion and job possibilities since both genders will contribute to the nation's GDP, which will boost economic growth and lower unemployment (Zehri et al. 2024).

SDG 10 (Reduced Inequalities) Industry-level study is sometimes overlooked while addressing issues of inequality of income at the societal level. Current government initiatives, such as progressive taxes and grants, allocate wealth after inequality is created but do not hold individual responsibility. By pricing income inequality, the concept encourages financial inclusion of Women and Minorities in the finance sector to proactively reduce inequality by offering incentives for ethical remuneration practices prior to the emergence of inequality (Husted and Salazar 2020) (Fig. 2).

2 The Impact of Learning and Acquiring Skills to the Financial Empowerment of Women

In the last few decades, advanced industrialized countries have been the main focus of educational research, which has addressed topics like improved information technology and conceptual comprehension in science and mathematics. However, this agenda has neglected important issues, including the requirements of the world's almost one billion illiterate people, who are mostly women and live mostly in the Asian continent. To end the cycle of poverty, their education is crucial. In addition to presenting an exploratory case study on a financial literacy tool intended to improve comprehension of long-term investment, this report seeks to emphasize the learning needs of these adults. The results reinforce the idea that learning is a socially situated behavior by emphasizing the use of local settings to develop successful teaching tools (Sharma and Johri 2014). According Kinyanjui and Ocholla suggest that financial literacy abilities are a crucial component in the economic empowerment of women. Social constructionism helps women entrepreneurs develop their financial literacy. Women entrepreneurs are unable to achieve economic empowerment due to a lack of financial literacy (Kinyanjui and Ocholla 2025).

The importance of the connection among financial awareness and economic independence for women in advancing economic growth and gender equality is becoming more widely acknowledged. FL plays an essential role in women financial inclusion as it enables women to effectively deal with risks and make a smart financial choice. It is a basis to a larger economic empowerment and social change. Advance financial services enhance women financial inclusion and entrepreneurship by enhancing women access to financial services using technology. Digital financial services are beneficial because they facilitate the economic activities of women, including management of their finances and the establishment of their businesses, making them more autonomous and empowered (Showkat et al. 2025).

2.1 The Foundation Capital of Education

Education is a crucial right that enhances society and economy. Women economic empowerment would be required in order to combat discrimination in the developing countries and to facilitate economic development. Women can be given equality and this helps to reduce poverty, inclusive development, and solve gender imbalance. People should be economically savvy in order to make prudent choices. Basic fiscal services would assist the poor and also improve the overall condition of the economy. Moreover, there is an absence of social connections that hinder cooperation and trust, which are the critical elements to the maximization of positive interactions (Fengwen and Ali 2023). A knowledge of these special problems is necessary to develop special ways to meet the special needs and cases of women in different situations (Choudhary and Jain 2023).

2.2 The Bridge to Usage: Financial Literacy (FL)

Other past studies show that women are more cautious in their investment than men and this is contrary to the conventional ideas in financial services industry. Even though the gender gap in pensions has been thoroughly researched, investment behavior is more recently analyzed. The differences have posed great public policy issues particularly as self-directed pension plans start to grow and the invention of the question of how to privatize Social Security. Since women live longer, their investment decisions have the potential to reduce the amount of retirement income, which could contribute to high poverty levels among women in old age. With this understanding of the subtle causes of these investment imbalances, it is important to devise capable policy interventions (Bajtelsmit and Bernasek 1996). This is perhaps the reason why women remain ignorant of financial services and products that can enable them be financially stable and empowered in the long run. They are enabled to bridge the ignorance and awareness divide by being financially literate. It enhances decision-making skills during pressure.

2.3 *Training on Digital Skills and Inclusion*

One must keep up with the curve in the current advanced world, especially with the replacement of the traditional banking system with the digital one. Involving advanced techniques, digital payment and mobile banking training is necessary but regularly overlooked in the context of financial emancipation of women. Through digital literacy, women who are not financially and digitally literate because of lack of access to technology or even internet connectivity can get jobs as digital literacy has the potential of providing them with new economic opportunities to start up small or online businesses. Governments and financial organizations should come up with programs to help women deal with the daily challenges that they face particularly financial illiteracy (Guruprasad et al. 2025).

3 The Impact of Norms on Women on Societal and Cultural Cognition and Use of Financial Goods and Services

The 2030 Agenda on the Sustainable Development Goals has identified women economic empowerment as a key to achieving gender equality and inclusive economic growth. Inequality in accessing financial resources, gender discrimination at the workplace, and the absence of decision-making powers are significant challenges causing high levels of extreme poverty in women. In the chapter, several structural and cultural problems, such as patriarchy, gender norms, illiteracy, and lack of social security, that limit the financial inclusion of women are discussed. The conclusion talks about empowerment strategies, including challenging the discriminatory norms, ensuring educational equity, broadening financial inclusion, supporting women entrepreneurs, etc. (Manhas 2025).

Inclusion in the financial sector is critical to socioeconomic development particularly in the emerging countries such as India since it ensures that the targeted financially vulnerable groups will access finance services. It is necessary to go over a lot of hurdles to reach complete financial inclusion, including the fact that the proportion of individuals without bank accounts is large, and financial literacy is high in remote locations. Worsening the scenario is a set of issues to do with the procedures, such as the stringent verification steps and the thorny paperwork, and strict risk assessment criteria that rejects a large number of individuals with a bad credit record. Women are also locked out of financial participation because of the socio-cultural obstacles, including gender prejudice and past traditions. Even though e-financial services may have a bright solution, the full potential of e-financial services cannot be achieved at the moment because of the insufficient level of confidence and the usage of online platforms (Sikka and Bhayana 2024).

3.1 Social Standards and Obstacles to Access

The empowerment of women is a critical development concern of the world as it is affected by economic, sociocultural, family, legal, political and psychological concerns. The dissonance between the time requirements of the profession and the historic role of women, which they have been assigned, influences the lack of the sufficient representation of women in the field of finance. Unlike their peers in the normal population and males in the financial sector, the women who venture into the profession are more achievement oriented and less conservative and conformist. This has a selection effect. Consequently, female financiers face some gender biases, and they successfully address them because they focus on achievement rather than tradition in comparison with their counterparts of the general population and the banking sector (Adams and Barber 2016).

3.2 Household Structure and the "Management Deficit"

Social standards surrounding gender roles are frequently reflected in how family chores are divided, which can result in an unequal distribution of obligations. Claim that adherence to gender-specific norms, rather than efficiency, is the reason why women are assigned more domestic chores. Traditionally viewed as masculine areas, financial tasks are frequently delegated to men, which may limit women's participation in financial management. This essay examines how this gender norms affect household welfare, arguing that keeping women out of financial decision-making may result in less consumption and worse-than-ideal decisions. The study examines how the gender of household economic decision-makers has changed, showing a notable trend toward women having more decision-making authority (Eds n.d.).

3.3 Actions that Change Rules and Regulatory Behavior

Since Women now work and continue to society, women's empowerment is crucial for both Economic growth and society. However, in order to overcome social and cultural barriers, it is necessary to change some laws and regulations that support financial inclusion and advance healthy gender equality. Women's roles, rights, and capabilities in terms of money and employment should be the main focus of these interventions. Through campaigns and training, the behavior should target and question oppressive gender stereotypes. Encourage other women in the industry by promoting female entrepreneurs, financial agents, and decision making in public settings. In an attempt to enhance equality between the genders and to support adjustments in the laws and regulations, it is necessary to establish gender-responsive policies (Swamy 2014).

4 The Studies of the Role of Gender-Responsive Policies in Empowering Women

Empowerment is a compound notion, which is influenced by socioeconomic, cultural, and demographic factors. Women empowerment has changed through the years to no longer be seen as a means of gaining essential rights like the right to work and to study but rather resolving problems that affect the society to place women in their proper positions. Such tools as Sustainable Development Goals set by the United Nations show that it is imperative to enable women in political participation to enable gender equality and effective governance. Women still have no representation in the position of powers and decision-making due to cultural traditions, political bodies, and lack of funds, although there were significant improvements over the past decades. The existing policies also undermine equal treatment of women in political representation due to the fact that in most cases, they do not have the resources and enforcement mechanisms to make tangible change (Kamila 2025). Women all over the world are faced with a lot of obstacles in joining the workforce, occupying leadership roles, and starting businesses, which hinders not only their economic self-sufficiency but also the growth of their nations. Against the backdrop of the worldwide competition in the area of investment, the current working paper explains why policymakers must ensure that women are put into consideration in development programs. It describes investment incentives that governments could offer to promote gender equality, and focuses on supply-side (access to jobs and funding) and demand-side (cost to firm to hire) impediments. The three major types of incentives outlined in the work are tax incentives, public procurement incentives and grants and subsidies. It warns though that such rewards need to be designed to ensure market inefficiencies and must meet specific gender equality restrictions and policy targets. The research gives illustrations of some countries that have succeeded in doing so (Kronfol 2019).

4.1 GRP in Legislative and Civil Changes

Gender Responsive Governance (GRG) entails the participation of both men and women stakeholders in the governmental levels to ensure social justice, gender equity and empowerment of women. It guarantees access to services, increased participation of women in governance and gives them the power to raise issues of injustice in unison and without any fear. Democracy is not merely constituted by the constitutional regulations but consists of a wide variety of institutions, political parties and organizations which is pluralism. This plurality means that these agencies are not reliant on the state to gain their legitimacy or authority. Equality and liberty which were considered synonymous since the 19 th century involve important notions such as individual development, tolerance of power relations, and right to express ideas. Inclusion of women in politics and empowerment of women is very important in promoting democratic governance. The political participation of women plays a very important role in the development of democracy and good governance which depends mainly on the political parties which determine the policy agendas and the candidates to be put forward. The mobilization of women in these parties also impacts a lot on their political empowerment. The growing attention of civil society, international agencies and development donors to women has brought about an awareness on the role that political parties can play in strengthening the women politically, but women remain locked out of the major structures of governance the world over (Sabo Ahmad et al. 2023).

4.2 Integration of Policies for Control and Access

Enforcement of different acts and legislation to ensure that women can access and control their financial resources. The private corporate sector should participate in the transition process in addition to the government sector. For low-income, non-working women without bank accounts or access to financial services and goods, governments ought to launch financial literacy campaigns. Laws that discriminate against women by requiring them to have a male representative in order for them to enjoy legal autonomy and power should be repealed.

4.3 The Triple Effect (FI, FL, GRP)

Improving women's status in today's society should be the aim of gender-responsive policies and financial inclusion. In addition to increasing women's independence, financial inclusion helps them get ready for unforeseen circumstances in the future. Making informed decisions and using financial services and products, such as bank accounts, credit, savings accounts, and insurance, is a result of financial inclusion and financial literacy. Women who are financially literate understand their position and rights regarding money and utilize it responsibly. Financial inclusion with gender-responsive policies results in sustainable access and legal security to boost women's engagement in the economic sector and increase employment. Financial knowledge promotes self-assured use and safeguards against exploitation. Programs for financial inclusion should be combined with instruction in digital skills so that women can be confident in utilizing financial resources.

5 Women Centric Schemes by Different Ministries of the Government of India (GOI)

See (Table 1).

6 Future Scope

Future research on the number of women with financial literacy and the use of financial resources to advance in the capital market should be both quantitative and qualitative. Researchers can include women who actively participate in digital financial literacy from both rural and urban areas. Women should have a practical understanding of how monetary policies operate, as financial knowledge alone is insufficient. Future studies should examine how women's knowledge levels may be impacted by the growing use of artificial intelligence and machine learning. The studies must focus on high-level digital financial literacy and cybersecurity among women so that they can enjoy their privacy and safety and operate with high-quality digital financial products without fear, but not on basic financial literacy as an access and management tool.

7 Conclusion

Conclusively, both financial inclusion and financial literacy are essential not only to women but also to society and the economy because women constitute a section of society and most of them are now gradually being gainfully employed in various sectors, which makes them part of the economic growth of the country. The administration has also introduced several women-focused programs implying that it is not only dealing with the gender inequality problem but also doing its best to empower women. Empowerment of women requires a holistic initiative beyond simple financial access. The problem that women cannot access financial resources is the culture of patriarchy that involves legal, social, and cultural barriers. This may have long term effects such as increased unemployment levels, stereotypical behavior and restriction of the independence of women. Financial inclusion entails the use of financial knowledge in real-life scenarios and learning about finance. Women are to be empowered, trained, and equipped with the requisite financial and digital literacy to enable them declare their autonomy and gain admission in society on equal grounds. The reason behind this is that restricting the access of women to financial literacy would affect society in a negative way. This study concludes that empowering women is possible through the use of financial inclusion, literacy, gender-responsive policies that enhance capacity, and proactive efforts to overcome social-cultural barriers.

References

Adams RB, Barber BM (2016) Family, values, and women in finance

Agarwal S, Amromin G, Ben-David I, Chomsisengphet S, Evanoff DD (2015) Financial literacy and financial planning: evidence from India. J Hous Econ 27:4–21. https://doi.org/10.1016/j.jhe.2015.02.003

Ahmad I, Sudha RT (n.d.) Women empowerment: a literature review. Int J Econ, Bus, Account, Agric Manag Sharia Adm (IJEBAS) 2(6):1353–1359

Bajtelsmit VL, Bernasek A (1996) Why do women invest differently than men? vol 970

Cabeza-García L, Del Brio EB, Oscanoa-Victorio ML (2019) Female financial inclusion and its impacts on inclusive economic development. Women's Stud Int Forum 77:102300. https://doi.org/10.1016/j.wsif.2019.102300

Chakraborty S, Chaudhuri T, Ghosh J (2025) Women empowerment by breaking barriers: gender equality in the workplace, leadership, and beyond. In: Bridging the leadership gap: policies and workplace equality, pp 71–101. https://doi.org/10.4018/979-8-3373-6345-5.ch003

Choudhary H, Jain H (2023) Addressing financial exclusion through financial literacy training programs: a systematic literature review. Empir Res Vocat Educ Train 15(1):8. https://doi.org/10.1186/s40461-023-00147-9

Eds NT (n.d.) Impact of patriarchy and gender stereotypes on working women exploring its past, present and future, vol 783031

Erman T, Kalaycıoğlu S, Rittersberger-Tılıç H (2002) Money-earning activities and empowerment experiences of rural migrant women in the city: the case of Turkey. Women's Stud Int Forum 25(4):395–410. https://doi.org/10.1016/S0277-5395

Fengwen W, Ali M (2023) Women's financial inclusion and women's financial literacy: Bibliometric analysis. In: Leong C-M, Ali M, Raza SA, Puah C-H, Eksi IH (eds) Financial inclusion across Asia: bringing opportunities for businesses. Emerald Publishing Limited. https://doi.org/10.1108/978-1-83753-304-620231001

Fisher PJ, Yao R (2017) Gender differences in financial risk tolerance. J Econ Psychol 61:191–202. https://doi.org/10.1016/j.joep.2017.03.006

Goel S, Sharma R (2017) Developing a financial inclusion index for India. Procedia Comput Sci 122:949–956. https://doi.org/10.1016/j.procs.2017.11.459

Gupta P, Aggarwal M, Bamba M (2025) Antecedents and measures of women's financial empowerment: a comprehensive and global literature review. J Int Women's Stud 27(4) https://www-scopus-com-geu.knimbus.com/inward/record.uri?eid=2-s2.0-105017890532&partnerID=40&md5=c398577ab4cd69786b7a78934b5ad856

Guruprasad LS, Ramasamy KRABA, Kulkarni S (2025) Gender disparities in financial inclusion: promoting women's economic empowerment in G20 countries for sustainable development. Digit Policy, Regul Gov. https://doi.org/10.1108/DPRG-10-2024-0269

Hibbert AM, Lawrence ER, Prakash AJ (2013) Does knowledge of finance mitigate the gender difference in financial risk-aversion? Glob Financ J 24(2):140–152. https://doi.org/10.1016/j.gfj.2013.07.002

https://www.ncw.gov.in/publications/women-centric-schemes-by-different-ministries-of-government-of-india-goi/

Husted BW, Salazar J (2020) Reducing inequalities: toward the development of a market for income inequality. J Clean Prod 245:118931. https://doi.org/10.1016/j.jclepro.2019.118931

Iqbal BA, Sami S (2017) Role of banks in financial inclusion in India. Contaduría Y Adm 62(2):644–656. https://doi.org/10.1016/j.cya.2017.01.007

Kamila DI (2025) Policy analysis of women' s empowerment in political and government participation 4(2):67–77

Kinyanjui JW, Ocholla D (2025) In: Ocholla D, Onyancha OB, Adesina AO (eds) Financial literacy amongst wsomen entrepreneurs for economic empowerment in Kenya BT-information, knowledge, and technology for teaching and research in Africa: information behavior in knowledge and economy. Springer Nature Switzerland, pp 53–83. https://doi.org/10.1007/978-3-031-65745-0_3

Kronfol, H. (2019) Women at work how can investment incentives be used to enhance economic opportunities for women?

Manhas S (2025) In: Koul R, Gupta R, Tandon B, Gupta R (eds) Structural and cultural barriers to the economic empowerment of women BT-women and empowerment: achieving sustainable development goal 5. Springer Nature Singapore, pp 91–107. https://doi.org/10.1007/978-981-96-2759-2_8

Parthiban U, Pannerselvam VD, Murugesan A, Ramasamy K (2025) She power chain: a blockchain-based platform for women's financial inclusion and empowerment. Asian Women 41(3):82–97. https://doi.org/10.64446/aw.2025.9.41.3.82

Perrin C, Weill L (2022) No man, no cry? Gender equality in access to credit and financial stability. Financ Res Lett 47:102694. https://doi.org/10.1016/j.frl.2022.102694

Sabo Ahmad H, Tahir Hamid K, Salisu Maigoshi Z, Lawan R, Shehu Aliyu F (2023) Gender responsive budgeting and women economic empowerment: a review. Int J Soc Sci Manag Rev:45–58

Sharma A, Johri A (2014) Learning and empowerment: designing a financial literacy tool to teach long-term investing to illiterate women in rural India. Learn Cult Soc Interact 3(1):21–33. https://doi.org/10.1016/j.lcsi.2013.10.003

Showkat M, Nagina R, Baba MA, Yahya AT (2025) The impact of financial literacy on women's economic empowerment: exploring the mediating role of digital financial services. Cogent Econ & Financ 13(1):2440444. https://doi.org/10.1080/23322039.2024.2440444

Sikka V, Bhayana P (2024) Barriers to comprehensive financial inclusion across the globe: from sociocultural norms to systemic challenges. In: Mansour N, Baral S, Garg V (eds) E-financial strategies for advancing sustainable development: fostering financial inclusion and alleviating poverty. Springer Nature Switzerland, pp 89–126. https://doi.org/10.1007/978-3-031-67523-2_7

Swamy V (2014) Financial inclusion, gender dimension, and economic impact on poor households. World Dev 56:1–15. https://doi.org/10.1016/j.worlddev.2013.10.019

Taghizadeh-Hesary F, Yoshino N, Otsuka M (2022) Sustainable development goals need and financial sector imperatives. In: Anbumozhi V, Kalirajan K, Kimura F (eds) Sustainable development goals and pandemic planning: role of efficiency based regional approaches. Springer Nature Singapore, pp 721–746. https://doi.org/10.1007/978-981-16-6734-3_19

Zehri C, Mohammed El Amin B, kadja A, Inaam Z, Sekrafi H (2024) Exploring the nexus of decent work, financial inclusion, and economic growth: a study aligned with SDG 8. Sustain Futur 7:100213. https://doi.org/10.1016/j.sftr.2024.100213

Corporate Social Responsibility as a Tool for Sustainable Development: Insights for Indian Government Regulations and Corporate Practices

Ajay Sharma and Ravi Kant Sharma

Abstract Corporate social responsibility (CSR) has changed from being an optional charity to a method of supposed sustainable development. The introduction of Company Act 2013 in India and government reforms in reporting rules have influenced companies to contribute funds to fulfil social and environmental goals. This chapter looks at India's legal and policy work to examine the connection between CSR and sustainability, challenges and suggestions for improving accountability. This chapter also examines India's legal and policy framework, the alignment of CSR with sustainability and sustainable development goals (SDGs), key achievements, challenges and recommendations for strengthening accountability.

Keywords Corporate social responsibility · CSR regulations in India · Sustainability · Sustainable development goals (SDGs)

1 Introduction

Corporate social responsibility (CSR) and sustainability have related concepts that can shape policies related to sustainable development. CSR reflects a company's obligation towards society beyond just making profit, while sustainability focuses and balances economic growth, social equality (Elkington 1997). More scholars view CSR as an advanced practical way for companies to contribute towards sustainability. This connection has become a format and mandatory CSR rules, turning corporate spending on social welfare into a legal pathway for promoting sustainable development (Ministry of Corporate Affairs 2025). In recent years, both researchers and policymakers have shown growing interest in how companies communicate their

A. Sharma (✉)
School of Management Studies, Graphic Era Hill University, Dehradun, India

R. K. Sharma
School of Business, Galgotias University, Greater Noida, India
e-mail: ravikantsharma@galgotiasuniversity.edu.in

A. Singh et al. (eds.), *Human-Centered Sustainability: Bridging Well-Being, Gender Equity, and Technology*, World Sustainability Series,
https://doi.org/10.1007/978-3-032-22020-2_27

efforts toward CSR and sustainability, especially in emerging economies. For many firms, these disclosures have evolved into a critical bridge connecting their business practices with broader national and global sustainability goals.

India presents a particularly distinctive case. The country's CSR–sustainability relationship has been influenced not only by legislation but also by long-standing developmental challenges such as poverty, environmental pressures, unequal income distribution, and infrastructural gaps. These realities shape how CSR is interpreted and implemented across Indian industries.

A defining feature of India's approach is its regulatory structure. Under the Companies Act 2013, eligible firms are required to allocate at least 2% of their average net profits to activities specified in Schedule VII. The Ministry of Corporate Affairs (MCA) plays a central role by outlining expectations, compliance norms, and reporting guidelines so that CSR funds are used meaningfully rather than treated as a procedural requirement. The latest updates (2024–25) have further strengthened this system by revising reporting formats, introducing mandatory CSR registrations, and tightening audit procedures. These reforms aim to improve transparency, strengthen accountability, and ensure that corporate contributions truly align with national sustainability priorities (ClearTax 2024; The Economic Times 2025a). Recent evidence suggests that emerging economies such as India follow a hybrid CSR model, where statutory requirements and voluntary ethical commitments operate together. The introduction of mandatory CSR under the Companies Act 2016 provided a legal backbone for corporate responsibility while simultaneously aligning business activity with national sustainability priorities. This blend of compulsion and discretion has encouraged firms not only to comply with minimum norms but also to experiment with innovative sustainability initiatives.

For instance, the CSR Rules (2016) include provisions that link corporate spending with SDG-7 by promoting investments in affordable, clean, and renewable energy. As a result, many companies have expanded efforts in areas such as renewable energy installations, waste reduction, conservation practices, and other environmentally responsible interventions. Amendments introduced in subsequent years further strengthened this ecological orientation and broadened expectations from firms (Jain and Kumar 2024).

Beyond environmental initiatives, CSR projects in India have increasingly targeted social domains—education, healthcare, gender equality, rural upliftment, and community welfare. These efforts play an important role in closing long-standing development gaps and contribute to more equitable social outcomes (Sharma 2023). Consequently, CSR in India has evolved from occasional charitable giving to a **strategic development instrument** with clear links to national priorities.

The India CSR Outlook Report 2024 (ICOR 2024) also shows how recent regulatory amendments are influencing the direction and spread of CSR activities. The report highlights sectoral trends, regional dispersal, and a growing alignment between CSR investments, the SDGs, and broader governmental development agendas. It notes especially strong synergy in areas such as health, environment, livelihood enhancement, and digital-enabled program monitoring. Together, these trends indicate that CSR in India is steadily transforming into a **catalyst for sustainable and inclusive development**.

2 Theoretical Background of CSR and Sustainability

The foundation of CSR and sustainability ground from a rich body of scholars who examine the evolving relationship between business and Society. One of the most influential perspectives is shareholder theory, which argues that corporations have obligations toward a diverse group of stakeholders not limited to shareholders but also including employees, consumers, suppliers, community and the natural environment. Freeman's early conceptualization has been explained in contemporary scholarship to emphasize that sustainability value creation requires balancing the interests of all stakeholders throughout long term ethical and transparent decision making (Freeman et al. 2022). This theoretical concept recognized CSR as an essential mechanism through which firms engaged with societal expectations and reduced negative externalities.

Institutional theory explains how organizations change their behavior in response to external pressure. Regulatory mandates, societal expectations and industrial norms influence how firms design to carry out CSR activities. The growing influence of sustainability reporting frameworks, environmental social grievances metrics. International agenda such as the Sustainable Development Goals (SDGs) has created a structured environment that encourages companies to integrate CSR into their long-term development strategies. In India, compulsory CSR spendings disclosure requirements under SEBI's business responsibility and sustainability reporting (BRSR) framework and sector-specific policies on renewable energy and climate adoption serve as a strong institutional mechanism guiding corporate sustainability requirements (Kumar and Taneja 2024).

Elkington's (1997) Triple Bottom Line-people, planet and profit illustrate CSR alignment with sustainability. CSR projects in the community safeguard ecosystem and Secure long-term profitability. Integrated CSR report improved legitimacy, employee's satisfaction and financial resilience. The sustainability is further informed by the triple bottom line TBL Framework which integrates economic performance, environmental protection and social equality. This model has been widely adopted by firms seeking operationalized sustainability within business strategies. Recent empirical evidence demonstrates that companies that adopt TBL principles achieve a better environment and performance, stronger community relationships and improve long term competitiveness (Singh and Dhil 2023). TBL principles are particularly relevant in the current era of climate change. Now business is expected to reduce emissions, conserve resources and contribute to a resilient social system. The SDGs provide a global framework for balancing social environment and economic objectives. CSR, when strategically aligned, enables businesses to contribute through investment in renewable energy, health Care, Education, poverty elevation and climate adoption (Nair and Madhavan 2015). The SDGs provide a global framework for balancing social environment and economic objectives. CSR, when strategically aligned, enables businesses to contribute through investment in renewable energy, health Care, Education, poverty elevation and climate adoption (Nair and Madhavan 2015). For example, CSR 2016 notification addresses SDG 7 (affordable and clean energy) and India's renewable energy targets. CSR activities also enhance environmental stewardship by adopting renewable energy and reducing and conserving their wastes, and in this way, they facilitate the ecological aspect of sustainability (Jain and Kumar 2024). Simultaneously, education, healthcare, gender empowerment and rural development CSR initiatives aim to solve urgent social requirements and decrease inequality and build positive growth (Sharma 2023). Therefore, CSR does not just act as philanthropy but as an important channel through which development is to be attained within the Indian context.

CSR mitigates reputational and regulatory risks while opening avenues for innovation in sustainable supply chains, inclusive Business models and technology. CSR simultaneously supports risk management and opportunity creation.

3 CSR and Sustainability in the Indian Context

India legislates CSR under sect. 135 of Companies Act, 2013. Firms meeting financial thresholds must allocate 2% of their average net profit to CSR activities aligned with schedule VII. Amendments in 2025 for tightened accountability through mandatory registration of implementing agencies and sticker disclosure norms (Khaitan & Co 2025; Ministry of Corporate affair 2025). Governmental alignment and CSR priorities aligned with the national Action Plan on climate change (NAPCC) and state level climate strategies. Corporate CSR projects in renewable energy, afforestation, and water management compliment National climate resilience while fulfilling compliance requirements (United Nations/GOI 2024). Indian firms, particularly in the energy and IT sector have moved towards long term CSR programs linked to sustainability.

There are many examples, including renewable micro-grids, sanitation projects and green skill training. Such initiatives demonstrate the shift from philanthropy to strategy-driven corporate decisions (The CSR Journal 2024). Globally CSR and sustainability conversation through an integrated ESG framework and standard like Global Revolving Initiative. Indian corporates, driven by regulatory mandates and investor expectations, are adopting these frameworks. All these measures are positioning CSR as a central pillar of sustainability strategy. (Carroll and Brown, 2018). Emerging economies such as Brazil and South Africa also regulate CSR, but India remains unique in mandating CSR by law, offering lessons for mobilizing corporate capital for sustainability (Gupta 2025).

3.1 Government Regulations and Corporate Social Responsibility Practices in India

Empirical studies indicate that CSR spending has grown steadily with investments connected to education, environmental sustainability and rural development. Challenges such as uneven regional distribution, limited evaluation of long-term impacts and persistent complaint-driven practices remain in India. This issue requires enhanced accountability, and engagement with stakeholders to ensure that CSR contributes meaningfully to National Development objectives. India has established a comprehensive framework to institutionalize corporate social responsibility and align it with the national development sustainable goals. According to Sect. 135 of the Companies Act, 2016, which mandates that companies must fulfil specific requirements of net worth turnover or profit allocation at least 2% of their average net profits to CSR activities. This provision has transformed CSR from a voluntary philanthropic activity into a structural, legally accountable mechanism which helps in mobilizing substantial corporate resources for social economic, and mental protection as well as in development.

Schedule VII of company, Act guides permissible CSR interventions encompass in areas such as education, health, equality environmental sustainability and rural development (Gupta 2023) to strength and accountability come on the company's policy rule 2014 along with subsequent amendments requires company to form a CSR committee of the board. It is mandatory to report activities through standardized forms CSR 1 and CSR 2 and register implementing agencies. These measures promote systematic planning monitoring and evaluation of CSR programs ensuring measurable outcome that rather than mare compliance.

Indian corporations have leveraged this regulatory framework to implement strategic CSR initiatives in area as related to sustainability and green energy. The TATA Group invests in renewable energy water conservation and rural livelihood through data trust integrating business availabilities with community needs. Reliance Industries Limited focus on education health technology driven rural development by using digital tools to monitor impacts on rural development. ITC limited promotes sustainable agriculture and rural livelihood through the each of all initiative while also investing in afforestation and water management projects similarly Mahindra and Mahindra supports girls' education through *Nanhi Kali* that promotes rural Entrepreneurship and vocational training.

In a recent development in area of CSR and sustainability, CSR has become a critical tool for climate change mitigation and environmental sustainability. Adani group invests in solar and wind energy and promotes energy access for rural populations. NTPC Limited engage in afforestation, water and Solar Energy programs. Suzlon Energy conducts ecosystem, restoration and give training to members in community, especially in rural areas for green jobs and renewal energy. Hindustan Unilever Ltd. focuses on water conservation son and environmental awareness campaign where as JSW Steel and Infosys Limited implement solar energy projects carbon reduction measures and sustainability education initiatives.

These are certain examples which highlights the key trends in Indian corporate houses increasingly adopting strategic CSR using technology and data analytics to monitor impact and design Program that promote inclusive growth and environmental sustainability. All together the regulatory framework and efforts of corporate shows that CSR potentials have become a strategic instrument for long term sustainable development that integrates corporate objectives with national priorities which align with sustainable development goals.

4 Integration of CSR with National and Global Sustainability Goals

The integration of corporate social responsibility with the National and global sustainability agenda has gained significant momentum in India over the past decade. Growing regulatory clarity combined with international commitments such as sustainable development goals and the Paris climate agreement. All this increased firms to reorient CSR activities towards long term sustainability priorities. Recent evidence shows that the rising proportion of CSR funds in India now aligned with SDG linked factors such as clean energy, climate adoption, skill development infrastructure making a shift from traditional philanthropic to strategic development contribution (India CSR Outlook report 2024).

This transition is partly driven by policy framework like national action plan on climate change and sectoral mission on energy efficiency, water conservation and sustainable agriculture. New policies provide structured pathways for corporate participation in national sustainability missions. Empirical studies further demonstrate a clear pattern of CSR-SDGs alignment, during and after pandemic, as companies began systematically making mapping their CSR interventions to SDGs such as good health and well being (STG 3) quality education, (STG 4) gender equality (STG 5), clean water and sanitation (STG 6) and climate action (STG 13) (Kumar et al. 2023). This indicates that CSR has evolved into a contemporary mechanism for supporting India's national development target especially in domains where public funding gaps persist. At the same time the emergence of environmental, and governance standards as further strengthen CSR sustainability integrations. Recent analysis highlights that many Indian forms are moving beyond compliance driven CSR and embedding environmental, social and governance (ESG) frameworks into corporate decision making, their investments in renewable Energy and reduction and circular economy innovations (Sharma and Mishra 2024).

Furthermore, CSR policy is increasingly aligned with National sustainability priorities such as the National Action Plan on climate change, increase in climate change mitigation, water management, and biodiversity protection. Help to improve CSR outcomes that support structural CSR initiatives.

5 Challenges in Aligning CSR with Sustainability

Although the company act 2016 and its later amendments provide formal foundation for connecting CSR initiatives with India's broader development. Actual alignment between CSR and long-term sustainability remains complicated. Many organizations still face structure and operation issues that limit the effectiveness of CSR efforts. Despite India has mandatory application CSR for corporate business firms but there are some challenges also, these are given below:

5.1 Fragmentation and Short-Term Focus

A major challenge is that many companies drive one or more community events, such as campus or donation derives, which while beneficial in the short term. But it fails to create sustainable impact on social development or environmental conservation. Still a question arises how they demonstrated a constant contrasting approach by implementing long term effective CSR program.

In Jharkhand and Odisha which span multiple years and integrated local community participation, there by outcomes. First smaller steps without strategic planning of one struggle to scale CSR projects leading to limited systematic impact.

5.2 Measurement and Transparency Gap

Another critical challenge is the absence of a sterile matrix to evaluate CSR outcomes. Report activities in quantitative terms surge number of beneficiaries reached or funds dispersed, without a clear indicator of long term social and environmental outcomes. For example, while industries have invested in rural education and health initiatives, the impact of these programs on community literacy or health is not consistently measured. With use of standard matrix CSR reports detail monitoring or evaluation meeting in difficult to assess whether resources translate into meaningful societal change (Sinha et al. 2024).

5.3 Greenwashing and Outcome Measurements

Greenwashing happens when companies mislead the public and stakeholders about their environmental responsibility. They might exhibit smart initiative, use unverified claims or highly positive actions while hiding harmful practices. This breach of behavior undermines the trust of the people, the overall attitude towards genius CSR programs and delays positive steps towards global sustainability. Greenwashing has increased tremendously as sustainability is included in corporate strategies. In order to guarantee the reputation of CSR activities, firms spearhead unbiased determination of performance measures associated with the sustainable development objectives. CSR activities can evolve into the obligatory approach that sustains sustainable development instead of optional charity through the regular monitoring and computer-linked evaluation systems.

5.4 Exclusion of MSMEs

There is a very important role of micro, small and medium (MSME) enterprises in the aspect of local employment and community development. However, these businesses are not subject to sect. 135 of the Companies Act, even though they play crucial roles in the creation of job opportunities in the local supply chain and community wellbeing. This omission narrows the general influence of CSR on sustainability at the grassroots level. Most required are CSR programs of micro and smaller enterprises that are located in regions where actual sustainability is done. For instance, a large company like ITC might implement watershed management in its supplies but smaller suppliers are not part of the regulatory framework, which creates a gap in environmental and social protection goals and results.

6 Recent Reforms in CSR (2024–25)

Amendments introduced new reporting forms CSR1/CSR2 mandatory registration for implementing agencies and sticker disclosure norms. These reforms seek to professionalize CSR implementation, improve oversight and enhance transparency (Ministry of Corporate Affairs 2025). New guidelines align CSR with National sustainability priorities, India's sustainability mission under the (NAPCC) create a framework for CSR investment in renewable energy, climate adaption, water management and biodiversity. CSR spending can thus be directed towards measurable sustainability outcomes. Governments complement CSR mandates with incentives, partnerships and campaigns. Updated registration and reporting mechanism strengthen state capacity to direct CSR towards national mission.

7 Measures to Achieve Sustainability Goals through CSR

CSR has increasingly become a tool for advancing sustainability in developing nations. Business firms can align their policies according to the objective of SDGs. Most of the leading companies in India following this path rather than contributing funds to second party, they are investing in research and development projects related to development of efficiency, waste management, use of renewable energy etc. the following measures highlight how CSR serves as a pathway for attainment sustainability. Sustainability goals can be achieved with the help of continues monitoring by authorities, media and stakeholders.

7.1 Social Responsibility Measurements

Social responsibility assessment evaluates how CSR activities improve community well being. These assessments often look at the scope and outcomes of initiatives like education programs, digital literacy, and skill development activities. Most of the companies are engaged in social activities like employment creation, improvement of livelihood and women empowerment that reflect the social value added by corporate involvement. Similarly, progress in sanitation facilities, access to safe drinking water and community infrastructure development are also necessary for the improvement of communities living in rural areas as well as in urban areas. The broader social changes reports supported by CSR activities.

7.2 Economics Sustainability Measurements

Economics sustainability is perceived by looking at CSR contribution to local economic growth. This may include increase in household income, support for micro enterprises and purchasing from local suppliers. Companies may also try to reduce the cost of environmentally friendly practices. This Shows how ecological efforts work alongside financial performance. Long term contributions such as developing self sustaining models for the highlight the lasting economic impact on the poor section of society.

Measuring the contribution of corporate social responsibility to sustainability goes requires a multiple dimensional assessment covering environment social, governance expects. On the environmental front, organizations can evaluate their progress by examining reduction in carbon.

7.3 Governance and Ethical Measurements

Governance related indicators evaluate the transparency, ethical behavior, accountability in CSR planning and exhibition. These measurements are helpful in the process of identifying organizational goals and outcomes. This is what makes companies adhere to responsible management practices and integrity of their sustainability commitments. Good reporting and external audit demonstrates the fact that a company is responsible in its governance. Interaction of stakeholders such as the existence of grievance, redressal system and community consultant are also indicators of inclusive CSR practices. Besides, the conformity to the international criteria, including ESG framework and SDG indicators, places CSR in an expanded sustainability framework.

7.4 Assessment and Valuation Matric

The effective assessment and the impact of CSR activities needs a powerful evaluation tool. This instrument will be able to guarantee the CSR programs result in significant and quantifiable transformations. They could apply such frameworks as theory of change, social and output development indicators to judge the achievement of initiatives in their objectives. Perpetuated effects are evidenced by long term monitoring of beneficiaries and manufacturing of environmental results and behavior change. All these come in handy in redefining CSR strategies.

7.5 Partnership and Collaboration

CSR initiatives become more effective when companies work with government and nonprofits. These types of partnership enhance scalability and ensure alignment with local and national environmental as well as developmental goals. Joint programs in sanitation, clean energy shows CSR role in advancing SDGs through Corporation among multiple stakeholders. Collaboration between the government, nonprofits and business entities has improved the contribution in sustainability. These kinds of joint efforts in CSR activities are beneficial for the development of the poor sector of society and the protection of the environment. Most of the NGOs in India work for the upliftment of poor people through plantation, education and employment generation. This also aligned with the agenda of the government; therefore, a partnership is required between both of them.

7.6 Sustainability Alignments with Business Models

The long-term impact of CSR is closely tied to how well it integrates into a company over all business strategy. When sustainability principles become part of core operations CSR shifts from a secondary activity to a driver of innovation and competitiveness. A good example is Mahindra and Mahindra whose investment in electric mobility and renewable energy shows how environmental responsibility can be built into the business model.

8 Conclusion

This chapter summarises that CSR is capable of greatly improving sustainability results. When company activities take care of long term needs of society in lieu of immediate short term compliance goals. In order to attain significant improvements, organizations should strive towards incessant checking of the system as well as incorporating powerful valuation processes. With the support of the culture and strategic choice which allows the company to implement sustainability values, the companies are likely to have long-term social benefits. This also enhances their competitiveness and publicity. Nevertheless, obstacles in this regard are still present such as imbalanced CSR spendings, ineffective implementation abilities, deficient evaluation, and risk of superficial CSR activities.

References

Ahmad S, Chauhan Y (2023) Corporate sustainability capabilities and competitive advantage: a resource-based perspective. J Sustain Manag 12(2):55–68

Carroll AB, Brown JA (2018) Corporate social responsibility: a review of current concepts, research, and issues. Bus Soc 57(7):1189–1199. https://doi.org/10.1177/0007650317745631

ClearTax (2024) Corporate social responsibility under Section 135 of the Companies Act, 2013. Retrieved from https://cleartax.in/s/csr-under-companies-act

CSRBOX (2024) India CSR outlook report 2024 (ICOR 2024). CSRBOX Impact

Elkington, J. (1997). Cannibals with forks: the triple bottom line of 21st century business. Oxford: Capstone

Freeman RE, Phillips R, Kujala J (2022) Stakeholder theory: future directions. Bus Soc 61(7):1513–1534

Gupta S (2023) Corporate social responsibility: issues and challenges in India. Recent Adv Bus Soc Sci

Gupta, S. (2025). Corporate social responsibility and sustainable development: status and working dimension of CSR in India. J Corp Soc Responsib. https://doi.org/10.xxxx/jcsr.2025

India CSR Outlook Report (2024) CSRBOX & NGOBOX

Jain S, Kumar R (2024) CSR and environmental sustainability: emerging perspectives from India. J Asian Bus Stud 18(3):421–438

Khaitan & Co. (2025). Companies (corporate social responsibility policy) amendment rules, 2025: commentary. Ministry of Corporate Affairs, Government of India. Retrieved from https://www.mca.gov.in

Kumar A, Taneja N (2024) Institutional drivers of CSR and sustainability reporting in India. Asian J Bus Account 17(1):112–130

Kumar, R., Sharma, A., Nair, P. (2023). Aligning corporate social responsibility with sustainable development goals in India across pandemic phases

Ministry of Corporate Affairs (2025) Companies (Corporate Social Responsibility Policy) Amendment Rules, 2025. Government of India Notification (7 July 2025)

Nair R, Madhavan A (2015) SDG3 and mandatory corporate social responsibility (CSR). J Sustain 14(2):115–128. https://doi.org/10.3390/su1402115

Sensharma, S. et al. (2023) Do Indian firms engage in greenwashing? evidence from corporate social responsibility reports. Afr Bus & Financ J

Sharma A (2023) CSR as a driver of community development in India. Int J Soc Responsib 9(1):101–118

Sharma R, Misra A (2024) Corporate social responsibility in India: evolution, compliance and emerging ESG focus. Afr J Bus Manag
Singh P, Dhir A (2023) Triple bottom line-oriented firms and sustainability performance. Corp Gov Rev 21(4):221–239
Sinha BC, Kumari A, Pandey Y, Tandon P, Agrawal B (2024) A critical analysis of corporate social responsibility in context to the provisions of the companies act, 2013 of India. Acad Mark Stud J 28(4):1–10
The CSR Journal (2024) Top 100 companies for CSR and sustainability in India. Retrieved from https://thecsrjournal.in
The Economic Times (2025a) New registration form for CSR entities. Retrieved from https://economictimes.indiatimes.com
The Economic Times (2025b) 30 companies penalised in three years over non-compliance with CSR rules. Retrieved from https://economictimes.indiatimes.com
United Nations/Government of India (2024) National Action Plan on Climate Change (NAPCC): updated missions and progress report. Retrieved from https://moef.gov.in
Zhang K, Hao X (2024) Corporate social responsibility as the pathway towards sustainability: a state-of-the-art review in Asia economics. Discov Sustain 5:348. https://doi.org/10.1007/s43621-024-00577-9

The manufacturer's authorised representative in the EU is Springer Nature Customer Service Centre GmbH, Europaplatz 3, 69115 Heidelberg, Germany. If you have any concerns regarding our products, please contact ProductSafety@springernature.com

Printed and bound by CPI Group (UK) Ltd, Croydon, CR0 4YY
17/09/2026
02218458-0012